Palaeohydrology

Palaeohydrology

Understanding Global Change

Edited by

K.J. Gregory
Department of Geography, University of Southampton, UK

and

G. Benito
Environmental Sciences Centre, CSIC, Madrid, Spain

WILEY

Other Wiley Editorial Offices

John Wiley & Sons Inc., 111 River Street, Hoboken, NJ 07030, USA

Jossey-Bass, 989 Market Street, San Francisco, CA 94103-1741, USA

Wiley-VCH Verlag GmbH, Boschstr. 12, D-69469 Weinheim, Germany

John Wiley & Sons Australia Ltd, 33 Park Road, Milton, Queensland 4064, Australia

John Wiley & Sons (Asia) Pte Ltd, 2 Clementi Loop #02-01, Jin Xing Distripark, Singapore 129809

John Wiley & Sons Canada Ltd, 22 Worcester Road, Etobicoke, Ontario, Canada M9W 1L1

Wiley also publishes its books in a variety of electronic formats. Some content that appears
in print may not be available in electronic books.

British Library Cataloguing in Publication Data

A catalogue record for this book is available from the British Library

ISBN 0-470-84739-5

Typeset in 10/12pt Times by Laserwords Private Limited, Chennai, India
Printed and bound in Great Britain by Antony Rowe Ltd, Chippenham, Wiltshire
This book is printed on acid-free paper responsibly manufactured from sustainable forestry
in which at least two trees are planted for each one used for paper production.

Contents

List of Contributors

K. Alverson — Pages International Project Office, Bärenplatz 2, 3011 Bern, Switzerland

V.R. Baker — Department of Hydrology and Water Resources, The University of Arizona, Tucson, Arizona 85721-0011 USA

G. Benito — CSIC-Centro de Ciencias Medioambientales, Serrano 115 bis, 28006 Madrid, Spain

O. Borisova — Institute of Geography, Russian Academy of Sciences, Staromonetny, 29, 109017 Moscow, Russia

J. Branson — GeoData Institute, University of Southampton, Southampton SO17 1BJ, UK

A.G. Brown — School of Geography and Archaeology, University of Exeter, Amory Building, Rennes Drive, Exeter EX4 4RJ, UK

Z. Cao — Department of Geography, University of Southampton, SO17 1BJ, UK

P. Carling — Department of Geography, University of Southampton, SO17 1BJ, UK

M.J. Clark — Department of Geography and GeoData Institute, University of Southampton, Southampton SO17 1BJ, UK

T.J. Cohen — School of Geosciences, University of Wollongong, New South Wales, 2522, Australia

C.J. Doyle — School of Geosciences, University of Wollongong, New South Wales, 2522, Australia

T. Edwards — University of Waterloo, Canada

K.J. Gregory — Department of Geography, University of Southampton, SO17 1BJ, UK

A. Gupta — School of Geography, University of Leeds, Leeds, LS2 9JT, UK

J. Herget — Geography Institute, Ruhr University, Bochum, Germany

M. Huisink — ICG, Vrije Universiteit, Institute of Earth Sciences, De Boelelaan 1085, 1081 HV Amsterdam, The Netherlands

V.S. Kale — Department of Geography, University of Pune, Pune 411 007, India

R. Kidson — Department of Geography, University of Cambridge, CB2 3EN, UK

J.C. Knox,   Department of Geography, University of Wisconsin Madison,
                       Madison, Wisconsin, 53706-1491 USA

E.M. Latrubesse   Laboratory of Geology and Physical Geography, IESA,
                       Federal University of Goiás, Goiânia, Brazil

S. Lewis   Department of Geography, Queen Mary, University of
                       London, Mile End Road, London, E1 4NS, UK

D. Maddy   Department of Geography, University of Newcastle, Daysh
                       Building, Newcastle upon Tyne NE1 7RU

G. Nanson   School of Geosciences, University of Wollongong, New South
                       Wales, 2522, Australia

T. Oguchi   Center for Spatial Information Science, The University of
                       Tokyo, c/o Department of Geography, Faculty of Science,
                       7-3-1, Hongo, Bunkyo-ku, Tokyo 113-0033, Japan

A. Panin   Geographical Faculty, Moscow State University, 119899
                       Moscow, Russia

D.G. Passmore   Department of Geography, University of Newcastle, Daysh
                       Building, Newcastle upon Tyne NE1 7RU

D.M. Price   School of Geosciences, University of Wollongong, New South
                       Wales, 2522, Australia

S.A. Schumm   Earth Resources Department, Colorado State University, Fort
                       Collins, 80523, Colorado, USA

A. Sidorchuk   Geographical Faculty, Moscow State University, 119899
                       Moscow, Russia

A.K. Singhvi   Physical Research Laboratory, Ahmedabad 380009, India

L. Starkel   Department of Geomorphology and Hydrology, Institute of
                       Geography, Polish Academy of Sciences, 31-018 Kraków,
                       ul. św. Jana 22, Poland

M.F. Thomas   University of Stirling, Stirling, FK9 4LA, UK

J.B. Thornes   King's College, University of London, WC2R 2LS, UK

M.B. Thorp   National University of Ireland, Dublin 4, Republic of Ireland

J. Vandenberghe   ICG, Vrije Universiteit, Institute of Earth Sciences, De
                       Boelelaan 1085, 1081 HV Amsterdam, The Netherlands

Preface

As this volume was being prepared for the publisher, there were reports of devastating floods in central Europe in August 2002 and of equally serious flood events in parts of Asia. In view of their terrible consequences, such events necessarily remind us of the question as to whether floods of this character and magnitude are to be expected in the course of the natural secular climatic sequence or whether they are triggered by the consequences of global change. Also at the time of writing, the agenda of the World summit meeting in Johannesburg included the consequences of climate change. Whatever the conclusion reached, it is increasingly appreciated that information from a long-term perspective is required and can address understanding of contemporary and future situations. The INQUA Commission on Global Continental Palaeohydrology (GLOCOPH) worked from 1991 to 2003, and in 2000 it was agreed that this volume, demonstrating the state-of-world knowledge of palaeohydrology in relation to the understanding of global change, should be produced. We appreciate that this can never be the last word on the subject and that many researchers, in addition to those who have contributed chapters, have added significantly to the progress that has been made in palaeohydrology and related fields, particularly in the period 1991 to 2003. In that time, we acknowledge the contribution made by earlier Presidents of GLOCOPH, Leszek Starkel (1991–1995), Victor Baker (1995–1999), and Ken Gregory (1999–2003), and also by Secretaries of the Commission, Tony Brown (1991–1999) and Gerardo Benito (1999–2003).

Our sincere thanks are due to Stan Schumm, a really influential founder of palaeohydrology, for contributing the Foreword, and to the authors of the chapters for their assistance. The award of a Leverhulme Emeritus Fellowship to K.J. Gregory (1998–2001) was extremely valuable during the development of this phase of GLOCOPH and the preparation of this volume.

The present phase of GLOCOPH research concludes at the INQUA Congress in Reno in 2003, but we are confident that work in this exciting multidisciplinary field will continue. Our sincere thanks are due to the researchers who have contributed so far and have set an admirable example to be followed by successive generations. GLOCOPH was able to take the lead as one of a consortium of six international commissions in successfully making a bid to ICSU to support collaborative research results on "Past Hydrological Events related to understanding Global Change", and that will certainly be one way in which the present research foundations progress and continue.

Ken Gregory
Gerardo Benito

Foreword

The word *'palaeohydrology'* appears to have been used first by Leopold and Miller in their US Geological Survey Water Supply Paper 1261 (1954) as follows:

"Consideration of the alluvial chronology brings to mind a host of questions concerning a comparison of the present with conditions which prevailed in the past. These questions are concerned not so much with past climate itself, but with the interaction of climate, vegetation stream regimen and runoff, which obtained under different climates from that of the present. That is, with hydrological rather than merely climatic factors. To describe this general subject the word 'palaeohydrology' is introduced."

Subsequent work using empirical relations among precipitation, temperature, runoff, and sediment yield was done in order to estimate past hydrologic conditions for geomorphic and stratigraphic purposes and to estimate the hydrologic conditions of small, ungauged watersheds for land management purposes. Only a few specialists were involved in this activity. However, with the advent of concerns about global climate change, the situation changed markedly with scientists with many different perspectives cooperating in the development of a broad-based and internationally oriented palaeohydrology. For example, in this volume, one of a series edited by Professor Gregory and his colleagues, eleven chapters are devoted to the palaeohydrology of various geographic regions of the globe. The remaining chapters deal with the interrelation of evidence for environmental change.

The purpose of this volume is to summarise developments in global continental palaeohydrology for the period 1991 through 2002, during which the INQUA Global Continental Palaeohydrology Commission was active, and to demonstrate how the results of a decade of activity can be used to contribute to the understanding of global climate change and its impact on all the topics listed by Leopold and Miller in the definition of palaeohydrology.

S.A. Schumm

PART 1   INTRODUCTION

1 Potential of Palaeohydrology in Relation to Global Change

K.J. GREGORY[1] AND G. BENITO[2]
[1]*University of Southampton, Southampton, UK*
[2]*CSIC-Centro de Ciencias Medioambientales, Madrid, Spain*

> *The past is not dead. It is not even past.* William Faulkner (Cited in Peter Carey
> *True History of the Kelly Gang.* Faber and Faber, London, 2001)

William Faulkner's sentiment is very apposite to the study of the earth's physical environment: "the present is the key to the past" has been an acclaimed dictum since the nineteenth century, but it is only very recently that the importance of past environmental change for understanding the present and the future has become more widely acknowledged. Greater attention to environmental processes (Gregory, 2000) focusing on techniques, process mechanics and modelling meant that research was developed independently from that on Quaternary environmental change. Once analysis of environmental processes was firmly established and short-term variations recognised, paralleled by questions raised by interpretations of Quaternary environmental change, the potential of a fruitful interaction was appreciated. Such interaction has become even more prescient with the advent of global change research in the last two decades of the twentieth century. Concern for future global change requires understanding of past as well as present environmental conditions so that the title for this volume "Palaeohydrology – Understanding Global Change" was chosen in order to expound one particular perspective. No single discipline is able to do justice to all aspects of global change, many new hybrid research fields having been generated at research frontiers where disciplinary boundaries intersect. Palaeohydrology is one such multidisciplinary field: its development, and the methods of approach employed, may be seen in relation to global change and scenarios developed, as a basis for potential links between palaeohydrology and global change, prior to showing how this volume is structured to address the potential research opportunities.

1 DEVELOPMENT OF PALAEOHYDROLOGY

Palaeohydrology is an excellent example of a research field that arose at the interface of several disciplines and is multidisciplinary in nature. Its development since 1954, when it was first formally defined (Leopold and Miller, 1954), has been chronicled (Gregory, 1983). Alluvial chronology was fundamental to the approach employed by Leopold and Miller (1954), and the global hydrologic cycle (Schumm, 1965), involving comparison of present and past water balance situations, was basic to the approach of Quaternary palaeohydrology specified by Schumm (1965). It was later supplemented by water

Palaeohydrology: Understanding Global Change. Edited by K.J. Gregory and G. Benito
© 2003 John Wiley & Sons, Ltd ISBN: 0-470-84739-5

quality and composition (Schumm, 1977), with important contributions developed from the investigation of underfit streams (Dury, 1964a; 1964b; 1965; 1977). Palaeohydrology could therefore appropriately be defined as "the science of the waters of the earth, their composition, distribution and movement on ancient landscapes from the beginning of the first rainfall to the beginning of continuous hydrological records" (Gregory, 1983). Subsequent research developments increased interdisciplinary effort (Baker, 1983; 1998), involved a broadening scope (Baker, 1995) encompassing ancient lake sediments and past groundwater levels. It visualised palaeohydrology as a new branch of the earth sciences (Starkel, 1995) with branches that included fluvial palaeohydrology (Branson et al., 1996) all focusing upon interpreting indices of past hydrological processes (Baker, 1996). One way in which this new branch of geoscience could be manifested was by applying models of hydrological cycles and water balance to the scenarios of the past (Grosswald, 1998) and speculations relating to the geological past had been presented (Schumm, 1968).

Palaeohydrological research is therefore less than 50 years old; it began with a global thrust, evolved and advanced by focus on particular areas such as the Temperate Zone (Starkel et al., 1991) and has now returned to a global focus supported by particular research themes (Gregory et al., 1995; Branson et al., 1996). Throughout this time, emphasis shifted from a conviction that predictions can be achieved to a more realistic view that acknowledges the difficulties of the past as well as that of the future "reconstructions". Contributions from international researchers were significant during the progress of International Geological Correlation Programme 158 A and B (Starkel et al., 1991) and from 1991 to 2003 by the Global Continental Palaeohydrology Commission of INQUA (GLOCOPH). Both these international research programmes culminated in collections of research papers, most recently concerned with environmental change (Benito et al., 1998), and with particular areas (Brown and Kadomura, 2001) in addition to the many productive international research meetings. The Commission on Global Continental Palaeohydrology was established by INQUA in Beijing in 1991 (see Chapter 2) in order *to undertake research on the global process of water for the continental areas of the Earth using evidence of past changes and emphasizing issues relevant to the human habitability of the planet.* The primary aim specified in 1994 was *to analyze the nature of global and zonal hydrological changes, fluxes and stores using the timescale 100–1,000 years with emphasis on those areas which hold the greatest human population and are most sensitive in terms of water resources and global changes.* Professor Leszek Starkel was President of the Commission from 1991 to 1995, Professor Vic Baker from 1995 to 1999, and Professor Ken Gregory from 1999 to 2003. Global palaeohydrological change was considered by the previous President (Baker, 1995), and a recent GLOCOPH volume emphasised palaeohydrology and environmental change (Benito et al., 1998), so that this book, at the end of the GLOCOPH period, is overtly constructed to focus attention upon the contributions that palaeohydrology can make.

External changes have necessarily affected the development of palaeohydrological research, but not perhaps always as rapidly as they might have done. They have included, first, the way in which philosophy has been more prominent in the scientific agenda (Baker, 1996), the techniques revolution increasingly reflected in research reported in palaeohydrological publications (Gregory, 1983; Wohl and Enzel, 1995), particularly the impact of information and database technology (Branson et al., 1995), culminating in admirable palaeohydrological reconstructions (Starkel, 1987; 1990; Starkel et al., 1991). Secondly, external changes have encouraged greater concern for relevance including research on global change, on the collation of world data, on the linkages from global circulation models to hydrological models, and on management of the present system from basin and river channel perspectives. Underpinning these emerging perspectives has been greater recognition of the benefits of knowledge of temporal change (see Chapter 21, Table 21.3).

2 APPROACHES EMPLOYED

Palaeohydrology when first defined was associated with retrodictive analyses based upon quantitative relationships between precipitation, runoff and sediment yield. The earliest approaches in palaeohydrology were founded on the relationships proposed by Langbein and Schumm (1958) and by Schumm (1965), subsequently developed by river metamorphosis equations (Schumm, 1969) from proposals by Lane (1955). As indicated in Chapter 17, a numerical approach was also involved in the analysis of underfit streams pioneered by Dury (1977), so that there emerged a major group of approaches based particularly upon studies of palaeochannels. A somewhat separate group of research investigations was founded using pollen analysis and other techniques for the reconstruction of environmental history and was able to develop interpretations of water balances in the past. Subsequently, a series of significant contributions came to be made from the study of palaeofloods. In the 1980s, under the aegis of IGCP 158, palaeohydrological research investigations were organized into those associated with lakes and mires (the *pollen* emphasis in subproject 158B) and those under the fluvial subproject (158A) in which investigations based upon *palaeochannels* were especially notable. By the end of the 1980s, these two subprojects saw the polarisation of approaches: one dedicated to environmental history and the water balance, and the other more concerned with river and channel behaviour. From this foundation, it was appreciated that analysis of the spatial pattern was necessary as a context for understanding the sequence of hydrological development. In theory, palaeohydrology is concerned with all components of the hydrological cycle, but in practice most research focuses on river channels and discharge, on lake level fluctuations, on fluctuations in groundwater levels and isotope chemistry and on proxy indicators of past precipitation characteristics such as tree rings, ice cores, pollen, or soils (Anthony and Wohl, 1998).

Significant developments have been achieved in the integrated global context since 1991 when GLOCOPH was established as a Commission of INQUA. At the four international conferences in Southampton in 1994, Toledo in 1996, Tokyo in 1998 and Moscow in 2000, 221 papers were presented and their subjects give clues to the approaches prevailing on those dates. It is not easy to categorise all the approaches adopted and the techniques employed because some of the intriguing technique developments included forest colonisation of river beds, lake ice phenology and the use of literary records; research included contributions from all continents except the Polar areas and the size of areas studied ranged from the Amazon and the Yenesei to small headwater basins in Crete, Japan and Poland. However, throughout all these GLOCOPH research investigations, it is notable that many were based upon analysis of data from sedimentary sequences, especially alluvial sediments and alluvial chronology. Of the papers presented at the four international conferences (Table 1.1), 29% were explicitly concerned with sediment-based investigations; if the papers that implicitly involved sediment analysis are included, then this accounts for 55% of all the papers presented. The rank order of the major research themes (from Table 1.1) is:

Table 1.1 Subjects of papers presented at GLOCOPH meetings

Dominant theme	Southampton and London 1994	Toledo, Spain 1996	Kumagaya, Japan 1998	Moscow 2000	Total
Glacial	0	5	6	2	13 (5.9%)
Techniques-based including palaeoecology, historical records	4	12	17	15	48 (21.8%)
Modelling	4	0	1	10	15 (6.9%)
Processes	3	6	15	9	33 (15.0%)
Sediment-based	15	19	23	7	64 (29.0%)
Basin components including palaeochannel, lakes, planform	11	17	4	4	36 (16.3%)
Drainage basin	2	5	3	2	12 (5.4%)

- *sediment-based* including slackwater deposits, sedimentary sequences and terrace deposits,
- *techniques-based* including palaeoecology and historical records,
- *basin components* including palaeochannels, lakes and planform,
- *processes* including volcanic eruptions, tectonics, groundwater, palaeofloods and water balance.

Therefore palaeohydrology is no longer founded upon the water balance and pollen, or upon the river and palaeochannels because other themes have become prominent.

The trend towards greater emphasis upon results from analysis of sedimentary data is very necessary because an increasingly diverse range of analytical techniques has been available to enable palaeohydrological reconstructions to be undertaken at a level that could not be anticipated in 1954 when palaeohydrology was first formally specified. However, relatively few research investigations, just 5% (Table 1.1), explicitly focused upon basin-wide changes or referred analysis to a basin framework, although major shifts of river channel pattern and glacial drainage changes have been investigated. This is perhaps inevitable in view of the way in which GLOCOPH has been able to advance palaeohydrology because it was necessary for emphasis to be placed upon use of a range of techniques, upon constructing different climatic scenarios and upon formulating palaeohydrological sequences for particular areas. However, as it is now appropriate to see how the results from GLOCOPH, together with those from the previous basin-based IGCP 158 research programme can be integrated and applied, reference to a basin context becomes essential to facilitate a link to hydrological modelling.

3   GLOBAL CHANGE SCENARIOS

The greenhouse effect has probably been recognised since 1827 (Jones and Henderson-Sellers, 1990), but it is only the last two decades of the twentieth century that saw substantial interest in global change; and only towards the end of that century has there been worldwide consensus on the direction of that change, namely, an increase of temperature resulting from an equilibrium doubling in greenhouse gases over pre-industrial levels (Henderson-Sellers, 1994). From the 1972 United Nations Stockholm conference up to the 1992 Rio Earth Summit, which focused on issues of climate change, biodiversity protection and sustainable development, debate about global change accelerated. The Agenda 21 Report produced for the 1992 meeting laid out an agenda for research and action over the next century. Subsequent world meetings in Kyoto (1997) and Buenos Aires (1998) were equally significant for furthering international discussion. The Intergovernmental Panel on Climate Change (IPCC) was established in 1988 to assess scientific information in relation to global warming and to suggest response strategies (Drake, 2000), three reports having already been published, engendering considerable international discussion and debate. Global environmental change was sometimes used in the sense of global aspects of natural environmental change, but this is no longer tenable (Mathews, 2001) because of the interaction of natural and anthropogenic processes. Global environmental change, often used synonymously with global change, can be thought of in two ways (Goudie, 2000): firstly, systemic global change including those global changes in climate brought about by atmospheric pollution; and secondly, change which is the consequence of areal or localized transformation processes so ubiquitous and pervasive that they cumulatively result in global change. In addition to the IPCC reports concerned primarily with climate change, there have been a number of international initiatives concerned with global change (see Chapter 3) including the International Geosphere–Biosphere Programme (IGBP) of the International Council of the Scientific Unions (ICSU). The six key questions posed in their Global Change report in 1990 notably did not include a specific hydrological question. Growing interest in global change research has been reflected in the creation of new journals such as *Global Environmental Change* (Pergamon, 1990), *Global Outlook Environment* (NY UNEP distributed by OUP, 1997) and *Global Ecology and Biogeography* (Blackwell, 1992); and also in the publication of encyclopaedic volumes such as *The Encyclopaedic Dictionary of Environmental Change* (Mathews, 2001) and the *Encyclopedia of Global Environmental Change* (Munn, 2001).

Despite greatly increased attention being given to global change, it has not been easy to raise general public awareness of the problems involved although a report *Global Environmental Outlook 2000* (United Nations, 1999) was widely reported as indicating that global warming will trigger a series of disasters with serious world implications. A major focus has been the contribution made by research on global climate models with knowledge and understanding of the mechanics of the world climate system being enhanced as a result of the building of sophisticated climate computer models, the vast amounts of remotely sensed data available and the establishment of a reliable calendar of geological events (Huggett, 1991). Some research has focused upon the ways in which climate change, as indicated by the outputs from Global Circulation Models (GCMs), might impact on the environmental system, exemplified by changes of surface air temperatures, precipitation rates and soil–water world patterns produced for different seasons (Henderson-Sellers, 1994).

Research undertaken on the hydrological consequences of global change has included exploration of aspects of greenhouse hydrology (Wilby, 1995), but GCMs are not well suited to answering questions concerning regional hydrologic variability (Xu, 1999), which are of primary interest to hydrologists. GCMs were originally designed to predict average synoptic scale general circulation patterns of the atmosphere, so their outputs cannot easily be harnessed by downscaling to provide inputs to hydrologic models at regional or local scales. Estimation of hydrology and water resource changes continues to pose problems for hydrologists (Nemec, 1995). Analyses of scenarios have been considered for river flow extremes and fluvial erosion in particular areas such as England and Wales (Newson and Lewin, 1991) and implications have also been explored for water resources in the United Kingdom (Arnell, 1996) and in Europe (Arnell, 1999). Against a background of methodologies for climate change impact assessments, techniques for developing climate change scenarios and hydrological models, Arnell (1996) utilised a case study of 21 catchments in the United Kingdom to explore changes in river flows and groundwater recharge in Britain that might occur by the 2050s, and analysed the possible impacts of the changes on water uses and the management of water resources. However, it is not simply an issue of documenting climate change and proposing appropriate mitigation measures; it is also important to consider how to adapt to climate change (Parry et al., 1998).

Whereas results from GCMs suggest that rising concentrations of greenhouse gases may have significance for global climate, it is less clear as to what extent local (subgrid) scale processes will be affected (Wilby and Wigley, 1997). Downscaling techniques emerged to bridge the gap between what climate modellers are able to provide and what impact assessors require. However, it is not only a question of downscaling spatially but also of recognising that the data analysed also has to be downscaled temporally – it is these problems of spatial and temporal scale that need to be addressed. A major limitation in undertaking downscaling and in using hydrological model approaches is the lack of sufficient data sets in a variety of climatic and physiographic regions related to a range of spatial and temporal scales (Xu, 1999) and it is towards such a paucity of data that palaeohydrological results may contribute.

4   POTENTIAL LINKS BETWEEN PALAEOHYDROLOGY AND GLOBAL CHANGE

Advances in the investigation of climate change and associated global change have not been accompanied by equivalent attention devoted to impacts on the hydrological system, despite the fact that in facets of hydrology major impacts of global change are really sustained. This disparity has arisen because attention had to be focused initially upon climate change and only once sufficient research had been achieved was it possible to consider detailed impacts such as hydrological ones. Additionally, the downscaling necessary to establish hydrological impacts cannot be completely achieved by reference to continuous hydrological records and contemporary timescales, so that it may be necessary to search longer timescales for analogous behaviour. Climate reconstructions of the past were achieved from the COHMAP (Cooperative Holocene Mapping Project) project (Wright et al., 1993) in which water balance changes were included, although it did not extend to include hydrology or palaeohydrology. In scrutinising aspects of climate change in the last 2,000 years, GCMs are insufficiently detailed so that a better understanding of not only what has happened over the past 2,000 years but also the history of forcing factors is vital to distinguish between natural variability and that resulting from anthropogenic influence on the climate system (Jones et al., 1994). Prior to the end of the twentieth century, the citation pattern, as demonstrated by content analysis, reflects the emphasis on palaeoclimate showing that palaeohydrological research has not been extensively quoted, so that Physical Geography Abstracts show Palaeoclimate, Palaeoenvironment, Palaeooceanography, and Palaeolimnology all cited more frequently than palaeohydrology (Table 1.2); and in books dealing with Quaternary environmental change, or with global change in general, palaeohydrology is seldom mentioned. This may be because development of palaeohydrology is recent although it has been argued that one of the major applications of research from GLOCOPH is to inform interpretations

Table 1.2 References in Physical Geography Abstracts

Term	1994	1995	1996	1997	Total
Palaeoclimate	38	43	43	51	175
Palaeoenvironment	55	40	59	72	226
Palaeooceanography	19	12	20	45	96
Palaeolimnology	35	25	15	15	90
Palaeohydrology	23	12	15	16	66
Palaeogeography	10	5	24	10	49
Palaeoclimatology	10	9	10	4	33
Palaeobotany	4	3	5	3	15
Palaeopedology	1	1	0	1	3
Palaeohydrology: database	101	85	57	22	265

of global climate change (Gregory, 1998). In specific terms, it has been shown how palaeoflood hydrology can assist in the design of hydraulic structures and in water resources management (Jarrett, 1996); and it has been noted that attention needs to be given to human adaptation to the changing global environment (Baker, 1998). The situation is changing so that books (e.g. Arnell, 2002) and dictionary entries (Anthony and Wohl, 1998; Mathews, 2001) now refer to the potential that palaeohydrology offers in the range of palaeosciences, defined as those branches of the natural environmental sciences that focus on the reconstruction and modelling of past events rather than direct observation and experiment (Mathews, 2001). It is presumptuous to assume that any one multidisciplinary discipline can, on its own, contribute to knowledge of global change because throughout many disciplines it is now appreciated (e.g. Cotton and Pielke, 1995) that scientists are grossly underestimating the complexity of interactions between the earth's atmosphere, ocean, geosphere and biosphere so that earth system science (Chapter 21) is being promulgated. Furthermore, when considering global change it has to be recognised that adjustments in water resources will be regional (with floods and droughts both projected to increase in intensity) or local (riparian zones) but, despite the sure knowledge that global change will occur (Perry, 1999), the exact nature of the consequential changes cannot be predicted with current limitations in our knowledge and computing power (Gleick, 1993). Indeed, Baker (1995) regretted how the international global change science initiative has made little appropriate use of the treasure of experience gained from past environmental changes.

Climate change analysed using GCMs (Henderson-Sellers, 1994) has been considered in relation to palaeohydrology (Arnell, 1996), but the question arises as to how far outputs from GCMs can be effectively coupled to hydrological models of an appropriate scale. It is generally agreed that regional climate prediction is not an insoluble problem, although it is characterized by inherent uncertainty that is derived from two sources: the unpredictability of the climatic system as a result of deterministic chaos and of the global system that renders climate predictions uncertain through unpredictability of external forcings superimposed on the climate system (Mitchell and Hume, 1999). It has been contended (e.g. Airey and Hulme, 1995) that model simulations of future changes to magnitude, timing and spatial pattern of global precipitation should be viewed as scenarios, not predictions. There remain, therefore, significant gaps relating to hydrological forecasts and to water resource estimation, including requirements for reliable techniques for downscaling climate model simulations to the catchment scale and for more integrated models (Arnell, 1996). To achieve relationships between global climate change investigations and impacts, it is necessary to complement inductive and deductive global change science approaches with retroductive global change science. Retroduction is a characteristic reasoning mode in Earth Sciences that involves synthetic reasoning often using analogies, applying the classical doctrines of commonsensism, fallibilism and realism (Baker, 1995) and because it emphasises deriving hypotheses from nature, it is very appropriate for palaeohydrology.

In considering potential links between palaeohydrology and global change, it is possible to identify several categories, to show how the subsequent chapters relate to these issues (Section 5) and demonstrate their pertinence in relation to river channel management (Chapter 21, Table 21.3), and then finally (Chapter 22) to suggest how conclusions from the chapters relate to the issues identified. Major issues relate to the following:

- Derivation of *data* to complement periods of continuous hydrological records relating to water balance, hydrological extremes, water quality, especially sediment involving historical sequences of change more clearly elucidated from the neolithic to the twentieth century.
- *Mechanics* of temporal change, including the significance of thresholds, sensitivity/hypersensitivity/undersensitivity, control by vegetation, human activity in this regard, debris stores and slugs, links with ecology including woody debris and planform.
- *Spatial contrasts* – differences between world zones, within zones and within basins, and their synchroneity.
- *Coupling* of Global Climate Change Models to hydrological models and the necessary forcing mechanisms.
- Construction of *new models* of a retroductive kind that may be non-linear, aided by techniques such as ^{137}Cs.

Examples of palaeohydrological contributions already made to these five categories (Table 1.3) are necessarily sequential with examples of the later ones depending upon further research. There will be the need to develop a conceptual basis as a framework of reference within which research results are presented in order to facilitate their application. Although research contributions focused upon the component elements of the palaeohydrological water balance, upon the balance itself, upon alluvial chronology and upon contributions in accord with the PAGES (see Chapter 3) framework are essential, it is also necessary to place research results in the context of a drainage-basin-based qualitative model of river channel structure. It is argued here that this is one necessary next step for palaeohydrology in order to facilitate further applications of research results; some progress has been made in this direction as indicated in Chapter 17. The need for a basin framework has been anticipated in some studies presented to GLOCOPH and it has been argued that, in the humid tropics, although considerable attention has been paid to studies of lake levels worldwide and to the palynology of mires, there has been a reluctance to translate these findings into interpretations or models of landscape response to climate change (Thomas, 1994), so that alternative models of stream response to climatic change are still necessary (Thomas, 1998). Some uncertainty remains an element in any forward estimation of future scenarios because of complex interactions of systems with large numbers of variables that may be non-linear and difficult to model, as well as the paucity of long-term data

Table 1.3 Examples of use of palaeohydrology results in relation to global change

Category of application	Example	Source
Derivation of *data* to complement periods of continuous hydrological records relating to water balance, hydrological extremes, water quality, especially sediment. Involving historical sequences of change more clearly elucidated particularly from the Neolithic to the twentieth century	Balance between precipitation and evaporation derived from past lake levels in closed basins by dating sedimentary deposits.	(Fontes and Gasse, 1991)
	Holocene flood chronology for the Upper Mississippi Valley from palaeochannel evidence.	(Knox, 1993)
	Dendrochronologic evidence for the frequency and magnitude of floods.	(Yanosky and Jarrett, 2002)
	Thousand-year record of channel change reconstructed for middle Trent and may be valuable for both model validation and planning purposes.	(Brown *et al*., 2001)
Mechanics of temporal change, including control by vegetation, woody debris, human activity, debris stores and slugs, the significance of thresholds, sensitivity/ hypersensitivity/ undersensitivity	Climatic and human controls on runoff and qualitative descriptions on torrential floods since the Bronze Age.	(Provansal, 1995)
	Impact of land use changes since the Neolithic on soil erosion and hydrology, with quantification of average rates of sediment yields at different periods.	(Jorda *et al*., 1991)
Spatial contrasts – differences in palaeohydrology and palaeoforms between world zones, within zones and within basins and their synchroneity	*Between zones*: Intracontinental runoff systems of North Asia and the influence of ice-dammed lakes.	(Rudoy, 1998)
	Within zones: Regional chronology of flood magnitude and frequency in small basins in western Arizona.	(House and Baker, 2001)
	Within basins: Palaeoflood hydrology of the Tagus river, Spain. Modelling fluvial dynamics along Allier/Loire, France.	(Benito *et al*., 1998; Veldkamp and Van Dijke, 1998)
Palaeohydrologic data used in *coupling* of Global Climate Change Models to hydrological models and the necessary forcing mechanisms	Lake and pollen observations of past climates are compared with climates simulated by numerical climate models.	(COHMAP Members, 1988)
	A simplified GCM is used to investigate the climate and hydrological regime sensitivity to the changes in insolation flux at the top of atmosphere, CO_2 variations, and topographic changes during the last 20 ka.	(Kislov and Surkova, 1998)
Construction of *new models*, possibly non-linear, of a retroductive kind, aided by techniques such as [137]Cs	Regional palaeoflood databases applied to flood hazards and palaeoclimate analysis.	(Diez-Herrero *et al*., 1998.)

and the fact that there may be factors previously ignored, with nature possibly having surprises up its sleeve in the form of catastrophic or extreme events (Goudie, 1993).

5 STRUCTURE

The chapters in this book are designed, in the light of the culmination of GLOCOPH Research (1991–2003), to demonstrate how palaeohydrology can be of value in the understanding of global change. A brief section of perspectives, therefore, provides some indication of the INQUA context (Chapter 2) and of the relationship to global change programmes (Chapter 3). The major sections of the volume are concerned with 10 major world areas, with explanation of the regional division that is adopted in Chapter 4. A second major section is devoted to recent progress interpreting evidence of environmental change and, like the regional section, cannot be comprehensive; topics are selected to focus upon issues subject to current research and capable of considerable application. In the final two chapters of the section, attention is directed to the significance of short-term changes (Chapter 20), a topic appearing as one of increasing recognition from many strands of palaeohydrological research; and to river channel management (Chapter 21), which is a major area for application.

Referring back to William Faulkner's sentiment: the past is certainly not dead, for scientists directly interested in palaeohydrological research and also for many others who might benefit from the results obtained.

REFERENCES

Airey, M. and Hulme, M., 1995. Evaluating climate simulations of precipitation: methods, problems and performance. *Progress in Physical Geography*, **19**, 427–448.

Anthony, D. and Wohl, E., 1998. Palaeohydrology. In R.W. Herschy and R.W. Fairbridge (eds), *Encyclopedia of Hydrology and Water Resources*. Kluwer Academic Publishers, Dordrecht, Boston, London, 508–511.

Arnell, N., 1996. *Global Warming, River Flows and Water Resources*. Wiley, Chichester.

Arnell, N., 1999. The effect of climate change on hydrological regimes in Europe: a continental perspective. *Global Environmental Change*, **9**, 5–23.

Arnell, N., 2002. *Hydrology and Global Environmental Change*. Prentice Hall, Harlow.

Baker, V.R., 1983. Large scale fluvial palaeohydrology. In K.J. Gregory (ed.), *Background to Palaeohydrology*. Wiley, Chichester, 453–478.

Baker, V.R., 1995. Global palaeohydrological change. *Quaestiones Geographicae, Special Issue 4, Late-Quaternary Relief Evolution and Environment Changes*. Poznan, 27–36.

Baker, V.R., 1996. Discovering earth's future in its past: palaeohydrology and global environmental change. In J. Branson, A.G. Brown and K.J. Gregory (eds), *Global Continental Changes: The Context of Palaeohydrology*. Special Publication No. 115, Geological Society, London, 73–83.

Baker, V.R., 1998. Palaeohydrology and the hydrological sciences. In G. Benito, V.R. Baker and K.J. Gregory (eds), *Palaeohydrology and Environmental Change*. Wiley, Chichester, 1–12.

Benito, G., Baker, V.R. and Gregory, K.J., (eds), 1998. *Palaeohydrology and Environmental Change*. Wiley, Chichester.

Benito, G., Machado, M.J., Perez-Gonzalez, A. and Sopena, A., 1998. Palaeoflood hydrology of the Tagus River, central Spain. In G. Benito, V.R. Baker and K.J. Gregory (eds), *Palaeohydrology and Environmental Change*. Wiley, Chichester, 317–333.

Branson, J., Brown, A.G. and Gregory, K.J., (eds), 1996. *Global Continental Changes: The Context of Palaeohydrology*. Special Publication No. 115, Geological Society, London.

Branson, J., Clark, M.J. and Gregory, K.J., 1995. A database for global continental palaeo-hydrology: technology or scientific creativity? In K.J. Gregory, L. Starkel and V.R. Baker (eds), *Global Continental Palaeohydrology*. Wiley, Chichester, 303–325.

Brown, A.G., Cooper, L., Salisbury, C.R. and Smith, D.N., 2001. Late Holocene channel changes of the middle Trent: channel response to a thousand-year flood record. *Geomorphology*, **39**, 69–82.

Brown, A.G. and Kadomura, H., (eds), 2001. Contributions to temperate and humid tropical palaeohydrology. *Geomorphology*, **39**, 1–82.

COHMAP Members, 1988. Climatic changes of the last 18,000 years: observations and model simulations. *Science*, **241**, 1043–1052.

Cotton, W.R. and Pielke, R.A., 1995. *Human Impacts on Weather and Climate*. Cambridge University Press, Cambridge.

Diez-Herrero, A., Benito, G. and Lain-Huerta, L. 1998. Regional palaeoflood databases applied to flood hazards and palaeoclimate analysis. In G. Benito, V.R. Baker and K.J. Gregory (eds), *Palaeohydrology and Environmental Change*. Wiley, Chichester, 335–347.

Drake, F., 2000. *Global Warming. The Science of Climate Change*. Arnold, London.

Dury, G.H., 1964a. *Principles of Underfit Streams*. US Geological Survey Professional Paper 452A.

Dury, G.H., 1964b. *Subsurface Exploration and Chronology of Underfit Streams*. US Geological Survey Professional Paper 452B.

Dury, G.H., 1965. *Theoretical Implications of Underfit Streams*. US Geological Survey Profes-sional Paper 452C.

Dury, G.H., 1977. Underfit streams: retrospect, perspect and prospect. In K.J. Gregory (ed.), *River Channel Changes*. Wiley, Chichester, 281–293.

Fontes, J.C. and Gasse, F., 1991. PALHYDAF (Palaeohydrology in Africa) pro-gram – objectives, methods, major results. *Palaeogeography, Palaeoclimatology, Palaeoe-cology*, **84**, 191–215.

Gleick, P., 1993. *Water in Crisis: A Guide to the World's Fresh Water Resources*. Oxford Uni-versity Press, Oxford.

Goudie, A.S., 1993. Environmental uncertainty. *Geography*, **78**, 137–141.

Goudie, A.S., 2000. Global environmental change. In D.S.G. Thomas and A. Goudie (eds), *The Dictionary of Physical Geography*. Blackwell, Oxford, 228.

Gregory, K.J., (ed.), 1983. *Background to Palaeohydrology*. Wiley, Chichester.

Gregory, K.J., (ed.), 1998. Applications of palaeohydrology. In G. Benito, V.R. Baker and K.J. Gregory (eds), *Palaeohydrology and Environmental Change*. Wiley, Chichester, 13–25.

Gregory, K.J., (ed.), 2000. *The Changing Nature of Physical Geography*. Arnold, London.

Gregory, K.J., Starkel, L. and Baker, V.R., (eds), 1995. *Global Continental Palaeohydrology*. Wiley, Chichester.

Grosswald, M., 1998. New approach to the ice age paleohydrology of Northern Eurasia. In G. Benito, V.R. Baker and K.J. Gregory (eds), *Palaeohydrology and Environmental Change*. Wiley, Chichester, 199–214.

Henderson-Sellers, A., 1994. Numerical modelling of global climates. In N. Roberts (ed.), *The Changing Global Environment*. Blackwell, Oxford, 99–124.

House, P.K. and Baker, V.R., 2001. Palaeohydrology of flash floods in small desert watersheds in western Arizona. *Water Resources Research*, **37**, 1825–1839.

Huggett, R.J., 1991. *Climate, Earth Processes and Earth History*. Springer-Verlag, Berlin, Hei-delberg, New York.

Jarrett, R.D., 1996. Palaeohydrology and its value in analyzing floods and droughts. *Water Resources Investigations Reports*, 95-4015, US Geological Survey, Denver, 13–14.

Jones, M.D.H. and Henderson-Sellers, A., 1990. History of the greenhouse effect. *Progress in Physical Geography*, **14**, 1–18.

Jones, P.D., Bradley, R.S. and Jouzel, J., (eds), 1994. *Climatic Variations and Forcing Mecha-nisms of the Last 2000 years*. Springer-Verlag, Berlin, Heidelberg, New York.

Jorda, M., Parron, C., Provansal, M. and Roux, M., 1991. Erosion et détritisme holocènes en Basse Provence calcaire. L'impact de l'anthropisation. *Physio-Géo*, **22–23**, 37–47.

Kislov, A.V. and Surkova, G.V., 1998. Simulation of the Caspian sea level changes during the last 20,000 years. In G. Benito, V.R. Baker and K.J. Gregory (eds), *Palaeohydrology and Environmental Change*. Wiley, Chichester, 235–244.

Knox, J.C., 1993. Large increases in flood magnitude in response to modest changes in climate. *Nature*, **361**, 430–432.

Lane, E.W., 1955. The importance of fluvial morphology in hydraulic engineering. *Proceedings American Society of Civil Engineers*, **81**, 1–17, Paper 745.

Langbein, W.B. and Schumm, S.A., 1958. Yield of sediment in relation to mean annual precipitation. *Transactions American Geophysical Union*, **32**, 347–357.

Leopold, L.B. and Miller, J.P., 1954. *Postglacial Chronology for Alluvial Valleys in Wyoming*. US Geological Survey Water Supply Paper 1261, 61–85.

Mathews, J.A., (ed.), 2001. *The Encyclopaedic Dictionary of Environmental Change*. Arnold, London.

Mitchell, T.D. and Hume, M., 1999. Predicting regional climate change: living with uncertainty. *Progress in Physical Geography*, **23**, 57–78.

Munn, T., (ed.), 2001. *Encyclopedia of Global Environmental Change*. Vol. 5, Wiley, Chichester.

Nemec, J., 1995. General circulation models (GCMS), climatic change, scaling and hydrology. In G.W. Kite (ed.), *Time and the River*. Water Resources Publications, Highland Ranch, CO, 317–356.

Newson, M.D. and Lewin, J., 1991. Climatic change, riverflow extremes and fluvial erosion – scenarios for England and Wales. *Progress in Physical Geography*, **15**, 1–17.

Parry, M., Arnell, N., Hume, M., Nicholls, R. and Livermore, M., 1998. Adapting to the inevitable. *Nature*, **395**, 741.

Perry, J.A., 1999. Water, water quality, water supply. In D.E. Alexander and R.W. Fairbridge (eds), *Encyclopedia of Environmental Science*. Kluwer Academic Publishers, Dordrecht, Boston, London, 674–682.

Provansal, M., 1995. The role of climate in landscape morphogenesis since the Bronze Age in Provence, southeastern France. *The Holocene*, **5**(3), 348–353.

Rudoy, A., 1998. Mountain ice-dammed lakes of Southern Siberia and their influence on the development and regime of the intracontinental runoff systems of North Asia in the Late Pleistocene. In G. Benito, V.R. Baker and K.J. Gregory (eds), *Palaeohydrology and Environmental Change*. Wiley, Chichester, 215–234.

Schumm, S.A., 1965. Quaternary palaeohydrology. In H.E. Wright and D.G. Frey (eds), *The Quaternary of the United States*. Princeton University Press, Princeton, 783–794.

Schumm, S.A., 1968. Speculations concerning palaeohydrologic controls of terrestrial sedimentation. *Bulletin Geological Society of America*, **79**, 1573–1588.

Schumm, S.A., 1969. River metamorphosis. *Proceedings American Society of Civil Engineers, Journal Hydraulics Division*, **95**, 255–273.

Schumm, S.A., 1977. *The Fluvial System*. Wiley, New York.

Starkel, L., 1987. *Evolution of the Vistula River Valley During the Last 15,000 years*, Part II, Polish Academy of Sciences, Geographical Studies, Institute of Geography, Special Issue 4, Warsaw.

Starkel, L., 1990. *Evolution of the Vistula River Valley During the Last 15,000 years*, Part III, Polish Academy of Sciences, Geographical Studies Institute of Geography, Special Issue 5, Warsaw.

Starkel, L., 1995. Introduction to global palaeohydrological changes. In K.J. Gregory, L. Starkel and V.R. Baker (eds), *Global Continental Paleohydrology*. Wiley, Chichester, 1–20.

Starkel, L., Gebica, P., Niedzialkowska, E. and Podgorska-Tkacz, A., 1991. *Evolution of Both the Vistula Floodplain and Late-Glacial – Early Holocene Palaeochannel Systems in the Grobla Forest (Sandomierz Basin), Evolution of the Vistula River Valley During the Last 15,000 years*, Part IV. Polish Academy of Sciences, Geographical Studies Institute of Geography, Special Issue 6, Warsaw, 87–99.

Starkel, L., Gregory, K.J. and Thornes, J.B., (eds), 1991. *Temperate Palaeohydrology: Fluvial Processes in the Temperate Zone During the Last 15,000 years*. Wiley, Chichester.

Thomas, M.F., 1994. *Geomorphology in the Tropics*. Wiley, Chichester.

Thomas, M.F., 1998. Late Quaternary instability in the humid and sub-humid tropics. In G. Benito, V.R. Baker and K.J. Gregory (eds), *Palaeohydrology and Environmental Change*. Wiley, Chichester, 247–258.

Veldkamp, A. and van Diijke, J.J., 1998. Modelling long term erosion and sedimentation processes in the late medieval climatic deterioration in Europe. In G. Benito, V.R. Baker and K.J. Gregory (eds), *Palaeohydrology and Environmental Change*. Wiley, Chichester, 53–66.

Wilby, R.L., 1995. Greenhouse hydrology. *Progress in Physical Geography*, **19**, 351–369.

Wilby, R.A. and Wigley, T.M.L., 1997. Downscaling general circulation model output: a review of methods and limitations. *Progress in Physical Geography*, **21**, 530–548.

Wohl, E.E. and Enzel, Y., 1995. Data for palaeohydrology. In K.J. Gregory, L. Starkel and V.R. Baker (eds), *Global Continental Palaeohydrology*. Wiley, Chichester, 23–59.

Wright, H.E., Kutzbach, J.E., Webb III, T., Ruddiman, W.F., Street-Perrott, F.A. and Bartlein, P.J., (eds), 1993. *Global Climates Since the Last Glacial Maximum*. University of Minnesota Press, Minneapolis, MN.

Xu, Chong-yu, 1999. From GCM's to river flow: a review of downscaling methods and hydrologic modelling approaches. *Progress in Physical Geography*, **23**, 229–249.

Yanosky, T.M. and Jarrett, R.D., 2002. Dendrochronologic evidence for the frequency and magnitude of palaeofloods. In P.K. House, R.H. Webb, V.R. Baker and D.R. Levish (eds), *Ancient Floods: Principles and Applications of Palaeoflood Hydrology*. American Geophysical Union, Washington, DC, 77–89.

PART 2   PERSPECTIVES

2 INQUA Research and Palaeohydrology

Institute of Geography and Spatial Organisation,
Polish Academy of Sciences, Cracow, Poland

1 PALAEOHYDROLOGICAL COMPONENTS IN QUATERNARY STUDIES

Water is a key resource for the habitability of the Earth. The hydrological cycle involves complex water-transfer stages in the oceans, atmosphere, lakes, glaciers, rivers and in groundwater. Patterns of water storage and water fluxes in time and in space are affected by climatic changes, as well as by simultaneously occurring modifications of climate and by the influence of the circulation of mineral and of organic matter. Water circulating on the Earth is analogous to blood circulating in the human body, transforming energy and mineral matter as well as supporting biomass production (Starkel, 1993). Every change in the water cycle can create a chain of environmental effects. Changes in precipitation, evaporation and runoff are reflected both in changes in interrelations between different types of storage (in oceans, lakes, glaciers, bogs, permafrost, groundwater) and also in all the geoecosystem components.

Reconstructions of past hydrological changes are based on an examination of various geological records, including those reflecting the changes in vegetation, soils, fluvial systems, aeolian activity, sea- and lake-level changes and variations in the volumes of ice sheets and of mountain glaciers. Therefore, various specialists have, quite rightly, made their input to palaeohydrology. The palaeohydrological approach is complementary to that of palaeoclimatology and is present, more or less implicitly, in all Quaternary palaeogeographic reconstructions as well as in global and regional climatostratigraphic subdivisions of the Quaternary. Most earlier reconstructions mainly considered fluctuations of temperature and precipitation; only in the 1960s did the attention turn to fluctuations in the water cycle and to water budget as a whole (Schumm, 1965).

2 PALAEOHYDROLOGICAL COMPONENTS IN THE ACTIVITIES OF INQUA COMMISSIONS

Temporal and spatial variations of elements of the water cycle are apparent in the considerations of problems undertaken by most of the INQUA Commissions. This component is relatively less clear in the studies by various teams of the Commission on Stratigraphy, although the facies sequence of sediments and biotic changes is in fact controlled by changes in water storage and fluxes. Commissions on lithology and

---

Palaeohydrology: Understanding Global Change. Edited by K.J. Gregory and G. Benito
© 2003 John Wiley & Sons, Ltd ISBN: 0-470-84739-5

the origin of sediments involve reconstructing the course of hydrologic processes; on glaciation, investigating the extension and decay of ice sheets and reconstructing changes in water storage; on loess and on palaeopedology, examining variations in both humidity and temperature. For several decades, the Commissions on sea-level changes, on shorelines, or on coastal evolution, took into consideration variations in global water storage and the extent of the seashore.

A second group of Commissions concentrates more specifically on the reconstruction of the elements of the hydrological cycle, most particularly the Commission on Palaeoclimatology, which uses various records for the reconstruction of precipitation as well as of the entire water budget, with significant use being made of records of palaeovegetation and of lake-level changes (Street-Perrott *et al.*, 1983; Mörner and Karlen, 1985). On a more regional scale and for shorter time units, similar reconstructions were undertaken from the 1960s to the 1970s by the Commission on the Holocene, where GLOCOPH really originated. Palaeohydrologic components are also very evident in the research by the Commission on the Palaeogeographic Atlas, which compiled maps of precipitation, ice sheets, permafrost and vegetation for several time slices of the upper Quaternary (Frenzel *et al.*, 1992). Among the more recent Commissions, the Carbon Commission should be mentioned because questions of CO_2 sinks and biomass production cannot be answered without an analysis of precipitation, evaporation and the other components of the water cycle (Adams and Faure, 1996).

Palaeohydrological problems became clearly evident after S.A. Schumm's presentation in 1965 at the INQUA Congress in Denver. At the subsequent Congress in Paris, the first session was devoted to geomorphology and palaeohydrology with emphases on rivers, lakes, and periglacial phenomena. At the Ottawa Congress in 1987, there were separate sessions on palaeohydrology and on the palaeoenvironments of various continents, as well as on palaeolimnology, glaciations, permafrost, monsoons and sea-level changes.

3 PALAEOHYDROLOGIC PROBLEMS IN IGCP AND OTHER PROJECTS

The simultaneous consideration of questions on changes in temperature and precipitation during the Quaternary raised awareness of the need for investigation of changes in the water cycle in their entirety; first realized by the CLIMAP Project (1976) and later by the COHMAP Project (1988); both were based on the comparison of observations and model simulations. Variations in precipitation also started to be reconstructed from ice cores (Oeschger and Langway, 1989). Separate working groups also involving reconstructions of hydrological components were formed by palynologists, palaeolimnologists and fluvial geomorphologists. In 1978, the IGCP project No. 158 "Palaeohydrology of the Temperate zone during the last 15,000 years" was established investigating fluvial systems and lake and mire environments in parallel. Among their results and the many conference proceedings and monographs, should be recognised *Background to Palaeohydrology* (Gregory, 1983), *Handbook of Holocene Palaeoecology and Palaeohydrology* (Berglund, 1986), *Temperate Palaeohydrology* (Starkel *et al.*, 1991) and *Palaeoecological events during the last 15,000 years* (Berglund *et al.*, 1996). Parallel to this IGCP Project was work by an active team concentrating on the reconstruction of extreme floods in the past (Baker *et al.*, 1988). In many other IGCP Quaternary projects, palaeohydrological changes were also considered including No. 253 Termination of the Pleistocene (Lundquist *et al.*, 1995);

No. 349 Desert margins and palaeomonsoons since 135 ka BP; No. 378 Karst process and carbon cycle; and No. 415 Glaciation and reorganisation of Asia's drainage in the Quaternary.

During the last decade of the twentieth century, the International Hydrological Programme started to devote more attention to changes in the water budget of the past, from the point of view of the long-term human impact on water resources (Issar, 1995; Issar and Brown, 1996). However, the most extensive programme considering palaeohydrological problems on a large scale was undertaken by PAGES (see Chapter 3, and Alverson *et al.*, 1995).

4   THE FOUNDATION OF GLOCOPH

During 1987 to 1988, very extensive interest by various interdisciplinary teams, especially during the very fruitful conclusion of IGCP Project No. 158, led to collaborative efforts to establish and organise a framework on global palaeohydrology (Starkel, 1989). After the Ottawa Congress in 1987, a Working Group on Global Continental Palaeohydrology was formed within the Holocene Commission of INQUA. This small team prepared a proposal for a new global-scale project (Starkel, 1990), which was submitted to the IGCP Board in 1989 but not accepted. This proposal underlined the need for a global framework in order to understand temporal changes in water storages and fluxes, proposing to concentrate research on a time resolution of 10^2 to 10^3 years. This project was proposed to integrate studies on all major elements of the hydrological cycle, among them the reconstruction of runoff, previously omitted in many palaeogeographic reconstructions. The research, like that in IGCP-158, was intended to concentrate upon selected rivers and key lake basins located in various climatic vegetation zones, and to recognise changes in storages and fluxes leading to reconstruction of water budgets (e.g. Kutzbach, 1980). The organisation of appropriate databases would assist in the construction of zonal and global models of past hydrological changes.

This project was slightly modified and later presented as the specific task for the proposal submitted as a new INQUA Commission on Global Continental Palaeohydrology, finally accepted at the Congress in Beijing in 1991. In this way, it was possible to fill a gap in the structure of the interdisciplinary commissions of INQUA.

5   THE IMPORTANCE OF PALAEOHYDROLOGY IN RELATION TO BETTER PRESENT-DAY AND FUTURE WATER MANAGEMENT

The volume of water on the globe exceeds $1,450 \times 10^5 \, \text{km}^3$ of which only $85 \times 10^6 \, \text{km}^3$ is stored on the land as groundwater, as ice and in lakes and rivers (Lvovitch, 1974; Shiklomanov, 1990). The annual circulation of water involves about $0.6 \times 10^6 \, \text{km}^3$, of which 21% falls on the continents. Reviewing the distribution of spatial water changes directs attention to areas with a surplus or deficit of precipitation, storage and runoff. Various types of human activity such as forest clearance, soil cultivation and overgrazing and, later, irrigation, exploitation of groundwater and construction of water reservoirs have changed the water cycle (Gregory and Walling, 1987) and have increased existing water deficits, especially in the arid belt. Currently at least $4,000 \, \text{km}^3$ of water per year is used for irrigation purposes (cf. Shiklomanov, 1990) and more than $1,000 \, \text{km}^3$ is used for other types of water consumption. Similar volumes of water are stored in reservoirs. More than half of all water used does not return to the cycle, and the various forms of human activity lead to lowering of groundwater tables, desiccation of lakes and rivers, and shifting of desert-steppe and steppe-forest ecotones.

Superimposed over this trend are changes connected with global warming caused by the rapid rise of CO_2 and other greenhouse gases released into the atmosphere (Brouwer and Falkenmark, 1989), reflected in the accelerated melting of glaciers and a higher frequency of extreme meteorological events with the consequence that we may expect rising sea levels and very variable runoff conditions.

In order to understand better the ongoing and future hydrological changes, we need to appreciate the mechanisms of past hydrological changes on longer timescales (Starkel, 1993; Benito *et al.*, 1998). Of particular importance is the recognition of short-term hydrological changes on timescales of decades, centuries and millennia; the role of intense heavy rains and extreme droughts and their clustering; and the rate of response of hydrological regimes and ecosystems, to changes in precipitation and land use. For these reasons, special attention should be concentrated on the period of the last millennia of accelerated circulation of water and sediment load (Starkel, 2002).

REFERENCES

Adams, J.M. and Faure, H., 1996. *Changes in Moisture Balance Between Glacial and Inter-glacial Conditions: Influence on Carbon Cycle Processes*. Special Publication 115, Geological Society, London, 27–42.

Alverson, K.D., Oldfield, F. and Bradley, R.S., (eds), 1995. Past global changes and their significance for the future. *Quaternary Science Reviews*, **19**, 1–5.

Baker, V.R., Kochel, R.C. and Patton, P.C., (eds), 1988. *Flood Geomorphology*. Wiley, Chichester.

Benito, G., Baker, V.R. and Gregory, K.J., (eds), 1998. *Palaeohydrology and Environmental Change*. Wiley, Chichester.

Berglund, B., (ed.), 1986. *Handbook of Holocene Palaeoecology and Palaeohydrology*. Wiley, Chichester.

Berglund, B.E., Birks, H.J.B., Ralska-Jasiewiczowa, M. and Wright, H.E., (eds), 1996. *Palaeoecological Events During the Last 15,000 years*. Wiley, Chichester.

Brouwer, F. and Falkenmark, M., 1989. Climate-induced water availability changes in Europe. *Environmental Monitoring and Assessment*, **13**, 75–98.

CLIMAP Project Members, 1976. The surface of the ice-age earth. *Science*, **191**, 1131–1144.

COHMAP Members, 1988. Major climatic changes of the last 18000 years: observations and model simulations. *Science*, **241**, 1043–1052.

Frenzel, B., Pecsi, M. and Velichko, A.A., 1992. *Paleogeographical Atlas of the Northern Hemisphere*. Hungarian Academy of Sciences, Budapest.

Gregory, K.J., (ed.), 1983. *Background to Palaeohydrology A Perspective*. Wiley, Chichester.

Gregory, K.J. and Walling, D.E., (eds), 1987. *Human Activity and Environmental Processes*. Wiley, Chichester.

Issar, A.S., 1995. *Impacts of Climate Variations on Water Management and Related Socio-Economic Systems*. IHP-IV Project H-2.1. UNESCO, Paris, 1–97.

Issar, A.S. and Brown, N., (eds), 1996. Water, environment and society in times of climatic changes. *Water Science and Technology Library*, 31, Kluwer Academic Publishers, Dordrecht.

Kutzbach, J.E., 1980. Estimates of past climate of palaeolake Chad, North Africa, based on a hydrological and energy-balance model. *Quaternary Research*, **14**, 210–223.

Lundquist, J., Saarnisto, M. and Rutter, N., (eds), 1995. IGCP-253 – termination of the Pleistocene – final report. *Quaternary International*, **28**.

Lvovitch, M.I., 1974. *Wodnyje resursy mira i ich buduscije (Water resources of the Earth and their future, in Russian)*. Mysl., Moskwa.

Mörner, N.A. and Karlen, V., (eds), 1985. *Climatic Changes on a Yearly to Millenial Basis*. D. Reidel, Dordrecht.

Oeschger, H. and Langway Jr., C.C., (eds), 1989. *The Environmental Record in Glaciers and Ice Sheets*. Wiley, Chichester.

Schumm, S.A., 1965. Quaternary paleohydrology. In H.E. Wright and D.G. Frey (eds), *The Quaternary of the United States*. Princeton University Press, Princeton, 783–794.

Shiklomanov, I.A., 1990. Global water resources. *Natural Resources*, **26**(3), 34–43.

Starkel, L., 1989. Global palaeohydrology. *Quaternary International*, **2**, 25–33.

Starkel, L., 1990. Global continental palaeohydrology. *Palaeogeography, Palaeoclimatology, Palaeoecology*, (Global Planetary Change Section). **82**, 73–77.

Starkel, L., 1993. Late Quaternary continental palaeohydrology as related to future environmental change. *Global and Planetary Change*, **7**, 95–108.

Starkel, L., 2002. Change in the frequency of extreme events as the indicator of climatic change in the Holocene (in fluvial systems). *Quaternary International*, **91**, 25–32.

Starkel, L., Gregory, K.J. and Thornes, J.B., (eds), 1991. *Temperate Palaeohydrology*. Wiley, Chichester.

Street, F.A. and Grove, A.T., 1979. Global maps of lake-level fluctuations since 30000 yr BP. *Quaternary Research*, **12**(1), 83–118.

Street-Perrott, A., Beran, M. and Ratcliffe, R., (eds), 1983. *Variations in the Global Water Budget*. D. Reidel, Dordrecht.

3 Palaeohydrology and International Global Change Programs

KEITH ALVERSON[1] AND TOM EDWARDS[2]
[1]*PAGES International Project Office, Berne, Switzerland*
[2]*University of Waterloo, Waterloo, Canada*

The balance of evidence, pundits tell us, suggests that human actions have had a discernable influence on the global average temperature. Current best estimates derived from instrumental data suggest that the northern hemisphere average temperature has increased over the past 100 years by approximately 1°C, and can be expected to rise at least another degree, perhaps as much as 5, over the next 100 years. In contrast, estimates based on palaeorecords suggest that the range of Northern Hemisphere average-temperature variability over the past millennium, which included periods such as the "Little Ice Age" and the "Medieval Warm Period," was less than one degree, indicating that the ongoing anthropogenic changes are indeed substantial when compared with this longer term record. Only a decade ago, palaeorecords such as snow accumulation rate and δ^{18}O-inferred temperature obtained from Greenland ice cores (Figure 3.1), were invoked to suggest that a relatively stable global climate existed during the Holocene as compared to the large oscillations associated with glacial periods (Dansgaard *et al.*, 1993). However, this polar-centric paradigm of Holocene stability has now been rejected in favor of a picture of large and sometimes abrupt, natural climatic oscillations during the Holocene, clearly of great relevance to modern concerns, largely based on evidence for pronounced hydrological variability at lower latitudes.

Although recent attention has been focused mainly on rising global temperatures, as a monitor of anthropogenic influence, changes in regional hydrological balance have had a more substantial and direct effect on human livelihoods over the past century. In addition, the more distant past record is replete with examples of sustained hydrological variability on all timescales. Figure 3.2 shows a highly condensed impression of global and regional hydrological variability over different timescales. On timescales of 100,000 years, for example, global sea level fluctuated on the order of 100 m in response to the waxing and waning of large continental ice sheets (Waelbroek *et al.*, 2002). These sea-level changes, at times extremely rapid – on the order of 1 m per century – reflect profound changes in the global hydrological cycle. Not surprisingly, regional water balance during glacial periods was in most cases also radically different from that of the present. This is not without relevance to many of today's concerns, since many of the aquifers currently being exploited in densely populated, semiarid regions were recharged during the glacial times and have insignificantly small recharge rates under modern conditions. Regional hydrological balance has also varied greatly on millennial timescales, as shown here by the example of 100-m excursions

Palaeohydrology: Understanding Global Change. Edited by K.J. Gregory and G. Benito
© 2003 John Wiley & Sons, Ltd ISBN: 0-470-84739-5

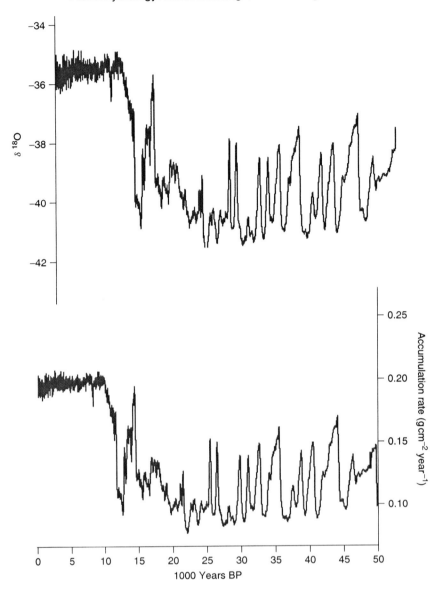

Figure 3.1 Accumulation and isotopically inferred temperature over the past 50,000 years as measured in Greenland ice cores (Dansgaard *et al.*, 1993). Figure generated from data stored at the World Data Center for Palaeoclimatology, from NOAA. http://www.ngdc.noaa. gov/paleo/

in the water level in Lake Abhé, Ethiopia, sustained over hundreds to thousands of years (Gasse, 2000). The magnitude of the highly publicized multiyear Ethiopian droughts of the last few decades, though devastating to the local population, was a minor event by comparison. At still finer temporal resolution, decadal to century-scale hydrological variability such as that manifested by the shifting level of Lake Naivasha, Kenya, has substantially influenced the history of local populations (Verschuren *et al.*, 2000). Annually resolved records of snow accumulation from high elevation ice cores can be used to extract detailed information about both anthropogenic influences and

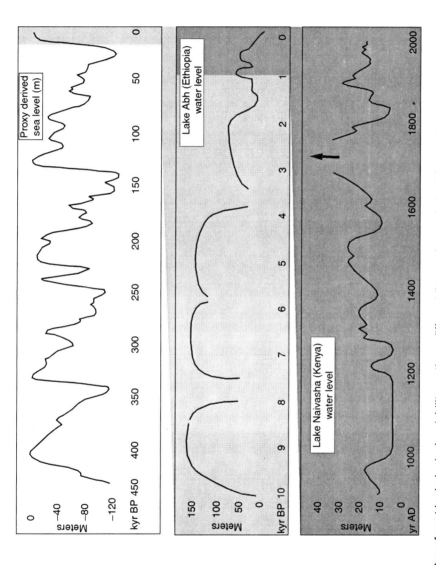

Figure 3.2 Examples of past hydrological variability on three different timescales as derived from palaeoproxy measurements. Sea-level variability associated with glacial cycles (Waelbroek *et al.*, 2002), lake-level changes in Lake Abhé, Ethiopia during the Holocene (Gasse, 2000) and lake-level changes in Lake Naivasha, Kenya over the last millennium (Verschuren *et al.*, 2000). Figure adapted from Oldfield, F. and Alverson, K., The social relevance of palaeoenvironmental research. In K. Alverson, R. Bradley and T. Pedersen (eds), *Palaeoclimate, Global Change and the Future*. Springer-Verlag 2003

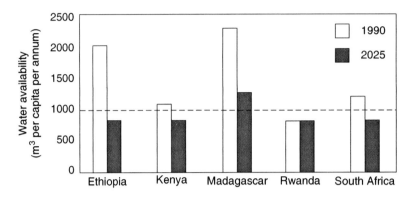

Figure 3.3 Water availability in selected African countries in 1990 and as projected for 2025. The indicated level of 1,000 m³ per capita per annum is a widely publicized level of "well being" in the industrialized world. Figure adapted from Beniston, M., 2000. *Environmental Change in Mountains and Uplands*. Arnold, London, 172, reprinted by permission of Hodder Arnold

long-term variability of naturally occurring climatic nodes with strong influences on regional hydrological balance such as El Niño and the Pacific Decadal Oscillation (Moore *et al.*, 2001; Moore *et al.*, 2002).

The sustainability of freshwater resources is a major and growing international concern. In Africa, 60% of the rural population and 25% of the urban population do not currently have access to safe drinking water and, due primarily to population growth, these percentages are expected to increase (Beniston, 2000; Figure 3.3). Changes in the hydrological balance due to greenhouse gas-induced climate change, although still poorly constrained by predictive models, are likely to compound this dire situation. Of even greater concern, however, is the prospect of enormous and potentially catastrophic hydrological changes such as those evident from the records shown in Figure 3.2 that represent abrupt step-shifts in local or regional water balance, possibly reflecting the crossing of critical thresholds during much more gradual climate change. Thus, detailed consideration of regional palaeohydrology is vital in any internationally coordinated approach for sustainable development of freshwater resources in at least two key ways: first, as a fundamental component of basic water resources assessment with particular focus on the understanding of groundwater recharge history and, second, as a guide to the sensitivity of regional water balance to ongoing and future climate change.

Despite the clear importance that palaeohydrological understanding must play in order to achieve sustainable development of freshwater resources, the number of international programs, purporting to deal with this subject, which actually includes palaeohydrological information in their considerations is astonishingly low. Explicitly palaeo-oriented programs, such as the International Quaternary Research Association (INQUA, see Chapter 2) and Past Global Changes (PAGES), have long promulgated the need to understand past environmental change, and especially palaeohydrologic change, in order to better constrain or predict future change (Alverson *et al.*, 2000; Alverson *et al.*, 2001; Alverson and Kull, 2003). Yet, this potential source of key insight into how the earth system functions has received remarkably little recognition within international programs devoted to the hydrological aspects of the earth system science and even less within the sustainable development community.

1   INTERNATIONAL EARTH SYSTEM SCIENCE PROGRAMS FOCUSING ON HYDROLOGY

Water, together with food and carbon, is one of three cross-cutting joint projects on global sustainability issues due to begin full operation in 2003 within the framework of the ICSU (International Council for Science) Earth System Science partnership. This partnership includes the World Climate Research Program (WCRP), International Geosphere Biosphere Programme (IGBP), the International Human Dimensions Programme on Global Environmental Change (IHDP) and Diversitas. The overall goal of this new Joint Water Project is to increase the scientific understanding, from an earth system perspective, of the interactions between human activities and the global environment inherent to global water systems, and thereby to contribute to the sustainability of these systems. Specifically, the project seeks to ascertain the impacts of global change on regional water systems and the subsequent feedback these changes have on the earth system. Within this context, the project aims to elucidate measures that are needed to achieve sustainability of these water systems, and the technical and institutional possibilities for meeting growing demands. To a large extent, this project will draw upon other well-established programs under the umbrella of the Earth System Science partnership, including the IGBP core project – Biospheric Aspects of the Hydrological Cycle (BAHC) and the WCRP-based Global Energy and Water Cycle Experiment (GEWEX). The relevance of palaeohydrological information to these water-focused Earth System Science partnership programs is clear. The development of adequate linkages with palaeoscience researchers, in order to ensure that long-term system variability is adequately addressed within these programs, is one task of the IGBP Past Global Changes (PAGES) project, the only program within the Earth System Science partnership explicitly concerned with palaeoenvironmental science.

2   INTERNATIONAL PROGRAMS IN SUSTAINABLE DEVELOPMENT OF WATER RESOURCES

An exhaustive consideration of international programs concerned with palaeohydrology and its relevance to sustainable development of water resources is beyond the scope of this short contribution. Instead, some are highlighted and web sites with appropriate links are provided in Table 3.1. Several programs and agencies within the United Nations System of Organizations (UNSO) have direct involvement in hydrology and global change science. These include, for example, the International Hydrological Programme (IHP), UNESCO's intergovernmental scientific cooperative program in water resources. The IHP is a vehicle through which UN Member States can enhance their knowledge of the water cycle and thereby increase their capacity to better manage and develop their own water resources. Its frame of reference includes improvement of the scientific and technological basis for rational and sustainable management of water resources in concert with protection of the environment. IHP is also one of many UN cosponsors of the World Water Assessment Programme (WWAP) for development, capacity building and the environment. The United Nations Environment Program (UNEP), as part of its various programs, has developed a web portal with links to extensive information on water resources. Unfortunately, and partly as a consequence of limitations on personnel and resources, these major international sustainability programs have so far taken very little account of palaeohydrological variability in their efforts to promote sustainable water resources development.

Table 3.1 International programs listed in the text and their Web sites

BAHC	Biospheric Aspects of the Hydrological Cycle	http://www.pik-potsdam.de/~bahc/
Diversitas		http://www.icsu.org/diversitas/
GASPAL	Groundwater as Palaeoindicator	
GEWEX	Global Energy and Water Cycle Experiment	http://www.gewex.com/
GNIP	Global Network for Isotopes in Precipitation	http://isohis.iaea.org/GNIP.asp
IAEA	International Atomic Energy Agency	http://www.iaea.org/
ICSU	International Council for Science	http://www.icsu.org/
IGBP	International Geosphere Biosphere Programme	http://www.igbp.kva.se/
IHDP	International Human Dimensions Programme on Global Environmental Change	http://www.uni-bonn.de/ihdp/
IHP	International Hydrological Program	http://www.unesco.org/water/ihp/
INQUA	International Quaternary Research Association	http://inqua.nlh.no/
ISOHIS	Isotope Hydrology Information System	http://isohis.iaea.org/
PAGES	Past Global Changes	http://www.pages-igbp.org
UNEP	United Nations Environment Program	http://freshwater.unep.org
UNESCO	United Nations Educational Scientific and Cultural Organization	http://www.unesco.org
UNSO	United Nations System of Organizations	http://www.unsystem.org/
WWAP	World Water Assessment Programme	http://www.unesco.org/wwap/partners/index.shtml
WCRP	World Climate Research Program	http://www.wmo.ch/web/wcrp/

One active water resources organization under the UN umbrella that does have a strong commitment to palaeoscience is the Isotope Hydrology Section of the International Atomic Energy Agency (IAEA), which specializes in the use of isotopic techniques in hydrology in support of water resources investigations in developing countries and, increasingly, in international climate and palaeoclimate research. Naturally occurring water isotope tracers are particularly useful for tracking and dating groundwater, as well as carrying primary quantitative information about palaeoenvironmental conditions prevailing at the time of aquifer recharge or the accumulation of snow subsequently incorporated into glacier ice. A central element of the IAEA's activities is the Global Network for Isotopes in Precipitation (GNIP), operated jointly since the 1960s with the World Meteorological Organization. The GNIP database provides the only comprehensive coverage of the distribution of isotopes in global precipitation, providing essential input functions for hydrologic and palaeohydrologic studies, a highly sensitive monitor of contemporary global climate variability and change, and a key source of information for calibrating isotope indicators of palaeoclimate.

The IAEA Isotope Hydrology Information System (ISOHIS), which includes the GNIP database, also serves as a permanent archive of crucial isotopic and chemical data from groundwater, collected worldwide by the IAEA and in various international initiatives. Examples of the latter include the European Community funded Groundwater as Palaeoindicator (GASPAL) project, carried out in collaboration with PAGES, which seeks to derive information about the age and history of groundwater in the arid Sahel region in order to improve the management of this critical resource.

3 CONCLUSIONS

This chapter briefly outlines some of the primary reasons why palaeohydrological information such as that considered in this volume has a vital role to play in achieving sustainable development of freshwater resources. A brief description of some relevant international global change research and water resources programs is also provided. Regrettably, although exceptions exist, many of the programs that are actively involved in water resource management currently take rather limited (or even no) account of the insight to be gained from palaeohydrologic investigations, whether it be to better assess the limits of the resource or to more accurately predict the nature and magnitude of future hydrological variability under the influence of climate change.

REFERENCES

Alverson, K., Bradley, R. and Pedersen, T. 2001. *Environmental Variability and Climate Change*. IGBP Science No. 3, IGBP Secretariat, Stockholm, 1–32.

Alverson, K., Oldfield, F. and Bradley, R. (eds), 2000. Past global changes and their significance for the future. *Quaternary Science Reviews*. Elsevier Science, 1–479.

Alverson, K. and Kull, C. 2003. Understanding future climate change using palaeorecords. In X. Rodo and F.A. Comin (eds), *Global Climate: Current Research and Uncertainties in the Climate System*. Springer-Verlag, 153–185.

Beniston, M., 2000. *Environmental Change in Mountains and Uplands*. Arnold, London, 105.

Dansgaard, W., Johnsen, S.J., Clausen, H.B., Dahl-Jensen, D., Gundestrup, N.S., Hammer, C.U., Hvidberg, C.S., Steffensen, J.P., Sveinbjornsdottir, A.E., Jouzel, J. and Bond, G. 1993. Evidence for general instability of past climate from a 250-kyr ice-core record. *Nature*, **364**, 218–220.

Gasse, F., 2000. Hydrological changes in the African tropics since the last glacial maximum. *Quaternary Science Reviews*, **19**, 189–211.

Moore, G.W.K., Holdsworth, G. and Alverson, K., 2001. Extra-tropical response to ENSO 1736–1985 as expressed in an ice core from the Saint Elias mountain range in northwestern North America. *Geophysical Research Letters*, **28**(18), 3457–3461.

Moore, G.W.K., Holdsworth, G. and Alverson, K., 2002. Climate change in the North Pacific region over the last three centuries. *Nature*, **420**, 401–403.

Oldfield, F. and Alverson, K., 2003. The social relevance of palaeoenvironmental research. In K. Alverson, R. Bradley and T. Pedersen (eds), *Palaeoclimate, Global Change and the Future*. Springer-Verlag, 1–11.

Verschuren, D., Laird, K.R. and Cumming, B.F. 2000. Rainfall and drought in equatorial east Africa during the past 1,100 years. *Nature*, **403**, 410–414.

Waelbroek, C., Labeyrie, L., Michel, E., Duplessy, J.C., McManus, J., Lambeck, K., Balbon, E. and Labracherie, M. 2002. Sea-level and deep water temperature changes derived from benthic foraminifera isotopic records. *Quaternary Science Reviews*, **21**, 295–306.

PART 3   WORLD AREAS

4 Introduction to Regional Palaeohydrologic Regimes and Areas

K.J. GREGORY[1] AND G. BENITO[2]
[1]*University of Southampton, Southampton, UK*
[2]*CSIC-Centro de Ciencias Medioambientales, Madrid, Spain*

Palaeohydrology involves the reconstruction of the main components of the hydrological cycle: precipitation, evaporation and runoff. Pollen concentration in lake sediments and peat deposits has been used for broad reconstructions of the vegetation and climatic history, as proxy information to obtain annual precipitation and evaporation, at particular sites (Wright *et al.*, 1993). Lake sediment records and dated shorelines can record lake volumes and fluctuations, which imply shifts in the precipitation/evaporation balance.

Runoff or stream flow is the most intractable component of the water cycle because it is affected by intrinsic parameters of the drainage basin including vegetation, geology, land use and soils. Reconstruction of runoff and palaeoflows (see Chapter 1) can be achieved through the analysis of: (1) landscape evolution at the catchment scale (e.g. Rinaldi *et al.*, 1995); (2) relict fluvial landforms at the channel–valley scale (Dury, 1977; Starkel, this volume Chapter 7; Sidorchuk *et al.*, Chapter 6); (3) alluvial sequences in fluvial terraces and floodplains analysing decadal to millennial fluctuations in fluvial activity (Nanson *et al.*, Chapter 14); or slackwater flood bars and benches deriving event-based stratigraphy (Knox, Chapter 10). Palaeohydrological interpretations analyse how changing patterns of flow/sediment discharge can be induced by climatic changes or by other factors acting over the drainage basin area such as land-use changes, vegetation changes, or permafrost conditions (Starkel, 1996) and including the impact of human activity.

Since the Last Glacial Maximum (LGM; $\sim$18 ^{14}C kyr BP or 21.5 cal kyr BP), river discharge and flow regimes have shown major changes. In some regions, such as northern and central Europe, past flow regimes were totally different from the present ones, whereas in others, such as some Mediterranean regions, hydroclimate-driven processes have changed only in intensity rather than in character. In order to demonstrate palaeohydrological variability in relation to global change, this volume has been organised into three broad latitudinal groups (high, mid and low latitudes), with further subdivisions being achieved according to continental or physical distribution (Figure 4.1). Since the atmosphere is an important component of the water cycle, climate variability and changes in the atmospheric circulation patterns are reflected in a suite of past hydrological responses according to the character of particular regions and their sensitivity to hydrological changes during the past. The variety of past changes

Palaeohydrology: Understanding Global Change. Edited by K.J. Gregory and G. Benito
© 2003 John Wiley & Sons, Ltd ISBN: 0-470-84739-5

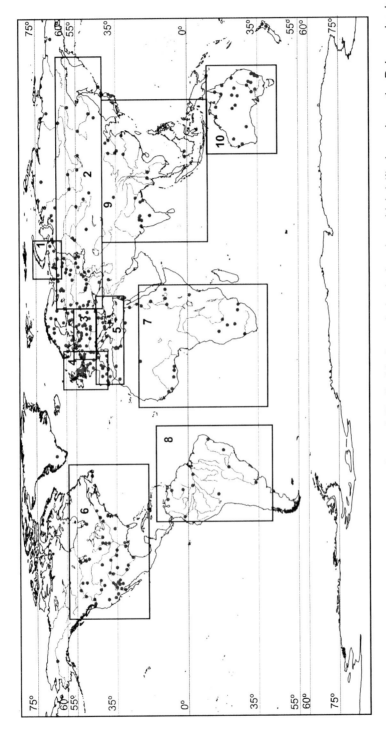

Figure 4.1 World division of palaeohydrological studies on latitudinal belts and physical regions: high-latitude regions: 1 – Polar and subpolar zone; mid-latitude regions: 2 – Northern Eurasia; 3 – Central Europe; 4 – Western Europe; 5 – Mediterranean basin; 6 – North America; low-latitude regions: 7 – Tropical and subtropical Africa; 8 – Central and South America; 9 – Southern Asia; 10 – Australia. Solid dots show areas of published research on palaeohydrology

and responses collectively provide a perspective on the range of multidisciplinary approaches to palaeohydrology, furnishing real scenarios to tune climatic models assessing the effects of future global change.

1 WORLD AREAS AND REGIMES

1.1 High-latitude Regions: Polar and Subpolar Regions

Polar and subpolar regions are defined as those extending north of 55° N (Figure 4.1). The region includes zones of polar desert, tundra and northern boreal forest (taiga) as well as ice-marginal zones (ice caps and glaciers). Present river regimes are characterised by a marked seasonality associated with winter freeze-up, followed by a high-magnitude late spring and early summer flow (Figure 4.2, Table 4.1).

Three basic catchment types of palaeohydrological regimes can be recognised in polar and subpolar regions: glaciated, extraglaciated or ice-marginal, and periglacial or non-glaciated permafrost catchments (Maziels, 1995). Many ice-marginal zones contained large ice-dammed glacial lakes that changed the drainage flow paths and eventually generated huge floods or jökulhlaups producing extensive geomorphic changes. Long-term fluvial activity and palaeohydrological changes also need to consider catchment changes affected by crustal uplift of glacio-isostatic origin which can induce incision, or submergence related to ice-ground melting leading to alluviation.

Whereas ice sheets of the last glaciation disappeared from subpolar regions during the Lateglacial (ca 15 kyr BP), they persisted in polar latitudes at least until the Pre-Boreal or Boreal (ca 10–8 kyr BP). The Holocene history seems to indicate that the appearance and disappearance of permafrost coincides with river instability, possibly

Table 4.1 Global classification of river regimes (Regime types after Haines *et al.*, 1988, with permission from Elsevier Science). World distribution of these types is shown in Figure 4.2

Groups	Classes	Regime types
1. Uniform flow regime	1. Uniform	1. Uniform
2. Summer flow dominance	2. Early summer flow	2. Mid-late spring
		3. Late spring–early summer
		4. Extreme early summer
		5. Moderately early summer
		6. Mid-summer
	3. Late summer flow	7. Extreme late summer
		8. Moderate late summer
		9. Early autumn
		10. Mid-autumn
3. Winter flow dominance	4. Autumn and spring flow	11. Moderate autumn
		12. Moderate winter
	5. Winter-spring flow	13. Extreme winter
		14. Early spring
		15. Moderate spring

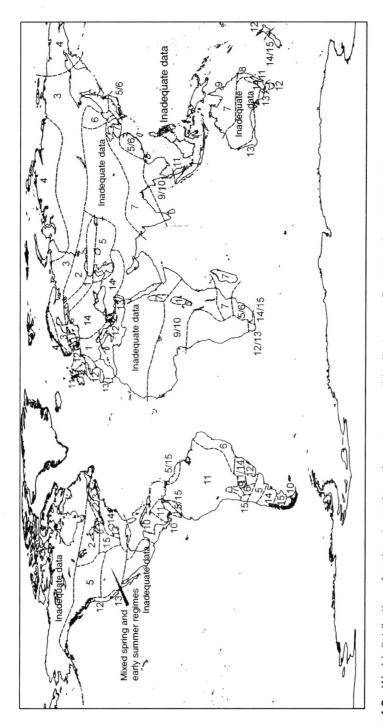

Figure 4.2 World distribution of regime types based on mean monthly discharges. Reprinted from *Applied Geography*, Volume 8, Haines, A.T., Finlayson, B.L. and McMahon, T.A., 1988. A global classification of river regimes, 255–272, with permission from Elsevier Science. Description of the 15 regime types is shown in Table 4.1

through changes in flow regime, vegetation and sediment load (Vandenberghe and Huisink, Chapter 5).

1.2   Mid-latitude Regions: Temperate Regions

These regions are defined as those extending between 55° and 35° in latitude, including zones of cold climates (D type in the Köppen–Geiger climatic classification) of the Russian Plain and Siberia, and temperate climates of central western Europe, eastern North America and south and southeastern Australia. River runoff regimes are characterised by uniform flows throughout the year in central Europe and south and western Russia, moderate to extreme winter discharges in the Atlantic areas of Europe, and autumn and/or spring flows in the rest of the area, including the Mediterranean (Figure 4.2; Table 4.1). Variability of annual precipitation and runoff increases southwards so in the Mediterranean region flood magnitudes can be more than 50 times the average annual flow.

Lateglacial palaeohydrology in northern Eurasia, the Russian plain and West Siberia, was characterised by very high annual discharges between 1.5 and 4 times greater than those at present, forming large channels with widths up to 15 times the present ones (Sidorchuck et al., Chapter 6). In addition, drainage from ice-dammed lake outbursts from ice-marginal zones (Baker et al., 1993; Grosswald, 1998) produced significant geomorphic impacts on particular fluvial systems, triggering valley erosion and/or degradation. During the Lateglacial, ice-dammed lakes drained ca 34,500 km^3 of water into the Aral Sea and, through the Caspian and Black seas, about 13,800 km^3 into the Mediterranean. Transformation of large periglacial channels into significantly smaller Holocene channels began ca 12 kyr BP, although discharges remained greater than modern ones until the beginning of the Boreal; discharges diminished during the first half of the Holocene and increased again from late Holocene to the present time, although the chronology of these changes varied considerably from one region to another (9 kyr BP; Sidorchuck et al., Chapter 6).

The *Central European* region covers the area between the Rhine and the Dnieper River, and from the Baltic to the Alps (47–55° N; 8–35° E). During the LGM, the Scandinavian ice sheet covered the northern part of Central Europe, including the site of present-day Berlin, whereas the southern zones were under continuous permafrost. Late glacial fluvial regimes were characterised by high flow seasonality, being especially high during heavy rains and snowmelt periods (probably late spring and summer types). Following the climate amelioration that started in the Bölling interstadial (ca 13–12 kyr BP), forest vegetation spread throughout the region producing a significant decrease of snowmelt runoff and sediment load (Starkel, 1991). At this time, the main fluvial systems shifted from braided to large palaeomeanders, and subsequently into small palaeomeanders at the beginning of the Holocene (Kozarski and Rotnicki, 1977). The late shift into small meanders was probably related to a change in the moisture source and the onset of the prevailing influence of the westerlies and cyclones carrying precipitation to the region. In Central Europe, a good correlation of the more distinct humid and usually cooler phases with a higher flood frequency and fluvial activity has been recognised (Starkel, 2002): 8.5–7.8, 6.6–6.0, 5.5–4.9, 4.5–4.0, 3.5–3.0, 2.8–2.7, 2.2–1.8 kyr BP, in the X–XI century, and the Little Ice Age.

The *Western European* zone (38–59° N; 10° W–8° E) now lies in the path of the westerly atmospheric circulation, with a cyclogenesis and cyclonic tracks related to the North Atlantic Oscillation (NAO) and the North Atlantic Deep Water (NADW).

River flows are usually highest in winter and discharge variance increases southwards. During the LGM, the north European ice sheet covered parts of the United Kingdom and Ireland; and the North Atlantic Ocean was ice-covered perhaps all the way south to the coast of Portugal during the winter season (Bjørn and Borns, 1994). During deglaciation, extensive areas free from ice sheets were transformed into areas with poor drainage forming lakes and peatbogs. Synchronous palaeohydrological patterns related to climate-driven forcing did not occur in the Late Quaternary, probably due to control of runoff by other factors such as geology, land-use patterns or catchment size (Brown, Chapter 8). Downcutting was dominant during the late Pleniglacial or Older Dryas, although in some rivers incision occurred later during the Younger Dryas. At the beginning of the Holocene, floodplain aggradation was scarce, probably due to the herbaceous and shrub ground cover (Vandenberghe, 1995). In the late Holocene, the 2,650-BP climatic event (van Geel *et al.*, 1998) produced significant alluvial records on the flood plains of the Severn, Avon, Ouse and Thames (Macklin and Lewin, 1993). Increased flood frequency during the Medieval Warm Period (AD 950–1200) and the warming phase of the Little Ice Age (ca AD 1600–1850; Le Roy Ladurie, 1967) is also indicated in some alluvial records of Western Europe (Brown, 1998).

The *Mediterranean region* (30–47° N; 10° W–35° E) is characterised by high-runoff seasonality and inter-annual variability. During the LGM, the region was deglaciated, except for the main mountain systems, the Pyrenees, Alps, and Pindus Mountains. At this period, a cooler climate, lower (total) precipitation and the development of steppe vegetation favoured aggradation along many Mediterranean rivers (Macklin *et al.*, 2002). This relation between cool climatic conditions and fluvial activity was not found in other Holocene climatic events such as during the Younger Dryas. Late Pleistocene palaeohydrology is not well recorded either because of the limited fluvial activity or due to the sparse numerical dating. In some areas of the Iberian Peninsula and Morocco, alluviation took place during the Lateglacial period. In the western Mediterranean, fluvial activity during the Holocene notably occurred at ~8.5–8.0 kyr BP (9.5–8.7 cal kyr BP), ~7–6 kyr BP (7.8–6.8 cal kyr BP), 4.2–3.5 kyr BP (4.7–3.7 cal kyr BP), 700–500 BC, AD 500–1000 and/or AD 1600–1700. A general trend of river channel metamorphosis from braiding to meandering throughout the Holocene has been reported in many areas of the northern Mediterranean region.

In *North America*, present river regimes (Figure 4.2; Table 4.1) are the result of rainfall, snowmelt, or combined rainfall and snowmelt patterns. Precipitation is connected to the general location of the middle latitude jet stream, and to the tropical and subtropical storms originating in the Gulf of Mexico and the Pacific, providing a mosaic of multi-type flow regimes across North America. This complexity is further amplified in the palaeohydrological regimes over the last 20,000 years. The maximum limit of the ice sheets was reached 21,000–18,000 years ago, although considerable readvance took place at 15,000–14,000 years ago (Bjørn and Borns, 1994). Late Pleistocene and early Holocene hydroclimatology of northern North America responded to the forcing effects of the disintegrating ice sheets with river discharges dominated by snowmelt patterns with spring and summer peaks. Probably the most remarkable fluvial changes during the Late Pleistocene were produced by large floods resulting from breaches of glacial-age ice dams that blocked the large mid-continent drainage systems along the Cordilleran Ice Sheet margin (e.g. Bretz, 1925; Baker, 1973; O'Connor and Baker, 1992), and along the margins of the Laurentide Ice Sheet (e.g. Thorson, 1989; Teller and Kehew, 1994). The cataclysmic outburst of glacial lake Missoula released 2,100 km³ of water at a peak discharge of about $17 \times 10^6 \, \text{m}^3 \, \text{s}^{-1}$, producing a suite of erosional and depositional flood features over 40,000 km² of Montana, eastern Washington and Oregon. During the same period, conditions cooler and wetter than those at present dominated over most of the southern part of North America in southwestern USA and Northern Mexico (Thompson *et al.*, 1993), with a major gradient of change across Mexico into dry conditions prevailing in the circum-Gulf of Mexico/Caribbean and other parts of Central America (Markgraf, 1993). Mid-late Holocene environments involved forcing factors that represent climate and hydrologic conditions that approximate to modern climates (Knox, Chapter 10). Palaeoflood records from many regions of North America show extreme hydrological events to be more sensitive to climate change than modal-range hydrological events (Knox, 2000). Palaeoflood chronologies for different regions of North America have been derived from slackwater flood deposits or by reconstructing overbank floods from competence-based equations, morphologic dimensions of palaeochannels preserved in floodplain or from stratigraphical sequences of floodplain and terrace alluvium. In a regional synthesis, Ely (1997) observed that the frequency of large floods increased throughout the region from 5,000 to 3,600 yr BP (5,800–4,200 cal yr BP) and increased again after 2,200 yr BP (2,400 cal yr BP) with maximum magnitude and frequency around 1100–900 yr BP (AD 900–1100), and after 500 yr BP (AD 1400). In southwest United States, most of the severe current floods are associated with winter storms and tropical cyclones, which are also the most probable causes of the palaeofloods over the last 5,000 years (Ely *et al.*, 1993 and Ely, 1997). In the Upper Mississippi Valley, using an excellent record of overbank flood deposits, Knox (1993) describes periods of small floods from 5,000 to 3,300 cal yr BP, and periods of average increased flood magnitudes between 7,000 and 5,000 cal yr BP, after 3,300 cal yr BP, and even larger floods from AD 1250 to 1450 at the end of the Medieval Warm Period. In-phase or out-of-phase flood cluster relations between regions of North America reflect the changing influence of the dominant atmospheric circulation patterns and storm tracks through time (Knox, Chapter 10).

1.3   Low-latitude Regions: Tropical and Subtropical Regions

Tropical and subtropical regions are defined as those extending between 0–25° and 25–35°, respectively, within both hemispheres. The region includes tropical rainforest, savanna and desert zones. River regimes are complicated since many of the major rivers such as the Nile, Parana or Zambeze rivers flow across different climatic regions. Most rivers, except those in the tropical rainforest, are characterised by strong seasonal patterns, which depend upon the movement of the Intertropical Convergence Zone (ITCZ), and the activity of the African and Asian monsoons. River discharges, therefore, increase during either autumn or summer, although some areas may present a second peak during wet years (Figure 4.2; Table 4.1). Rivers at latitudes between 15 and 35° (desertic zones of both hemispheres) have an ephemeral regime with a lack of discharge for most of the year. Throughout tropical areas, LGM (5°C cooler) palaeohydrology was characterised by dry climatic conditions and low river discharges and, therefore, by a hydrological regime notably different from that of the present. The largest decline in discharge probably occurred in southeast Asia, associated with a weakened summer monsoon (Kale *et al.*, Chapter 13).

   In *Tropical Africa*, the return to humid conditions is not evident until 12,500 yr BP (ca. 15 cal kyr BP) (Kadomura, 1995). There is evidence that a Younger Dryas arid interval, intervening before the main pluvial phase of the early Holocene occurred after 10 kyr BP (11.5 cal kyr BP), lasting probably until 6–5 kyr BP (Thomas and Thorp, 1995). In Africa, mid-Holocene alluvial sediments 4.5–4 kyr BP (5.1–5.7 cal kyr BP) are scarce or absent indicating arid conditions and low fluvial activity (Thomas and Thorp, 1995; Chapter 11). In northern Ethiopian Highlands, palaeoenvironmental reconstruction based on infilled valley deposit sequences suggests that the past 4,000 years comprised three major wetter periods at ca. 4,000–3,500, 2,500–1,500 and 1,000–960 yr BP, and two degradation episodes at ca. 3,500–2,500 and 1,500–1,000 yr BP (Machado *et al.*, 1998). Flow stages recorded in the Roda Nilometer show high floods occurring in the early 800s, around AD 1100, and also between AD 1350–1470 (Hassan, 1981; Said, 1993).

In *South and Central America*, low discharges associated with the LGM prevailed at least until 13 kyr BP, with discharges 7–10 times smaller than present (Dumont *et al.*, 1992). In the Parana basin, there is evidence pointing to a wetter climate at the early Holocene (ca 8,000–3,500 yr BP), producing a change from braided to anastomosed patterns as well as floodplain aggradation (Stevaux, 2000). A short dry episode between 3.5 and 1.5 ka affected several river basins including the Parana and the Chaco system (Stevaux, 2000; Latrubesse, Chapter 12). Since 1,500 BP the tropical fluvial systems of South and Central America have not experienced significant changes.

In *Southern Asia*, the LGM was cool, dry and variable (Thompson *et al.*, 1997) associated with a weakening of the summer monsoon. Consequently, river discharge experienced a significant reduction in water and sediment discharge, which led to highly seasonal or ephemeral rivers (Kale *et al.*, Chapter 13). Climate amelioration after about 13 kyr BP, together with strengthening of the summer monsoon, was accompanied by increases in flow and sediment discharges. During this period, there is multiple evidence of huge floods in Asian rivers related both to extreme rainfalls and rapid release of water stored behind natural dams, produced by landslides, debris flows, or within glaciers due to moraine or ice dams (e.g. Coxon *et al.*, 1996; Wohl and Cenderelli, 1998). During the mid-late Holocene (between 6–3 kyr BP), a progressive weakening of the monsoon increased aridity and led to a subsequent drop in water and sediment discharge in Asian rivers. In the last 3,000 years, fluvial activity responses to alternate humid and dry phases were not synchronous across Asia. Sequences of extreme floods occurred at distinct periods over the last 2,000 years (Ely *et al.*, 1996; Kale *et al.*, 2000) with extreme ones occurring between ca AD 400–1000 and after the 1950s (Kale, 1999).

Australia is a hydroclimatically diverse continent extending from the tropics (10° S) to latitudes near 45° S. The northern part is dominated by a tropical climate with strong influence of maritime air masses and a monsoon regime that produces an extreme late summer river discharge. The central and west-central parts of the continent comprise extensive areas of desert climate, which makes Australia the world's driest continent, excluding Antarctica, with only 45 mm average annual continental runoff, being 1/4 that of Africa or 1/14th that of South America (McMahon *et al.*, 1992). The southern part (south of 30°) is dominated by a temperate climate under the influence of westerly winds and a winter rainfall system that produces extreme winter river flow patterns. Some southeast regions follow an equinox pattern with discharge maxima during either autumn or spring (Figure 4.2; Table 4.1). Most of the available palaeohydrological information refers to the southeast, located within the limit of the temperate regions of the continent. In general terms, Late Quaternary climate changes in Australia are marked by rainfall variability, and in this respect river flows are very sensitive to these global changes (Nanson *et al.*, 1992). During the LGM, the high fluvial activity in northern Australia contrasts with the absence of fluvial activity in the southeast (Nott *et al.*, 2002). This dryer period is also recorded in lake levels of the region that dried out between 19,000 and 16,000 yr BP (Harrison and Dodson, 1993), favouring the extension of the desert limits into the northern (tropical) areas. In southeast Australia, the post LGM Yanco Phase (20–12 TL ka; equivalent to 17–10 kyr BP) of fluvial activity was characterized by much greater runoff and alluvial activity than in the late Holocene (Nanson *et al.*, Chapter 14). This interval of wetter conditions was also recorded in high lake levels between 15,000 and 13,000 yr BP, with a maximum at 12,000 yr BP. During the very late Pleistocene or with the onset of the Holocene, channels changed from the wide, large-amplitude meander systems to smaller, deeper channels with smaller meander amplitudes and wavelengths (Baker *et al.*, 1995; Nanson *et al.*, Chapter 14). Sediment loads also changed from a gravel regime to mixed load regime and sedimentation was confined between the Pleistocene terraces. This decrease in fluvial activity in southeast Australian rivers is ascribed (Nanson *et al.*, Chapter 14) to flows less active than those occurring during the Pleistocene but more active than those of today, whereas in contrast, the Holocene Climatic Optimum (8–5 ka) heightened fluvial activity in northern Australia. Rivers of coastal southeastern Australia decreased activity from about 4,000 years to the present, with a change of fluvial sedimentation into vertically accreting floodplains.

2 CONCLUDING REMARKS

At present, only 2.5% of the earth's water is in the freshwater phase, 68.6% is frozen into polar ice caps and a further 30.1% is contained in shallow groundwater aquifers, leaving only 1.3% of the earth's freshwater mobile in the surface and atmospheric phases of the hydrological cycle. During the LGM, long-term storage of water in the ice sheets accounted for a sea-level lowering between 127 and 163 m, together with a consequent diminution of precipitation, evaporation and even runoff, at least in areas without permanent permafrost conditions. As climate started to ameliorate towards the Lateglacial, river discharges increased, first in the temperate zones of North America, Europe, Asia and southeast Australia (16–10 kyr BP). Palaeohydrological evidence indicates that the scale and rate of Late Pleistocene hydrological processes, such as large-scale proglacial lakes or huge jökulhlaups draining several million cubic metres per second, may not have present analogues. In the Tropical zones, dry conditions prevailed until 12 kyr BP, followed by humid conditions between the onset of the Holocene until 7–6 kyr BP. Present-day river regimes prevailed for at least the last 4,000 years, with some minor abrupt climatic events producing changes in mean discharge and flood magnitude and frequency. These changes are not synchronous in different parts of the world, although some teleconnections have been established.

  Palaeohydrological research results have to be collated according to regions and for the sake of convenience these have been organised into the three broad latitudinal groups explained above with further subdivisions made according to continent or physical character. Palaeohydrological approaches in each of the areas (Chapters 5–14) necessarily vary according to the characteristics of the area, to features of environmental history, and to the character of the research that has been undertaken.

REFERENCES

Baker, V.R., 1973. *Paleohydrology and Sedimentology of Lake Missoula Flooding in eastern Washington*. Geological Society of America Special Paper 144, 79.

Baker, V.R., Benito, G. and Rudoy, A.N., 1993. Paleohydrology of late Pleistocene superflooding, Altay Mountains, Siberia. *Science*, **259**, 348–350.

Baker, V.R., Bowler, J.M., Enzel, Y. and Lancaster, N., 1995. Late Quaternary palaeohydrology of arid and semi-arid regions. In K.J. Gregory, L. Starkel and V.R. Baker (eds), *Global Continental Palaeohydrology*. Wiley, Chichester, 203–231.

Bjørn, G.A. and Borns, H.W., 1994. *The Ice Age World*. Scandinavian University Press, Oslo, 1–208.

Bretz, J H., 1925. The Spokane flood beyond the Channeled Scabland. *Journal of Geology*, **33**, 97–115, 236–259.

Brown, A.G., 1998. Fluvial evidence of the medieval warm period and the late medieval climatic deterioration. In G. Benito, V.R. Baker and K.J. Gregory (eds), *Palaeohydrology and Environmental Change*. Wiley, Chichester, 43–52.

Coxon, P., Owen, L.A. and Mitchele, W.A., 1996. A late Quaternary catastrophic flood in the Lahul Himalaya. *Journal of Quaternary Science*, **11**, 495–510.

Dumont, J.F., Garcia, F. and Fournier, M., 1992. Registros de cambios climáticos para los depositos y morfologías fluviales en la Amazonia Occidental. In L. Ortlieb and J. Marcharé (eds), *Paleo ENSO Records. International Symposium*, Extended Abstracts, ORSTOM, Lima, 87–92.

Dury, G.H., 1977. Underfit streams: retrospect, perspect and prospect. In K.J. Gregory (ed.), *River Channel Changes*. Wiley, Chichester, 281–293.

Ely, L.L., 1997. Response of extreme floods in the Southwestern United States to climatic variations in the late Holocene. *Geomorphology*, **19**, 175–201.

Ely, L.L., Enzel, Y., Baker, V.R. and Cayan, D.R., 1993. A 5000-year record of extreme floods and climate change in the Southwestern United States. *Science*, **262**, 410–412.

Ely, L.L., Enzel, Y., Baker, V.R., Kale, V.S. and Mishra, S., 1996. Changes in the magnitude and frequency of late Holocene monsoon floods on the Narmada River, central India. *Geological Society of America Bulletin*, **108**, 1134–1148.

Grosswald, M.G., 1998. New approach to the ice age palaeohydrology of Northern Eurasia. In G. Benito, V.R. Baker and K.J. Gregory (eds), *Palaeohydrology and Environmental Change*. Wiley, London, 199–214.

Haines, A.T., Finlayson, B.L. and McMahon, T.A., 1988. A global classification of river regimes. *Applied Geography*, **8**, 255–272.

Harrison, S.P. and Dodson, J., 1993. Climates of Australia and New Guinea since 18,000 yr B.P. In H.E. Wright, J.E. Kutzbach, T. Webb III, W.F. Ruddiman, F.A. Street-Perrot and P.J. Bartlein (eds), *Global Climates Since the Last Glacial Maximum*. University of Minnesota Press, Minnesota, MN, 265–293.

Hassan, F.A., 1981. Historical Nile floods and their implications for climate change. *Science*, **212**, 1142–1145.

Kadomura, H., 1995. Palaeoecological and palaeohydrological changes in the humid tropics during the last 20,000 years, with reference to equatorial Africa. In K.J. Gregory, L. Starkel and V.R. Baker (eds), *Global Continental Palaeohydrology*. Wiley, Chichester, 177–202.

Kale, V.S., 1999. Late Holocene temporal patterns of palaeofloods in central and western India. *Man and Environment*, **24**, 109–115.

Kale, V.S., Singhvi, A.K., Mishra, P.K. and Banerjee, D., 2000. Sedimentary records and luminescence chronology of late Holocene palaeofloods in the Luni River, Thar Desert, northwest India. *Catena*, **40**, 337–358.

Knox, J.C., 1993. Large increases in flood magnitude in response to modest changes in climate. *Nature* **361**, 430–432.

Knox, J.C., 2000. Sensitivity of modern and Holocene floods to climate change. *Quaternary Science Reviews*, **19**, 439–457.

Kozarski, S. and Rotnicki, K. 1977. Valley floors and changes of river channel patterns in the North Polish Plain during the late Würm and Holocene. *Quaestiones Geographicae*, **4**, 51–93.

Le Roy Ladurie, E., 1967. *Historie du climat depuis l'an mil*. Flammarion, Paris, 2nd ed., 1983.

Macklin, M.G., Fuller, I.C., Lewin, J., Maas, G.S., Passmore, D.G., Rose, J., Woodward, J.C., Black, S., Hamlin, R.H.B. and Rowan, J.S., 2002. Correlation of fluvial sequences in the Mediterranean basin over the last 200 ka and their relationship to climate change. *Quaternary Science Reviews*, **21**, 1633–1641.

Macklin, M.G. and Lewin, J. 1993. Holocene river alleviation in Britain. *Zeitschrift für Geomorphologie N. F.*, Supplementband **88**, 109–122.

Markgraf, V., 1993. Climatic history of Central and South America since 18,000 yr BP: comparison of pollen records and model simulations. In H.E. Wright, J.E. Kutzbach, T. Webb III, W.F. Ruddiman, F.A. Street-Perrot and P.J. Bartlein (eds), *Global Climates Since the Last Glacial Maximum*. University of Minnesota Press, Minnesota, MN, 357–385.

Maziels, J.K., 1995. Palaeohydrology of polar and subpolar regions over the past 20,000 years. In K.J. Gregory, L. Starkel and V.R. Baker (eds), *Global Continental Palaeohydrology*. John Wiley, 258–299.

McMahon, T.A., Finlayson, B.L., Haines, A.T. and Srikanthan, R., 1992. *Global Runoff. Continental Comparisons of Annual Flows and Peak Discharges*. Catena Verlag, Cremlingen, Germany, 1–166.

Machado, M.J., Pérez-González, A. and Benito, G., 1998. Paleoenvironmental changes during the last 4000 yr in the Tigray, Northern Etiopía. *Quaternary Research*, **49**, 312–321.

Nanson, G.C., Price, D.M. and Short, S.A., 1992. Wetting and drying of Australia over the past 300 ka. *Geology*, **20**, 791–794.

Nott, J., Price, D. and Nanson, G., 2002. Stream response to Quaternary climate change: evidence from the Shoalhaven River catchment, southeastern highlands, temperate Australia. *Quaternary Science Reviews*, **21**, 965–974.

O'Connor, J.E. and Baker, V.R., 1992. Magnitudes and implications of peak discharges from glacial Lake Missoula. *Geological Society of America Bulletin*, **104**, 267–279.

Said, R., 1993. *The River Nile. Geology, Hydrology and Utilization*. Pergamon Press, Oxford, 1–320.

Starkel, L., 1991. The Vistula River Valley: a case study for Central Europe. In L. Starkel, K.J. Gregory and J.B. Thornes (eds), *Temperate Palaeohydrology*. John Wiley, Chichester, 171–188.

Starkel, L., 1996. Palaeohydrological reconstruction: advantages and disadvantages. In J. Branson, A.G. Brown and K.J. Gregory (eds), *Global Continental Changes: The Context of Palaeohydrology*. Special Publication No. 115, Geological Society of London, London, 9–17.

Starkel, L., 2002. Change in the frequency of extreme events as the indicator of climatic change in the Holocene (in fluvial systems). *Quaternary International*, **91**, 25–32.

Stevaux, J.C., 2000. Climatic events during the late Pleistocene and Holocene in the upper Parana River: correlation with NE Argentina and South-Central Brazil. *Quaternary International*, **72**, 73–85.

Teller, J.T. and Kehew, A.E., 1994. Introduction to the late glacial history of large proglacial lakes and meltwaters runoff along the Laurentide ice sheet. *Quaternary Science Reviews*, **13**, 795–799.

Thomas, M.F. and Thorp, M.B., 1995. Geomorphic response to rapid climatic and hydrologic change during the late Pleistocene and early Holocene in the humid and sub-humid tropics. *Quaternary Science Reviews*, **14**, 193–207.

Thompson, R.S., Whitlock, C., Bartlein, P.J., Harrison, S.P. and Spaulding, W.G., 1993. Climatic changes in the Western United States since 18,000 yr BP. In H.E. Wright, J.E. Kutzbach, T. Webb III, W.F. Ruddiman, F.A. Street-Perrot and P.J. Bartlein (eds), *Global Climates Since the Last Glacial Maximum*. University of Minnesota Press, Minnesota, MN, 468–513.

Thompson, L.G., Yao, T., Davis, M.E., Henderson, K.A., Thompson, E., Lin, P.N., Beer, J., Synal, H.A., Cole-Dai, J. and Bolzan, J.F., 1997. Tropical climate instability: the last glacial cycle from a Qinghai-Tibetan ice core. *Science*, **276**, 1821–1825.

Thorson, R.M., 1989. Late Quaternary paleofloods along the Porcupine River, Alaska: implications for regional correlation. *U.S. Geological Survey Circular*, **1026**, 51–54.

Van Geel, B., Van der Plicht, J., Kilian, M.R., Klaver, E.R., Kouwenberg, J.H.M., Ressen, H., Reynaud-Farrera, I. and Waterbolk, H.T., 1998. The sharp rise of δ ^{14}C at ca. 800 cal. BC. Possible causes, related climatic teleconnections and the impact on human environments. *Radiocarbon*, **40**, 335–350.

Vandenberghe, J., 1995. The role of rivers in climatic reconstruction. *Palaeoclimate Research*, **14**, 11–20.

Wohl, E.E. and Cenderelli, D., 1998. Flooding in the Himalaya Mountains. In V.S. Kale (ed.), *Flood Studies in India*. Memoir 41, Geological Society of India, Bangalore, 77–99.

Wright Jr., H.E., Kutzbach, J.E., Webb III, T., Ruddiman, W.F., Street-Perrott, F.A. and Bartlein, P.J., (eds), 1993. *Global Climates Since the Last Glacial Maximum*. University of Minnesota Press, Minneapolis, MN, 1–569.

Section I   High-latitude Regions

5 High-latitude Fluvial Morphology: The Example from the Usa River, Northern Russia

J. VANDENBERGHE AND M. HUISINK
Vrije Universiteit, Amsterdam, The Netherlands

1 INTRODUCTION

This chapter discusses the potential impact of several climatic and environmental factors, including vegetation type and density, river gradient, annual flood, permafrost occurrence and sediment load on the river channel pattern. Braided river types were often assumed to be the typical style for permafrost regions. However, there appears to be a wide morphological variety of rivers in such regions (Vandenberghe, 2001). Meandering and anabranching rivers, and rivers that are annually shifting between meandering and braided now frequently occur next to braided rivers in cold regions such as northern Canada (Vandenberghe, 2001; Figure 5.5). Also fluvio-morphological patterns and deposits from former permafrost regions demonstrate a similar wide variety of river styles. In addition, the evolution of former periglacial river systems has often shown a complex response of these systems to climatic changes (e.g. McCann *et al.*, 1971; Starkel, 1991; Vandenberghe *et al.*, 1994; Bridgland and Allen, 1996; Huisink, 1997; Mol *et al.*, 2000; Vandenberghe, 2002).

The sensitivity for fluvial changes to occur in response to climatic changes in cold regions may be analysed to consider the impact of *future* climatic changes on the river systems in those areas. Scenarios of future temperature and precipitation changes can be used in hydrological runoff models to calculate the corresponding mean annual or peak discharges. These discharges in turn, form the starting point to predict the possible changes in channel style.

In this chapter, the rich fluvio-morphological diversity of an arctic (subarctic) river system is discussed in the light of its environmental conditions. In addition, the probability for potential future changes in the different parts of the system are analysed on the basis of reconstructed Holocene changes. Publications in English on fluvial morphology in Russia are scarce, but exceptions are the papers by Alabyan and Chalov (1998), Bareiss *et al.* (1999), Panin *et al.* (1999), Rachold *et al.* (1996), Sidorchuk and Borisova (2000), Sidorchuk *et al.* (in press) and Yamskikh (1998). These papers describe a wide variety of river types, but most of the research sites are not located in northernmost Russia.

The fluvial morphology of arctic rivers may be exemplified by the Usa catchment in northern Russia, which, like many (sub)arctic river systems shows a wide variety of channel patterns. Furthermore, some parts of the catchment showed a change in

Palaeohydrology: Understanding Global Change. Edited by K.J. Gregory and G. Benito
© 2003 John Wiley & Sons, Ltd ISBN: 0-470-84739-5

river channel pattern one or more times during the Holocene. Therefore, the fluvial characteristics and evolution of the Usa catchment in the north Russian lowlands was analysed in the framework of the multi-disciplinary research programme TUNdra Degradation in the Russian Arctic (TUNDRA) (Huisink *et al.*, 2003) (Figure 5.1). This programme on (TUNDRA) studied the effect of global change on a large arctic catchment, and more specifically, collected information on present and past rivers in the catchment (Oksanen *et al.*, 2001; Huisink *et al.*, in prep.). The results are compared with descriptions from present-day river morphology in Canada and other Russian catchments.

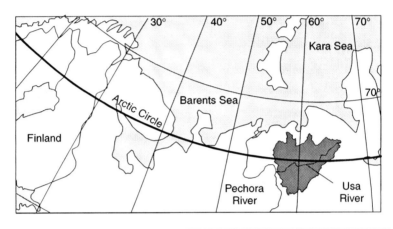

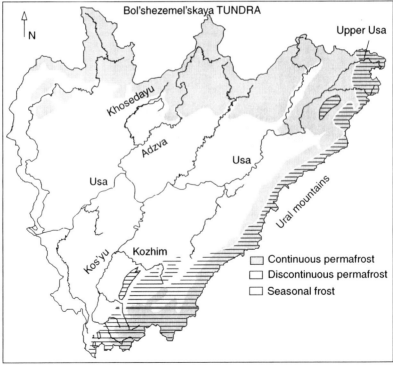

Figure 5.1   Location map of the Usa catchment in northern Russia

2 PRESENT-DAY FLUVIAL MORPHOLOGY IN THE USA CATCHMENT

The Usa River is a tributary of the Pechora River, which is one of the major south-north flowing rivers in European Russia. The 93,000 km^2 Usa catchment is located in the European Russian Plain. The Upper Usa originates in the Ural Mountains, which is in the northernmost part of the catchment. Various other rivers originate in the Ural Mountains, including the Kos'yu and Kozhim rivers, coming together with the Upper Usa River in the central part of the basin. Other tributaries, the Khosedayu and Adzva rivers, drain the gently undulating northern and eastern Bol'shezemel'skaya tundra plains. The Usa basin is located in the zones of continuous and discontinuous permafrost as well as in environments with deep seasonal frost. The area is very remote, and apart from isolated towns and oil wells, human settlement is virtually absent. This results in an undisturbed, pristine landscape, where river morphology is not disturbed or altered by man. The vegetation cover is dense and changes from taiga forests with swamps and mires in the south to tundra in the north. Slopes vary from steep in the Ural Mountains (up to 240 cm km^{-1}) to very low in the Bol'shezemel'skaya plains (less than 10 cm km^{-1}). The infiltration capacity is very low because of the frozen soils that favour very irregular, peaked discharges, characteristic of the nival discharge regime according to Church (1974).

A wide variety of river types occur in northern Russia varying from braiding through intermediate stages between braiding and meandering to meandering (Huisink *et al.*, in press). In contrast to what might be expected, the dominant river type in this (sub)arctic area is not the braided type. Instead, meandering or anabranching patterns occur more frequently. This variety is illustrated in Figure 5.2 showing an aerial photograph in the middle reach of the catchment. The tributary at the bottom left of the picture is a clearly meandering one with a highly sinuous course. The Usa River is straight or even braided at present but in the central lower part of the picture large palaeo-meanders are clearly visible and represent the evidence of a past meandering Usa River.

The geology, and more specifically the lithology, of the subsoils is an independent factor that influences fluvial action and morphology since it determines longitudinal gradient and sediment yield. The Ural Mountains and the foothills consist of hardrock in which narrow valleys have been incised. The other part of the catchment is underlaid by unconsolidated sediments and the rivers are less confined and have wider flood-plains. Braided rivers occur within the Ural Mountains and at their rims where the river gradients exceed 1.30 m km^{-1}. Most of the catchment is located in the gentler sloping taiga and tundra lowlands where braided rivers are scarce. It is striking, however, that in regions with similar river gradients, very different river types may be observed, varying from meandering to anabranching (Huisink *et al.*, in press). Meandering is considered traditionally as a typical pattern for relatively constant river discharges. Therefore, it is unexpected to find that the very irregular, peaked discharges during the snowmelt season, when large amounts of water are discharged in a very short period, did not prohibit the formation of a large number of meanders. However, it should be realized that, when discharges are sufficient to transport all the sediment that is supplied, the determining factor in the establishment of a specific river pattern is not the discharge variability, but the sediment supplied in terms of volume and grainsize. Irregular discharges often provoke the removal and transport of much heterogeneous material. But this is not the case in the lowland reach of the Usa River where the very dense vegetation is largely responsible for increased bank stability, thus diminishing the sediment load in the river. In addition, the frost in the soil at the time

Figure 5.2   Aerial photograph of the middle reach of the Usa River

of snowmelt, the "break up", also prohibits a large sediment yield towards the river. Thus, it appears that vegetation and soil frost diminish the availability of river sediment load, and consequently, also hinder the establishment of a braided river type, as Vandenberghe (2001) found for a wide range of present-day and past fluvial systems.

A detailed spatial analysis of river types in relation to vegetation belts demonstrates that each of the river types occurs in all the distinguished vegetation zones (Huisink *et al.*, 2003, Figures 7 and 8). This shows that there is no relation between river pattern and vegetation type. In fact, it is not the vegetation *type* that determines sediment removal, but it is the vegetation *density* that is the determining factor for sediment removal (Huisink *et al.*, in press). It is indeed the vegetation density that controls riverbank cohesion and stability and sediment supply throughout the catchment.

The peak floods in spring, in combination with ice jams, promote locally abrupt chute cut-offs of the large meander bends and sudden channel abandonment. This results in the formation of a river system with multiple meandering channels and stable islands as is the case with the Kos'yu River (Figure 5.2; Huisink *et al.*, 2003). These properties are characteristic of an anabranching river type (Nanson and Knighton, 1996). On a large timescale, this is a river system that may be considered as intermediate between meandering and braided.

3 PAST CHANGES IN FLUVIAL MORPHOLOGY

Figure 5.2 shows the changes in fluvial style of the Usa river during the Holocene, varying between sinuous, braided or straight at some reaches. Human impact has been and is still very limited in this remote area. Thus, the observed changes in fluvial styles may be most likely attributable to climate changes. The impact of Holocene climate changes on the rivers in the Usa basin is complex (Figure 5.3). In some valleys, the channel pattern did not change (Khosedayu River, Figure 5.3), while in other valleys the channel pattern changed from highly sinuous via low sinuous to straight and meandering (Usa River). The Khosedayu and middle Usa valleys show identical terrace stratigraphies that are related to distinct and simultaneous phases of increased incision (Figure 5.3).

The terrace stratigraphies and climate evolution are described in detail by Huisink *et al.* (in prep.). A major terrace level is distinguished in both the Khosedayu and Usa River valleys at 7 to 15 m above the present floodplains. Dating of the organic material in the fluvial sediments suggests that incision took place at the very beginning of the Holocene, certainly before 8.9 ka BP. Afterwards, fluvial deposition and reworking took place until around 7.3 ka BP in the Usa valley and 5.7 ka BP in the Khosedayu valley. A second incision phase occurred around 6–7 ka BP that resulted in a second, much smaller terrace level. This was accompanied by a change in channel style.

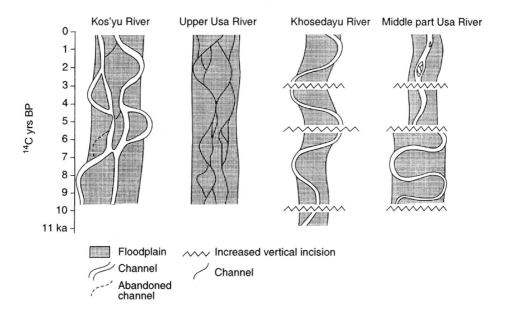

Figure 5.3 Examples of Holocene morphological change from four rivers in the Usa catchment: the anabranching Kos'yu River; the braiding upper Usa River; the meandering Khosedayu River; and the middle part of the Usa River

Information on the age of the incision that led to the formation of the present-day floodplains is scarce, but this final incision had definitely taken place before 2.8 ka BP.

The incision phases coincide with major changes in vegetation or permafrost conditions. The early Holocene incision is related to the establishment of a denser vegetation cover after the last glacial (Oksanen *et al.*, 2001). Reduced sediment load to the rivers lowered the energy needs of the rivers leading to lower gradients and incision. This is a phenomenon commonly recognised in postglacial fluvial development (e.g. references in Vandenberghe, 1995). The incision around 6–7 ka BP seems to coincide with the first establishment of permafrost after the Holocene climatic optimum in the Usa Basin (Kuhry, pers. comm.; Huisink *et al.*, in prep.). Some 3000 years ago, permafrost disappeared again in the region (Kuhry, pers. comm.), simultaneously with the last incision phase in the Usa catchment. Obviously, there is a coincidence between permafrost establishment and degradation and river activity. The formation/degradation of permafrost may possibly influence vegetation (sediment load) and discharges (soil water storage capacity) temporarily so much that a phase of instability occurs. This may trigger rivers to incise or even change river style. Although the processes inducing the incisions in the Usa catchment are not clearly known, it should be noted that phases of instability at times of climatic *change* have previously been discussed and modelled in other areas (Vandenberghe, 1993; Bogaart and van Balen, 2000). The functionality between channel patterns and soil frost, vegetation and discharge in the Usa catchment is in accordance with the general relationships found in arctic regions (Vandenberghe, 2001).

4 SENSITIVITY OF (SUB)ARCTIC FLUVIAL MORPHOLOGY TO GLOBAL CHANGE, EXEMPLIFIED BY RIVERS FROM THE USA BASIN

An attempt has been made to indicate possible channel planform changes as a consequence of globally changing environmental conditions. The estimation of the sensitivity of channels to change their morphological patterns is based on the potential specific stream power concept described by Van den Berg (1995). This concept uses the valley slope and bankfull discharge or mean annual flood to calculate the potential stream power (ω_v). Plotting this specific potential stream power against the median grain size of the riverbed sediment enables a distinction to be made between single- and multi-thread rivers. The threshold value underlying this distinction is also a function of the three parameters: valley slope (gradient), mean grainsize of the bedload and the mean annual flood. Van den Berg (1995) calibrated the threshold value against the three parameters worldwide for a data set of 228 rivers with wide alluvial floodplains like those of the northern Russian rivers.

Figure 5.4 shows the calculated potential stream power for several river segments in the Usa basin. Slope values were determined from 1:50,000 topographical maps. For the present Kos'yu, Khosedayu and Usa (Makarikha station) rivers, the other two parameters are available from hydrological measurements and bedload sampling. For the Kozhim River, the bed grainsize was estimated by analogy with the nearby Kos'yu field station. The latter station is at similar distance to the Ural Mountains, the sediment source area, so that the same median bed grain size may be assumed. For the Adzva River and the Usa River (Adzva station), an intermediate grain size between the up- and downstream sampled measurements is taken. After calculating the specific potential stream power, this value was plotted against the median bed grainsize. It appears that the observed present channel morphology (single- versus multichannel) of the Usa basin segments generally corresponds to the channel morphology that should follow from the diagram constructed by Van den Berg (Table 5.1). An exception is the anabranching Kos'yu River segment that may be considered as a kind of transitional form between braided and meandering and thus has no real place in the diagram by Van den Berg. The calculated value of the Kos'yu River plots in the single-thread domain, but very near to the threshold line, while it is certainly not a meandering river.

From Figure 5.4, it appears that three river segments are located deep in the single-thread domain (Numbers 3, 4 and 5). These river sections are thus stable meandering rivers and they are not expected to change their pattern in response to small external changes. This is confirmed by the Holocene reconstructions of their evolution. The

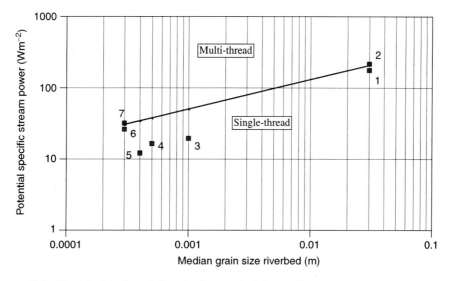

Figure 5.4 Fluvial style in relation to the potential specific stream power using mean annual discharge. The numbers correspond to the rivers listed in Table 5.1

Table 5.1 Data used in Figure 5.4 – rivers, mean annual flood, valley slope, median grain size of bed material, potential specific stream power, the critical potential specific stream power, the calculated channel type (single-thread or multi-thread), the observed channel type and the corresponding points in Figure 5.4

	Mean annual flood $(m^3 s^{-1})$	Valley slope (m/m)	Median grain size (m)	ω_v	$\omega_{v\ critical}$	Channel type calculated	Channel type observed	Number in Figure 5.4
Kos'yu	1,339	0.001470	0.03	178	206	Single	Multi	1
Kozhim	1,584	0.001662	0.03	218	206	Multi	Multi	2
Khosedayu	565	0.000390	0.001	19	49	Single	Single	3
Adzva	2,106	0.000170	0.0005	16	37	Single	Single	4
Usa at Adzva	9,209	0.00006	0.0004	12	37	Single	Single	5
Usa at Makarikha	10,372	0.00015	0.0003	32	30	Multi	Multi	6
Usa – 30% at Makarikha	7,000	0.00015	0.0003	26	30	Single	Single	7

Khosedayu River, for instance, (Number 3 in Figure 5.4) is a highly sinuous mean-dering river at present, but was also meandering during Holocene times (Figure 5.3). For the Adzva and Usa River (station Adzva, Numbers 4 and 5 in Figure 5.4) that are meandering at present, no historical reconstructions are available. But topographical maps and satellite images do not show any indication of former channel patterns that are distinct from the present one.

The Usa River (Makarikha station) plots in the multi-thread domain, which is con-sistent with its contemporary braided planform. However, the data plots very close to the threshold line. Slightly different environmental conditions during the Holocene may have instigated pattern changes. The occurrence of large palaeo-meanders along most of the Usa River, and also at Makarikha, confirms this assumption. These palaeo-meanders were active during the early and middle Holocene (Figure 5.3; Huisink *et al.*, in prep.). A mean temperature 3°C higher during the Holocene optimum may be assumed (Oksanen *et al.*, 2001). Introducing this temperature in a hydrological model results in a reduction of the mean annual flood by some 30% (Van der Linden, personal communication) and an equivalent reduction in specific potential stream power. This should bring the value plotted for the Usa at Makarikha below the threshold line, thus into the single-thread domain, during the middle Holocene.

5   POTENTIAL FOR PREDICTION OF FUTURE RIVER PATTERN CHANGES AS A CONSEQUENCE OF ENVIRONMENTAL CHANGES

The proximity of the present-day channel pattern to the thresholds for change of that pattern is thus very important. This explains why some palaeo-rivers adapted their pat-terns to external forcing, while other rivers in nearly similar situations did not adapt to the same forcing conditions. An illustrative example is the reaction of different Euro-pean rivers to the Younger Dryas cooling (Vandenberghe *et al.*, 1994). The use of the potential specific stream power concept of Van den Berg (1995) offers an appropriate way of more precisely calculating the specific position of a river with regard to the threshold value as a basis for estimating the change in its pattern. This may be of great significance for predicting the future reaction of river morphology as a response to changing external conditions, including, for instance, changes in temperature, precip-itation or land use. In the case of the Usa basin, therefore, it should not be expected that, after some global climate change, the Khosedayu, Adzva and Usa (station Adzva) would change their pattern. But the braided Usa River near Makarikha will reach its threshold after relatively slight changes. For instance, according to hydrological mod-elling, a temperature lowering of one to two degrees will result in a reduction of the annual flood by about 10 to 20%. That could be sufficient for an enormous change of the pattern for that specific river section: from multi- to a single-thread channel. It may be concluded that this river section is highly susceptible to relatively small changes in temperature or precipitation.

Using the above-mentioned method requires, firstly, data on bankfull discharge or mean annual flood magnitude. However, such hydrological information is often hard to obtain in arctic regions. Secondly, knowledge of the bedload grainsize requires sediment sampling. Thirdly, the river gradient may be derived from topographical maps, although sufficiently detailed topographical maps are not always available. Furthermore, the gradient estimate may be quite approximate, and for rivers with a fine-grained bedload, a slight change of the gradient may result in relatively large changes of the specific potential stream power and thus, the crossing of threshold values. Such changes in annual flood may be induced by changes in temperature, permafrost occurrence or depth, while vegetation cover or land use may also influence the grainsize of the bedload.

6 CONCLUSIONS

Channel patterns in the Usa basin vary between braided, meandering and anabranching. Some river segments changed their pattern during the Holocene whereas others did not. It appears that the density of the vegetation cover plays a major role in defining the channel pattern, rather, the vegetation type and the river gradient. Detailed analysis of the Holocene responses of the Usa River to climatic changes provides analogues for potential future climatic changes. The Holocene history seems to indicate that the appearance and disappearance of permafrost coincides with river instability, possibly through changes in flow regime, vegetation and sediment load. The sensitivity of the river pattern to respond to such changes is dependent on the river gradient, the grainsize of the bedload and the mean annual flood. The relative proximity of these factors especially determines whether a change in channel pattern is, or is not to be expected within the constraints of a certain external forcing. In the case of a specific river network (for instance the Usa), the river gradient will not change dramatically, but most change may be expected from changing discharges (annual flood) and – to a lesser extent–bedload grainsize. The sensitivity of (sub)arctic fluvial morphology to external changes is therefore important for the prediction of channel pattern changes that may occur as a response to possible future climatic environmental changes.

ACKNOWLEDGEMENTS

We thank S. van der Linden for providing the hydrological data used to calculate the present and future potential specific stream power values.

REFERENCES

Alabyan, A.M. and Chalov, R.S., 1998. Types of river channel patterns and their natural controls. *Earth Surface Processes and Landforms*, **23**, 467–474.

Bareiss, J., Eicken, H., Helbig, A. and Martin, T., 1999. Impact of river discharge and regional climatology on the decay of sea ice in the Laptev sea during spring and early summer. *Arctic, Antarctic, and Alpine Research*, **13**, 214–229.

Bogaart, P.W. and van Balen, R.T., 2000. Numerical modelling of the response of alluvial rivers to Quaternary climate change. *Global and Planetary Change*, **27**, 147–163.

Bridgland, D.R. and Allen, P., 1996. A revised model for terrace formation and its significance for the early middle Pleistocene terrace aggradations of North-East Essex, England. In C. Turner (ed.), *The Early Middle Pleistocene in Europe*. Balkema, Rotterdam, 121–134.

Church, M., 1974. Hydrology and permafrost with reference to Northern North America. *Proceedings Workshop Seminar on Permafrost Hydrology*, Canadian National Committee, IHD, Ottawa, 7–20.

Huisink, M., 1997. Late glacial sedimentological and morphological changes in a lowland river as a response to climatic change: the Maas, The Netherlands. *Journal of Quaternary Science*, **12**, 209–223.

Huisink, M., Kultti, S., Kuhry, P. and Oksanen, P., Holocene periglacial river dynamics in relation to vegetation and palsa-peat in the Eurasian Usa catchment; in prep.

Huisink, M., de Moor, J.J.W., Kasse, C. and Virtanen, T., 2003 Factors influencing periglacial morphology in the northern European Russian tundra and taiga. *Earth Surface Processes and Landforms*, **27**, 1223–1235.

McCann, S.B., Howarth, P.J. and Cogley, J.G., 1971. Fluvial processes in a periglacial environment. *Transactions Institute of British Geographers*, **55**, 69–82.

Mol, J., Vandenberghe, J. and Kasse, C., 2000. River response to variations of periglacial climate. *Geomorphology*, **33**, 131–148.

Nanson, G.C. and Knighton, A.D., 1996. Anabranching rivers: their cause, character and classification. *Earth Surface Processes and Landforms*, **21**, 217–239.

Oksanen, P.O., Kuhry, P. and Alekseeva, R.N., 2001. Holocene development of the Rogovaya river peat plateau, European Russian Arctic. *The Holocene*, **11**, 25–40.

Panin, A.V., Sidorchuk, A.Y. and Chernov, A.V., 1999. Historical background to floodplain morphology: examples from the East European plain. In S.B. Mariott and J. Baker (eds), *Floodplains: Interdisciplinary Approaches*. Special Publications 163, Geological Society, London, 217–229.

Rachold, V., Alabyan, A., Hubberten, H.-W., Korotaev, V.N. and Zaitsev, A.A., 1996. Sediment transport to the Laptev sea – hydrology and geochemistry of the Lena river. *Polar Research*, **15**, 183–196.

Sidorchuk, A.Y. and Borisova, O.K., 2000. Method of paleogeographical analogues in paleohydrological reconstructions. *Quaternary International*, **72**, 95–106.

Sidorchuk, A.Y., Panin, A.V., Borisova, O.K., Elias, S.A. and Syvitski, J.P., Channel morphology and river flow in the Northern Russian plain in the late glacial and Holocene. *International Journal of Earth Sciences*, in press.

Starkel, L., 1991. Long-distance correlation of fluvial events in the temperate zone. In L. Starkel, K.J. Gregory and J.B. Thornes (eds), *Temperate Palaeohydrology*. Wiley, Chichester, 473–496.

Van den Berg, J.H., 1995. Prediction of alluvial channel pattern of perennial rivers. *Geomorphology*, **12**, 259–270.

Vandenberghe, J., 1993. Changing fluvial processes under changing periglacial conditions. *Zeitschrift für Geomorphologie*, Supplementband **88**, 17–28.

Vandenberghe, J., 1995. Postglacial river activity and climate: state-of-the-art and future prospects. In B. Frenzel, J. Vandenberghe, C. Kasse, S. Bohncke and Gläser, B. (eds), *European river activity and climatic change during the Lateglacial and early Holocene*. Paläoklimaforschung, G. Fischer Verlag, Stuttgart, **14**, 1–9.

Vandenberghe, J., 2001. A typology of Pleistocene cold-based rivers. *Quaternary International*, **79**, 111–121.

Vandenberghe, J., 2002. The relation between climate and river processes, landforms and deposits during the Quaternary. *Quaternary International*, **91**, 17–23.

Vandenberghe, J., Kasse, C., Bohncke, S. and Kozarski, S., 1994. Climate-related river activity at the Weichselian-Holocene transition: a comparative study of the Warta and Maas rivers. *Terra Nova*, **6**, 476–485.

Yamskikh, A.A., 1998. Late Holocene soil formation in the valley of the river Yenisei; Central Siberia. *Catena*, **34**, 47–60.

Section II Mid-latitude Regions

6 The Lateglacial and Holocene Palaeohydrology of Northern Eurasia

ALEKSEY SIDORCHUK, ANDREY PANIN[1] AND OLGA BORISOVA[2]

[1]*Moscow State University, Moscow, Russia*
[2]*Institute of Geography, Moscow, Russia*

1 INTRODUCTION

Extreme floods on rivers fed by melt water from glaciers or glacial lakes have been well studied in the past periglacial areas (e.g. Baker, 1973; Baker *et al.*, 1993). However, the palaeohydrology of rivers in cold climates with widespread permafrost with no source of glacial melt water has attracted less attention, though such rivers drained an area over $30 \times 10^6 \, \text{km}^2$ in the Northern hemisphere at the termination of the Last Glaciation. At that time, the periglacial zone with continuous permafrost and low soil permeability extended over the northern Russian Plain and the West Siberian Plain as far south as 49° N (Velichko, 2002). The relics of large palaeochannels are found on the lower levels of river terraces and on the flood plains throughout the former periglacial zone. The majority of these palaeochannels had a meandering pattern and their widths exceeded those of the recent channels by up to 15 times (Panin *et al.*, 1999). Palaeolandscape reconstructions (Sidorchuk *et al.*, 2001b) show that the modern climatic and hydrological analogues of the periglacial conditions are found in the western margins of the Altai Mountains (closest climatic conditions) and in the tundra zone of the northern Russian Plain and West Siberia (closest hydrologic regime). The palaeohydrology of the periglacial rivers can be reconstructed on the basis of their morphology using such palaeogeographical analogues.

East Siberia is situated at present within the permafrost zone with continuous or discontinuous permafrost. Therefore, the type of hydrological regime did not change dramatically at the Lateglacial/Holocene transition, and the contemporary regional relationships between the flow hydraulics and channel morphology can potentially be used as a basis for palaeohydrological reconstructions. The river valleys in this tectonically active region are mainly incised so that the morphology of the confined channels has much more complicated relationships with the channel-forming discharge than in the case of free alluvial rivers. Accordingly, the use of the hydrological regime equations in this region is limited to channel sections with free alluvial meanders.

2 METHODS OF PALAEOHYDROLOGICAL AND PALAEOLANDSCAPE RECONSTRUCTION

River channel morphology can be used for palaeohydrological reconstruction as it strongly relates to the channel-forming hydrological regime. There are two main ways

Palaeohydrology: Understanding Global Change. Edited by K.J. Gregory and G. Benito
© 2003 John Wiley & Sons, Ltd ISBN: 0-470-84739-5

of calculating palaeodischarges: using (1) hydraulic or (2) regime equations. Dury (1964; 1965) was the first to use both approaches for palaeohydrological reconstructions. The hydraulic approach requires detailed field investigations. The longitudinal profile and cross-section morphology of each channel, as well as the texture and lithology of the infilling sediments must be carefully studied (Rotnicki, 1991). Of the fluvial relief forms, riffles (crosses) of the meandering palaeochannels are the best for the reconstruction of water discharge Q (m^3 s^{-1}). The Chezy–Manning formula can be applied for such calculations:

$$Q = A \frac{D^{1/6}}{n} \sqrt{SD} \tag{1}$$

The boundaries of a palaeochannel cross-section correspond to the top of the basal layer of coarse alluvium. The cross-section area A (m^2) and its mean depth D (m) can be calculated for different water stages. The slope S is presumably stage-independent and equal to the longitudinal slope of the mean surface of the infilling deposits. The Manning roughness coefficient n (s m$^{-1/3}$) can be estimated from the modern fluvial analogue of the palaeochannel. The palaeochannel lifetime and the return period of the extreme discharge are calculated using the meander evolution model. Mean maximum and annual discharges with different return periods can be calculated with the use of the two-parametric gamma-distribution and Gudrich distribution (Evstigneev, 1990), the distribution parameters being obtained from the modern fluvial analogues. This approach was discussed in detail by Sidorchuk and Borisova (2000).

The relationships between channel morphology and flow hydrology are used in the regime equations approach. In this investigation, the relationships between the mean annual discharge (Q) and the bankfull channel width (W_b) were established for 185 sections of meandering rivers in the Russian Plain, West and East Siberia:

$$Q = 0.012 y^{0.73} W_b^{1.36} \tag{2}$$

The parameter y is related to the seasonal flow variability and represents the ratio between the annual discharge Q and the mean maximum discharge Q_{max}:

$$y = 100(Q/Q_{max}) \tag{3}$$

An increase in the flow variability generally causes an increase in the floodplain height and flow concentration in a single channel with larger bankfull width. Flow variability depends on the basin area F (km^2):

$$y = aF^{0.125} \tag{4}$$

Parameter a in Formula (4) reflects the geographical distribution of the flow variability and can be estimated for each palaeochannel using the recent fluvial analogues. It is then possible to calculate y with Formula (4), the mean annual discharge Q from the palaeochannel width with Formula (2), and the mean maximum discharge Q_{max} with Formula (3). Flood runoff depth h_p (mm) can be calculated from Q_{max} and F with the formula

$$h_p = \frac{Q_{max}(F + b)^N}{K_0 F} \tag{5}$$

as in Evstigneev (1990). Parameters b, N and K_0 should be estimated for the region-analogue.

To use both methods of palaeohydrological reconstruction, it is crucial to determine the recent region-analogue and the fluvial analogue. The palaeogeographical analogue method (Sidorchuk and Borisova, 2000) is based on the assumed possibility of transforming the present-day spatial (geographical) relationships in hydrology into temporal relationships at certain locations. Kalinin (1966) demonstrated that such an assumption is based on the ergodic theorem. Broad experience of interpolation and extrapolation of hydrological variables in space and time suggests that it is possible to use this theorem in palaeohydrology. Geographical influences on river flow bring about similarity of hydrological regimes for rivers in similar landscapes (Evstigneev, 1990). Geographical controls over river flow and their applications to palaeohydrology lead to the principle of palaeogeographical analogy:

- Similar hydrological regimes were characteristic of palaeorivers in similar palaeolandscapes.
- The hydrological regime of a palaeoriver within a palaeolandscape would be similar to that of a present-day river within the same type of landscape.

The second statement forms the basis of the method of palaeogeographical analogy. Therefore, for a palaeohydrological reconstruction, the reconstruction of the palaeolandscape is required. The hydrological regime of modern rivers in a certain type of landscape can be used for estimations of the palaeohydrological regime in the same type of palaeolandscape.

To reconstruct the landscape and climatic conditions that existed at various stages of palaeochannel development, palynological studies of dated alluvium, lake and peat sediments are necessary. The use of palaeobotanic data for palaeoclimatic and palaeolandscape reconstruction implies that the flora of a particular region directly reflects the influence of the natural environment and of the climate in particular. The method of reconstructing vegetation and climate from the composition of fossil florae was developed by Grichuk (1969), who used a concept derived from Szafer (1946). Geographical analysis of the modern spatial distribution of all the plants of a certain fossil flora allows finding the location of the closest modern floristic analogue to the past vegetation at the site. By identifying the region where the majority of plant species found in a fossil flora grows at the present time, it is possible to determine the closest modern landscape and climatic analogue to the past environment under consideration. Usually, the conditions suitable for all the species of a given fossil flora can be found within a comparatively small area. The present-day features of plant communities and the main climatic indices of such a region-analogue would be close, if not identical, to those that existed at the site in the past.

River channel morphology is characterised by significant inertia, so that the sizes of alluvial relief forms of the lowland sandy rivers do not change much because of the short-term (several decades) oscillations of the river flow. A measurable transformation of the channel width and meander wavelength usually takes hundreds of years. Accuracy of calculations with hydraulic and regime equations is relatively low, the error being about 40% (Sidorchuk and Borisova, 2000). Therefore, palaeohydrological reconstructions using palaeochannel morphology reveal only long-term changes of a relatively high magnitude.

3   SURFACE RUNOFF CHANGES IN NORTHERN EURASIA DURING THE LATEGLACIAL AND THE HOLOCENE

3.1   The Russian Plain

The Lateglacial

On the basis of the detail coring of the sediments infilling several palaeochannels, the geometry of cross-sections of these palaeochannels was reconstructed. Calculations with Chezy–Manning formula (Figure 6.1) for the palaeochannels of the Protva River (Site 1 – the Oka River basin, mixed broadleaved-coniferous forest zone), the Seim

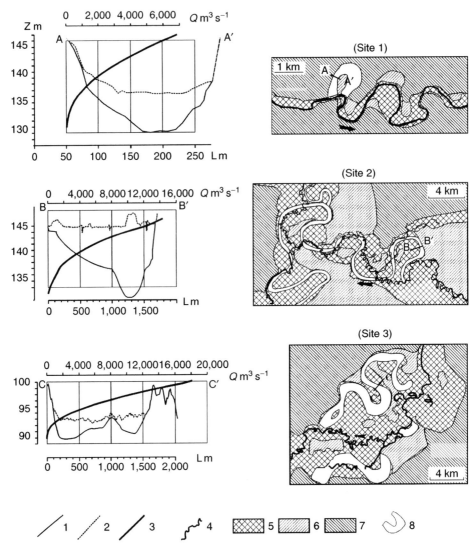

Figure 6.1 Morphology of macromeanders, cross-section geometry and calculated stage discharge curves for the palaeochannels of (Site 1) Protva, (Site 2) Seim and (Site 3) Khoper rivers. Key: 1 – the reconstructed bed of a palaeochannel; 2 – surface of the infilling deposits; 3 – stage discharge relationships; 4 – modern channel; 5 – floodplain; 6 – low terraces; 7 – high terraces and 8 – palaeochannel

River (Site 2 – the Dnieper River basin, forest-steppe zone), and the Khoper River (Site 3 – the Don River basin, steppe zone) show that the reconstructed extreme and mean maximum discharges (MMD) were up to 14 times greater than the modern ones (Table 6.1) and close to MMD of much larger modern rivers. For comparison, the modern MMD of the Oka River near Kaluga (basin area of 54,900 km^2) is 4820 m^3 s^{-1}, the MMD of the Dnieper River near Kiev (328,000 km^2) is 7500 m^3 s^{-1}, and the MMD of the Don River near Razdorskaya (378,000 km^2) is 5610 m^3 s^{-1}.

All these large palaeorivers were active in the Lateglacial. The beginning of infilling of the Protva River large channel with sediments occurred approximately 12,700 ± 110 yrs BP (Ki-7312). The large channel of the Khoper River was active 14,430 ± 110 yrs BP (Ki-7694). Its filling began about 12,000 yrs BP (11,900 ± 120 yrs BP, Ki-5305; 11,325 ± 120 yrs BP, Ki-7680). The large palaeochannels of the Seim and Svapa rivers were abandoned about 14,000 years ago (13,800 ± 85, Ki-6984; 14,030 ± 70, Ki-6997; 13,510 ± 85, Ki-6991). All these and subsequently quoted dates are given as uncalibrated ^{14}C yrs BP.

There are more than 70 sites with well-preserved ancient palaeochannels in the Russian Plain. Although most of them were not cored to establish the cross-section geometry, their widths and meander wavelengths are available from large-scale maps, air photos and space images (Sidorchuk et al., 2001b). Mean annual and MMD for these palaeorivers were calculated with the Formulae (2) to (5). Parameters $b = 1$, $N = 0.17$ and $K_0 = 0.004$ were estimated for the tundra region-analogue. The spatial distribution of the surface flow can be illustrated by the flood runoff values from Formula (5). Unlike the modern latitudinal distribution of the runoff depth on the Russian Plain, its distribution in the Lateglacial generally followed the shape of the ice sheet margins (Figure 6.2). The latter had the northeastern direction in the northwest of the Russian Plain and the meridional direction in the east of the Plain, near the slopes of the Ural Mountains. The maximum runoff depths existed in the areas adjacent to the ice sheets although none of the rivers used in our calculations were fed by glacier melt water. An excess of the flood water flow above the modern one can be explained by both greater snow depth and a greater runoff coefficient value. Flood runoff depth reached 450 to 600 mm in the basins of the Mezen' and upper Pechora rivers. It was about 300 to 400 mm in the basins of the Oka and upper Volga rivers. Minimum flow

Table 6.1 Extreme flood discharges at several Lateglacial rivers of the Russian Plain

River	Protva River	Seim River	Khoper River
Morpho-hydrological characteristics	mod/palaeo	mod/palaeo	mod/palaeo
Basin area (km^2)	2,170/2,170	10,700/10,700	19,100/19,100
Channel width (m)	80/180	60/1,000	60/1,400
Meander wavelength (m)	760/1,600	780/5,600	720/5,000
Longitudinal slope	4.1 × 10^{-4} /3.24 10^{-4}	1.3 × 10^{-4} /7.5 × 10^{-5}	6.3 × 10^{-4} /1.54 × 10^{-4}
Roughness coefficient n (s m$^{-1/3}$)	0.028/0.028	0.024/0.024	0.029/0.029
Mean discharge (m^3 s^{-1})	11.6/240	37.1/490	67.8/450
Mean maximum discharge (m^3 s^{-1})	250/3,100	575/6,400	991/5,800
Extreme discharge (m^3 s^{-1})	507/7,000	1,920/14,500	2,910/13,200

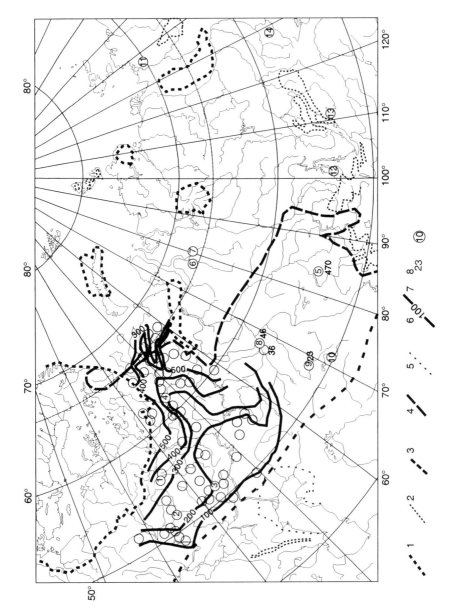

Figure 6.2 Lateglacial flood runoff depth in northern Eurasia. Key: 1 – the ice sheet margins; 2 – mountain glacier margins; 3 – the southern boundary of permafrost at the maximum stage of the last glaciation; 4 – Palaeo-Caspian sea coast line (1 to 4 after Velichko, 2002, © GEOS); 5 – the modern boundary of the permafrost; 6 – reconstructed flood runoff-depth contours (mm); 7 – reconstructed flood runoff depths at West Siberia (mm) and 8 – sites mentioned in the text, © Atlas-Monograph

for this area was calculated for the basins of the upper Severnaya Dvina (250 mm) and Vychegda (350–400 mm) rivers.

Flood runoff depth decreased with the distance from the edge of the glaciers, presumably with the reduction of precipitation and runoff coefficients. It was about 200 to 250 mm in the Seim, upper and middle Don, and Khoper River basins. Even smaller water flow values were reconstructed for the rivers of the lower Dnieper and Don, middle Volga and lower Kama basins, where flood runoff depths were about 100 to 200 mm.

The spatial variability of the runoff depth in the Lateglacial was significantly lower than the modern situation. In the north of the Russian Plain, the flood runoff at that period was two to three times greater than the modern values. In the south of the Plain, a relatively low flood runoff was then three to five times greater than the modern amounts. The reduced spatial variability of the runoff was caused by more uniform (Velichko, 1973) landscape conditions of flow generation in the periglacial zone compared to the present-day pattern of latitudinal landscape zones. Nevertheless, the geographical pattern of runoff was rather distinctive, corresponding largely to the shape of the main glacial and orographic boundaries.

On the basis of the reconstructed annual runoff depths, it is possible to estimate the annual flow volumes for the river basins in the central and eastern regions of the Russian Plain (Table 6.2). At the time of existence of the macromeanders, the annual flow from the northern megaslope of the Plain was about 380 km^3 or 1.5 times greater than the modern amount from the same area. An almost two-fold increase of water flow took place in the Pechora and Mezen' river basins. The annual surface runoff volume from the Volga River basin reached 585 km^3; this amount, being more than twice the modern value, can provide an explanation for the high level of the Caspian Sea during the Late Khvalyn transgression. A major part of the runoff volume from the Volga River basin was supplied by the Oka and Kama rivers. Their input was 3 to 3.5 times greater than that at present. The biggest ratio between the Lateglacial and modern annual flow volumes (almost 4 to 1) is reconstructed for the Don River basin. It is worth noting that none of these rivers were fed with glacial melt water.

Table 6.2 Annual water flow volume of the Lateglacial rivers in the Russian Plain and in West Siberia

River basin	Basin area (10^3 km^2)	Annual water flow volume (km^3)	
		Lateglacial	Modern
Severnaya Dvina	380	115	107
Mezen'	78	45	20
Pechora	322	220	126
Upper Volga	220	93	85
Oka	245	147	41
Kama	507	260	88
Middle and lower Volga	249	85	40
Don	422	110	28
Pur	95	50	28
Taz	100	38	34

The Holocene

At the end of the Last Glacial epoch, degradation of permafrost and increase of soil permeability in spring, combined with increasing groundwater flow in summer, brought about dramatic changes in the hydrologic regime and morphology of river channels. These factors caused a decrease of runoff coefficients, flood flow and MMD during the snow thaw period. Changes in the groundwater regime during the summer caused an increase in the basic flow and vegetation spread over the bare floodplains and sandy bars. Large periglacial channels were abandoned and transformed into floodplain lakes and fens. New channels were formed under conditions of lower annual flow and much steadier flow regime. Flood waves became lower and less steep. Flooding of the floodplains covered with dense vegetation caused a significant reduction of the flow velocities and rates of channel erosion during the high water stages. In contrast, the water level and flow velocities in the channel in summer became greater, so that the channel erosion by the low water became more important for channel morphology. Therefore, the newly formed channels had much smaller widths and meander lengths than the periglacial ones. These transformations began about 12 kyrs BP in the southern part of the Russian Plain and continued until approximately 8.5 kyrs BP in the northern part.

The degree of channel transformation varied significantly over the Russian Plain due to different magnitudes of changes in water discharge at the Lateglacial/Holocene transition (Figure 6.3). Rivers in the tundra zone still exist in conditions similar to periglacial ones, with runoff coefficients of 0.9 to 1.0. Consequently, modern river flow in the tundra is close to periglacial values. In the taiga zone, the recent annual flow is about 80 to 85% of the periglacial amount in the east and 30 to 60% in the west of the region. In the broadleaved forest, it is about 40 to 50% of the periglacial amount in the east and 20 to 25% in the west. In the steppe and forest steppe, the modern annual flow is about 40 to 60% of the periglacial value in the east and only 10% in the west of the region. Presumably, spatial variability in the decrease of the runoff coefficients at the Lateglacial/Holocene transition was the main cause of this

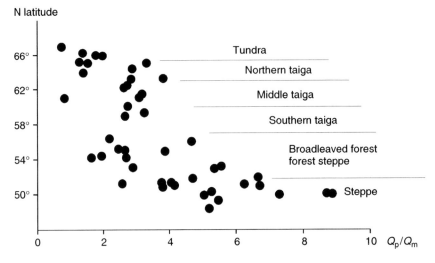

Figure 6.3   Changes in water discharges in the rivers of the Russian Plain and West Siberia at the Lateglacial/Holocene transition. Lateglacial/Recent annual discharges ratio Q_p/Q_m versus northern latitude

different degree of change in annual flows. The reduction was greatest in the steppe, where percolation in spring and evapotranspiration in summer increased substantially with permafrost degradation and climate warming. In the taiga zone, it was less evident due to deeper seasonal freezing of soils and low soil permeability during the flood period. In the northern taiga and tundra, the soil permeability during the flood period in the Holocene remained at generally the same level as in the Lateglacial.

The overall decrease of water flow during the Holocene was not uniform. For example, the palaeochannels on the low terraces and floodplain of the lower Vychegda River (Site 4) show considerable morphological change during the Holocene (Sidorchuk et al., 2001a). In the early Boreal, the palaeochannels were larger than the modern ones. They became significantly smaller than the modern ones during the late Boreal–Atlantic and then increased again, beginning from the Subboreal. The MMD of palaeo-Vychegda changed in a similar way; it was greater than the modern value at the beginning of the Boreal, became half the present-day value during the late Boreal–early Atlantic, and since the end of the Atlantic has increased up to its modern value.

Similar changes in the river morphology were reconstructed for the central part of the Russian Plain. The channel of the Protva River (Site 1) was deeper than the modern one during the period from 10 to 6.5 kyrs BP. Later (6.5–1.0 kyrs BP) the channel depth decreased significantly, presumably owing to lower floods. During the last 1,000 years spring floods increased, which caused an increase of the river depth and incision of the channel so that the base of the coarse channel alluvium became 2 to 2.5 m lower than before.

A similar sequence of periods with low and high floods was characteristic for the rivers in the Oka, middle Dnieper and middle Volga basins. A well-developed horizon of the zonal or floodplain-type soils was formed on the floodplains in this region 2.4 to 1.1 kyrs BP. This soil horizon was later buried under floodplain deposits with the earliest radiocarbon dates about 1.1 kyrs BP (Butakov et al., 2000).

3.2 The West Siberian Plain

The Lateglacial

The palaeochannel of the Berd' River near Novosibirsk (Site 5), described by Volkov (Zykina et al., 1981), gives clear evidence of the high surface runoff in the Ob' River basin. This meandering palaeochannel, 300 to 500 m wide, is situated on the first terrace of the Berd' River (Figure 6.4), with meander wavelength of 4 to 6 km. The modern river channel is 40 to 60 wide, indicating that the palaeochannel was formed by much higher runoff. Our calculations with Formulae (2) to (5), using the tundra region-analogue show that MMD was then about 400% greater than at present, and the annual discharge was 450% greater. Three radiocarbon dates (12,450 ± 55 yrs BP, SOAN-411; 11,100 ± 30 yrs BP, SOAN-112; 12,820 ± 500 yrs BP, SOAN-11) obtained on the peat layers in the lower part of the palaeochannel infill indicate the time of the palaeochannel abandonment. The palaeochannel was still flooded during the Holocene, as the horizontally bedded floodplain alluvium filling in the channel is dated to 6.8 to 4.4 kyrs BP on the basis of the peat interlayers (6,780 ± 145 yrs BP, SOAN-9; 5,930 ± 100 yrs BP, SOAN-113; 4,400 ± 340 yrs BP, SOAN-114).

The large palaeochannels situated on the first terraces and floodplains are known all over West Siberia from the semi-desert zone in the south to the tundra in the north. Although largely not dated, these palaeochannels, presumably (Volkov, 1994) were active during the Lateglacial. In the tundra, the palaeochannel on the first terrace of

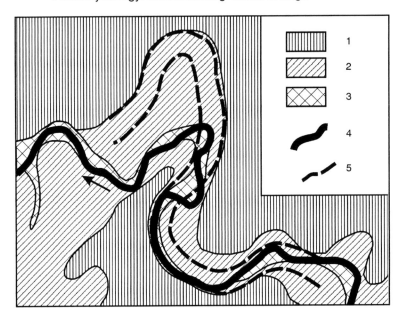

Figure 6.4 The Lateglacial palaeochannel of the Berd' River (Ob' River basin, West Siberia), after Zykina *et al.*, 1981, simplified. Key: 1 – interfluvial areas and high terraces; 2 – the first terrace; 3 – flood plain; 4 – the modern channel and 5 – palaeochannel on the first terrace, © Nauka

the lower Pur River (Site 6) was formed by the annual flow of $50 \, \text{km}^3$ (Korotaev *et al.*, 1999), and the one on the first terrace of the lower Taz River (Site 7) – by the annual flow of $38 \, \text{km}^3$ (Table 6.2). In the dry steppe, the palaeochannel on the first terrace of the Tobol River (Site 8) was 800 to 1000 m wide (5 times the width of the recent channel) with a meander wavelength about 10 to 12 km. In the semi-desert, the palaeochannel on the first terrace of the Ishim River (Site 9) was 200 to 300 m wide (also 5 times the width of the recent channel) with a meander wavelength of about 4 km. Calculations with Formulae (2) to (5) show that MMD of the lateglacial Ishim River was 450% and the annual discharge was 350% of the recent values. The same indices for the Tobol River were 350% and 200%, respectively. The annual flows of the Taz and Pur rivers were 110% and 180% of the recent ones. These estimates suggest that the spatial distribution of the surface flow in the Lateglacial was more uniform than at present. During the Lateglacial, rivers in the southeastern part of West Siberia (including the Ob' River headwaters) were possibly fed partly by melt water from mountain glaciers. It is also possible that their morphology was influenced to some degree by catastrophic floods that might have resulted from blowouts of the glacial lakes (Baker *et al.*, 1993).

The Holocene

The calculations discussed above show that during the Holocene the river flow reduced by 3 to 4 times in the southern part, and by 1.1 to 2 times in the northern part of West Siberia. The causes of these changes were generally the same as in the Russian Plain. The extent of channel transformation decreased from the south to the north according to the degree of permafrost degradation. The most significant changes in the river flow regime took place in the south of West Siberia and in northern Kazakhstan, where the flow decrease caused disjunction of the river net. For example, the Ishim River basin area reduced by 30%, and the upper Nura River (Site 10) is lost at present in the salt lakes (Volkov, 1960).

The morphology of the palaeochannels on the upper and middle Ob' River floodplain (Chernov and Garrison, 1981) shows significant long-term changes in the hydrological regime during the Holocene. About 6.5 to 3 kyrs BP, the upper Ob' River was characterised by a meandering pattern with meander wavelengths 1.2 times smaller than those of the recent channel. At the same time, the zonal chernozem and grey forest soils were formed on the floodplain (terrace at that period) of the middle Ob' River. The aeolian processes and dunes formation were common on its surface. During the last 3 kyrs, the meandering channel of the upper Ob' River was transformed into a braided one with several meandering sections, which had larger meander wavelengths than previously. In the middle Ob' valley, the zonal soil horizon was overlain by flood deposits. Therefore, the river flow in the southern part of West Siberia was relatively low at the end of the Atlantic and in the Subboreal, but increased substantially in the Subatlantic.

3.3   East Siberia

Rivers of East Siberia in the basins of the Yenisei, Lena, Yana and Indigirka rivers have mainly deeply incised valleys with relatively rare alluvial sections. The macromeanders found in some of the incised valleys do not have the same palaeohydrological significance as those in free alluvial channels. For example, the main feature of the channel pattern of the lower Yana River (Site 11) is a series of 11 macromeanders with wavelengths from 10.8 to 30.4 km, which is 25 to 60 times the average channel width (Matveev et al., 1994). This size corresponds to the characteristic diameter of the dome-like geological structures at the Kular Ridge, crossed by the Yana River antecedent valley. Thus, the development of such channel macrobends was controlled initially by the tectonic structure of the bedrock. In the course of subsequent downstream migration of the macrobends, they acquired a characteristic asymmetry similar to that of free alluvial meanders. The slopes of the convex banks of the macrobends are gentle and contain a sequence of the Late Neogene/Quaternary alluvial terraces and relatively wide segments of the youngest terrace and floodplain. Such a slope profile indicates a general increase in the amplitude of channel macrobends during the last one million years (Sidorchuk and Panin, 1996). The opposite slopes of the concave banks of the channel bends are steep. The main cause of such meander-like development of structurally controlled bends is the influence of the valley curvature on the second-order channel relief forms. The ordinary meanders and braids are shifting down the macrobends at a rate of 0.5 to 1.5 m per year, that is, 10 to 100 times faster than macrobends. During their movement, the zone of active erosion is shifted alternately from the concave to the convex sides of macrobends. Nevertheless, because of the influence of macrobend curvature, the cumulative erosion at their concave sides is greater than at the convex ones. The origin and significance of this influence on the river valley asymmetry can be compared to that of the Coriolis force. The palaeohydrological signal in such geologically and hydraulically controlled river valleys is relatively weak.

The use of the hydrological regime equations in East Siberia is limited to the rather rare channel sections with free alluvial meanders. The existence of periods with greater or smaller flood levels in the past for incised river channels is indicated in some locations by characteristic changes in the alluvium stratigraphy. Such evidence is available from the tributaries of the Angara River and the Yenisey River upstream from the Angara mouth, for the rivers of the Baikal Lake region, and for the Aldan River basin.

The Lateglacial

According to Vorob'eva *et al.* (1992), the floodplain facies of alluvium on the low terrace (6.3–7.1 m above the low water level) of the Belaya River (Site 12) contains three interlayers of coarse sand, separating culture horizons of the Mesolithic age. The upper culture horizon dated to 11.9 kyrs BP is overlain by sand with numerous pebbles and cobbles. These horizons of coarse floodplain alluvium indicate high levels of spring floods during the Lateglacial, presumably related to the mountain glaciers melting in the Sayany Mountains. At approximately 10 kyrs BP, the level of spring floods decreased by at least 3 m.

The high spring flood flow during the Lateglacial was reconstructed for the rivers in the northeastern part of the Baikal Lake drainage basin (Site 13). Palaeomeanders of this age were on average 1.9 times larger than the modern ones (Antoshchenko-Olenev, 1982). In the Aldan River basin (Site 14), 14 to 12 kyrs BP the flood levels were also high: the alluvium of this age was deposited at the elevation of 14 m above the modern low water level (Mochanov, 1977).

The Holocene

In the Holocene, some changes in the river channel morphology can be traced in the wider sections of the floodplains and river deltas. For example, free palaeomeanders of the lower Yana River developed on the floodplain within the confined macrobends are smaller than the recent ones (Figure 6.5). As indicated by the radiocarbon dating of

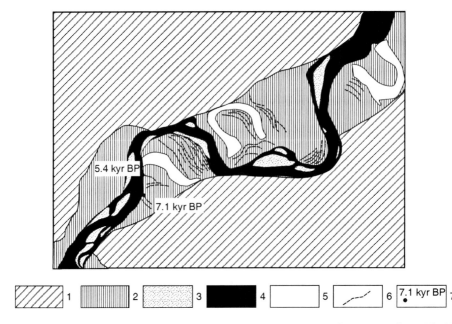

Figure 6.5 The Holocene palaeochannels on the Yana River floodplain (Site 11). Key: 1 – interfluvial areas and high terraces; 2 – flood plain; 3 – the islands and the bars in the channel; 4 – the modern channel; 5 – palaeochannel on the floodplain; 6 – natural levees and 7 – sampling sites and ^{14}C dates

the natural levees on the floodplain, these meanders were formed at a stage of slightly lower water flow about 5 kyrs BP. Estimations based on the size of palaeomeanders show that the annual discharges in two main delta branches of the Yana River were 1,500 and 550 m³ s⁻¹ at the period 0.9 to 1.1 kyrs BP (Sidorchuk, 1975). That was twice the modern annual flow at the head of the delta.

The highest floods in the Belaya and Kan rivers (Yenisei River basin) occurred in the Atlantic, about 6.7 kyrs BP (Vorob'eva *et al.*, 1992). A considerable increase of the flood levels is indicated for the Belaya River in the last 1,000 years, as a horizon of floodplain deposits containing artefacts of the Iron Age (2.5–1 kyrs BP) is buried there by a layer of alluvial silt 20 to 30 cm thick (Vorob'eva *et al.*, 1992).

In the northeastern part of the Baikal Lake drainage basin, the meanders formed during the Atlantic were 1.2 to 1.3 times larger than the modern ones (Antoshchenko-Olenev, 1982). The Subboreal and earlier part of Subatlantic were characterised in this region by reduced runoff and low levels of the spring floods.

In the Aldan River basin (site 14), the intervals from 10.5 to 9.5 kyrs BP and 3.9 to 1.1 kyrs BP were characterised by low flood levels: under 10 m and 11 m, respectively, as indicated by buried soil horizons in alluvial sequences (Mochanov, 1977). During the last 1,000 years, the flood levels increased substantially as indicated by a layer of sandy clay, which covers both the floodplain deposits of various age and soil horizons up to the elevation of 12 to 13 m above the modern low water level. The age of sediments at the base of this layer is 0.8 to 0.6 kyrs BP. At present, the maximum flood levels in some parts of the Aldan valley reach 14 to 15 m.

4 DISCUSSION AND CONCLUSION

During the Lateglacial, under conditions of cold periglacial climate, widespread permafrost and low soil permeability, rivers of the western part of northern Eurasia (the Russian Plain and West Siberia) formed large channels having widths up to 15 times greater than the present ones. As periglacial conditions do not persist in the major part of this region at present, reconstructions of the palaeodischarges of these rivers are based on the concept of the palaeohydrological analogue. The tundra zone of the northern Russian Plain and West Siberia (the region with the closest hydrologic regime to that of the past) was used to estimate the parameters in the regime-type formulae, which relate channel morphology with the water flow characteristics.

Large sizes of the periglacial river channels indicate that the maximum and annual discharges of those rivers were high during the Lateglacial. The annual flow volume from the Severnaya Dvina, Mezen', Pechora, Volga and Don river basins was about 1,100 km³, thus reaching about 200% of recent values. Changes of annual flow in the Ob' River basin were probably of similar magnitude. The MMD of the rivers in periglacial conditions was up to 14 times greater than the recent ones. The main part of the annual flow was presumably discharged during the spring flood. The spatial distribution of the surface runoff depth was more uniform than at present, due to the relative homogeneity (Velichko, 1973) of the periglacial hyperzone. At the same time, the shape of the runoff isolines generally reflected the boundaries of the continental ice sheets, mountain glaciers and major orographic features, as well as an increase in the continentality of climate towards the east.

The available radiocarbon dates suggest that transformation of the large periglacial channels into significantly smaller Holocene channels began approximately 12 to 13 kyrs BP in the southern part of the periglacial hyperzone of the Russian Plain and West Siberia. This process developed mainly because of the degradation of permafrost, decrease in rainfall and surface runoff depth. Later, the broad belt of channel transformations gradually shifted towards the north. In the Boreal, some rivers in the taiga zone still remained larger than the recent ones. Rivers in the tundra did not change substantially compared to the Lateglacial.

The overall decrease in surface runoff did not occur gradually over time. The surface runoff reached its maximum during the Lateglacial and remained greater than present at the beginning of the Boreal. It generally decreased during the first half of the Holocene and increased again later, although the chronology of these changes varied considerably in different regions. Thus, the lowest runoff in the northern Russian Plain occurred in the early Atlantic, while in the central part of the Plain and in West Siberia the minimum runoff corresponded to the late Atlantic. Runoff on the Russian Plain and in West Siberia increased in the Subboreal compared to the Atlantic low flow interval, but remained lower than present. Runoff changes in the Subatlantic were similar all over northern Eurasia, the beginning of this period being relatively dry, with subsequent increase of the flood levels during the last 1,000 years.

East Siberia, with its extremely continental climate, is situated at present within the permafrost zone. In this region, there are no indications of large-scale changes in river channel morphology during the Lateglacial and the Holocene that are comparable to those registered for the Russian Plain and West Siberia. This difference can be partly explained by a greater stability of the confined channels, which is characteristic of the East Siberia area. In the southern regions of East Siberia, the Lateglacial was characterised by relatively high flood levels, although it is possible to attribute such an increase in water flow at least partly to the melting of glaciers in the mountains of southern Siberia. The water flow changes through the Holocene were different in the several regions of this large territory. In the southern part of East Siberia, the late Atlantic (6–5 kyrs BP) was characterised by the highest flood levels for the entire Holocene. This period was followed by a substantial decrease of runoff in the Subboreal. In the northern part of East Siberia, the surface runoff during the late Atlantic (5 kyrs BP) was lower than at present. Here, the rise of the water flow was indicated in the Subatlantic time.

The temporal and spatial patterns of changes in runoff over the Russian Plain, West and East Siberia during and since the Lateglacial in relation to landscape evolution demonstrate how an understanding of palaeohydrological change can provide useful background information for studies of global change.

ACKNOWLEDGEMENTS

The research was funded by Russian Foundation for Basic Research grants 95-05-14,435 and 97-05-64,708.

REFERENCES

Antoshchenko-Olenev, I.V., 1982. *The History of the Natural Environment and Tectonical Movement at the Late Cenozoic in the Western Transbaikal Region*. Nauka, Novosibirsk, 155 (in Russian).

Baker, V.R., 1973. *Palaeohydrology and sedimentology of Lake Missoula flooding in Eastern Washington*. Geological Society of America, Special Paper 144.

Baker, V.R., Benito, G. and Rudoy, A., 1993. Palaeohydrology of late Pleistocene superflooding, Altai Mountains, Siberia. *Science*, **259**, 348–350.

Butakov, G.P., Kurbanova, S.G., Panin, A.V., Perevoshchikov, A.A. and Serebrennikova, I.A., 2000. Human induced sedimentation of the floodplains of the Russian Plain rivers. In R.S. Chalov (ed.), *Erosion and Channel Processes*. Vol. 3. Izdatelstvo Moskovskogo Universiteta, Moscow, 78–92 (in Russian).

Chernov, A.V. and Garrison, L.M., 1981. Palaeogeographical analysis of the channel deformation evolution of the rivers with the wide floodplains in the Holocene (upper and middle Ob' River case study). *Bull. MOIP*, geologiya, **56**(4), 97–108 (in Russian).

Dury, G.H., 1964. *Principles of Underfit Streams*. US Geological Survey Professional Paper 452-A, Washington, DC, 67.

Dury, G.H., 1965. *Theoretical Implications of Underfit Streams*. US Geological Survey Professional Paper 452-C, Washington, DC, 43.

Grichuk, V.P., 1969. An attempt of reconstruction of certain climatic indexes of the northern hemisphere during the Atlantic stage of the Holocene. In M.I. Neustadt (ed.), *Golotsen*. Nauka, Moskva, 41–57 (in Russian).

Evstigneev, V.M., 1990. *River Flow and Hydrological Calculations*. Izdatelstvo Moskovskogo Universiteta, Moscow (in Russian).

Kalinin, G.P., 1966. Space – temporal analysis and ergodicity of hydrological elements. *Vestnic Moskovskogo Universiteta, ser. Geografiya*, **5**, 6–12 (in Russian).

Korotaev, V.N., Sidorchuk, A. Yu. and Tarasov, P.E., 1999. Palaeogeomorphological analysis of the river deltas of the Taz estuary. *Geomorfologiya*, **2**, 78–84 (in Russian).

Matveev, B., Panin, A. and Sidorchuk, A., 1994. Rates of formation of forms in a river channel hierarchy: the case of the River Yana in Northeast Russia. In L.J. Olive, R.J. Loughram and J.A. Kesby (eds) *Variability in Stream Erosion and Sediment Transport*. IAHS Publ. No. 224, 181–186.

Mochanov, Yu.A., 1977. *The Oldest Stages of the Human Population in the North-Eastern Asia*. Nauka, Novosibirsk (in Russian).

Panin, A.V., Sidorchuk, A.Yu. and Chernov, A.V., 1999. Historical background to floodplain morphology: examples from the East European Plain. In S.B. Marriott and J. Alexander (eds), *Floodplains: Interdisciplinary Approaches*. Special Publication No. 163, Geological Society, London, 217–229.

Rotnicki, K., 1991. Retrodiction of palaeodischarges of meandering and sinuous rivers and its palaeoclimatic implications. In L. Starkel, K.J. Gregory and J.B. Thornes (eds), *Temperate Palaeohydrology*. John Wiley & Sons, Chichester, 431–470.

Sidorchuk, A.Yu., 1975. The main stages of the Yana River delta evolution. In A. Ivlev (ed.), *Geomorfologiya i Paleogeografiya Dal'nego Vostoka*. Khabarovsk, Geographical Society 166–180 (in Russian).

Sidorchuk, A. and Panin, A., 1996. Water supply from the Yana River basin since late Pliocene. *Terra Nostra*, **9**, 97.

Sidorchuk, A.Yu. and Borisova, O.K., 2000. Method of paleogeographical analogues in paleohydrological reconstructions. *Quaternary International*, **72**(1), 95–106.

Sidorchuk, A., Panin, A., Borisova, O. and Kovalyukh, N., 2001a. Lateglacial and Holocene palaeohydrology of the lower Vychegda river, Western Russia. In D. Maddy, M.G. Macklin and J.C. Woodward (eds), *River Basin Sediment Systems: Archives of Environmental Change*. Swets & Zeilinger B.V., Amsterdam, 265–296.

Sidorchuk, A., Borisova, O. and Panin, A., 2001b. Fluvial response to the late Valdai/Holocene environmental change on the East European plain. *Global and Planetary Change*, **28**, 303–318.

Szafer, W., 1946. Flora pliocenska w Kroscienku nad Dunajcam. *Rozprawy Wydzialu Matematyceno-przyrodniczego*, Polska academia nauk, **72**(B. 1–2), 98.

Velichko, A.A., 1973. *Natural Process in the Pleistocene*. Nauka, Moscow (in Russian).

Velichko, A.A. (ed.), 2002. *Evolution of the Landscapes and Climates of Northern Eurasia. The Late Pleistocene – Holocene – Elements of Forecasts)*. Atlas-Monograph, Vol. 2, GEOS, Moscow (in Russian).

Volkov, I.A., 1960. About the recent past of the rivers Ishim and Nura. *Trudy laboratorii aerometodov AN SSSR*, **9**, 15–19 (in Russian).

Volkov, I.A., 1994. Climate change and landscape evolution during the Sartansk cold period and the Holocene according to geological and geomorphological information (upper near-Ob' region case study). *Geologiya i Geofizika*, **35**(10), 14–24 (in Russian).

Vorob'eva, G.A., Goryunova, O.I. and Savel'ev, N.A., 1992. Chronology and palaeogeography of the Holocene of the southern Middle Siberia. In *Geochronology of the Quaternary*. Nauka, Moscow, 174–181 (in Russian).

Zykina, V.S., Volkov, I.A. and Dergacheva, M.I., 1981. *Upper-Quaternary Deposits and Fossil Soils of Novosibirsk near-Ob' region*. Nauka, Moscow, 203 (in Russian).

7 Palaeohydrology of Central Europe

L. STARKEL
Institute of Geography and Spatial Organisation,
Cracow, Poland

1 SPECIFIC FEATURES OF CENTRAL EUROPE

Central Europe will be considered as the area between the Rhine in the west, the Niemen and the Dnieper in the east, and extending from the Baltic and the North Sea southwards to the Alps, the Carpathians and the Pannonian basin. Information on hydrological and climate changes are registered in fluctuations of lake levels; advances and recessions of mountain glaciers and associated fluctuations of parallel vertical zones such as the upper tree line, the snow line, and the solifluction limit; the rate of peat growth; the transformation of plant communities; fluvial accumulation and erosion; precipitation of calcareous tufa and speleothem growth; the dynamics of mass movements, soil profiles, permafrost extent; with further evidence provided by molluscs cladocera and rodents as well as from continuous records of annually laminated lake sediments and tree rings.

The area of central Europe is characterised by a great variety of morphotectonic units and by very frequent palaeogeographic changes as a result of being open to various shifts of climatic zones both in the west-to-east direction as well as in the north-to-south direction. In this paper, special attention will be concentrated on the period of the last 20 ka BP, starting from the last advance of the Scandinavian ice sheet and covering the whole postglacial period.

Central Europe is located at the junction of the three main tectonic units of the European subcontinent: in the northeast – the stable and flat Fennoscandian shield and Russian platform; in the west – the uplands and depressions of the old Hercynian orogenic system, rejuvenated in the Tertiary; and in the south – young mountain chains of the Alpine orogeny separated by the submontane and intermontane tectonic depressions (Figure 7.1). Central Europe is therefore characterised by highland groups in the south, lowlands in the north, and is separated from Scandinavia by shallow inland seas, the North Sea and the Baltic. Most of the area falling towards the north is drained by parallel rivers: the Rhine, Weser, Elbe, Oder, Vistula and Niemen. In contrast, the mountains and depressions of the Alpine mountain system are drained by the extensive catchment of the Danube river flowing to the Black Sea, and the Dnieper and Dniester at the eastern margin (Figure 7.2).

The vertical zonality connected with orography creates distinct diversity in the rainfall pattern and the regime of rivers, with headwaters in the mountains being controlled by snowmelt and by heavy summer floods. Most of these floods also affect the intra- and submontane depressions. The vast lowlands located to the north of the Harz, Sudetes and Carpathians, invaded several times by the Scandinavian ice sheets,

Palaeohydrology: Understanding Global Change. Edited by K.J. Gregory and G. Benito
© 2003 John Wiley & Sons, Ltd ISBN: 0-470-84739-5

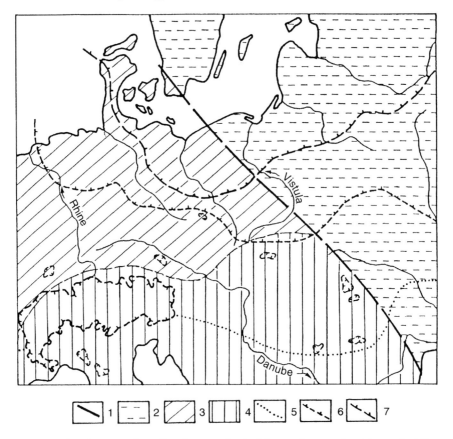

Figure 7.1 Morphotectonic units, rivers, extent of ice and permafrost in central Europe. 1 – limit between platformed and orogenic Europe; 2 – shield and platform; 3 – Hercynian orogenic system; 4 – Alpine orogenic system; 5 – maximum extent of permafrost 20 ka BP; 6 – limit of last ice sheet and Alpine glaciation; 7 – maximum extent of Scandinavian ice sheets

are covered by a thick blanket of glacial and glacifluvial deposits, frequently 100 to 200 m deep, overlying the Tertiary lacustrine and fluvial deposits. Several groundwater horizons occur in both units.

A west–east transect of central Europe shows a decrease in precipitation and an increase in continentality towards the east, the latter expressed in severe winters, freezing of the ground, long duration of snow cover and more frequent snowmelt floods with ice jams. This west–east trend is also reflected in the composition of plant communities, the eastern limits of beech, hornbeam and fir and the western limit of spruce across the territory of Poland. A very flexible circulation pattern in these latitudes is responsible for unstable hydrological conditions, great fluctuations from year to year being reflected in the frequency of extreme rainfalls (Starkel, 1996), floods and in water storage.

During the maximum extent of the Scandinavian ice sheet, the northern part of central Europe was under ice (Figure 7.1) and piedmont glaciers extended at the outlets of Alpine valleys. Most of central Europe at that time was in the belt of continuous permafrost; only the southern part of the Pannonian basin as well as the Black Sea coastal zone had discontinuous permafrost. The late Vistulian warming with several

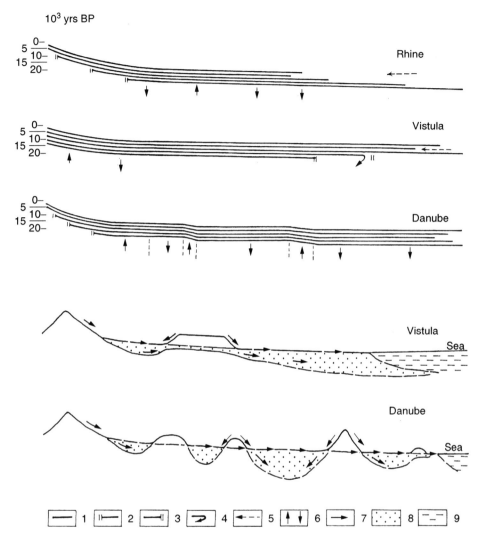

Figure 7.2 Schematic changes in river length (above) and longitudinal profiles across mountains, plateaus and sedimentary basins (below). 1 – existing river length at 5 ka intervals; 2 – river beginning as glacial melt water; 3 – river blocked by ice sheet; 4 – river changed direction; 5 – trend of transgression or regression; 6 – trend of uplift or subsidence; 7 – direction of outflow; 8 – sedimentary basin; 9 – sea basin

distinct rises in temperature was accompanied by precipitation increase and melting of permafrost, which allowed a very rapid spread of forest communities, mainly from the southeast and from refuges in the mountain foothills (Ralska-Jasiewiczowa, 1989; Starkel, 1977). In the north, the new topography exposed after the melting of the ice sheet created new conditions for water storage and runoff and for establishing a new valley and drainage network. The relatively stable fluvial regime that was obtained under dense vegetation cover showed only centennial and millennial fluctuations in the water balance. With increasing human activity – starting from the late Bronze Age or the Roman period and becoming especially intensive during the last millennium – evaporation declined, groundwater levels rose and in hilly areas, it was responsible for accelerating runoff and increasing sediment loads (Starkel, 1988). Recent centuries are characterised by direct human intervention in the water cycle including regulation of river channels and construction of reservoirs.

2   REGIONAL CHANGES

The characteristics of hydrological changes during the last 20 ka is presented for seven different types of regions:

Mountain areas with vertical zonality
Submontane and intramontane depressions
Former periglacial uplands
Karstic areas
Former periglacial lowlands
Deglaciation zone of last ice sheet
Coastal areas.

2.1   Mountain Areas with Vertical Zonality

These are characterised by existence of vertical belts and by surplus of precipitation. Runoff is especially high during heavy rains and snowmelt periods. For vertical belts in the mountains, of special significance are the upper forest limit, the snowline, the extent of permafrost and glaciers. The upper treeline limits the zone of restricted runoff and slope wash and delayed snowmelt. It is located in the northern Alps at an elevation of about 2,000 to 2,100 m a.s.l. and in the Tatra Mts. only 1,550 m a.s.l. Up to the snowline (in the Northern Alps 3,000 m a.s.l., in the Tatra Mts. practically non-existent) is the belt of cryogenic processes, snowmelt and intensive runoff. Above the snowline is an area of permanent water storage in snow, which, in favourable conditions is transformed into ice. Rivers fed by melting snow and ice have a typical pattern of annual discharge (Parde, 1955).

   During the last cold stage, in addition to the precipitation decrease of 20 to 50% (cf. Frenzel et al., 1992), the snowline was lowered to 1,800 m a.s.l. in the Alps (Patzelt, 1977) and 1,500 m a.s.l. in the Tatra Mts. The Alps were completely under snow – Alpine glaciers 1 to 2 km thick filled the valleys extending as piedmont glaciers in the foreland, all controlling the summer high discharges of the Danube and the Rhine. Storage in the High Tatra was much lower, with glaciers 5 to 12 km long and up to 230 m thick (Klimaszewski, 1987) and in the other parts of the Carpathians, Sudetes, Bayerisches Wald and Harz, even smaller in amount. Therefore, their impact on the fluvial regime was more restricted.

   The Lateglacial warming caused the gradual melting of the ice, rising of the snow-line, retreat of permafrost and expansion of forest communities, which practically reached the present-day positions in the early Holocene (Patzelt, 1977; Bortenschlager, 1982; Zoller, 1977; Burga, 1988; Röthlisberger, 1986). In the Tatra Mountains, the last remnants of corrie glaciers melted in the Younger Dryas (Klimaszewski, 1987) and new research suggests that the melting continued to the Venediger phase ca 8,500 to 8,000 BP (Baumgart-Kotarba and Kotarba, 2001).

   During the Holocene, several advances of Alpine glaciers were recognised, dated in Austria and Switzerland at about 8.6–8.0 ka BP, 6.5–6.0, 5.3–5.0, 4.7–4.4, 3.3–3.0, 2.8–2.3, 1.8–1.3, about 1.0 ka BP and in XVII–XIX ca AD (Patzelt 1977). More detailed examination of some glaciers, such as the Aletsch, has shown a distinct coincidence of wetter and cooler phases with declines in solar radiation: 1050–1130 AD, 1300–1350 AD, 1450–1500 AD, 1550–1650 AD and 1750–1850 AD (Flohn, 1993). Distinct phases of recession are interpreted as periods of warmer and drier climate (Hormes *et al.*, 2001). Most of these fluctuations are also expressed in the shift of the upper tree line, usually by not more than 50 to 100 m (e.g. Burga, 1988), in reactivation of slope debris by solifluction (Furrer *et al.*, 1975; Matthews *et al.*, 1993), and lake-level fluctuations in the Swiss Plateau and Jura Mountains (Magny, 1993). More attention was given to the great variety of mass movements in the Carpathians, which correlate with events described in the Alps (Figure 7.3). Frequent deep landslides in the Flysch Carpathians indicate the phases with wet years and continuous rains (Starkel, 1997; Alexandrowicz, 1996; Margielewski, 2000), and debris flows in the cryogenic zone were instigated by frequent high-intensity heavy

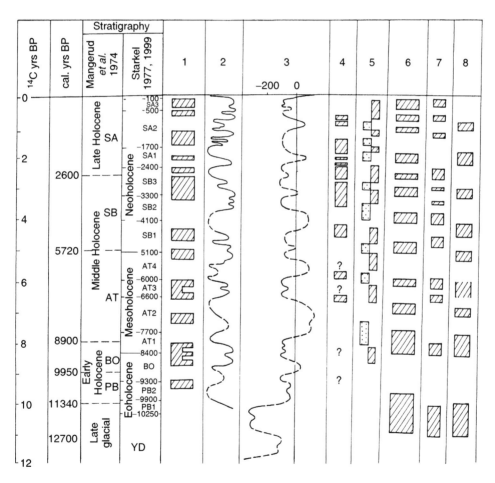

Figure 7.3 Cooler and wetter phases in the Alps and Carpathians. 1 – glacial advances in the Austrian Alps (Patzelt, 1977); 2 – glacial advances in the Swiss Alps (Röthlisberger, 1986); 3 – snow line fluctuations (Patzelt, 1977 and others); 4 – lowering of solifluction zone in the Alps (Matthews et al., 1993); 5 – high lake level in the Jura Mountains (Magny, 1993); 6 – landslide phases in the Carpathians (Starkel, 1997; Alexandrowicz, 1996; Margielewski, 2000); 7 – debris flows in the Tatra Mountains (Kotarba, 1998); 8 – flood phases in the Carpathian foreland (Starkel, 1983; 2001a; Kalicki, 1991 and others)

downpours (Kotarba and Baumgart-Kotarba, 1997). A good indicator of increase in humidity in the early Subboreal is the rate of peat deposition and culmination of Picea pollen at several localities in the Carpathians and Sudetes (Szczepanek, 1989; Madeyska, 1989; Starkel, 1995; Rybnickova and Rybnicek, 1996).

2.2 Submontane and Intramontane Depressions

Submontane and intramontane depressions formed mainly during the Neogene and Quaternary subsidence providing natural storage areas of water and sediments (Figure 7.2). Fluctuations in precipitation and runoff in the surrounding mountains are reflected in the type of sediments and morphological features in the depressions, with fluvial systems registering variations in the frequency and character of extreme hydrological events (Starkel, 1983; Figure 7.4).

In the subalpine depressions of the Danube basin, several steps in glacifluvial fans and cut and fill terraces (Troll, 1957; Fink, 1977; Brunnacker, 1978) have long been

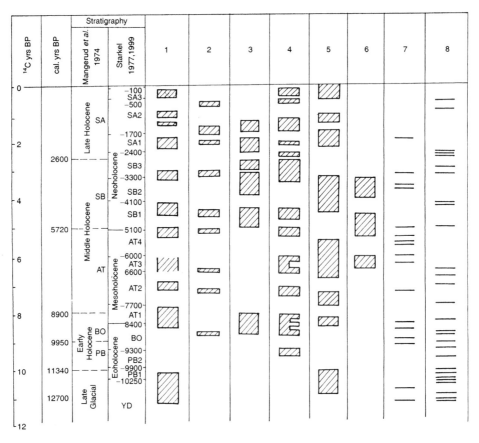

Figure 7.4 Phases of increased fluvial activity in central Europe during the Holocene. 1 – upper Vistula basin (Starkel, 1994; Kalicki, 1991; Starkel *et al.*, 1996a); 2 – right bank tributaries of the upper Danube (Brunnacker, 1978; Schreiber, 1985); 3 – subfossil oaks in southern Germany (Becker, 1982); 4 – advances of glaciers in the eastern Alps (Patzelt, 1977; 1993); 5 – upper Main (Schirmer, 1983); 6 – flood phase in the Rhine delta (van der Woude, 1981); 7 – age of abandoned channels in the Lodz region (Turkowska, 1988; Kamiński, 1993); 8 – age of abandoned channels in the Warta (Kozarski, 1991) and Prosna (Rotnicki, 1991) valleys

recognised. Connection with climatic fluctuations has been documented by horizons of subfossil oaks buried in alluvia and dated dendrochronologically (Becker, 1982) at 8.7–8.0, 5.0–4.3, 3.9–3.1, 3.0–2.7, 2.0–1.7 and 1.4–1.2 ka BP. The transition Atlantic–Subboreal is especially marked by erosion and channel avulsions (Schreiber, 1985; Wohlfarth and Ammann, 1991), although a different opinion has been expressed (Buch, 1988) explaining differences in alluvial fill age in various sections of the longitudinal profile of the Danube valley by self-regulation of the fluvial system. In the Morava basin, greater diversity during the Subboreal followed by vertical accretion starting from the late Bronze or early Mediaeval period, connected with increasing deforestation, has been documented (Havlicek, 1991). In the neighbouring Czech basin of the upper Elbe catchment after a dry early Holocene with chernozem formation, the alluvia of the humid Boreal–Atlantic transition was followed by several dry spells dated 4.5–3.7 ka BC, 2.5–2.0 ka BC and 1.25–0.7 ka BP (Ložek, 1991; Ruzickova and Zeman, 1994; Cilek, 1997).

Over the extensive Pannonian basin, loess and cryogenic deposits were recognised, indicating cold and dry climate during the upper Plenivistulian with the presence of permafrost (Frenzel *et al.*, 1992), also documented by missing groundwater which should be dated between 25 and 13 ka BP. (Stute and Deak, 1989). In that area, the early Holocene was still dry, a solonchak lake ending with the expansion of forest at the beginning of the Atlantic (Ando and Mucsi, 1967). At the Boreal–Atlantic transition, the shift to more humid climate is reflected in channel deposits up to 8 m thick with tree trunks in the Danube valley dated between 9 and 8.1 ka BP (Kvitkovič, 1993). From the Tista valley, several generations of paleochannels were described, abandoned at the Atlantic–Subboreal and Subboreal–Subatlantic transitions (Borsy and Felegyhazi, 1983). The slackwater deposits of late Bronze age flood core recorded at Iron Gate (Brunnacker, 1971). Those short humid phases alternated with much drier ones, documented by precipitation of protodolomite in the Balaton lake (Müller and Wagner, 1978). Reconstruction of precipitation changes based on dated palaeochannel parameters (Gabris, 1985) is not correct, because the abandoned forms were created by discharges during drier stable phases before avulsions.

The most complete sequence of changes for the last 13 ka BP was documented in the Subcarpathian basins, in the Vistula valley and its tributary the Wisłoka (Starkel, 1983; Kalicki, 1991; Starkel *et al.*, 1982; 1991; 1996a). Several fills with abandoned systems of meandering channels were recognised, larger in the Lateglacial and smaller in the Holocene (Figure 7.5). More than 150 radiocarbon dates and 500 dendrochronologically dated oak trunks (Krąpiec, 1992; Starkel *et al.*, 1996a; Figure 7.6) gave a solid chronological background. Each of the phases with higher flood frequency (dated: 8.5–7.7, 6.5–6.0, 5.5–4.9, 4.4–4.1, 3.5–3.0, 2.7–2.6 ka BP, 225 BC–300 AD, 450–575 AD, 900–1150 AD and after 1500 AD) started with overbank deposition, widening and straightening of channel, followed by a tendency to braiding and channel avulsion (Starkel, 1983; 1995; Figure 7.6). Of special importance is the phase about 8.5 to 7.8 ka BP, older than the Neolithic forest clearance, having thick alluvial fans reflecting about 100 events (Starkel *et al.*, 1996a) as well as aggradation from mediaeval times, followed by a tendency to braiding during the Little Ice Age (Klimek and Starkel, 1974; Szumański, 1977). This incision started with nineteenth century channel regulation, continued in the twentieth century up to 2 to 4 m (Klimek, 1987) at the mountain margin, causing the lowering of groundwater levels, and reduced the discharge of "underground rivers" draining the 20- to 30-m-thick alluvial fills. (Klimek and Starkel, 1974; Szumański, 1977).

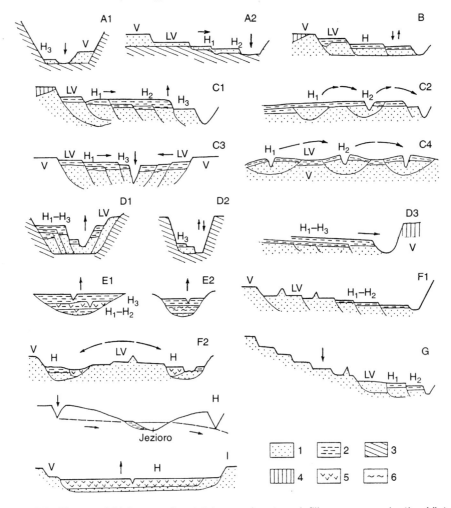

Figure 7.5 Types of Holocene floodplains and cut and fill sequences in the Vistula catchment (after Starkel, *et al.*, 1991). A – incised mountain valleys, narrow (A1) or wide (A2); B – mountain foothills; C – submontane basins, various tendencies of shifting and avulsion indicated; D – gaps across uplands; E – flat valleys of former periglacial zone; F – reaches of Vistula valley upstream of those invaded by ice sheet; G – reaches of Vistula incised after deglaciation; H – long profile of new river draining the dead-ice depressions; I – abandoned ice marginal streamway. Shading: 1 – alluvia of channel facies; 2 – overbank facies; 3 – substratum; 4 – loess; 5 – organic deposits; 6 – calcareous tufa. Abbreviations: V- Vistulian (Periglacial); LV – late Vistulian; H_1–H_3 – from eo to neo Holocene

2.3 Former Periglacial Uplands

The system of uplifted old mountain blocks and platforms in central- and southern Germany and southern Poland was totally in the zone of continuous permafrost during the upper pleniglacial (25–14 ka BP) and had been partly covered by older Scandinavian ice sheets. Hilly slopes are therefore covered by a blanket of periglacial solifluction debris, formed before the late Vistulian (Starkel, 1998; Völkel and Leopold, 2001). The lower parts of plateaus and higher terraces are buried by a thick mantle of loess, the product of a dry periglacial climate, which was deposited after 25 ka BP and before 14 BP (Maruszczak, 1987; Alexandrowicz, 1989). The presence of four

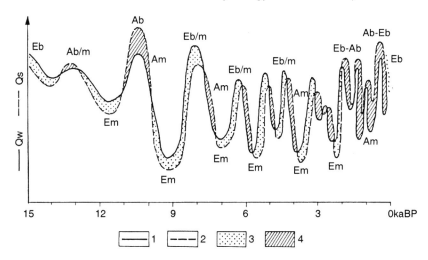

Figure 7.6 Fluctuations of channel-forming discharges (Q_w) and sediment load (Q_s) during the last 12 to 15 ka BP (modified after Starkel, 1983)

subfacies of loess on the Polish transect from dry in the east to humid in the west were distinguished on the basis of their texture and $CaCO_3$ content (Jersak, 1973), indicating the distinct gradient in rainfall totals and humidity even on the transect distance of 400 to 500 km. The transition from the Lateglacial to the Holocene is expressed in peat deposition due to the rise of groundwater table (Andres and Geyh 1970; Mausbacher et al., 2001).

In the German part, fluctuations in precipitation and runoff were documented in the valley fills of the Weser (Lüttig, 1960), upper Main (Schirmer, 1983) and Rhine (Brunnacker, 1978) all synchronised with the glacial advances in the Alps (Figure 7.4). Each younger fill has a weaker developed soil horizon. These phases were also proved by clusterings of subfossil oaks in the Rhine–Main catchment (Becker, 1982; Delorme and Leuschner, 1983). The floodplains of most of the valleys are covered by 1- to 5-m-thick cover of alluvial loams connected with anthropogenically accelerated runoff and soil erosion (Mensching, 1957; Richter, 1965; Starkel, 1988). The beginning of this soil erosion is well dated by buried archaeological sites and organic horizons. Erosion acceleration started in the Leine valley about 4 to 5 ka BP (Mayer et al., 1963), but is better exposed during late Bronze, Roman and Mediaeval periods (e.g. Jäger, 1962). Phases with higher precipitation and runoff were recorded in the Holzmaar lake (Zolitschka and Negendank, 1998). The content of Δ^2H (deuterium) in tree rings was used for reconstruction of fluctuations in rainfall and humidity (Frenzel, 2000). Some of the humid phases distinguished were 100 to 500 years long coinciding with flood phases (4.1–4.5 ka BP, 5.0–5.2 ka BP) whereas others did not. One phase dated 2,500 to 1,800 cal yrs BC is not reflected in the increase of soil erosion due to decreased cultivation by late Neolithic tribes.

A similar sequence of changes is observed in the uplands of Southern Poland after the retreat of permafrost. New organised subsurface drainage facilitated the gullying and piping in the loess deposits (Śnieszko, 1995), whereas dry valleys cut in the permeable sandy deposits stabilised. The rise of groundwater tables caused paludification of floodplains in the early Holocene (Nakonieczny, 1975; Onieszko, 1985). Therefore, the thanatocoenoses of floodplains reflect the treeless landscape (Alexandrowicz, 1995). An outstanding question is the origin of early Holocene chernozem, which may have been formed under the open woodland before the first humid phase at the end of the Boreal (Lożek, 1975; Śnieszko, 1995).

Accelerated soil erosion is well documented at several archaeological sites starting with early Neolithic slope wash at Pleszów dated about 6.0 ka BP (Wasylikowa *et al.*, 1985) and at several late Neolithic sites (Bronocice – by Kruk *et al.*, 1996), where this late Neolithic phase saw associated alluviation in larger valleys (Śnieszko, 1985; 1995), although this was much more extensive in the Roman period and later from the tenth to the fifteenth centuries, especially in totally deforested loess areas (Śnieszko, 1995).

2.4 Karstic Areas

Karstic areas give a unique opportunity to measure and date the intensity of karstic processes after the retreat of permafrost (Figure 7.7). The precipitation of stalagmites

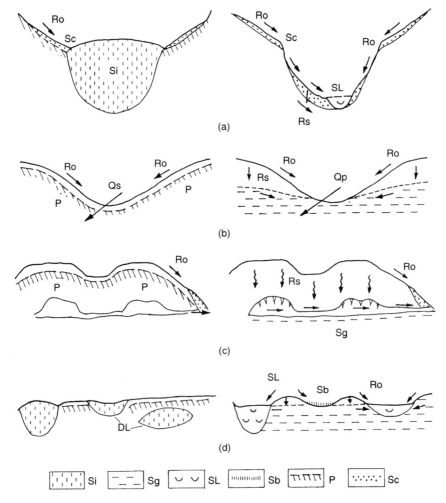

Figure 7.7 Schemes of water circulation in different landscapes during the last glacial maximum (left) and in the Holocene (right). (a) mountain valley; (b) hilly area; (c) karstic plateau; (d) deglaciated area; Si storage in ice; Sg groundwater storage; Sl lake storage; Sb bog storage; Sc storage in colluvia; P permafrost; Di dead ice; Ro overland runoff; Rs infiltration and subsurface runoff; Qs seasonal discharge; Qp perennial discharge

in central European caves started at lower elevations in the Lateglacial, but more generally in the Preboreal (Herzman, 1991). In the Hölloch cave in the Bavarian Alps, the rate of deposition above the date $8,530 \pm 160$ BP increased from $1.2 \, \text{mm yr}^{-1}$ to $2.8 \, \text{mm yr}^{-1}$ (Wurth *et al.*, 2000). This coincided with a high rate in Slovenian caves between 8,350 to 8,080 BP (Franke and Geyh, 1971) as well as with deposition of travertines in the Czech karstic region (Ložek, 1991; 1975) and southern Poland (Pazdur *et al.*, 1988). At Sieradowice between 8.7 and 8.1 ka BP, the rate of deposition increased from 0.1 mm to $0.25 \, \text{mm yr}^{-1}$ (Śnieszko 1995). At that time, foam sinters were formed at low elevations in the Slovak Carpathians, nowadays characteristic for higher belts with annual precipitation above 900 mm (Ložek, 1991). Detailed dating of calcareous tufa in the Cracow Upland helped distinguish phases with high deposition rates: 8.6–8.1, 6.8–6.0, 5.5–5.0, 4.5–3.0, 2.5–2.2 ka BP (Pazdur *et al.*, 1988). Their age differs from the speleothems, which show maxima about 8, 5.0–4.5 and 2.5–2.2 ka BP (Pazdur, 2000; cf. Figure 7.15), times when barriers built of tufa were dissected.

The age of higher rates of calcareous deposition in spring cupola peatbogs 8.0–7.5, 6.7–6.5, 6.0–5.6 and 2.5–1.7 ka BP shifts slightly (Dobrowolski, 1998) connected with intensified circulation of underground waters during warm and humid phases.

The transition to Subboreal in karstic areas is marked by breaks in deposition of calcareous tufa and the presence of rockfalls and debris talus at the sides of canyons in German and Czech karstic regions (Jäger and Ložek, 1983; Ložek, 1997; Brunnacker, 1975). The transition to Subatlantic from 1250 to 700 BC (Ložek, 1997) was especially dry and cool. A good indicator of the reactivation of karstic processes at the Boreal–Atlantic transition is the sinkholes formed in buried gypsum beds near Staszów in Southern Poland (Szczepanek, 1971).

2.5   Former Periglacial Lowlands

These lowlands comprise a zone 100 to 500 km wide between the maximum extent of the last ice sheet and a belt of uplands in the south, which is built of periglacial and glacifluvial sediments of older glaciations (mainly Warthe and Oder) transformed by the activity of rivers flowing from the south and later by periglacial processes (Klatkowa, 1994). During the interpleniglacial, this was a zone of discontinuous permafrost with tundra and forest–tundra vegetation but with the transgression of Scandinavian ice between 25 and 20 ka BP, this zone had continuous permafrost again and changed into arctic desert conditions (Dylik, 1967; GoYdzik, 1995; Vandenberghe and Pissart, 1993) so that subsurface drainage was blocked (Andres and Geyh, 1970). Rivers changed their character from anabranching to braided (Mol *et al.*, 2000), although in smaller valleys, this change followed after 20 ka (Turkowska, 1995). The strong aeolian activity is reflected in the presence of stony pavements with aeolian ventifacts (Manikowska, 1995). The first warming and appearance of patches of tundra vegetation with shrubs at the retreat of the Pomeranian phase, after 15 ka BP, facilitated the formation of low dunes, and the oldest palaeomeanders from that time in the Ner valley are dated $13,800 \pm 200$ BP (Turkowska, 1995). In the larger transit valleys, large palaeomeanders occurred when expansion of vegetation followed in the Bölling. The first abandoned meanders started to be filled in the early Alleröd (Kozarski *et al.*, 1988). Simultaneous with the melting of permafrost, groundwater reservoirs started to form and small lakes and swamps appeared in depressions (Nowaczyk and Okuniewska–Nowaczyk, 1999; Ralska-Jasiewiczowa *et al.*, 1998).

Aeolian activity was strong, the formation of dune systems in wide valley floors during the Older and Younger Dryas being connected with barriers created by shrubs and trees (Dylikowa, 1969; Manikowska, 1995). During the Alleröd and Younger Dryas, there were higher proportions of steppe species in the east and tundra species in the west (Ralska-Jasiewiczowa, 1989; Madeyska, 1995), indicating a drier and more continental climate to the east. The rebuilding of discontinuous or sporadic permafrost during the Younger Dryas was mainly restricted to Western Europe (Vandenberghe and Pissart, 1993).

Hydrological changes during the Holocene in the lowlands were small and the forest communities were relatively stable. Bogs reflected fluctuations in the rate of growth including distinct rises at the Boreal–Atlantic transition and about 5 to 4.5 ka BP (Żurek and Pazdur, 1999), as well as several rises in humidity separated by humic Grenzhorizonts dated between 4 and 1 ka BP (Behre et al., 1996).

Some transitional river valleys included several generations of paleochannels and alluvial fills, sometimes separated by organic horizons. Channels were abandoned in the Warta valley 8.3–8.1, 4.2–4.0 and after 2.5 ka BP (Kozarski, 1991), at similar periods in the subcarpathian basins, with similar phases documented for small river valleys near Łódź (Turkowska, 1988). For the Prosna valley, Rotnicki (1991) reconstructed the bankfull and mean annual discharges. He stated that the Qb during the Alleröd–Younger Dryas was 5 to 7 times higher and during more stable phases of the Holocene either up to 2 times higher or 2 times lower than at present (Rotnicki, 1991). This change reflected the shift from snowmelt floods to rainy floods.

Anthropogenic impact is registered in the superposition of younger alluvial loam with buried soils, dated in the Weisse Elbe and Mulde river valleys 5.7 to 5.1 and 3.3 to 2.9 ka BP (Hiller et al., 1991). To the east in Byelorussia, similar buried horizons interpreted as indicators of frequent floods were dated in Berezyna valley at $1,000 \pm 50$ BP and in the upper Dnieper at 940 ± 90 BP (Kalicki, 1991).

2.6   Deglaciation Zone of the Last Ice Sheet

The deglaciation zone of the last ice sheet experienced dramatic hydrological changes, followed, in the northern part of central Europe, by invasion by the last Scandinavian ice sheet between 22 and 14 ka BP (Figures 7.2, 7.7–7.10). The organic sediment near Łeba at the Baltic coast was dated $22,300 \pm 700$ BP and the maximum extent has been reached between 20, 5 and 20 ka BP (Rotnicki and Borówko, 1994; Kozarski, 1995). The advance of about 300 km in 2,000 years was a very rapid movement, thus blocking many river valleys. The course of deglaciation can be reconstructed from the sediments and morphological forms. During the maximum extent of the ice sheet (Leszno or Brandenburg Phase), the outflow of the Oder river was to the Elbe and North Sea. In the Warsaw basin, a small, dammed lake existed (Baraniecka and Konecka-Betley, 1987; Starkel, 1990), but there is no evidence supporting the hypothesis on the transfluence of Vistula waters towards the Warta and Oder at that time (cf. Wioniewski and Andrzejewski, 1994; Starkel, 2001a). The rate of ice-sheet retreat differed in the period between 20 and 16 ka BP when it was highest in the segment between the Oder and Vistula (above $50 \, \mathrm{m\,yr^{-1}}$) as a result of a different type of deglaciation. Instead of frontal recession, areal deglaciation characterised by many kame plateau and dead-ice depressions (Roszko, 1968) prevailed.

The climate of that phase of deglaciation was very cold and dry according to documentation from the extension of permafrost over the areas left free of ice (Kozarski, 1993; 1995; Liedtke, 1993; Böse, 1991; Figure 7.8). Permafrost aggradation caused

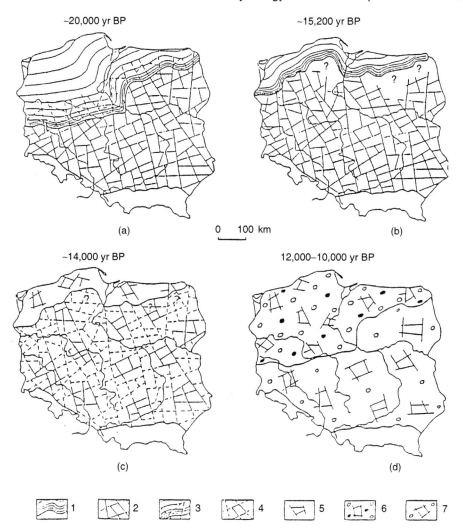

~20,000 yr BP

~15,200 yr BP

(a)

0   100 km

(b)

~14,000 yr BP

12,000–10,000 yr BP

(c)

(d)

1   2   3   4   5   6   7

Figure 7.8 Stages of change in permafrost extent and character in Poland outside mountains between 20 and 10 ka BP (after Goździk, 1995). 1 – ice sheet; 2 – continuous permafrost; 3 – ice sheet covering partly degraded permafrost; 4 – permafrost with thick active layer and possible taliks; 5 – discontinuous or sporadic permafrost; 6 – discontinuous permafrost with thermokarst depressions and buried dead ice; 7 – discontinuous permafrost with thermokarst depressions. Reproduced by permission of Lódzkie Towarzystwo Naukowe

the formation of cold-based ice beneath the ice margin, and many sites were found with syngenetic and epigenic ice-wedge casts and frost cracks with a polygonal pattern over outwash plains and sand wedge polygons over till, indicating a very cold and dry climate. These features extend up to the margin of the Pomeranian phase dated about 16.2 ka BP. The depth of the active layer at 0.5 to 1.2 m is connected with their decalcification and enrichment in $CaCO_3$ over the former permafrost table (Kowalkowski, 1990). Strong aeolian activity is documented by the presence of ventifacts, loess and cover sand patches (Kozarski and Nowaczyk, 1991). In the valley floors and over outwash plains, oriented icing depressions and widenings were observed connected with thermokarst (Kozarski, 1995).

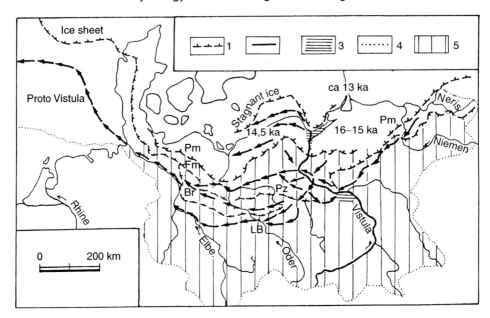

Figure 7.9 Changes of river courses and ice-sheet retreat in the Elbe-Oder-Vistula-Nieman river catchments (modified after Starkel *et al.*, 1991). 1 – position of ice-sheet front during various phases; 2 – direction of outflow (indicating marginal streamways); 3 – proglacial dammed lakes; 4 – watershed of proto-Vistula; 5 – catchment of great proto-Vistula during Pomeranian Phase ca16 ka BP. Br(Le) Brandenburg or Leszno Phase; Fm(Pz) Frankfurt or Poznan Phase; Pm Pomeranian Phase

During deglaciation, the main transformation was of the valley network (Galon, 1961; Wiśniewski, in Starkel (ed.) 1990). The Nieman, Vistula and Oder waters were trying to find new outflow across the dead-ice blocks (Figure 7.10). Finally, during the Pomeranian Phase (about 16 ka BP), the Toruń-Eberswalde ice marginal streamway was formed more than 20 km wide at some reaches, collecting the waters from Neris and Nieman up to the Elbe as well as outwash water from the ice margin, forming the largest catchment in Europe (beside Volga and Danube Figure 7.9). The ice retreat from the marginal zone of the Pomeranian Phase was also very rapid and the last moraines at the Baltic coast of the Gardno Phase dated 14,5–14,3 ka BP (Rotnicki and Borówko, 1994; Rotnicki, 2001) preceded even faster melting of ice. At the bottom of the Baltic Sea, peat layers were found dated in the Pomeranian Bay between 14 and 13 ka BP and in the Gulf of Gdańsk more than 12 ka BP (Mojski, 1995). This melting coincided with a distinct warming of Epe interstadial between 15 and 14 ka BP. Below sea level, marginal forms are preserved as well as large ice marginal streamways passing from the Gulf of Gdańsk to the west up to the Bornholm Depth, indicating the avulsion of Vistula waters to the Gulf of Gdańsk at that time (Figures 7.9 and 7.10).

The gradual lowering of the base level provoked incision upstream. The retreat from the margin of Poznan phase (18.8 ka BP) caused incision in the Vistula valley at Kamion reaching the depth of the present-day floodplain before 14.5 ka BP (Manikowska, 1995). Downcutting downstream of the Toruń–Eberswalde marginal streamway was of the order of 40 to 50 m. Between the lower Oder, Vistula and Nieman, there remained flat reaches of marginal streamways, later covered by bogs with underfit streams (Żurek, 1975). The level of the Lateglacial Baltic Lake was at about

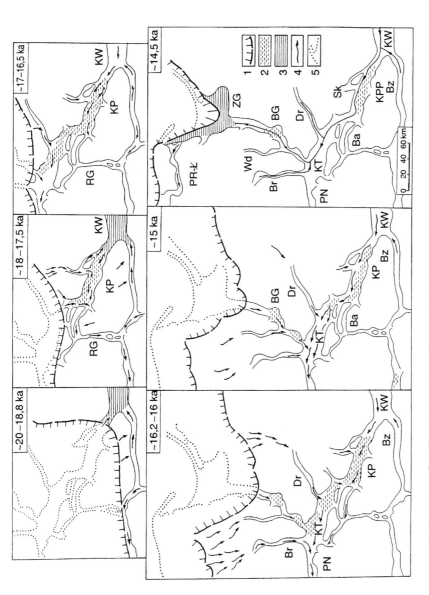

Figure 7.10 Ice retreat and directions of outflow in the lower Vistula basin between 20 and 14 ka BP (modified from Wiśniewski in Starkel (ed.) 1990). 1 – ice margin; 2 – dead-ice blocks occurring in the Vistula valley; 3 – ice-dammed lakes; 4 – drainage directions; 5 – present-day contours of the Vistula valley and Baltic Sea; KW Warsaw; KP Płock B; KT Toruń B; BG Grudziadz B; ZG Gdanskie damm lake, ice marginal streamways; PN Notec; PR-L Reda-Leba; RG Goplo. Other abbreviations indicate names of tributary valleys

80 m below sea level and the Vistula in the reach of the present-day delta was incised
up to −35 m (Mojski, 1990) and the Oder up to −60 m (Brose and Präger, 1983).

After the recession of the exposed undulating dead-ice topography along with permafrost that had expanded from the south, both underwent gradual melting (Figures 7.7
and 7.8). Melting of deeper dead-ice blocks started with the Lateglacial warming during
the Bölling and Alleröd (Niewiarowski, 1990; Kozarski, 1995). This time of formation of new groundwater reservoirs and of lakes was when a new drainage pattern was
being established connecting either with the deeply incised main river valleys or with
unrejuvenated sections of ice marginal streamways and outwash plains (Starkel, 1990).
Linear subglacial channels, many shallow depressions over the morainic plateau and
outwash plains overgrown firstly by peat have later been overdeepened by up to several tens of meters (e.g. Więckowski, 1978; Niewiarowski, 1990). Most of the lakes
did not have (and frequently still do not have) a connection with the valley network.
These lakes underwent gradual change and reduced their extent so that only one-third
of former lake surfaces now exist (Kalinowska, 1961). The main reasons for lake-
level changes were the climatic fluctuations, but their decrease in space and volume
was due to gradual infilling by sediments, overgrowth by peat and incorporation into
drainage basins so that the sequence of lake-level fluctuation may differ from lake to
lake (Ralska-Jasiewiczowa and Latałwa, 1996; Figure 7.11).

High water levels are recorded mostly before and during the beginning of melting
of dead-ice blocks (in Alleröd or Younger Dryas) followed by a rapid fall in water
level of several meters (Figure 7.11). The first Holocene rise about 8.5 to 8.0 ka BP is
limited to some lakes (Ralska-Jasiewiczowa and Latałowa 1996; Niewiarowski, 1995;
Starkel et al., 1996b; Wojciechowski, 1999), whereas several others show a continuous
lowering, which can be attributed to organisation of the new drainage system as well
as to increased evaporation during the Boreal–Atlantic warming (Kondracki, 1971;

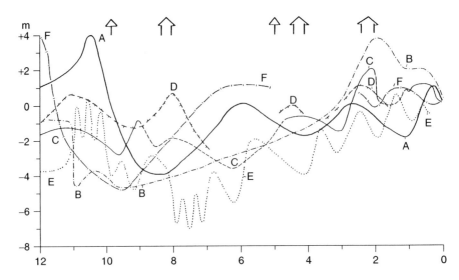

Figure 7.11 Fluctuations of water level in the lakes of North Poland. A – Masurian lakes
(after Kondracki, 1971); B – Lake Strazym (after Niewiarowski, 1988); C – Lake Biskupin
(after Niewiarowski, 1995); D – Lake Gosciaz (after Starkel et al., 1996b); E – Kornik-Bnin
lake (after Wojciechowski, 1999); F – Lake Jasien (after Florek et al., 1999); arrows indicate main phases of high water level in Polish lakes (after Ralska-Jasiewiczowa, 1989
and others)

Niewiarowski, 1990; Nowaczyk *et al.*, 1999). During the Atlantic, the water level in some basins was so low that they had no outflow (Figure 7.12). Starting from about 6 ka BP till present, a general tendency of water-level rise is observed, connected with gradual filling by sediments (Figure 7.13) and with increased drainage from lakes (Starkel *et al.*, 1996b; Ralska-Jasiewiczowa *et al.*, 1998; Wojciechowski, 1999). Shallow lakes react more to fluctuations in rainfall pattern and evapotranspiration, and

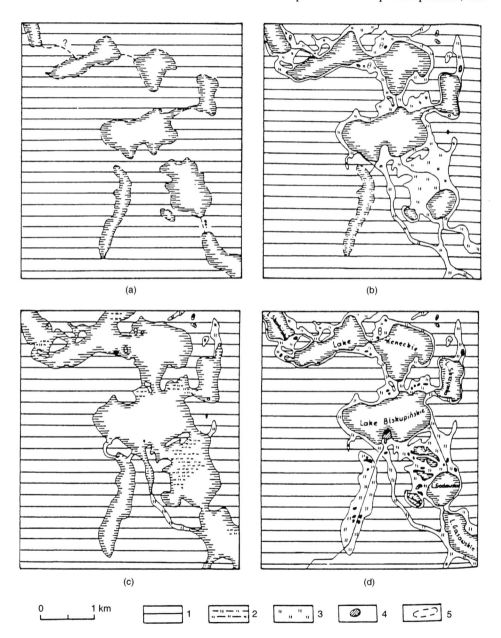

Figure 7.12   Palaeohydrology of the Biskupin region (after Niewiarowski, 1995) (a) about 6 ka BP; (b) about 3 ka BP(late Bronze Age); (c) about 2 ka BP; (d) present hydrology; 1 – land; 2 – swamps; 3 – bogs; 4 – settlement of Lusatian culture; 5 – presumed extent of lakes. Reproduced by permission of W. Niewiarowski, 1995

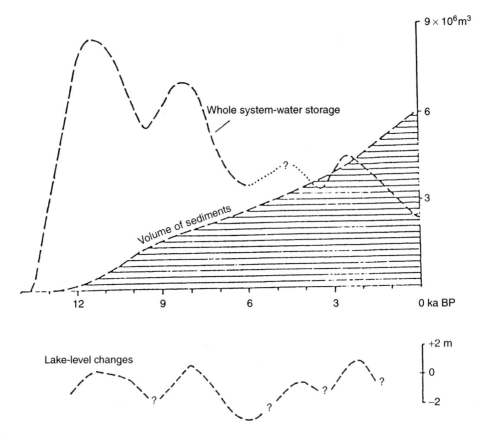

Figure 7.13 Changes of water storage (volume) and deposition in Lake Gosciaz during the last 12 ^{14}C ka BP (after Starkel *et al.*, 1996b)

the blanket of impermeable lake deposits separates lake waters from groundwater in glaciofluvial deposits (Figure 7.14). Superimposed on this trend, humid phases 4.5 to 4.1 ka especially 2.5 to 2.0 ka BP are expressed in water-level rises, as well as in higher rates of peat growth (Żurek and Pazdur, 1999). Flooding of late Bronze settlements of Lusatian culture is documented at many sites (Niewiarowski, 1990; 1995 Figure 7.12). The 2-m rise at Biskupin lake after 2,750 cal yrs BP was the basis for calculation of water-balance change (Skarżyńska, 1965), with precipitation rise evaluated from 413 to 658 mm and surface storage with outflow rise from 60 to 240 mm. The Little Ice Age rise was not so pronounced, and the fall in the twentieth century is caused by artificial drainage of lakes and bogs.

The diversity of results (Figure 7.11) is also connected with methods applied for reconstructions. For Gościąż Lake, we obtained two different pictures: by analysing changes of water volume reflected in the relationships of stable isotopes (Pazdur *et al.*, 1999) and by levelling of sediments and forms in the littoral zone (Starkel *et al.*, 1996b). Anthropogenic factors started to play an increasingly significant role after the late Bronze Age and Roman times. On the shores of Biskupin lake and in small kettles over the morainic plateaus, thick slope wash deposits accumulated (Sinkiewicz, 1995). Deforestation and soil cultivation about 2 ka BP reduced the evapotranspiration rate, and the rise of groundwater table caused the formation of shallow lakes south

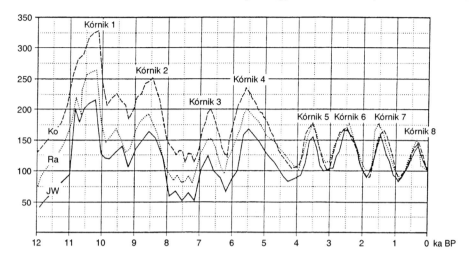

Figure 7.14 Changes in water resources of selected Kornik-Zaniemysl lakes during the last 12 ka expressed as a percentage of present-day water volume (after Wojciechowski, 1999). Ko Lake Kornik; Ra Lake Raczynskie; JW Lake Jeziory Wielkie. Reproduced by permission of Komitet Badan Czwartorzedu

of Poznan (Borówka, 1990). In the Kashubian Lakes, the part mineral components during the Subatlantic increased by up to 60 times (Gołębiewski, 1981).

2.7 Coastal Areas

During the first stages of the Flandrian transgression, extensive areas of the North Sea shelf were covered by water followed between 8,600 and 7,100 BP (Merkt *et al.*, 2001) by an especially rapid rise (from 45 to 15 m below sea level). The Baltic Sea had a separate history; but after the Ancylus Lake phase, the Litorina transgression was simultaneous and its several fluctuations caused the formation of Rhine and Vistula deltas, coastal barriers as well as estuaries (Figures 7.2 and 7.9). Stages of lowering of the level of the North Sea coincided with thicker and coarser members in deltaic deposits (6.4–6.5; 5.3–4.6; 4.4–3.4 ka BP) being synchronous with more humid phases and with advances of Alpine glaciers (van der Woude, 1981; Figure 7.4). In the lowland Peene valley, the influence of sea-level rises has been registered by water-level fluctuations in bog sequences and transfluent lakes up to 90 to 120 km upstream (Richter, 1968). Many small bays were closed by marine barriers and the coastal lakes and bogs register the ingressions of the Baltic Sea about 8.2 to 7.8, 6.4 to 6.0 and after 4 ka BP (Rotnicki and Borówka 1995; Tobolski, 1991).

3 VALLEYS OF RIVERS DRAINING SEVERAL REGIONS

Large rivers, both flowing to the north and those belonging to the Danube catchment, flow through areas of various rainfall characteristics and water balance. The hydrological regime of major rivers is usually controlled by headwater areas located in the mountains, where the waves of rain and snowmelt floods originate (Figure 7.2). These flood waves determine the water storage in the submontane depressions. Farther downstream, tributaries of draining neighbouring areas start to play a leading role again.

In the case of catchments drained northwards, the melting of river ice during cold winters proceeds from the upper part downstream. Therefore, in the east the ice-jam floods combined with frozen ground are more frequent and cover the extensive floodplains, especially in the former ice marginal valleys drained by underfit streams. In the case of the Vistula River valley, the presence of frequent floods between the first and third centuries AD has been documented along the whole longitudinal profile. In the lower course, organic horizons were buried at that time under overbank loams (Tomczak, 1987; Starkel, 2001a).

Construction of embankments along the channels reduced the flooded areas and by contrast, increased the water level variations, leading to catastrophic flooding during breaks of embankments (Gębica and Sokołowski, 2001). Incision of river beds after the regulation of river channels (Klimek, 1987) combined with density of the drainage networks is responsible for the lowering of groundwater levels and reductions in water storage. As a result of this, flood waves originating in the mountain ranges reach the sea more rapidly and during the dry season the discharges are lower. Construction of water reservoirs works in a contrary direction.

4   CONCLUSIONS

Contrasting with present-day diversity in water resources over the area of central Europe, three general stages of palaeohydrologic evolution can be distinguished.

During Stage I (20–14 ka BP), glaciated areas, with water storage in the glacial ice, occurred in areas occupied by the Scandinavian ice sheet and also in the high mountains, existing in parallel with the periglacial zone with permafrost and seasonal runoff. The glaciated area supplied the periglacial zone with meltwaters, and in the case of lowlands ice sheets blocked the older river courses.

Stage II (14–10 ka BP) was partly diachronous with Stage I because the melting of ice and expansion of permafrost in the north and its retreat in the south started before 14 ka BP (Kozarski, 1995; GoYdzik, 1995). During the late Vistulian, it followed the formation of groundwater reservoirs, evapotranspiration increased and dense vegetation with trees developed inspired by a rise both in temperature and precipitation. This was a stage of transformation in the type and frequency of floods. In the areas left ice-free, the dead-ice blocks had melted and a new topography developed with surficial storage in lakes and bogs, so that a drainage system dominated by rivers was created.

Stage III (10–0 ka BP) was the period of expansion of forest communities, which outside the high mountain area controlled the water cycle. During the Holocene, the more humid phases (and usually cooler) with frequent extreme events usually 300 to 500 years long were initiated separated by longer phases, relatively dry and with less frequent floods (Starkel, 1983; Starkel et al., 1991). Those phases are reflected not only in fluctuations of glaciers and vertical belts in the mountains, in alluvial cuts and fills but also in mass movements, fluctuations of lake levels and other phenomena (Figure 7.15). The main cause of these fluctuations seems to be variations in the solar radiation and superimposed volcanic activity reflected in the precipitation regime (Magny, 1993; Starkel, 1998; 1999; Frenzel, 2000). Some of them are registered throughout central Europe 8.5 to 8.0 ka BP (Starkel, 1999) at the Atlantic–Subboreal transition (Starkel, 1995), after the late Bronze (van Geel et al., 1996) as well as the Little Ice Age.

In these fluctuations, slight differences are visible both between north and south, and west and east. The latitudinal difference is shown by comparison of lake levels (Gaillard, 1985) as well as by regional dendrochronological standards, which give much better correlation in latitudinal, rather than in longitudinal transect (Kr1piec, 1992). On the contrary, analysing the character of winter seasons and snowmelt flood

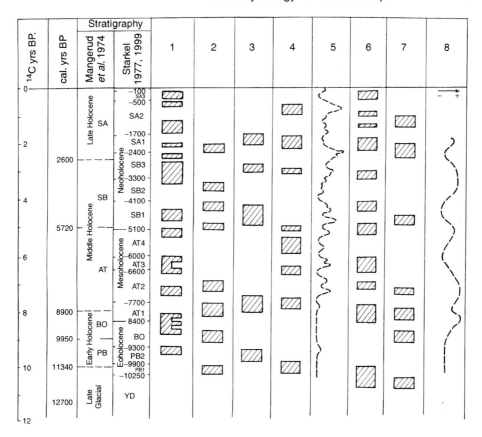

Figure 7.15 Palaeohydrological and other changes reflected in various sediments. 1 – advances of alpine glaciers (Patzelt, 1977 and others); 2 – phases of abrupt transformation of vegetation (after Ralska-Jasiewiczowa, 1989); 3 – phases of higher lake water level (Ralska-Jasiewiczowa, 1989); 4 – calcareous tufa in spring peatbogs (Dobrowolski, 1998); 5 – rate of stalagmite growth (Pazdur, 2000); 6 – phases of high flood frequency (Starkel, 1983; 2001a; Kalicki, 1991; Starkel et al., 1991); 7 – high rate of peat growth (Żurek and Pazdur, 1999); 8 – deposition rate of calcareous tufa (Pazdur et al., 1988)

frequency in a longitudinal transect, we find their much greater extent to the west during the Little Ice Age (Brazdil et al., 1999; Starkel 2001b).

Superimposed upon these fluctuations are natural or anthropogenic regional and local tendencies. The high surficial storage of water over the dead-ice topography is slowly declining with the silting, overgrowing and drainage of lake depressions (Figure 7.13). Deforestation and cultivation of soil generally accelerated the runoff and at the same time caused the increase of ground storage by reducing the evapotranspiration. These changes were evidenced in sediments and forms beginning either from the late Bronze Age or from Roman times. Direct regulation of runoff both on slopes and river channels has frequently totally transformed the water circulation. The present-day landscape of central Europe, therefore, now presents a mosaic of areas in different stages of transformation of water circulation and water resources. It is only by appreciating the palaeohydrology of these areas that the sequence of environmental changes can be understood, providing insight into some of the adjustments that may be possible in the future.

REFERENCES

Alexandrowicz, S.W., 1989. Stratigraphy and malacofauna of the Upper Vistulian and Holocene of Szklarka stream valley (Cracow Upland). *Bulletin Polish Academy of Sciences, Earth Sciences*, **37**, 59–67.

Alexandrowicz, S.W., 1995. The malacofauna preserved in Late Glacial and Holocene deposit within the Vistula catchment. *Geographical Studies*, spec(5), 178–184.

Alexandrowicz, S.W., 1996. Holoceńskie fazy intensyfikacji procesów osuwiskowych w Karpatach. *Geologia*, **22**(3), 223–262.

Ando, M. and Mucsi, M., 1967. Widerspiegelung der jungpleistozänen und holozänen Klimarhytmen in den Ablagerungsverhältnissen der sodahaltigen Teiche und periodischen Wasserdeckungen im Donau-Theiss-zwischenstromland. *Acta Geographica Szeged*, **7**(1–6), 43–53.

Andres, G. and Geyh, M.A., 1970. Paläohydrologische Studien mit Hilfe von [14]C über den pleistozänen Grundwasserhaushalt in Mitteleuropa Südliche Frankenalb. *Naturwissenschaften*, **59**, 418.

Baraniecka, M.D. and Konecka-Betley, K., 1987. Fluvial sediments of the Vistulian and Holocene in the Warsaw Basin. *Geographical Studies*, spec(4), 151–170

Baumgart-Kotarba, M. and Kotarba, A., 2001. Deglaciation in the Sucha Woda and Pańszczyca Valleys in the Polish High Tatras. *Studia Geomorphologica Carpatho-Balcanica*, **35**, 7–38.

Becker, B., 1982. Dendrochronologie und Paläoëkologie subfossiler Baumstämme aus Flussablagerungen, ein Beitrag zur nacheiszeitlichen Auenentwicklung imsüdlichen Mitteleuropa. *Mitteil*. Vol. 5, der Kommission für Quartärforschung, Österreich. Akademie der Wissenschaften, Wien, 1–120.

Behre, K.E., Brande, A., Küster, H. and Rosch, M., 1996. In B.E. Berglund, H.J.B. Birks, M. Ralska-Jasiewiczowa and H.E. Wright (eds), *Palaeoecological Events During the Last 15000 Years*. Wiley, Chichester, 507–551.

Borówka, K., 1990. Late Vistulian and Holocene denudation magnitude in morainic plateaux: case studies in the zone of maximum extent of the last ice sheet. *Quaternary Studies in Poland*, **9**, 5–31.

Borsy, Z. and Felegyhazi, E., 1983. Evolution of the network of water courses in the North-Eastern part of the Great Hungarian plain from the end of the Pleistocene to our days. *Quaternary Studies in Poland*, **4**, 115–124.

Bortenschlager, S., 1982. In M.J. Gaillard (ed.), *Glacial Fluctuations and Changes in Forest Limit in the Alps*. IGCP 158, Lundqua Report 27, Lund University, 43–45.

Böse, M., 1991. A palaeoclimatic interpretation of frost-wedge casts and aeolian sand deposits in the lowlands between Rhine and Vistula in the upper pleniglacial and late glacial. *Zeitschrift für Geomorphologie*, Supplementband **90**, 15–28.

Brazdil, R., Glaser, R., Pfister, Ch., Dobrovolny, P., Antoine, J.M., Barriendos, M., Camuffo, D., Deutsch, M., Enzi, S., Guidoboni, E., Kozyta, O. and Rodrigo, F.S., 1999. Flood events of selected European rivers in the sixteenth century. *Climatic Change*, **43**, 239–285.

Brose, F. and Präger, F., 1983. *Regionale Zusammenhänge und Differenzierungen der holozänen Flussgenese im nordmitteleuropaischen Vergletscherungsgebiet*. Ergänzungsheft Nr 282 zu *Petermanns Geographische Mitteilungen*, Gotha, 164–175.

Brunnacker, K., 1971. Geologisch-pedologische Untersuchungen in Lepensky Vir am Eisernen Tor. *Fundamenta Köln Series A*, **3**, 20–32.

Brunnacker, K., 1975. Aktivierungen des fluviatilen Geschehens im Holozän nördlich der Alpen. *Biuletyn Geologiczny U.W.*, **19**, 149–156.

Brunnacker, K., 1978. Der Niederrhein im Holozän. *Fortschritte der GeologieRheinland und Westfalen*, **28**, 399–440.

Buch, M.W., 1988. Spätpleistozäne und holozäne fluviale Geomorphodynamik im Donautal zwischen Regensburg und Straubing. *Regensburger Geogr. Schriften*, **21**, 1–197.

Burga, C.A., 1988. Swiss vegetation history during the last 18000 years. *New Phytologist*, **110**, 581–602.

Cilek V. (1997). Mid-Holocene dry spells in Bohemia, Central Europe. *Third Millennium B.C. Climate change end Old World Collapse*, NATO ASI Series 1, 49, NATO, Springer-Verlag, 309–320.

Delorme, A. and Leuschner, H.H., 1983. Dendrochronologische Befunde zur jüngeren Flussgeschichte von Main, Fulda, Lahn und Oker. *Eiszeitalter und Gegenwart*, **33**, 45–57.

Dobrowolski, R., 1998. *Strukturalne uwarunkowania rozwoju współczesnej rzeźby krasowej na międzyrzeczu środkowego Wieprza i Bugu*. Wydawnictwo UMCS, Lublin, 88.

Dylik, J., 1967. The main elements of upper Pleistocene paleogeography in Central Poland. *Biuletyn Peryglacjalny*, **16**, 85–115.

Dylikowa, A., 1969. Le probleme des dunes interieures en Pologne a la lumiere des etudes de structure. *Biuletyn Peryglacjalny*, **20**, 45–80.

Fink, J., 1977. Jüngste Schotterakkumulationen im österreichischen Donauabschnitt. Dendrochronologie und postglaziale Klimaschwankungen in Europa, *Erdwiss. Forschung*, **13**, 190–211.

Flohn, A., 1993. Climate evolution during the last millennium: what can we learn from it? In J.A. Eddy and H. Oeschger (eds), *Global Changes in the Perspective of the Past*. John Wiley, Chichester, 295–316.

Franke, H.W. and Geyh, M.A., 1971. ^{14}C-Datierungen von Kalksinter aus slovenischen Höhlen. *Der Aufschluss*, **22**(7/8), 235–237.

Frenzel, B., 1983. On the Central-European water budget during the last 15000 years. *Quaternary Studies in Poland*, **4**, 45–59.

Frenzel, B., 2000. Datiert der klimawirksame Eingriff des Menschen in den Haushalt der Naturgerst aus dem beginnenden Industriezeitalter? *Rundgespräche der Kommission für Ökologie*. Band 18, Entwicklung der Unwelt seit der letzten Eiszeit, München, 33–46.

Frenzel, B., Pecsi, M. and Velichko, A.A., 1992. *Paleogeographical atlas of the Northern Hemisphere*. Hungarian Academy of Sciences, Budapest.

Furrer, G., Leuzinger, H. and Amman, K., 1975. Klimaschwankungen während des alpinen Postglazials im Spiegel fossiler Böden. *Vierteljahrschr Naturf. Ges. Zürich*, **120**(1), 15–31.

Gabris, G., 1985. An outline of the paleohydrology of the Great Hungarian plain during the Holocene. In M. Pecsi (ed.), *Environmental and Dynamic Geomorphology*. Budapest, 61–75.

Gaillard, M.J., 1985. Postglacial paleoclimatic changes in Scandinavia and Central Europe: a tentative correlation based on studies of lake level fluctuations. *Ecologia Mediterranea*, **11**, 235–237.

Galon R., 1961. Morphology of the Noteć-Warta (or Toruń-Eberswalde) ice marginal streamway. *Geographical Studies*, **29**, 1–98.

Gębica, P. and Sokołowski, T., 2001. Sedimentological interpretation of crevasse splays formed during extreme 1977 flood in the upper Vistula river valley (South Poland). *Annales Societatis Geologorum Poloniae*, **71**, 53–62.

Glaser, R., 1992. The temperatures of Southwest Germany since 1500 – the examples of lower Frankonia and Northern Wittenberg. *Palaeoklimaforschung*, **7**, 51–64.

Gołębiewski, R., 1981. Kierunki i intensywność denudacji na obszarze zlewni górnej Raduni w późnym würmie i holocenie. *Zeszyty Nauk.Uniw.Gdańskiego*. seria Rozpr. i Monogr. 26.

Goździk, J.S., 1995. Permafrost evolution and its impact on deposition conditions between 20 and 10 ka BP in Poland. *Biuletyn Peryglacjalny*, **34**, 53–72.

Hammer, C.U., Clausen, H.B. and Dansgaard, W., 1980. Greenland ice sheet evidence of postglacial volcanism and its climatic impact. *Nature*, **288**, 230–255.

Havlicek, P., 1991. The Morava river basin during the last 15000 years. In L. Starkel, K.J. Gregory and J.B. Thornes (eds), *Temperate Palaeohydrology*. Wiley, Chichester, 319–341.

Hiller, A., Litt, T. and Eissman, L., 1991. Zur Entwicklung der jungquartären Tieflandstäler im Saale-Elbe-Raum unter besonderer Bericksichtigung von C-14 Daten. *Eiszeitalter und Gegenwart*, **41**, 26–42.

Hjelmroos-Ericsson, M., 1981. *Holocene Development of Lake Wielkie Gacno Area, Northwestern Poland*. Thesis 10, Department of Quaternary Geology, Lund University, Lund.

Hormes, A., Müller, B.U. and Schlüchter, Ch., 2001. The Alps with little ice: evidence for eight Holocene phases of reduced glacier extent in the Central Swiss Alps. *The Holocene*, **11**(3), 255–265.

Jäger, K.D., 1962. Über Alter und Ursachen der Anelehmablagerung Thűringischer Flűsse. *Praehistorische Zeitschrift*, **40**(1/2), 1–59.

Jäger, K.D. and Ložek, V., 1983. Paleohydrological implications on the Holocenedevelopment of climate in Central Europe based on depositional sequence of calcareous fresh-water sediments. *Quaternary Studies in Poland*, **4**, 81–89.

Jersak, J., 1973. Litologia i stratygrafia lessu wyżyn Południowej Polski. *Acta Geographice. Lodzensia*, **32**, 139.

Kalicki, T., 1991. The evolution of the Vistula river valley between Cracow and Niepołomice in late Vistulian and Holocene time. Evolution of the Vistula river valley during the last 15000 years, part IV. *Geographical Studies*, Special Issue 6, 11–37.

Kalicki, T., 1995. Lateglacial and Holocene evolution of some river valleys in Byelorussia. *Paläoklimaforschung*, **14**, 89–100.

Kamiński, M., 1993. *Późnopleistoceńska i holoceńska transformacja doliny Moszczenicy jako rezultat zmian Środowiska naturalnego oraz działalności człowieka*. Doctor Thesis, Department of Geography, Łódź University, Łódź.

Kalinowska, K., 1961. Zanikanie jezior polodowcowych w Polsce. *Przegląd Geograficzny*, **33**(3), 511–518.

Klatkowa, H., 1994. Evaluation du role de l'agent periglaciaire en Pologne Centrale. *Biuletyn Peryglacjalny*, **33**, 79–100.

Klimaszewski, M., 1987. The geomorphological evolution of the Tatra Mountains of Poland. *Zeitschrift fűr Geomorphologie*, Supplementband **65**, 1–34.

Klimek, K. and Starkel, L. (1974). *History and Actual Tendency of Flood-Plain Development at the Border of the Polish Carpathians*. Nachrichten Akad. Gőttingen, Rep. of Comm. on present-day Processes IGU, 185–196.

Klimek, K., 1987. Vistula valley in the Oswiecim Basin in the Upper Vistulian and Holocene. *Evolution of the Vistula River Valley during the last 15000 years*, Vol. II, Geographical Studies, Warsaw, Special Issue 4, 13–29.

Kondracki, J. (1971). Changes of the lake levels as a result of the climatic variations during the Holocene period. Études sur le Quaternarie dans le Monde VIII Congrés INQUA, Vol. 1, Paris, 1969, 119–120.

Kordos, L., 1978. A sketch of the vertebrate biostratigraphy of the Hungarian Holocene. *Kűlőunyomat a Főldrajzi Kőzlemenyek*, **1–3**, 222–229.

Kotarba, A. (1998). Lacustrine deposits in the Tatra Mountains as evidence of Late Vistulian and Holocene events related to global climate change. *Abstracts of XVI Congres*. Carpathian-Balkan Geological Association, Vienna, 291.

Kotarba, A. and Baumgart-Kotarba, M., 1997. Holocene debris-flow activity in the light of lacustrine sediment studies in the High Tatra Mountains, Poland. *Paleoklimaforschung*, **19**, Special Issue ESF Project, 147–158.

Koutaniemi, L. and Rachocki, A., 1981. Paleohydrology and landscape development in the middle course of the Radunia basin, North Poland. *Fennia*, **159**(2), 335–342.

Kowalkowski, A., 1990. Evolution of Holocene soils in Poland. *Quaestiones Geographicae*, **11/12**, 93–120.

Kozarski, S., 1991. Warta- a case study of a lowland river. In L. Starkel, K.J. Gregory and J.B. Thornes (eds), *Temperate Palaeohydrology*. John Wiley, Chichester, 189–215.

Kozarski, S., 1993. Late Vistulian deglaciation and the expansion of the periglacial zone in NW Poland. *Geologie en Mijnbouw*, **72**, 143–157.

Kozarski, S., 1995. The periglacial impact on the deglaciated area of northern Poland after 20 kyr BP. *Biuletyn Peryglacjalny*, **34**, 73–102.

Kozarski, S., Gonera, P. and Antczak, B., 1988. Valley floor development and paleohydrological changes: the late Vistulian and Holocene history of the Warta River (Poland). In G. Lang and Ch. Schlűchter (eds), *Lake, Mire and River Environments*. Balkema, Rotterdam, 185–203.

Kozarski, S. and Nowaczyk, B. 1991. Lithofacies variation and chronostratigraphy of late Vistulian and Holocene aeolian sediments in northwestern Poland. In S. Kozarski (ed.) *Late

Vistulian (=Weichselian) and Holocene aeolian phenomena in Central and Northern Europe. Zeitschrift fur Geomorphologie Supplementband **90**, 107–122.

Krąpiec, M., 1992. Skale dendrochronologiczne późnego holocenu Południowej i centralnej Polski. *Geologia*, **18**(3), 37–119.

Kruk, J., Milisauskas, S., Alexandrowicz, S.W. and Śnieszko, Z., 1996. *Osadnictwo i zmiany środowiska naturalnego wyżyn lessowych*. Inst. Archeologii i Etnologii PAN, Kraków, 1–139.

Kvitkovič, J., 1993. Intenzita vertikalnych tektonickych pohybov zemskiej kory Nizinach Slovenska v holocene. *Geograficky Časopis*, **45**(2–3), 213–232.

Liedtke, H., 1993. Phases periglaziärgeomorphologischer Prägung während der Weichselkaltzeit im nordischen Tiefland. *Zeitschrift für Geomorphologie*, Supplementband **93**, 69–94.

Ložek, B., 1975. Zur Problematik der landschaftsgeschichtlichen Entwicklung in verschiedenen Höhenstufen der Westkarpaten während des Holozäns. *Biuletyn Geologiczny U.W.*, **19**, 79–92.

Ložek, V., 1991. Palaeogeography of limestone areas. In L. Starkel, K.J. Gregory and J.B. Thornes (eds), *Temperate Palaeohydrology*. Wiley, Chichester, 413–429.

Lüttig, G., 1960. Zur Gliederung des Auelehmes in Flussgebiet der Weser. *Eiszeitalter und Gegenwart*, **11**, 39–50.

Madeyska, E., 1989. Type region P.f. Sudetes Mts – Bystrzyckie Mts Environmental changes recorded in lakes and mires of Poland during the last 13000 years. Part 3, In M. Ralska-Jasiewiczowa (ed.), *Acta Palaeobotanica*, **29**(2), 37–41.

Madeyska, T., 1995. Roślinność Polski u schyłku części ostatniego zlodowacenia. *Przegląd Geologiczny*, **43**(7), 595–599.

Magny, M., 1993. Holocene fluctuations of lake levels in the French Jura and sub-Alpine ranges and their implications for past general circulation pattern. *The Holocene*, **3**(4), 306–313.

Manikowska, B., 1995. Aeolian activity differentiation in the area of Poland during the period 20-8 ka BP. *Biuletyn Peryglacjalny*, **34**, 125–165.

Margielewski, W., 2000. Landslide phases in the Polish outer Carpathians. In E. Bromhead, N. Dixon and M-L Ibsen (eds), *Landslides in Research, Theory and Practice, Proceedings of 8th International Symposium on Landslides*. T. Telford Publ., London, 1010–1016.

Maruszczak, H., 1987, 1986. Loesses in Poland, their stratigraphy and paleogeographical interpretation. *Annales UMCS*, **41**(2), 15–54.

Matthews, J.A., Ballantyne, C.K., Harris, Ch. and McCarroll, D., 1993. Solifluction and climatic variation in the Holocene: discussion and synthesis. In B. Frenzel (ed.), Special Issue ESF Project, *European Paleoclimate and Man*. Ak. Wiss., Mainz, 339–361.

Mensching, H., 1957. Soil erosion and formation of haugh-loam in Germany. *Extrait des Cemp. R. et Rapports*. Vol. I, Union Hydrologic, Toronto, 174–180.

Merkt, J., Müller, H. and Streif, H., 2001. Kurzfristige Klimaschwankungen in Quartär Klimaweissbuch. *Terra Nostra*, **7**, 84–93.

Mojski, J.E., 1990. The Vistula river delta. Evolution of the Vistula river valley during the last 15000 years. In L. Starkel (ed.), *Geographical Studies*, Special Issue 5, 126–141.

Mojski, J.E., 1995. An outline of the evolution of the Southern Baltic area at the end of the last glaciation and beginning of the Holocene. *Biuletyn Peryglacjalny*, **34**, 167–176.

Mojski, J.E. 1995. *Atlas poludniowego Baltyku*. Sopot, Warszawa.

Mol, J., Vandenberghe, J. and Kasse, C., 2000. River response to variations of periglacial climate in mid-latitude Europe. *Geomorphology*, **33**, 131–148.

Müller, G. and Wagner, F., 1978. Holocene carbonate evolution in Lake Balaton (Hungary): a response to climate and impact of man. *Modern and Ancient Lake Sediments*. Vol. 2, Spec. Publ. Inter. Assoc. Sedimentology, 57–81.

Nakonieczny, S., 1975. The development of river valleys of the Lublin upland during the Holocene. *Biuletyn Geologiczny*, **19**, 219–222.

Niewiarowski, W., 1988. Levels in subglacial channels and their significance in determining the channel origin and evolution. *Geographia Polonica*, **55**, 113–127.

Niewiarowski, W., 1990. Hydrological changes in the light of palaeolake studies. In L. Starkel (ed.), *Evolution of the Vistula Valley*. Vol. III, Geographical Studies, Special Issue 5, Warsaw, 170–178.

Niewiarowski, W. (red.), 1995. *Zarys zmian środowiska geograficznego okolic Biskupina pod wpływem czynników naturalnych i antropogenicznych w późnym glacjale i holocenie*. Turpress, Toruń, 1–290.

Panin, N., Panin, S., Herz, N. and Noakes, J.E., 1983. Radiocarbon dating of Danube delta deposits. *Quaternary Research*, **19**, 249–255.

Parde, A., 1955. *Fleuves et Riviers*. 3rd edn. Arman Colin, Paris.

Patzelt, G., 1977. Der zeitliche Ablauf und das Ausmass postglazialer Klimaschwankungen in den Alpen. In B. Frenzel (ed.), *Dendrochronologie und postglaziale Klimaschwankungen in Tirol*. Vol. 67, Veröff. des Museum Ferdinandeum, Mainz, 93–123.

Pazdur, A. 2000. Radiocarbon in freshwater carbonates as tool of late Quaternary studies. *Geologos*, **5**, 135–154.

Pazdur, A., Goslar, T., Gradziński, M. and Hercman, H., 1999. Zapis zmian hydrologicznych i klimatycznych w obszarach krasowych Polski Południowej na podstawie badań izotopowych. In A. Pazdur, A. Bluszcz, W. Stankowski and L. Starkel (eds), *Geochronologia górnego czwartorzędu Polski*. Inst. Fizyki Politechniki Śląskiej, Gliwice, 157–177.

Pazdur, A., Pazdur, M., Starkel, L. and Szulc, J., 1988. Stable isotopes of the Holocene calcareous tufa in Southern Poland as paleoclimatic indicators. *Quaternary Research*, **30**(2), 177–189.

Plichta, W., 1970. Wpływ wieku na stopień zbielicowania gleb wytworzonych z piasków wydmowych mierzei Świny. *Studia Soc. Sc. Torunensis*, **7**(3), 1–64.

Ralska-Jasiewiczowa, M. (ed.), 1989. Environmental changes recorded in lakes and mires of Poland during the last 13000 years part III. *Acta Palaeobotanica* **29**(2), 1–120.

Ralska-Jasiewiczowa, M. and Latałowa, M., 1996. In B.E. Berglund, H.J.B. Birks, M. Ralska Jasiewiczowa and H.E. Wright (eds), *Palaeoecological Events During the Last 15000 Years*. John Wiley, Chichester, 403–472.

Ralska-Jasiewiczowa M., Goslar, T., Madeyska, T. and Starkel, L. (eds), 1998. *Lake Gościąż, Central Poland, a Monographic Study*. W. Szafer Institute of Botany, Polish Academy of Science, Kraków.

Richter, G., 1965. Bodenerosion. Schäden und gefährdete Gebiete in der Bundesrepublik Deutschland. *Forschungen z. Deutschen Landeskunde*, **152**, 1–592.

Richter, G., 1968. Fernwirkungen der litorinen Ostseetransgression auf tiefliegende Becken und Flusstäler. *Eiszeitalter und Gegenwart*, **19**, 48–72.

Roszko, L., 1968. Recesja ostatniego lądolodu z terenu Polski. *Prace Geograficzne IG PAN*, Warszawa, 31–52.

Rotnicki, K., 1991. Retrodiction of palaeodischarges of meandering and sinuous alluvial rivers and its palaeohydroclimatic implications. In L. Starkel, K.J. Gregory and J.B. Thornes (eds), *Temperate Palaeohydrology*. John Wiley, Chichester, 431–471.

Rotnicki, K. and Borówka, R.K., 1994. Stratigraphy, palaeogeography and dating of the North Polish stage in the Gardno-Łeba coastal plain. In K. Rotnicki (ed.), *Changes of the Polish Coastal Zone*. A. Mickiewicz University, Poznań, 84–88.

Röthlisberger, F. 1986. *10000 Jahre Gletschergeschichte der Erde*. Verlag Sauerländer, Aarau.

Ruzickova, E. and Zeman, A. (eds), 1994. *Holocene Floodplain of the Labe River*. Report on Grant 31305, Geol. Inst. Academy of Sciences of Czech Rep., Praha, 116.

Rybnickova, E. and Rybnicek, K., 1996. Czech and Slovak Republics. In B.E. Berglund, H.J.B. Birks, M. Ralska-Jasiewiczowa and H.E. Wright (eds), *Palaeoecological Events During the Last 15000 Years*. John Wiley, Chichester, 473–505.

Schirmer, W., 1983. Die Talentwicklung an Main und Regnitz seit dem Hochwűrm. *Geologisches Jahrbuch*, **A71**, 11–43.

Schreiber U., 1985. Das Lechtal zwischen Schongau und Rain im Hoch-, Spät- und Postglazial. *Sonderverofentlichungen*, **58**, 1–192.

Sinkiewicz, M., 1995. Przeobrażenia rzeźby terenu i gleb w okolicy Biskupinawskutek denudacji antropogenicznej. In W. Niewiarowski (ed.), *W: Zarys zmian środowiska geograficznego okolic Biskupina w późnym glacjale i holocenie*. Toruń, Turpress, 247–279.

Starkel, L., 1977. The palaeogeography of mid- and east Europe during the last cold stage, with west European comparison. *Philos. Trans. R. Soc. London, Ser. B*, **280**, 351–372.

Starkel, L., 1983. The reflection of hydrological changes in the fluvial environment of the temperate zone during the last 15000 years. In K.J. Gregory (ed.), *Background to Palaeohydrology*. John Wiley, Chichester, 213–237.

Starkel, L., 1987. Anthropogenic sedimentological changes in Central Europe. *Striae*, **26**, 21–29.

Starkel, L., 1988. Stratigraphy of the periglacial zone in Poland during the maximum advance of the Vistulian ice sheet. *Geographia Polonica*, **55**, 151–163.

Starkel, L., (ed.), 1990. Evolution of the Vistula river valley during the last 15000 years, *Geographical Studies IG PAN*. Vol. III, Special Issue 5.

Starkel, L., 1994. Frequency of floods during the Holocene in the Upper Vistula basin. *Studia Geomorphologica Carpatho-Balcanica*, **27–28**, 3–13.

Starkel, L., 1995. Reconstruction of hydrological changes between 7000 and 3000 BP in the upper and middle Vistula river basin, Poland. *The Holocene*, **5**(1), 34–42.

Starkel, L., 1996. Geomorphic role of extreme rainfall in the Polish Carpathians. *Studia Geomorphologica Carpatho-Balcanica*, **30**, 21–38.

Starkel, L., 1997. Mass movements during the Holocene: the Carpathian example and the European perspective. *Paläoklimaforschung*, **19**, 401–408.

Starkel, L., 1998. Frequency of extreme hydroclimatically – induced events as a key to understanding environmental changes in the Holocene. In A.S. Issar and N. Brown (eds), *Water, Environment and Society in Times of Climatic Changes*. Kluwer Academic Publishers, 273–288.

Starkel, L., 1999. 8500–8000 yrs BP humid phase – global or regional? Science Reports of Tohoku University, 7th Series, *Geography* **49**(2), 105–133.

Starkel, L., 2001a. Historia doliny Wisły (od ostatniego zlodowacenia do dziś). *Monografie Geograficzne IGiPZ PAN*, **2**, 1–263.

Starkel, L., 2001b. Extreme rainfalls and river floods in Europe during the last millennium. *Geographia Polonica*, **74**(2), 69–79.

Starkel, L., Klimek, K., Mamakowa, K. and Niedziałkowska, E., 1982. The Wisłoka river valley in the Carpathian foreland during the Lateglacial and Holocene. *Prace Geograficzne IGiPZ PAN*, Special Issue 1, 41–56.

Starkel, L., Gębica, P., Niedziałkowska, E. and Podgórska-Tkacz, A., 1991. Evolution of both the Vistula floodplain and lateglacial – early Holocene palaeochannel systems in the Grobla Forest (Sandomierz Basin). *Evolution of the Vistula River Valley During the Last 15000 Years, Part IV*. Geographical Studies, Special Issue 6, 87–99.

Starkel, L., Kalicki, T., Kr1piec, M., Soja, R., Gębica, P. and Czyżowska, E., 1996a. *Hydrological Changes of Valley Floors in the Upper Vistula Basin During Latevistulian and Holocene*. Geographical Studies, Special Issue 9, 7–128.

Starkel, L., Pazdur, A., Pazdur, M.F., Wicik, B. and Więckowski, K., 1996b. Lake-level and groundwater-level changes in the Lake Gościąż area, Poland: palaeoclimatic implications. *The Holocene*, **6**(2), 213–224.

Stute, M. and Deak, J., 1989. Environmental isotope study (^{14}C, ^{13}C, ^{18}O, O, D, noble-gases) on deep groundwater circulation systems in Hungary with reference to paleoclimate. *Radiocarbon*, **31**(3), 902–918.

Sundborg, A. and Jansson, M., 1991. Hydrology of rivers and river regimes. In L. Starkel, K.J. Gregory and J.B. Thornes (eds), *Temperate Palaeohydrology*. John Wiley, 13–29.

Szumański, A., 1977. Zmiany układu koryta dolnego Sanu w XIX i XX wieku oraz ich Wpływ na morfogenezę tarasu tęgowego. *Studia Geomorph. Carpatho-Balcanica*. Vol. XI, Kraków, 139–154.

Szczepanek, K., 1971. Kras Staszowski w świetle badań paleobotanicznych. *Acta Palaeobotanica*, **12**(2), 140.

Szczepanek, K., 1989. Late glacial and Holocene pollen diagrams for Jassiel in the low Beskid Mts (the Carpathians). *Acta Palaeobotany*, **27**, 1.

Szeroczyńska, K., 1998. holoceńska historia jezior Lednickiego Parku Krajobrazowego na podstawie kopalnych wioślarek. *Studia Geologica Polonica*, **112**, 29–103.

Snieszko, Z., 1985. Paleogeografia holocenu w dolinie Sancygniówki. *Acta Geographica Lodziensia*, **51**, 1–106.

Snieszko, Z., 1995. Ewolucja obszarów lessowych wyżyn Polskich w czasie ostatnich 15000 lat. *Prace Naukowe Uniw. Śląskiego*, **1496**, 5–122.

Tobolski, K., 1991. Dotychczasowy stan badań paleobotanicznych i biostratygraficznych Lednickiego Parku Krajobrazowego. *Wstęp do paleo-ekologii Lednickiego Parku Krajobrazowego*. Wydawnictwo Naukowe UAM, Poznań, 11–34.

Tomczak, A., 1987. The evolution of the Vistula valley in the Toruń Basin in the Late Glacial and Holocene. In *Evolution of the Vistula River Valley During the Last 15000 Years*. Geographical Studies, Vol. II, Special Issue 4, 207–231.

Troll, C., 1957. Tiefenerosion, Scitenerosion und Akkumulation der Flüsse im fluvioglazialen und periglazialen Bereich. *Geomorph. Studien*. Machatchek Festschrift, Gotha.

Turkowska, K., 1988. Rozwój dolin rzecznych na wyżynie Łódzkiej w czwartorzędzie. *Acta Geographica Lodziensia*, **57**, 1–135.

Turkowska, K., 1995. Recognition of valley evolution during the Pleistocene – Holocene transition in non-glaciated regions of the Polish lowland. *Biuletyn Peryglacjalny*, **34**, 209–227.

Vandenberghe, J. and Pissart, A., 1993. Permafrost changes in Europe during the Last Glacial. *Permafrost and Periglacial Processes*, **4**, 121–135.

van der Woude, J.D., 1981. *Holocene Paleoenvironmental Evolution of a Perimarine Fluviatile Area*. Vrije Universitet, Amsterdam, 1–112.

Van Geel, B., Baurman, J. and Waterbolk, H.T., 1996. Archaeological and palaeoecological indications of an abrupt climate change in the Netherlands and evidence for climatological teleconnections around 2650 BP. *Journal of Quaternary Science*, **11**(6), 451–460.

Voznyachuk, L.N. and Walczyk, M.A., 1977. Terraces of the Niemen river, its age and relations to the shorelines of glacial lakes and the Baltic Sea. *Baltica*, **6**, 193–209.

Volkel, J. and Leopold, M., 2001. Zur zeitlichen Einordnung der jüngsten periglazialen Aktivitätsphase im Hangrelief zeutraleuropäischer Mittelgebirge. *Zeitschrift für Geomorphologie*, **45**(3), 273–294.

Wasylikowa, K., Starkel, L., Niedziałkowska, E., Skiba, S. and Stworzewicz, E., 1985. Environmental changes in the Vistula valley at Pleszów, caused by Neolithic man. *Przeglld Archeologiczny*, **33**, 19–55.

Więckowski, K., 1978. Bottom deposits in lakes of different regions of Poland, their characteristics, thickness and rates of accumulation. *Polish Archive Hydrobiology*, **25**(1–2), 483–489.

Wisniewski, E. and Andrzejewski, L., 1994. The problem of the Warsaw ice-dammed lade drainage through the Warsaw-Berlin Pradolina at the last ice-sheet maximum. *Zeitschrift für Geomorphologie N.F*, **38**(1), 141–149.

Wojciechowski, A., 1999. Late Glacial and Holocene lake-level fluctuations in the Kórnik – Zaniemyśl lakes area, Great Poland Lowland. *Quaternary Studies in Poland*, **16**, 81–101.

Wohlfarth, B. and Ammann, B., 1991. The history of the Aare river and the forealpine lakes in Western Switzerland. In L. Starkel, K.J. Gregory and J.B. Thornes (eds), *Temperate Palaeohydrology*. John Wiley, Chichester, 301–318.

Wurth, G., Niggemann, S., Frank, N., Mangini, A. and Richter, D.K., 2000. A late Glacial to Holocene stalagmite from Hőlloch cave (Gottesacker area, Bavarian Alps, Germany), *Paleoenvironmental implications. Climate Change, the Karst Record II*. Abstracts of Conference, Kraków, 104–106.

Zolitschka, B. and Negendank, J.F.W., 1998. A high resolution record of Holocene palaeohydrological changes from Lake Holzmaar, Germany. In B. Frenzel (ed.), *Paläoklimaforschung*, **25**, 37–52.

Zoller, H., 1977. Alter und Ausmass postglazialer Klimaschwankungen in den Schweizer Alpen. *Erwissenschafliche Forschung*, **13**, 271–281.

Żurek, S., 1975. Geneza zabagnienia Pradoliny Biebrzy. *Prace Geograficzne IG PAN*, **110**, 1–107.

Żurek, S. and Pazdur, A., 1999. Zapis zmian paleohydrologicznych w rozwoju torfowisk Polski. In A. Pazdur, A. Bluszcz, W. Stankowski and L. Starkel (eds), *Geochronology of the Upper Quaternary in Poland*. Gliwice, 217–228.

8 Global Environmental Change and the Palaeohydrology of Western Europe: A Review

A.G. BROWN

University of Exeter, Exeter, UK

1 WESTERN EUROPE AND THE GLOBAL CLIMATE SYSTEM

Continental NW Europe and the British Isles lie in the path of the mid-latitude westerly atmospheric circulation with the frequency of high pressure increasing eastwards and south into eastern and southern Europe. Oscillations in the NW European climate are related to the North Atlantic Oscillation (NAO) and the North Atlantic Deep Water (NADW) circulation.

During the period of weather recording, daily synoptic conditions have been analysed for several areas of Europe including the British Isles using the Lamb airflow classification system (Lamb, 1972), and elsewhere in NW Europe, using the circulation type system based mainly on the source of air masses – examples include Spain (Benito *et al.*, 1996), France (Pfister and Bariss, 1994) and Switzerland (Pfister, 1994). This has allowed palaeoclimatic reconstructions for NW Europe for historical forcing events such as the Maunder Minimum (Wanner *et al.*, 1995). These studies show a strong linkage between solar forcing, the North Atlantic sea surface temperatures (SST), pressure reversals and air-mass source. The most common situation being a centennial–millennial scale oscillation from high-pressure systems over northern NW Europe, a weakening of the Azores High and high frequencies of northeasterly continental airflows (negative NAO regime, Kapsner *et al.*, 1995) to a strong subtropical high-pressure system and deeper than the normal Icelandic low increasing the frequency of strong winds and warm wet winters (positive NAO regime). The historical record reveals the highly complex regional patterns within NW Europe caused by the distribution of land, sea and mountains within a zone of competition between maritime and continental air masses. A particularly notable feature of the synoptic climatology of NW Europe is the frequency of blocking and deflection of westerly air masses by northern and central European high-pressure systems. The scale of regional synoptic climatology contrasts across NW Europe can be seen from Figure 8.1 (Mayes and Wheeler, 1997). In situation (a), an area of intense high pressure was located over Russia and a deep depression was moving northeastwards from Iceland producing a vigorous southwesterly flow from the British Isles to Scandinavia and in situation (b), a meso-scale depression was located over Scotland associated with a low-pressure system in western Russia. This synoptic intraregional spatial variability in climate is further increased by orographic effects induced by the western uplands of the United Kingdom, the Scandinavian Massif, Alps, Pyrenees and Cantabrian Mountains.

Palaeohydrology: Understanding Global Change. Edited by K.J. Gregory and G. Benito
© 2003 John Wiley & Sons, Ltd ISBN: 0-470-84739-5

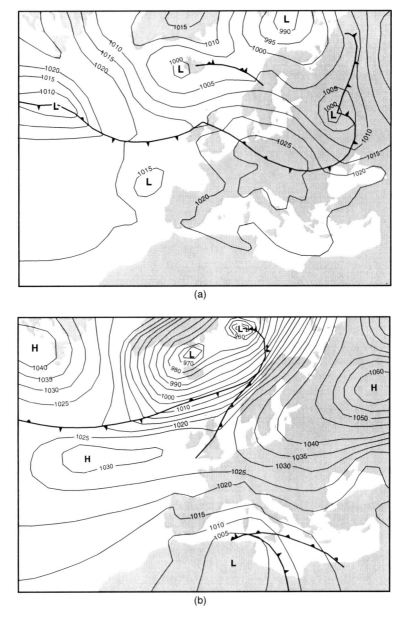

Figure 8.1   (a) Synoptic chart for 1200 GMT on March 13, 1995 and (b) synoptic chart for 1200 GMT on September 18, 1985 (both adapted from Mayes, J. and Wheeler, D., 1997 The anatomy of regional climates in the British Isles. In D. Wheeler and J. Mayes (eds), *Regional Climates of the British Isles*. Routlege, London, 9–46, by permission of Routlege

There are several implications for palaeohydrological studies of this intraregional complexity. Most obviously, areas may be out of phase as seems the case between the western and eastern Mediterranean; secondly, sensitivity to land–ocean coupling will vary. Therefore, as scale is important, one approach is to define and use hydro-climatologically uniform areas (e.g. FRIEND) in order to partition the data from large catchments so as not to confuse spatial differences of fluvial response, which quite easily could cancel each other out. Indeed, Benito has shown out-of-phase relationships between Northern and Southern Spain (Benito *et al.*, 1996). A second approach, discussed later, is to compare medium to small catchments from different hydroclimatic areas. However, although the long-term stability of these areas is unknown and although they are the result of geographical and topographic factors, they are the product of the interaction of these factors with the synoptic environment of NW Europe. Indeed, what we may look for in the palaeorecord is periods of disruption with persistent continental conditions prevailing as far west as the British Isles and persistent maritime conditions penetrating into central Europe.

The traditional fluvial approach has been to define discontinuities, something that fluvial archives are replete with, and correlate them from one area to another. There are several problems with this approach. Firstly, there are serious, if not insurmountable problems of dating discontinuities accurately and with enough precision to allow correlation to be done objectively. Secondly, correlation may suggest climatic linkages across Europe (teleconnections) but gives little information on the nature of the climatic fluctuations and the climatic environment. An alternative that can link discontinuities to climatic phenomena is correlation with long and continuous palaeoclimatic records such as ocean floor cores. NW Europe is not particularly well located in this respect, with a wide continental shelf made up of Tertiary erosional surfaces and Pleistocene fluvio-marine deposits. However, a North Atlantic sediment core has recorded changes in SST caused by variations in NADW circulation (Chapman and Shackleton, 2000). This core, from the Garder Drift 1,000 km SW of Iceland, reveals high-frequency fluctuations with periodicities of 550, 1,000 and 1,600 years. This has been correlated with the GISP II δ ^{18}O record and suggests that the NADW circulation is a significant factor in centennial-to-millennial-scale Holocene climatic variability in the North Atlantic (Chapman and Shackleton, 2000). Cores from further south off NW Iberia reveal alternations in the strength of the Azores High and therefore, the intensity and moisture content of westerly winds blowing off the Atlantic across southern Europe. Comparison of planktonic δ ^{18}O with arboreal pollen in the same core (core MD95-2042) illustrate that variation in the NW-Iberian tree pollen closely tracks millennial-scale climatic variability at least between 63 and 9 kyr BP (Roucoux *et al.*, 2001). Cacho *et al.* (2001) also report a correlation of SST profiles across the western Mediterranean with the North Atlantic record although the Younger Dryas seems to be of shorter duration and diachronous E–W. Though the ocean and ice core records are extremely important, Broecker (2000) has pointed out that it is not clear how such relatively weak-forcing pacemakers of climate that are recorded can produce such large changes in the atmospheric system. He concludes that the reason is the existence of extremely powerful and probably non-linear feedbacks in the system. This provides an important reason for studying the spatio-temporal record of hydroclimatology that is recorded in fluvial, lacustrine and mire systems. Although inferior, there are some semi-continuous terrestrial proxy data sources from NW Europe. These include long mire sequences and loess sequences. The best example of the former is the record from the La Grande Pile peat bog in the Vosges Mountains of NE France (Woillard, 1978). This record can also be correlated with the marine record (Woillard and Mook, 1982) but would seem to reveal only the larger D-O cycles, that is, interstadials and stadials. This filtering effect caused by the location of vegetation refugia and re-migration effects, may, however, not accurately reflect changes in precipitation.

Palaeohydrological studies in NW Europe have predominantly fallen into the stratigraphic, palaeobotanical or documentary categories. In comparison with the New World or tropics, relatively little palaeohydraulic or palaeoflood work has been published. This is almost certainly largely due to climatic-geomorphic factors. The sediments, which have proved most rewarding for palaeohydrological modelling, so called SlackWater Deposits (SWD) occur in bedrock gorges and are preserved and more visible under relatively arid conditions. However, the possible occurrence of these sediments in Japan (Jones *et al.*, 2001) suggests that they could occur in temperate NW Europe but that bioturbation would prevent the preservation of flood laminae. In SW Europe, Benito *et al.* (1998) have studied the SWD of the middle Tagus, which reveal three periods of late Holocene increased flood frequency and magnitude. These periods are from 1150 to 1290 AD, which is at the end of the Medieval Warm Period (MWP), or Late Medieval Climatic Deterioration (LMCD) *sensu* Brown, 1996, from 1400 to 1500 AD and 1850 to 1910 AD, which is at the end of the Little Ice Age (LIA). This technique allows the estimation of extreme values for engineering purposes and in this case a re-estimation of the 1947 flood. We are likely to see more SWD studies in western Europe in the next few years and these will add some badly needed quantification to the palaeohydrological record.

Relatively little regime-based palaeohydrological modelling of past discharges from alluvial palaeochannels has been attempted since the 1980s (see Gregory, 1983) because of the magnitude of errors involved, however, both Herget (1998) and Brown *et al.* (2001) have used this approach to discriminate between single and multiple functioning channel patterns.

2   EVENTS IN THE PALAEOHYDROLOGICAL RECORD

2.1   The Lateglacial (Oxygen Isotope Stage 2)

The recent trend in studies of Lateglacial fluvial change has been to use multiple palaeoenvironmental indicators alongside the sedimentology and dating. These have included macrofossils, pollen and spores, beetles and most recently, chironomid (Ruiz *et al.*, in prep.). This approach has been applied to several sites in the Trent Basin, UK. From a site on the River Idle, Howard *et al.* (1999) have described a braided river plain with organic accumulation in a scour hollow that commenced ca 13,500 cal BP. The pollen and coleopteran indicate a slow moving or stagnant water body fringed by reeds, sedges and moss in an open environment. This is probably the result of the reduction in river discharge that would have occurred at the end of the Dimlington Stadial. The organic deposits are covered by a cryoturbated and podzolised sandy clay, itself buried by coversand probably deposited during the Loch Lomond Stadial. The position of the sequence on a low terrace implies incision during the Loch Lomond Stadial. At Hemington, recent work has revealed organic deposits at the base of the sub-alluvial gravel dated to $12,530 \pm 70$ uncal. BP to $11,725 \pm 80$ uncal. BP (Brown and Salisbury, in press). The highly humified sandy peat lenses were probably the remains of a channel fill formed during the early Windermere Interstadial or on the declining limb of the Bølling–Allerod temperature curve (Greenwood and Smith, in press). The pollen spectra are dominated by *Betula*, Cyperaceae, Poaceae, *Filipendula, Thalictrum, Galium* t., *Artemisia* and *Salix* – typical of intermediate Stadial–Interstadial conditions in the Lateglacial. The coleopteran and caddis fly fauna contain species now limited to the montane area in the British Isles and the tundra and forests of the high latitudes in Europe. The sediments also constrain the window for incision; after the Dimlington Stadial but before the Windermere Interstadial (i.e. between 18,000 and 13,000 BP). They also indicate reworking during the early Loch Lomond Stadial (Younger Dryas) and the aggradation of up to 4 m of braided river gravel. Environmental data from a small valley in the same region at Croft revealed a multi-period fill including a clast of peat from an Iron Age channel with a radio-carbon age of 10,700 to 9,300 cal BC (Smith *et al.*, in press). The pollen spectra are dominated by Poaceae and Cyperaceae with *Pinus, Betula, Helianthemum* and other types of spectra typical of the Lateglacial, and more specifically, of the beginning of the Younger Dryas. In the Severn–Wye Basin, recent work at Wellington in the lower Lugg valley offers an unusually high-resolution picture of Lateglacial vegetation and fluvial change (Brown *et al.*, in press). The accelerator mass spectrometry (AMS) dat-ing is consistent with a Devensian date, probably between 18,000 and 14,000 years BP for the deposition of the terrace into which the palaeochannel was cut (Figure 8.2). The pollen diagram is one of the few in England, which span the later Dimlington Stadial/older Dryas – Windermere Interstadial boundary at ca 13,800 to 12,600 cal BC, although, typically of this period the dating is problematic (Brown *et al.*, in press). The dates also suggest that incision of the river forming the Holocene floodplain probably occurred during the Younger Dryas in contrast to the record at Hemington. There are indications of changing wetland conditions and extent in the early part of the time window and scrub invasion and heathland development in the later period associated with ameliorating climatic conditions. The pollen indicates a valley floor covered by a relatively open species-rich grassland that was probably maintained by natural grazing.

From a recent re-evaluation and collation of the chronostratigraphic and environmen-tal data from the Wetter Valley in Germany, Houben (in press; Figure 8.3) has shown high reach-scale variability in sedimentation at the commencement of the Younger Dryas. He suggests that this variability is greater than that during the Pleniglacial and is caused by a combination of configurational state, local climatic factors and the disequilibrium between climate and vegetation cover. The importance of the con-figurational state or inheritance (*sensu* Brown, 1990) is only now beginning to be appreciated by process geomorphologists (Lane and Richards, 1997). Richards (1999) has suggested that it poses serious questions for regime-based palaeohydrological studies. However, it does not follow that such techniques are invalid but simply that they can only be applied when the configurational state allows, that is, when the process response is externally driven. As the previous studies suggest, fluvial response to Lateglacial climate change was both variable between and within basins. An important question remains why did incision occur in the late Pleniglacial or Older Dryas in some basins but not until the Younger Dryas in others? In order to answer this a large dataset will be required from which possible factors can be evaluated including basin size, sediment supply and configurational state. These rela-tively high-precision records of fluvial change can be compared to the model proposed by Vandenberghe including the predicted incision at the beginning of the Holocene caused by a combination of low evapotranspiration due to the lack of woodland cover and low sediment supply due to herbaceous and shrub ground cover (Vandenberghe, 1995). Subsequently, as evapotranspiration increases discharge decreases and Van-denberghe suggests this will result in valley infilling. However, with low sediment supply, channel infilling and stability would result and with probably little or no over-bank alluviation. Indeed, this is the pattern that is generally recorded and alluvial sediments of the Pre-Boreal (early Mesolithic) age are rarely encountered except in association with lakes or on calcareous lithologies in which tufa and marl formation is common (e.g. Thatcham on the Thames in the United Kingdom and in the Seine Basin, France).

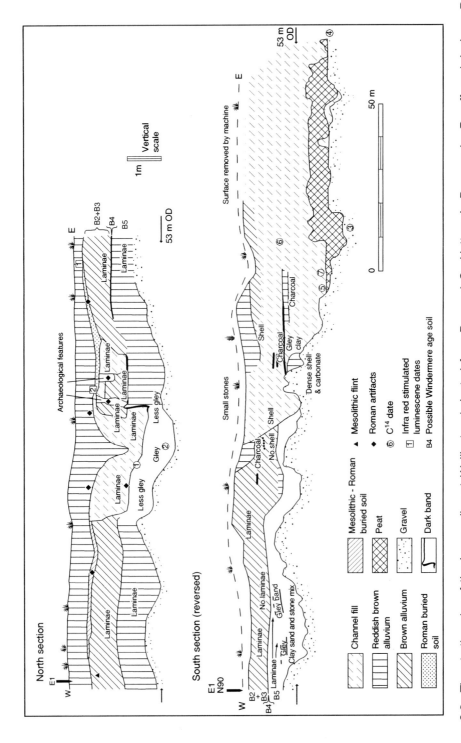

Figure 8.2 The stratigraphy of the Lugg valley at Wellington (adapted from Brown, A.G., Hatton, J., Pearson, L., Roseff, R., and Jackson, R., In M. Brickley and D. Smith (eds), *Lateglacial-Holocene Floodplain Palaeoenvironments in the Lugg Valley, UK.* Oxbow Books, Oxford; in press)

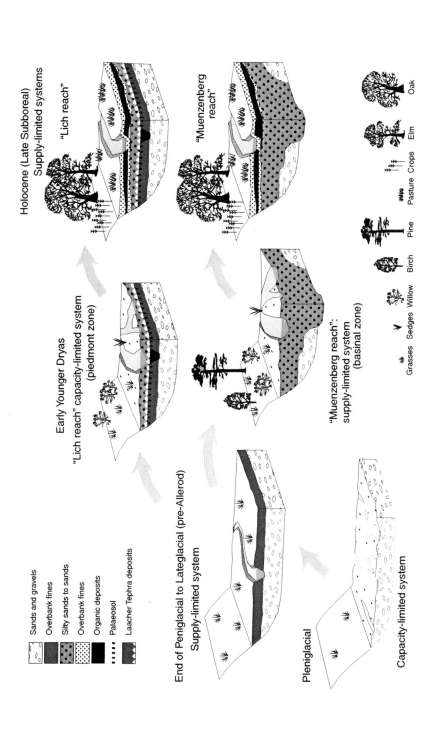

Figure 8.3 Descriptive model of the changing system-state in the Wetter valley during the Lateglacial and the Holocene (reproduced by permission from Houben, in press)

2.2   Holocene Climatic Events and Fluvial Response (Oxygen Isotope Stage 1)

All attempts to correlate the palaeohydrological record with other proxy climate data suffer from the "suck-in and smear" effects, *sensu* Baillie (1995) – the "suck-in effect" or the temptation to correlate dates with key events of the "smear effect" the impossibility of proving contemporaneity. As the number of Holocene climatic discontinuities from the Δ ^{14}C records derived from raised mires, tree rings, varves and other high-resolution records increase, this problem only magnifies. Indeed, it is possible that given dating precision it may be virtually impossible not to obtain a correlation; however, this does not necessarily imply causality. With this problem in mind only the most widely accepted discontinuities are discussed below.

The 8,200 BP Event

This climatic "event" is recorded from the GISP II record (Chapman and Shackleton, 2000). However, as yet it has proved elusive in NW European fluvial sequences. In a new synthesis of the British data, Macklin and Lewin (in press) show an increase in the frequency of ^{14}C-dated alluvial units ca 7,520 to 8,100 BP. The reason this climatic event seems to have had little fluvial effect is probably due to the stability of naturally vegetated slopes, floodplains and channel banks coupled with low sediment supply.

The 2,650 BP Event

This climatic event is now widely recognised across Europe (Geel *et al.*, 1996) and is believed to have been caused by solar forcing (van Geel *et al.*, 1999). It was a change in NW Europe from relatively warm and continental conditions to more ocean climates. It is argued that in the Netherlands the groundwater table rose (Geel *et al.*, 1996) and fens and bogs expanded. The event is well known to archaeologists as the Late Bronze Age – early Iron Age climatic "downturn" with both economic and cultural effects. Isarin and van der Beek (in press) has suggested it may be the cause of a lack of early Iron Age settlements in the Delta region of the Netherlands and a shift in channel patterns. It has also been associated with increased mass movement in high relief areas (Ballantyne, 1993). In terms of the fluvial record, changes in fluvial stratigraphy from many parts of the United Kingdom have been correlated with this event including increases in alluviation on the floodplains of the rivers Severn, Avon, Ouse and Thames (Macklin and Lewin, 1993). In a more recent analysis of the UK alluviation data (Macklin, in press), this discontinuity was again highlighted as one of the most significant in the fluvial record in the British Isles. There are, however, problems with using a dataset with variable chronological accuracy, resolution and precision. An obvious example is the alluvial record from a small catchment in the Severn Basin called Ripple studied by Brown and Barber (1985). The change in sedimentation traced along the Ripple floodplain was dated at one cross-section with four dates, the uppermost being $2,350 \pm 50$ uncal. BP (SRR-1907). This calibrates at 2σ to an envelope of 725 to 212 cal BC, although 84% of the curve falls within the period 544 to 353 cal BC. This would appear to be part of a regional pattern of increased alluviation in the English Midlands of around 2,500 to 2,700 BP as originally identified by Shotton (1978) and discussed further in this chapter. It is obviously tempting to correlate this change with the 2,650 event, but, as the pollen record reveals, this change is also correlated with a major change in the vegetation of the catchment, which could not have been caused by the 2,650 event. As mentioned earlier we see a smear-type correlation problem as even the most reliable dates linked to this event range in age from 2,450 BP to 2,750 BP (van Geel *et al.*, 1996).

In a multimedia CD-ROM publication of their work on the Rhine–Meuse delta, Berendsen and Stouthamer (2001) have described and modelled a Holocene channel change. The record is characterised by development of an early Holocene-incised system (10–8 kyr BP), sea-level induced aggradation (8–5 kyr BP) and more complex phases associated with environmental change (5 kyr BP to present). Although a river as large as the Rhine–Meuse can be affected by many factors including tectonics, Berendsen and Stouthamer (2001) suggest there are two periods of palaeohydrological change. The first is the well-documented decrease in discharge at the end of the last Glacial (Weichselian) and the second is ca 2,800 BP. The later change was associated with an increase in meander wavelength, amplitude, the number of functioning channels and the avulsion frequency. Increased flooding around this time appears to be common throughout Europe (Macklin and Lewin, 1993; Starkel, 1995; Kalicki, 1996) and at least as far east as Romania (Howard *et al.*, in press). This period, ca 500 to 800 BC, the Late Bronze–early Iron Age is also the period of population growth, and extensification of agriculture creating the open mixed landscapes of much of NW Europe.

The 535/536 AD Event

Tree-ring evidence for a climatic event in NW Europe was first published by Baillie in 1988 (Baillie, 1995). It is believed to have been caused by a volcanic eruption, possibly a proto-Krakatoa. Interestingly there is little evidence for any increase in alluviation around this date from the United Kingdom (Macklin and Lewin, 1993; Macklin, 1999). There does, however, appear to be an increase in flood magnitude and frequency recorded in the uplands around 400 to 600 AD. An example of this is the incision of 5 out of 11 sites in the Tyne Basin (Figure 8.4) as recorded by Passmore and Macklin (1997) and Macklin (1999). However, improved dating is clearly required to verify any correlation in dating the incision to 535 AD or immediately after.

The Little Ice Age (ca AD 1600 to 1850)

The LIA is well known from a wide variety of data sources and looked for in the fluvial records by Rumsby and Macklin (1996) and Brown (1996). Rumsby and Macklin (1996), suggest that many river basins in NW and Central Europe experienced enhanced fluvial activity (increased flood frequencies) between AD 1750 and AD 1900 associated with the later phase of the LIA. Brown (1996) came to similar conclusions using different data sources including historical hydrological records. Again we see the problem of correlating imprecisely dated fluvial records with proxy records that themselves vary in the LIA date. Rumsby and Macklin (1996) also distinguish an earlier Late Medieval phase in increased flooding ca AD 1250 to AD 1550. However, using high-precision (near calendrical) dating from dendrochronology and documentary sources, Brown (1998) has suggested that large floods were common throughout both the LMCD and the preceding MWP, (ca AD 950–AD 1250; Brown *et al.*, 2001) probably reflecting relatively little change in winter temperatures and/or increased precipitation during the MWP. Alluvial response, specifically accelerated valley sedimentation, has been found in the Mediterranean as far east as Crete (Maas *et al.*, 1998) and in many catchments it can probably be equated with the Younger Fill (*sensu* Vita-Finzi, 1969).

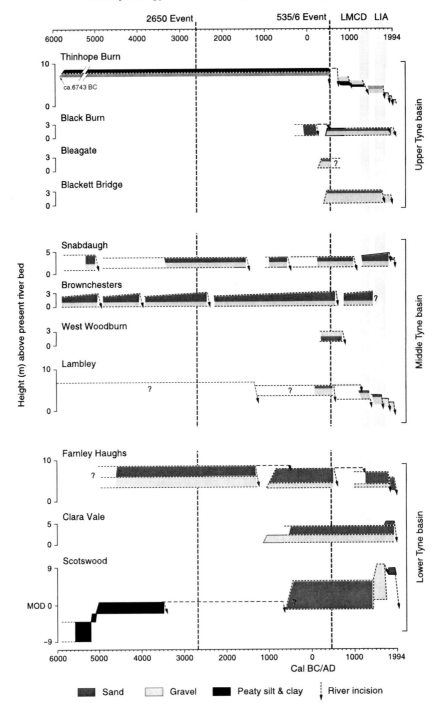

Figure 8.4 Age-depth diagrams for Holocene alluvial units and incision events in the upper and middle reaches of the Tyne Basin, United Kingdom (adapted from Macklin, M.G., 1999. Holocene river environments in prehistoric Britain: human interactions and impact. *Quaternary Proceedings*, **7**, 521–530)

3 SMALL CATCHMENT PALAEOHYDROLOGY

One of the ways of trying to investigate the causal relationship between climate change and the fluvial archive is by investigating small catchments. In the United States, this approach has been highly successful in investigating the role of land-use history in the sensitizing of basins to climatic events, particularly droughts (Trimble, 1981). In a major study of agricultural catchments in Lower Saxony, Bork (1989) has shown that soils had been truncated and slopes had been lowered by over 2 m during the historical period. In a detailed examination of the stratigraphy of gullies, Bork (1989) has revealed the full sequence near the deserted village of Drudeevenshusen (Figure 8.5). A Holocene soil developed on loess under forest from the early Holocene to the fifth and sixth centuries AD. Then, in the very late thirteenth century, there was disastrous soil erosion creating gullies 5-m deep on agricultural slopes and 10-m deep in the headwater valley floor. These gullies started to infill in the second decade of the fourteenth century. This and other sites in the region show a sudden and extreme erosional cycle over a few decades during the LMCD. There is evidence that another less extreme cycle occurred during the LIA between 1750 and 1850. The obvious question is why was the late thirteenth/early fourteenth century event so severe? In climatic terms, the evidence would suggest that the fall in temperatures and increase

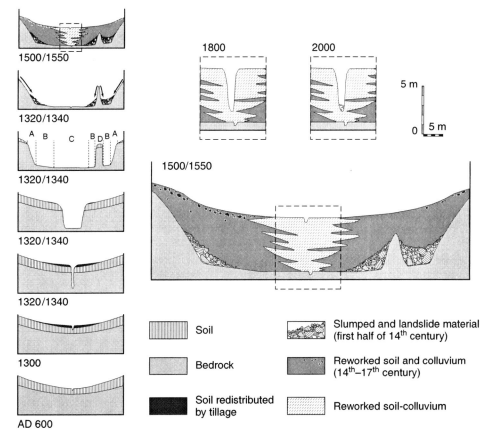

Figure 8.5 The environmental sequence for a headwater valley in Lower Saxony over the last millennia (adapted from Bork, 1989)

in flooding was no greater than during the LIA. Documentary evidence can, however, provide valuable evidence. In the eighteenth century, the erosion was minor on fields with winter crops, greater on fields with summer crops and most severe on fallow land with no vegetation cover, whereas in the fourteenth century, soil erosion was severe on all arable land but did not occur under woods or on pasture (Bork, 1989). Bork (1989) concludes that the cause of the fourteenth century erosion must have been extraordinary rainfall event, or events, increasing mean annual soil erosion from a typical arable/fallow maximum of around $50\,t\,ha^{-1}$ to several thousand tons per hectare and that the event might have been a multiple of the one thousand year rainfall event. The severity of this short-term climatic perturbation is illustrated elsewhere in continental Europe. Pörtge and Deutsch (2000) report the highest flood magnitude on the River Main near Würzburg of $3300\,m^3\,s^{-1}$ that occurred on July 22, 1342.

3.1 Lowland versus Upland in the United Kingdom

Because of the incised nature of most upland, first and second-order streams, exposures of terraces, boulder berms and colluvium are relatively common. Most studies have dated alluvial sequences in upland reaches using ^{14}C from peat or wood (Richards, 1981) or for the historical period lichenometry (Merrett and Macklin, 1999). In the Howgill Fells (Harvey et al., 1981; Harvey et al., 1984; Harvey, 1985) and in the Lake District, a phase of gulleying occurred in the tenth century AD, which was associated by Harvey et al. (1981) with the Viking settlement and the introduction of extensive sheep grazing. Harvey points out the regional diachrony in the record for gulleying, fan development and alleviation in northern upland Britain, which he associated with the spatially variable settlement and land use history. But as Macklin et al. (1992) have shown that although the major flood event will probably have been recorded within individual basins, these can be just as diachronous due to the combination of regional climatic factors, sub-regional meteorological factors and variations in basin sensitivity. In the Yorkshire Dales, Merrett and Macklin (1999) found three periods of increased flood magnitude and frequency, ca 1750 to 1800, ca 1850 to 1900 and ca 1940 to 1990. The earliest period caused aggradation but the later two caused incision suggesting non-climatic factors were also involved.

One of the reasons these upland catchments and others elsewhere in NW Europe seem to be sensitive to abrupt decadal-scale climate changes (Merrett and Macklin, 1999) is that they have all lost their natural vegetation and are covered by a plagio-climax under heavy grazing pressure. This may also explain why earlier prehistoric terraces and flood deposits are rare; alternatively, they may have been eroded out, or it may be a function of the pattern of climatic change during the Holocene, as Macklin (in press) has argued that the changes experienced during the last millennium were of larger magnitude than during any other periods in the Holocene.

Studies from lowland catchments are far less common due to the buried nature of the stratigraphic record. However, recent palaeoenvironmental work associated with archaeological investigations in central England have started to provide Holocene fluvial chronologies. At Kirby Muxloe, the infill of a headwater valley in the Trent Basin, UK, there is multi-proxy evidence of ecological and geomorphological changes, and transformation of a small lowland catchment in the valley floor from the Neolithic to the Iron Age. The palaeoecological record represents local conditions and can be compared with on-site and off-site archaeological evidence of human activity in the catchment and proxy record of climate change. Pollen analysis of the earliest organic sediments from a Neolithic palaeochannel indicate an open deforested landscape with spectra dominated by grasses and herbs, which are indicative of open, probably pastoral land use. There are no indications of arable cultivation from these samples; however, because of the low concentrations of all except the palaeochannel this cannot be regarded as conclusive evidence of absence. A second organic channel infill produced a continuous pollen and macrofossil record of vegetation change lasting approximately a thousand radiocarbon years from the Bronze Age into the Iron Age. The diagram has many interesting features, the most obvious is that it illustrates a complete transformation of the immediate landscape around the site during the Iron Age. The Neolithic deposits produced coleoptera associated with mature woodland including a species associated with *Tilia* (Lime), whereas the Bronze–Iron Age peats produced very few insects associated with woodland and was dominated by dung beetles indicative of clearance and grazing (Greenwood and Smith, in press). Although from the pollen evidence the first impact is seen in the Late Bronze/early Iron Age, this increases in the middle Late Iron Age around 700 cal BC immediately after which there is a change from organic sediments to silt and clay. Indeed, the calibrated date, at both two sigma and one sigma, overlap the 2,650 climatic event. This situation is remarkably like the Ripple Brook mentioned earlier and presents the same problem. The sudden change in the nature, and probably the rate of valley-floor sedimentation can be correlated with the ca 2,650 climatic event; however, both clearly also occur after deforestation of almost the entire catchment and the extensive adoption of arable cultivation. At Kirby Muxloe the archaeology can help, as the cultural information clearly indicates the laying out of an agricultural landscape with ditches and agricultural buildings in the eighth century BC. This would have greatly increased the sensitivity of the catchment to heavy rainfall events, which are assumed to have been of increased frequency around 2,700 to 2,600 cal BP. With the increasing use of multiple AMS dates on single macrofossils the temporal relationships between catchment change and climatic forcing should be revealed.

Further south, work in the United Kingdom at Slapton Ley in Devon by Foster *et al.* (2000), identified a change in sedimentation in the stratigraphy of a small-valley bottom wetland. The confining dates of the colluvial inwash at ca 950 uncal BP to 750 uncal. BP calibrate at 2σ between AD 1010 and 1175 and AD 1215 and 1325 or AD 1340 and 1390 and were correlated with climatic deterioration at the end of the Late Medieval Climatic Optimum (LMCD). Pollen evidence indicated that agricultural intensification had occurred prior to this date, however, at this time the area was under intensive arable agriculture sensitising the basin to extreme rainfall events, particularly those occurring in winter and early spring. The work on the Holocene palaeohydrological history of small catchments is in its infancy but it has the potential to integrate the land use and abuse signal with climatic forcing.

4   CONCLUSIONS: CLIMATIC IMPLICATIONS OF THE FLUVIAL RECORD

This review has attempted to show why, at a variety of spatial scales, fluvial response to both climatic and catchment-driven forcing may be diachronous and so why interregional correlation may be difficult. This is not to say regional patterns do not exist – they do, but that they are the result of a combination of factors including geologically controlled sensitivity, and regional land use patterns as well as climatological patterning. This thesis is examined in relation to major Late-Quaternary climatic events. Some of the diachrony may simply be related to catchment size. In the case of the United Kingdom, the tributaries of the Trent and Severn discussed here underwent incision during the Younger Dryas, however, both the Middle Trent and Lower Severn clearly show aggradation during this period. This is part of the complex response of rivers to climate change, which is common during the Lateglacial. The fluvial record of the 2,650 BP solar-forced climatic event is examined partly in order to highlight the problem of correlating the fluvial record with proxy palaeoclimatic data and the problem of causality. This climatic event occurs during a period of cultural and landscape change in NW Europe as Iron Age agriculture is undergoing both extensification and intensification in response to increasing population pressure in the Middle Iron Age. The result will be regional patterns in the nature and degree of fluvial response and there is evidence that this is also the case with the response of rivers to the LIA. The evidence of a response to the 535/536 event is equivocal largely because of the near impossibility of matching dating precision. If there was a response it would appear to have been in the uplands rather than the lowlands of Europe. This review concludes with a number of studies of small catchments, the reason being that within these catchments land-use changes can be quantified and typically affect the majority of the contributing area. Those studied so far illustrate how the sensitivity of these catchments change during the late Holocene effectively preconditioning to climate change, which will produce large, if not disproportionate response to changes in the frequency and magnitude of storms. Dramatic changes in land use in the English Midlands during the early-middle Iron Age amplifies the effect of climatic fluctuations in the third millennium BP producing a regional pattern of alluviation. Whereas in the English uplands and SW England, catchment conditions only appear to have amplified climatic forcing during the last millennium. This illustrates the point that the regional pattern of palaeohydrological response to climatic forcing is the result of the interaction of internal factors such as land-use history with climate.

REFERENCES

Ballantyne, C.K., 1993. Holocene mass movements on Scottish mountains: dating, distribution and implications for environmental change. In B. Frenzel (ed.), *Solifluction and Climatic Variation in the Holocene*. Special Issue 11, European Social Fund Project 'European Palaeoclimate and Man' Palaeoclimate Research, Fischer Verlag, Stuttgart, 71–85.

Baillie, M.G.L., 1995. *A Slice Through Time*. Batsford Ltd, London.

Benito, G., Machado, M.J. and Pérez-González, A., 1996. Climate change and flood sensitivity in Spain. In J. Branson, A.G. Brown and K.J. Gregory (eds), *Global Continental Changes: The Context of Palaeohydrology*. Special Publication No. 115, Geological Society, London, 85–98.

Benito, G., Machado, A., Pérez-González and Sopeña, A., 1998. Palaeoflood hydrology of the Tagus river, Central Spain. In G. Benito, V.R. Baker and K.J. Gregory (eds), *Palaeohydrology and Environmental Change*. Wiley, Chichester, 317–334.

Berendsen, H.J.A. and Stouthamer, E. 2001. *Physiographic Development of the Rhine-Meuse Delta*. Assen Koninklijk Van Gorcum, CD, Utrecht.

Bork, H-F., 1989. Soil erosion during the past millennium in Central Europe and its significance within the geomorphodynamics of the Holocene. In F. Ahnert (ed.), *Landforms and Landform Evolution in West Germany*, Catena Supplement, **15**, 121–132.

Broecker, W.S., 2000. Abrupt climate change: causal constraints provided by the paleoclimate record. *Earth Science Reviews*, **51**, 137–154.

Brown, A.G., 1990. Holocene floodplain diachrony and inherited downstream variations in fluvial processes: a study of the river Perry, Shropshire, England. *Journal of Quaternary Science*, **5**, 39–51.

Brown, A.G., 1996. Human dimensions of climate change. In J. Branson, A.G. Brown and K.J. Gregory (eds), *Global Continental Changes: The Context of Palaeohydrology*. Special Publication No. 115, Geological Society, London, 57–72.

Brown, A.G., 1998. Fluvial evidence of the medieval warm period and the late medieval climatic deterioration. In G. Benito, V.R. Baker and K.J. Gregory (eds), *Palaeohydrology and Environmental Change*. John Wiley, Chichester, 43–52.

Brown, A.G. and Barber, K.E., 1985. Late Holocene palaeoecology and sedimentary history of a small lowland catchment in Central England. *Quaternary Research*, **24**, 87–102.

Brown, A.G., Cooper, L., Salisbury, C.R. and Smith, D.N., 2001. Late Holocene channel changes of the middle trent: channel response to a thousand year flood record. *Geomorphology*, **39**, 69–82.

Brown, A.G., Hatton, J., Pearson, L., Roseff, R. and Jackson, R., In M. Brickley and D. Smith (eds), *Lateglacial-Holocene Floodplain Palaeoenvironments in the Lugg Valley, UK*. Oxbow Books, Oxford; in press.

Brown, A.G. and Salisbury, C.R., The geomorphology of the Hemington reach. In L. Cooper and S. Ripper (eds), *The Hemington Bridges: The Excavation of Three Medieval Bridges at Hemington Quarry, Castle Donington, Leicestershire*. University of Leicester Archaeological Services/English Heritage, Leicester, in press.

Cacho, I., Grimalt, J.O., Canals, M., Sbaff, L., Shackleton, N.J., Schönfeld, J. and Zahn, R., 2001. Variability of the Western Mediterranean Sea surface temperature during the last 25,000 years and its connection with the Northern Hemisphere climate changes. *Palaeoceanography*, **16**, 40–52.

Chapman, M.R. and Shackleton, N.J., 2000. Evidence of 550-year and 1000-year cyclicities in North Atlantic circulation patterns during the Holocene. *The Holocene*, **10**, 287–291.

Foster, I.D.L., Mighall, T.M., Wotton, C., Owens, P.N. and Walling, D.E., 2000. Evidence for mediaeval soil erosion in the South Hams region of Devon, UK. *The Holocene*, **10**, 261–271.

Geel, A., van Buurman, J. and Waterbolk, H.T., 1996. Archaeological and palaeoecological indications of an abrupt climate change in the Netherlands, and evidence for climatological teleconnections around 2,650 BP. *Journal of Quaternary Science*, **11**, 451–460.

Geel, B., van Raspopov, O.M., Renssen, H., van der Plicht, J., Dergechev, V.A. and Meijer, H.A.J., 1999. The role of solar arcing upon climate change. *Quaternary Science Reviews*, **18**, 331–338.

Greenwood, M. and Smith, D., 1983. *A Gazetteer of Coleoptera from Palaeo-Deposits from the Trent Valley*. A Volume in Honour of the retirement of S. Limbrey, M. Brickley and D. Smith (eds), Oxbow Books, Oxford.

Gregory, K.J. (ed.), 1983. *Background To Palaeohydrology*. Wiley, Chichester.

Harvey, A.M., Oldfield, F., Baron, A.F. and Pearson, G.W., 1981. Dating of post-glacial landforms in the central Howgills. *Earth Surface Processes and Landforms*, **6**, 401–412.

Herget, J., 1998. Anthropogenic influences on the development of the Holocene terraces of the River Lippe, Germany. In G. Benito, V.R. Baker and K.J. Gregory (eds), *Palaeohydrology and Environmental Change*. John Wiley, Chichester, 167–179.

Howard, A.J., Bateman, M.D., Garton, D., Green, F.L., Wagner, P. and Priest, V., 1999. Evidence of late Devensian and early Flandrian processes and environments in the Idle Valley at Tiln, North Nottinghamshire. *Proceedings of the Yorkshire Geological Society*, **52**, 383–393.

Howard, A.J., Macklin, M.G., Bailey, D.W., Mills, S. and Andreescu, R., Late glacial and Holocene river development in the Teleorman Valley on the Southern Romanian plain. *The Holocene*; in press.

Houben, P., Spatio-temporally variable response of fluvial systems to Late-Pleistocene climatic change: a case study from central Germany. *Quaternary Science Reviews*; in press.

Kalicki, T., 1996. Climate or anthropogenic alluviation in Central European valleys during the Holocene? In J. Branson, A.G. Brown and K.J. Gregory (eds), *Global Continental Changes:*

The Context of Palaeohydrology. Special Publication 115, Geological Society of London, London, 205–216.

Isarin, R.F.B. and van der Beek, H., The add-on value of fluvial geomorphology in the Betuwer-oute archaeology project: high resolution landscape reconstruction from the Rhine-Meuse delta. In A.D. Howard, M.G. Macklin and D.G. Passmore (eds), *The Alluvial Archaeology of North-West Europe and the Mediterranean*. Balkema, Rotterdam; in press.

Jones, A.P., Shimazu, H., Oguchi, T., Okino, M. and Tokutake, M., 2001. Late Holocene slackwater deposits on the Nakagawa river, Tochigi Prefecture, Japan. *Geomorphology*, **39**, 39–52.

Kapsner, W.R., Alley, R.B., Shuman, C.A., Anandakrishnan, S. and Grootes, P.M., 1995. Dominant influences of atmospheric circulation on snow accumulation in Greenland over the past 18,000 years. *Nature*, **373**, 52–54.

Lamb, H.H., 1972. British Isles weather types and a register of the daily sequence of circulation patterns 1861–1971. *Meteorological Office Geophysical Memoir*, **116**, 1–85.

Lane, S.N. and Richards, K.S., 1997. Linking river channel form and process, time, space and causality revisited. *Earth Surface Processes and Landforms*, **22**, 249–260.

Maas, G.S., Mackling, M.G. and Kirkby, M.J., 1998. Late Pleistocene and Holocene river development in Mediterranean steepland environments, southwest Crete, Greece. In G. Benito, V.R. Baker and K.J. Gregory (eds), *Palaeohydrology and Environmental Change*. Wiley, Chichester.

Macklin, M.G., 1999. Holocene river environments in prehistoric Britain: human interactions and impact. *Quaternary Proceedings*, **7**, 521–530.

Macklin, M.G., The condition of British alluvial geoarchaeology: progress, constraints and opportunities. In A.D. Howard, M.G. Macklin and D.G. Passmore (eds), *The Alluvial Archaeology of North-West Europe and the Mediterranean*. Balkema, Rotterdam; in press.

Macklin, M.G. and Lewin, J. 1993. Holocene river alleviation in Britain. *Zeitschrift für Geomorphologie N. F.* Supplementband **88**, 109–122.

Macklin, M.G. and Lewin, J., in press. River sediments, great floods and centenial-scale Holocene climate change. *Journal of Quaternary Science*.

Macklin, M.G., Rumsby, B.T. and Heap, T., 1992. Flood alleviation and entrenchment: Holocene valley-floor development and transformation in the British uplands. *Geological Society of America Bulletin*, **104**, 631–643.

Mayes, J. and Wheeler, D., 1997. The anatomy of regional climates in the British Isles. In D. Wheeler and J. Mayes (eds), *Regional Climates of the British Isles*. Routlege, London, 9–46.

Merrett, S.P. and Macklin, M.G., 1999. Historic river response to extreme flooding in the Yorkshire Dales, Northern England. In A.G. Brown and T.A. Quine (eds.), *Fluvial Processes and Environmental Change*. Wiley, Chichester, 345–360.

Passmore, D.G. and Macklin, M.G., 1997. Geoarchaeology of the Tyne basin: Holocene river valley environments and the archaeological record. In C. Tolan-Smith (ed.), *Landscape Archaeology in Tynedale. Solway Ancient and Historic Landscapes Research Programme Monograph 1*, University of Newcastle, 11–27.

Pfister, C., 1994. Spatial patterns of climate change in Europe AD 1675–1715. In B. Frenzel, C. Pfister and B. Glaeser (eds), *Climatic Trends and Anomalies in Europe 1675–1715*. Fischer Verlag, Stuttgart, 287–316.

Pfister, C. and Bareiss W., 1994. The climate in Paris between 1675 and 1715 according to the meteorological journal of Louis Morin. In B. Frenzel, C. Pfister and B. Glaeser (eds), *Climatic Trends and Anomalies in Europe 1675–1715*. Fischer, Stuttgart, 151–172.

Pörtge, K.-H. and Deutsch, M., 2000. Hochwasser in vergangenheit und gegenwart. *Rundgespräche der Kommission für Ökologie*, Supplementband **18**, 139–151.

Richards, K.S., 1981. Evidence of Flandrian valley alleviation in Staindale, N. York. *Earth Surface Processes and Landforms*, **6**, 183–186.

Richards, K.S., 1999. The magnitude-frequency concept in fluvial geomorphology, a component of a degenerating research programme? *Zeitschrift für Geomorphologie N.F.* Supplementband **118**, 1–15.

Roucoux, K.H., Shackleton, N.J., de Abreu, L., Schönfeld, J. and Tzedakis, P.C., 2001. Combined marine proxy and pollen analysis reveal rapid Iberian vegetation response to North Atlantic millennial-scale climatic oscillations. *Quaternary Research*, **56**, 128–132.

Rumsby, B.T. and Macklin, M. G., 1996. River response to the last neoglacial (the 'Little Ice Age') in Northern, Western and Central Europe. In J. Branson, A.G. Brown and K.J. Gregory (eds), *Global Continental Changes: The Context of Palaeohydrology*. Special Publication No. 115, Geological Society, London, 217–233.

Shotton, F.W., 1978. Archaeological inferences from the study of alluvium in the lower Severn-Avon valleys. In S. Limbrey and I.G. Evans (eds), *Mans Effect on the Landscape: The Lowland Zone*. Report 21, Council for British Archaeology Research, London, 27–32.

Smith, D. *et al.*, *Essays in honour of Susan Limbrey. Association of Environmental Archaeologists*. Oxbow Books, Oxford; in press.

Starkel, L., 1995. Reconstruction of hydrological changes between 7000 and 3000 BP in the upper and middle Vistula river basin, Poland. *The Holocene*, **5**, 34–42.

Trimble, S.W., 1981. Changes in sediment storage in the Coon Creek basin, Driftless Area, Wisconsin, 1853 to 1975. *Science*, **214**, 181–183.

Vandenberghe, J., 1995. The role of rivers in climatic reconstruction. *Palaeoclimate Research*, **14**, 11–20.

Van Geel, B., Buurman, J. and Waterbolk, H.T., 1996. Archaeological and palaeoecological indications of an abrupt climate change in the Netherlands and evidence for climatological teleconnections around 2650 BP. *Journal of Quaternary Science*, **11**, 451–460.

Vita-Finzi, C., 1969. *The Mediterranean Valleys*. Cambridge University Press, Cambridge.

Wanner, H., Pfister, C., Brázdil, R., Frich, P., Frydenahl, J., Jónsson, T., Kington, J., Lamb, H.H., Rosenørn, S. and Wishman, E., 1995. Wintertime European circulation patterns during the late Maunder minimum cooling period (1675–1704). *Theoretical and Applied Climatology*, **51**, 167–175.

Woillard, G.M., 1978. Grande pile peat bog: a continuous pollen record for the last 140,000 years. *Quaternary Research*, **9**, 1–21.

Woillard, G.M. and Mook, W.G., 1982. Carbon-14 dates at Grande pile: correlation of land and sea chronologies. *Science*, **215**, 159–161.

9 Palaeohydrological Changes in the Mediterranean Region during the Late Quaternary

G. BENITO
CSIC-Centro de Ciencias Medioambientales,
Madrid, Spain

1 INTRODUCTION

Mediterranean regions are highly sensitive to short-term climatic or environmental changes due to their fragile hydrologic, climatic and biological equilibrium (Vita-Finzi, 1969; Butzer, 1974; 1975; Lewin *et al.*, 1995). This sensitivity is shown in the temporal and spatial variability of natural indicators such as desertification, floods, droughts and biodiversity, which are closely related to global change factors of climate and human activity. Geographic location at the southern limit of influence of the polar front in the subtropical-sensitive Mediterranean zone provides great potential for reflecting any short-term global scale changes. The hydrological history of the Mediterranean basin appears to have been a complex alternation of wet and dry episodes with abrupt transitions related to the impact of changes in atmospheric circulation and oceanic currents. However, to decipher palaeohydrological changes in the Mediterranean basin from alluvial sequences is a complex task because of problems in the identification of successive phases of local cyclic events, their internal dating and external correlation, and uncertainties about their causation in a region with a long history of human intervention (Butzer, 1980; Macklin *et al.*, 1995). In fluvial systems, the best indicators of such variations are either sequences of cut and fill in the alluvial plains, which may record changes in the discharge–load relationships over long time periods (0.1–1 ka), or sedimentary records of individual events, such as floods, storms, droughts or human impacts on the environment.

This chapter aims at describing the long-term hydrological changes (over the last 20,000 years) recorded in the Mediterranean regions, demarcated here by river catchments within zones of Mediterranean climate, following the criteria by Pardé (1950) and Guilcher (1979) (Figure 9.1). These changes will be explained in the context of global change. Mediterranean palaeohydrology has contributed to the study of global change in two main ways: first, through the detection of past climatic changes and the corresponding response of fluvial systems; and second, by quantitatively estimating the hydrological effects of past global changes produced over extensive time periods (Late Pleistocene–Holocene). The latter is essential for the prediction of potential of water resource and hydrological risk problems arising from an eventual climatic change.

Palaeohydrology: Understanding Global Change.   Edited by K.J. Gregory and G. Benito
© 2003 John Wiley & Sons, Ltd   ISBN: 0-470-84739-5

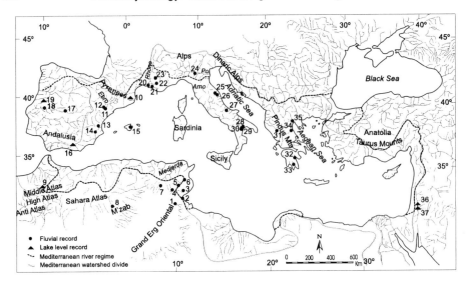

Figure 9.1 Major rivers and streams of the Mediterranean region and watersheds of catchments draining into the Mediterranean Sea. Numbers refer to the location of sites cited in Table 9.1 and described in the text

2   MEDITERRANEAN HYDROCLIMATOLOGY

The region lies between 30 and 47° N and between 10° W and 35° E (Figure 9.1), and is a zone of transition between the continental influences of Europe, Asia and the North African desert, and the oceanic effects of the Atlantic. In particular, the area lies within the boundary between subtropical and mid-latitude atmospheric patterns, which results in alternate wet and dry (summer) seasons. In terms of total annual rainfall, the Mediterranean region spans a wide range from arid (<200 mm) to humid (>1,000 mm), depending on location and orographic characteristics.

In summer, the western basin is mainly under the influence of the Azores subtropical high, whereas the eastern part falls under a low-pressure area extending from the Persian Gulf northwest towards the eastern Mediterranean basin (as far as Greece) and is associated with the Indian summer monsoon (Gat and Magaritz, 1980; Wigley and Farmer, 1982). During autumn, as the westerlies extend southwards, there is an increase in W, NW and SW circulation types. In the western Mediterranean, the SW types are more frequent in late autumn, corresponding to an increased blocking of circulation and the subsequent genesis of cold pools producing catastrophic flooding in eastern Spain, southern France and Italy. In winter, the western Mediterranean basin is normally dominated by a high frequency of zonal circulation in altitude producing the low-pressure conditions associated with the invasion of maritime polar air masses. Wintertime in the eastern part is affected by the Siberian high and associated polar continental air masses. Transitory excursions of the polar front jet, modified by the land–sea temperature contrast, favour cyclogenesis and winter rains in the central and eastern parts of the Mediterranean basin (Wigley and Farmer, 1982). In spring, as well as in late winter, with the expansion of the jet and the domain of the undulating circulation pattern, there is a change in the main flow type with increasing southerly and southwesterly flows, responsible for important precipitation volumes in Morocco, Algeria, E and SE Spain, and S Italy.

Mediterranean rivers comprise a large variety of regimes and seasonal discharge distribution (Pardé, 1950; Masachs, 1950) related to the above seasonal climatic characteristics. Mediterranean hydrology is characterised by maximum discharge during the cold season (October to April), minimum discharge in summer, extreme seasonal and annual discharge variability, severe floods with peak discharges more than 50 times the average discharge and high rates of sediment transport.

In the Mediterranean rivers, high flow stages are produced between September and May depending on the rainfall occurrence, with single or double maxima (usually in spring and autumn). Maximum rainfall amounts are produced during September and October in NE and E Spain, SE France and central N Italy; during November in Sardinia, Liguria (NW Italy), Andalusia (S Spain), W Anatolia (Turkey), North and Central Greece, Albania and Montenegro; in December in NW Africa and Sicily; and in January in Israel, Lebanon, Rousillon region (SE France) and Noire Montagne (central France). Double discharge maxima following the equinoxes are related to the influence of high pressure over Mediterranean regions during January and February (e.g. Arno River, Italy), or by snow accumulation in the mountains during winter months. Hydrological regimes of rivers draining high mountain areas (Atlas, Pyrenees, Alps, Dinaric Alps, Pindus and Taurus Mountains) are dominated by winter snow and by summer evapotranspiration, although high intensity rainfall may occur during the summer (Beckinsale, 1969). In these mountain regions, the hydrological regime is moderate pluvio-nival or nivo-pluvial spring maximum (March to May), followed by a slight low water in August or September.

The lack of rainfall during the summer is the common characteristic of the Mediterranean regions. Rainfall period may last only two to three months in some regions of North Africa and SE Spain or it may extend from autumn through to spring in France, N Italy and the Balkan region. Low latitude rivers frequently dry out during the summer months even in drainage basins larger than 1,000 km^2.

3 ALLUVIAL MORPHOGENESIS AND CLIMATE CHANGE

Glacial/interglacial cycles (10^5 yr in scale) are composed of minor "climate events" or abrupt climate changes at millennial, centennial or decadal scales. Many of these abrupt climate changes have been revealed by marine sediments associated with a layer of ice-rafted debris in the North Atlantic (Heinrich events; Heinrich, 1988; Bond et al., 1992), in the Greenland ice cores (Dansgaar-Oeschger events; Dansgaard et al., 1993), as well as in terrestrial records (Grimm et al., 1993).

In the Mediterranean basin, the hydrological and geomorphic responses to these changes are still poorly understood. It is clear that climatic changes may produce considerable effects on fluvial dynamics and morphology, namely aggradation–incision phases, channel pattern modification and changes in sediment load (Schumm, 1977). However, a change in the seasonal pattern of precipitation would have been more effective in terms of geomorphic work than changes in total annual rainfall (Vita-Finzi, 1969). In addition, areas where the vegetation cover protects the soil incompletely, changes in the number, duration and timing of rain showers of different sizes and intensities can have profound effects on the erosive power of runoff and on stream regimes (Vita-Finzi, 1969). Furthermore, it is difficult to discern the role played by climate and human activity in changes observed in fluvial channels. Since most of the studied sites and datable materials are related to archaeological sites, human activity is singled out as the main driving force of alluvial aggradation and incision (van

Andel and Zangger, 1990; Barker and Hunt, 1995). Other researchers consider human activities to enhance the climatic morphogenetic effects on fluvial systems by triggering environmental instability and shortening aggradation–incision periods (Ballais, 1995). According to Ballais (1995), bioclimatic variations during historical times are claimed to be effective if the geosystem has been considerably modified by human activity. The regionalisation of processes and parallel landform evolution may constitute an argument in favour of climatic changes being responsible for a particular morphogenetic period (Gutiérrez-Elorza and Peña-Monné, 1998).

Different conceptual models have been described to relate climatic changes to aggradation–incision phases. In moderately sized rivers, major phases of aggradation are usually associated with periods of cooler climate, lower (total) precipitation and the development of steppe vegetation (Macklin *et al.*, 2002). It is assumed that during cold periods, high rates of sediment delivery to trunk rivers from hill slopes and tributary streams resulted from greater winter runoff from slopes dominated by steppe vegetation (Fuller *et al.*, 1998). During transition to interglacial and interstadial periods, the return of tree cover in response to increasing temperatures reduced hillslope sediment delivery to rivers with a subsequent increase in channel incision.

Small size tributary rivers and/or alluvial infilled valleys may be more sensitive to local changes in climate, vegetation or land use. In these small valleys, environmental deterioration appears to have occurred in a punctuated manner, with relatively short periods of pronounced slope instability and valley alluviation, alternating with periods of relative stability and soil formation (Butzer, 1980; Pope and van Andel, 1984; Abbott and Valastro, 1995). These periods of instability have been interpreted as being triggered by climatic mechanisms (e.g. Vita-Finzi, 1969; Bintliff, 1975; Devereux, 1982), but many authors preferred an anthropogenic explanation (e.g. Brückner, 1986; Butzer, 1980). In this model, alluvial fills were associated with transition to drier climate conditions in which sediments that accumulated on slopes under a more extensive vegetation cover were eroded and transported into the river channels and valley bottoms (Knox, 1984).

In this chapter, the chronology is mainly based on radiocarbon ages (^{14}C yr BP), some U/Th dating, as well as thermoluminescence (TL) dating and related techniques such as optically stimulated luminescence (OSL) and infrared stimulated luminescence (IRSL). When available, published radiocarbon dates are used and expressed in ^{14}C yr BP or in kyr BP (10^3 ^{14}C yr BP). In addition, most of the radiocarbon dates have been converted into calendar estimates using the CALIB 3.0 program (Stuiver and Reimer, 1993) back to 18,000 ^{14}C yr BP. In the text, calibrated years are expressed either as cal. yr BP or in cal kyr BP (10^3 cal yr BP). Dates obtained by other methods (e.g. TL, OSL, IRSL, U/Th datings) were referred to as "years ago" or as ka ("thousands of years before present").

4   NORTH AFRICAN MEDITERRANEAN REGION

Early work on Mediterranean valleys by Vita-Finzi (1969) points out the development of two main aggradational phases in the Mediterranean region since the Late Pleistocene. The "older fill" was dated as the Late Pleistocene (between 30–15 ^{14}C kyr BP or perhaps even later). The older fill is tentatively explained in terms of an increase in frost weathering in the uplands during the Last Glaciation, and by the seasonal incidence of more intense rains throughout the area. The older fill was incised during most of the Holocene until post-Roman times, when Mediterranean streams began to aggrade. This medieval accumulation, known as the "younger fill", was explained as a result of tectonic and anthropogenic causes. In recent centuries, the younger fill was incised, which is still in operation in many Mediterranean streams. This model proposed in Vita-Finzi's classic work has been improved by new chronological and stratigraphical findings, which have led to advances in fluvial palaeohydrology and stratigraphy in the Mediterranean basin, although it can still be considered to be at an early stage of knowledge.

4.1 Late Pleistocene

In most of the Maghreb zone, between 40 ka and 10 ka, two or three successive episodes of silt and sand deposition, reddish to pink in colour and >6 m in thickness can be recognised (Rognon, 1984). This fine material contrasts with the large quantities of coarse material deposited during the Mousterian period (100,000 to 40,000 years ago) and these are interpreted as indicators of probably a colder climate and/or more frequent flood waters (Rognon, 1984). In the Atlantic Atlas of Morocco, the oldest silt unit overlies gravel units and was deposited before 38 ^{14}C kyr BP (Delibrias *et al.*, 1976) by successive flooding. This oldest silt unit was followed by extensive colluvial accumulation (ca 26 ^{14}C kyr BP) probably associated with a humid period and a vegetated landscape (Weisrock *et al.*, 1985). Glacial periods are considered to be influenced by the development of a fixed anticyclone over the north European ice sheet and by lower sea-surface temperatures (Rognon, 1987) that produced colder and drier conditions, but increased the seasonality of precipitation (Prentice *et al.*, 1992).

A second silt and sandy-silt accumulation took place between 26 and 13.9 ^{14}C kyr BP (Delibrias *et al.*, 1976) during a relatively wet period with frequent but low-intensity rain. These fine sediments have been interpreted as the result of deposition from suspension during flood events (Rognon, 1984). During this period (22–12 ^{14}C kyr BP), the Sahara region appears to have remained inhospitable and unoccupied, at least until ca 12 ^{14}C kyr BP (14 cal kyr BP) or later (Burke *et al.*, 1971; Servant, 1973), and the humans and savanna animals moved out into the regions bordering on the Mediterranean in the north (Clark, 1980).

A third and upper pink silt unit was deposited between 15 to 10 ^{14}C kyr BP (17.9–11.5 cal kyr BP) in SW Morocco. This unit also contains sands and clays, which were derived both from the erosion of old soils and from residual weathered material from the High Atlas and Anti-Atlas regions (Weisrock, 1998). At the margin of the Sahara, the Sebkha Mellala record (Site 8, Figure 9.1 and Table 9.1) shows a first increase in precipitation in the M'zab heights or in the Sahara Atlas mountains, centred at 13.9 ^{14}C kyr BP (16.4 cal kyr BP); further short-term wet pulses occurred at ca 12.7 ^{14}C kyr BP and 12 ^{14}C kyr BP (15 and 14 cal kyr BP), before an arid period that lasted from 10.6 to 9.3 ^{14}C kyr BP (12.3–10.3 cal kyr BP). This preceded the onset of Holocene permanent lacustrine conditions, with its maximum at 8.6 ^{14}C kyr BP (~9.6 cal kyr BP) (Gasse *et al.*, 1990). The regional data indicate that humid conditions had already commenced by ca 14.5 cal kyr BP (12.3 ^{14}C kyr BP), following full glacial hyperarid conditions during the latest Pleistocene (Street and Grove 1976; COHMAP Members, 1988), and were at their maximum during the Holocene African Humid Period (9–6 cal kyr BP after Ritchie *et al.*, 1985).

Table 9.1 Sites included in Figure 9.1 with bibliographic references

No.	Site name	Reference(s)
1	Abdallah & Makhrerouga Wadis	Petit-Maire *et al.*, 1991
2	el Akarit Wadi	Zouari, 1988; Ballais, 1991; 1995
3	Leben & Ben Sellan Wadis	Ouda *et al.*, 1998
4	Bir Oum Ali Wadi	Ballais, 1991
5	Kasserien Wadi	Ballais, 1995
6	es Sgniffa Wadi	Ballais, 1995
7	Chéria-Mezeraa Wadi	Farrand *et al.*, 1982; Ballais, 1995
8	Sebkha Mellala	Gasse *et al.*, 1990
9	Tigalmamine Lake	Lamb *et al.*, 1989
10	Banyoles and Salines Lakes	Roca and Juliá, 1997; Juliá *et al.*, 1998; Valero-Garcés *et al.*, 1998
11	Guadalupe and Bergantes rivers	Fuller *et al.*, 1996; 1998; López-Avilés *et al.*, 1998; Macklin *et al.*, 2002
12	Regallo River	Macklin *et al.*, 1994
13	River Jucar	Butzer *et al.*, 1983; Carmona González, 1995
14	Rivers Vilanopo, Mula, Lorquí, Nogalte, Guadalentín	Cuenca Payá and Walker, 1995
15	Calo d'es Cans	Rose and Meng, 1999
16	Lake Padul	Pons and Reille, 1988
17	Tagus River, Puente Arzobispo	Benito *et al.*, 1998; 2002
18	Tagus River, Alcantara	Benito *et al.*, 2002
19	Lagoa Comprida	van der Brink and Janssen, 1985; Van der Knaap and van Leeuwen, 1995
20	Rhone delta	Arnaud-Fassetta and Provansal, 1999; Bruneton *et al.*, 2001; Arnaud-Fassetta, 2002
21	Arc River	Jorda and Provansal, 1990; Provansal, 1995a,b
22	Durance River	Jorda, 1992
23	Sainte-Victoire Mountain	Ballais and Crambes, 1992
24	Po river	Cremaschi *et al.*, 1994
25	Cesano and Misa rivers	Nesci and Savelli, 1986; Coltorti, 1991; 1997
26	Musone valley	Baldetti *et al.*, 1983; Coltorti, 1991
27	Biferno valley	Barker and Hunt, 1995
28	Bradano River	Neboit, 1984; Brückner, 1986; Abbot and Valastro, 1995
29	Basento River	Neboit, 1984; Abbot and Valastro, 1995
30	Cavone River	Abbot and Valastro, 1995
31	Voidomatis River	Lewin *et al.*, 1991; Woodward *et al.*, 2001
32	Argolid valley	Bintliff, 1975; 1982; Pope and van Andel, 1984
33	Sparta, Gythion-Helos valleys	Bintliff, 1975; 1982
34	Larissa Plain, Thessaly	Demitrack, 1986 in Lewin *et al.*, 1991
35	Thessaloniki valley	Astaras and Sotiriadis, 1988
36	Dead Sea	Yechieli *et al.*, 1993; Frumkin *et al.*, 1991
37	Lake Lisan	Begin *et al.*, 1985; Geyh, 1994; Bartov *et al.*, 2002

4.2 Early Holocene

The African lakes were highest around 9,000 years ago and the vegetation changes during the early Holocene amelioration – linked to the intensification of the African monsoon (DeMenocal *et al.*, 2000) – instigated the large grazing and repopulation of desert zones (Clark, 1980). However, the major source of precipitation in northern Africa is cyclonic, and there is little correlation between modern temporal patterns of rainfall on the northern and southern margins of the Sahara (Roberts *et al.*, 1994). At the millennial scale, however, it is clear that a broad correlation exists between times of greater and lesser precipitation across the Saharan region for at least the major climatic cycles (Swezey *et al.*, 1999).

There is not much information on Holocene fluvial records and chronologies in the western Maghreb. However, in the eastern Maghreb, at least two Holocene terraces along the watercourses can be recognised (Ballais, 1995). The "lower Prehistoric Holocene terrace" (ca 3–5 m above the thalweg, exceptionally up to 15 m) was formed during a climatic period of greater humidity, which lasted from the beginning of the early Holocene (~8,300 BP) to the end of the Climatic Optimum (~5,000 BP) (Ballais, 1991; 1995). In northern Tunisia, this terrace aggradation probably started during the Late Pleistocene (Ballais and Heddouche, 1991). Water discharges increased during this period and gastropods lived in freshwater pools in places where annual rainfall is now approximately 50 mm (Petit-Maire *et al.*, 1991). During this humid period, vegetation returned to the slopes and predominant fine-grained alluvial sediments point to a low stream competence. At that time, wadi flow was probably maintained during part of the year, with construction of travertines ($6,970 \pm 70$ [14]C yr BP) at some wadis presently draining into the Grand Erg Oriental ca 7,788 cal yr BP (Site 1; Petit-Maire *et al.*, 1991). A decrease in stream discharge was identified at around 7.5 to 7 ^{14}C kyr BP (8.3–7.8 cal kyr BP) at least in the south of Tunisia (Site 2; Zouari, 1988; Ballais, 1991), although aggradation of fluvial silts continued until at least 6.4 cal kyr BP in southern Tunisia (Petit-Maire *et al.*, 1991) and until 4 cal kyr BP in the north of the country (Site 6; Zouari, 1988; Ballais, 1991). This Prehistoric Holocene terrace was compared by Ballais (1995) to the main Holocene fill in the southern French Alps (*remblaiement Holocène principal*; Jorda, 1985), the "*basse terrasse holocène préhistorique*" in Sainte-Victoire Mountain in Lower Provence (Site 23; Ballais and Crambes, 1992), and the Bradano in Basilicate in Italy (Site 28; Brückner, 1986). These aggradation chronologies have been confirmed in recent works carried out in the Wadi Leben and Wadi Ben Sellan in the Maknassy basin (central Tunisia; Site 3) by Ouda *et al.* (1998) where palustrine carbonate and gastropod shells dated with U-Th and radiocarbon techniques provided ages between 7 and 10 ka for black clayey silts and sandy silts of this terrace. Incision of this lower prehistoric terrace began around 4 cal kyr BP (3.6 ^{14}C kyr BP) representing a return to more arid conditions (Brun, 1989; Ballais, 1995), which are related to aeolian accumulations in the region (Ballais and Heddouche, 1991; Swezey *et al.*, 1999) and low lake levels in the Sahara and NW Africa (Gasse and Van Campo, 1994).

4.3 Middle-Late Holocene

The "very low Historic Holocene terrace", composed of fine sediments with either beige or grayer colours, is ubiquitous along the northern parts of the eastern Maghreb, and less developed in the south of Tunisia and in Libya (Ballais, 1995). The timing of this aggradation period, however, is not accurately constrained. Isotopic dating at Chéria-Mezeraa Wadi provided an age of $1,350 \pm 70$ [14]C yr BP (Site 7; Farrand *et al.*, 1982; 1,287 cal yr BP); $1,470 \pm 190$ [14]C yr BP at el Akarit Wadi (Site 2; Page, 1972; 1,348 cal yr BP); $2,050 \pm 58$ [14]C yr BP (malacofauna) at the Sgniffa Wadi terrace (Site 6; Ballais, 1995; 1,995 cal yr BP); and from $1,992 \pm 200$

to $3{,}270 \pm 100$ ^{14}C yr BP at Leben and Ben Sellan Wadis (Site 3; Ouda et al., 1998; 1,946–1,930 cal yr BP to 3,471 cal yr BP). However, inclusion of Roman pottery within the alluvial sediments in the Kasserien Wadi (Site 5) or its accumulation over a Roman town (Sgniffa Wadi) suggest a younger age (Ballais, 1995), certainly after the creation of Africa Novus (46 BC) and probably towards the end of Antiquity (third to fifth centuries AD) and the early Middle ages (Vita-Finzi, 1969; Brückner, 1986; Ballais and Crambes, 1992). The extensive occurrence of this terrace can be best explained by climatic factors as driving forces, although enhanced by human modifications in an already fragile environment (Ballais, 1995). Following this terrace, incision occurred until the present time, interrupted by two short minor aggradational episodes. The earlier (and larger) of these episodes, known as the very low Post-Islamic Holocene terrace, contained Islamic pottery from the tenth to eleventh centuries and has been radiocarbon dated as 610 ± 110 ^{14}C yr BP at el Akarit Wadi (Ballais, 1995). Climatically, this period corresponds to the transition between the Medieval Optimum and the beginning of the Little Ice Age (LIA) (Le Roy Ladurie, 1967). A terrace of the same age is widespread in the Mediterranean basin (Vita-Finzi, 1969; Ballais and Crambes, 1992). Following this aggradation phase, two benches were developed either cut on the previous terrace or as a coarse or fine-grained aggradation several decimetres thick. The second episode could be very recent (twentieth century) and produced by considerable bedrock erosion. At present, watercourses in North Africa continue to incise with channel bedload having grains coarser than those found in the Holocene terraces and even in the Late Pleistocene deposits.

5 NORTHERN MEDITERRANEAN REGION

5.1 Late Pleistocene

The palaeoenvironment during the Late Pleistocene consisted of a vegetation association of Pinus and Juniperus, together with steppe-like vegetation, which is especially characteristic of the Tardiglacial in the Mediterranean region (Guiot et al., 1993). There is also evidence of reduced vegetation on slopes during the Last Glacial Maximum (LGM), so that there was a steady supply of sediment to the rivers. At the LGM the shoreline was situated at ca − 100/120 m, with a constant rise of up to −40 m by ca 9 kyr BP (Shackleton and van Andel, 1985). The late Pleistocene–Holocene marine transgression resulted in the loss of a Peniglacial coastal strip some 10 to 20 km wide in areas of the Valencia Gulf and about 5 km wide along the shore of eastern Andalusia (Aura et al., 1998).

The most effective river activity was recorded during the LGM (21–26 ka) (Oxygen Isotope Stage 2; OIS), during which floodplain aggradation predominated, creating a fluvial terrace that was widespread in the Mediterranean region (Nesci and Savelli, 1986; Coltorti, 1991; Fuller et al., 1998; Rose and Meng, 1999; Macklin et al., 2002). In the Guadalupe river (NE Spain; Site 11), major aggradation took place between 24 and 22 ka (luminescence dating; Fuller et al., 1998) associated with the prominent cool episode of the Greenland Ice Core Project (GRIP) between interstadials 2 and 3 (Dansgaard et al., 1993). Despite lower moisture levels during cold climatic episodes, river activity and aggradation processes are considered to be due to rapid runoff across unvegetated surfaces with thin soils and bare rock surfaces (Lewin et al., 1995).

From the LGM until the beginning of the Holocene, at ∼11.5 ka, region-wide alluviation events are recorded at 16–19 ka and 12.5–13 ka (luminescence dating; Macklin et al., 2002; Sites 11, 12, 31), these being correlated with abrupt decreases in sea surface temperature in the northeast Atlantic (Broecker, 1998; Stocker, 1999). During these cold periods, increased seasonality of precipitation gave rise to flooding and to increased flood peaks. Slackwater flood deposits in the Pindus Mountains (NW Greece; Site 31) show increased flood peaks between $13,960 \pm 260$ ^{14}C yr BP ($17,550–15,950$ cal yr BP) and $14,310 \pm 200$ ^{14}C yr BP ($17,850–16,450$ cal yr BP) (Woodward *et al.*, 2001). In the eastern Mediterranean, high lake levels existed at between 15,000 and 13,000 BP (Bølling/Allerød interstadial), falling abruptly between 11,000 and 10,500 BP (Begin *et al.*, 1985; Yechieli *et al.*, 1993; Bartov *et al.*, 2002; Sites 37 & 36).

5.2 Early Holocene

The latest extreme cold phase, the Younger Dryas (about 8°C cooler than today), was characterised by the predominance of blocking at middle latitudes, which forced any progressive system to pass through very low latitudes (Lamb, 1971). In northwestern Spain (Allen *et al.*, 1996), pollen-based palaeoclimatic reconstructions indicate a dry and cool climate with strong seasonality for the Younger Dryas ($10,700–9,800$ ^{14}C yr BP; 13 to 11.7 cal kyr BP). In the Mediterranean basin, there is a lack of fluvial records associated with this climatic period, probably due to the severe dry conditions and a decrease in runoff.

The climatic change experienced after the Younger Dryas period is well represented in the palaeoflood record (Figure 9.2) with an anomalous number of very large flood events in the Tagus River, central Spain (Sites 17 & 18), with at least 12 extreme flood events occurring in ~200 years (from 9,440 to 9,210 ^{14}C yr BP; $10,490–10,060$ cal yr BP; Benito *et al.*, 1998; 2002). Because of the large size of the Tagus river catchment (ca $81,947\,km^2$), years with the occurrence of such extreme floods are usually related to anomalously wetter conditions. Lake-level records and pollen analysis of Lagoa Comprida (Serra de Estrela, Portugal; Site 19) showed wetter conditions after the Younger Dryas, suggesting an increase in winter precipitation on the Atlantic side of Iberia during the Lateglacial interstadial (Van der Brink and

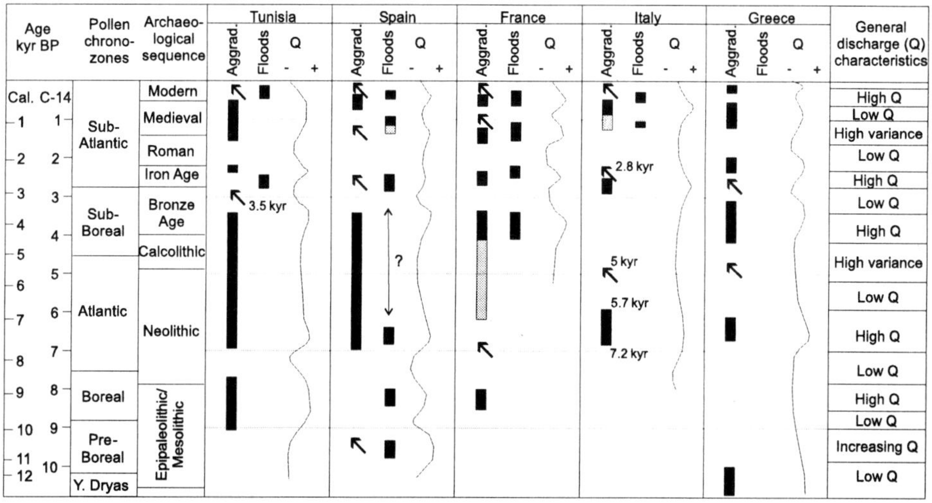

Figure 9.2 Fluvial activity through time in different Mediterranean countries. Aggradation and flood periods are composite data of Sites 2, 4, 5 and 6 for Tunisia; Sites 11, 12, 13, 14, 17 and 18 for Spain; Sites 21, 22 and 23 for France; Sites 27, 28, 29 and 30 for Italy; and Sites 31, 32, 34 and 35 for Greece. See Table 9.1 for references

Janssen, 1985; Van der Knaap and Van Leeuwen, 1995). These wetter conditions are also recorded in Banyoles and Salines lakes (NE Spain; Site 10), (Roca and Juliá, 1997; Juliá *et al.*, 1998; Valero-Garcés *et al.*, 1998).

In southern France (Arc River valley and Durance River tributaries; Sites 21 & 22), the oldest and most extensive depositional phase began during the Boreal period (ca 8,500–8,000 BP; 9.5–8.7 cal kyr BP), before the first evidence of human soil disturbance (Provansal, 1995a; Jorda, 1992). This is consistent with dates reported by Fuller *et al.* (1996) and López–Avilés *et al.* (1998) in the Guadalupe–Bergantes river system for early Holocene river terraces, which provided ages of 9.6 ± 1.0 ka and $9.0 \pm$ ka (IRSL datings). In the Tagus River (central Spain), frequent flooding was identified from 8,500 to 8,000 ^{14}C yr BP (9.5–8.7 cal kyr BP; Benito *et al.*, 2002), which is consistent with high lake level and forest development in the Iberian Peninsula and Morocco (Pons and Reille, 1988; Lamb *et al.*, 1989; Sites 16 & 9). A similar pattern of anomalously high numbers of extreme events between 8,500 and 8,000 ^{14}C yr BP has been described in other parts of the world such as central Europe (Starkel, 1999; 2002). This period was followed by dry climatic conditions ca 8,000 to 7,000 ^{14}C yr BP (8.7–7.8 cal kyr BP) reported from pollen and lake-level studies in the Iberian Peninsula (Pons and Reille, 1988; Pérez-Obiol and Juliá, 1994; Valero-Garcés *et al.*, 1998; Davis, 1994; Burjachs-Casas *et al.*, 1996), and in the eastern Mediterranean (Frumkin *et al.*, 1991; Geyh, 1994). There is no evidence of extensive fluvial aggradation during this period in the Mediterranean. A later aggradation phase was found at 7,200 to 6,000 ^{14}C yr BP (6.8–8 cal kyr BP) in other Mediterranean rivers of Spain (Fuller *et al.*, 1998; Site 11), S Italy (Neboit, 1984; Site 28) and Greece (Demitrack, 1986 in Lewin *et al.*, 1991, Site 31).

5.3 Middle–Late Holocene

At 6 kyr BP, winter temperatures were 1 to $3°$K greater than at present in the far N and NE Europe, but 2 to $4°$K colder than at present in the Iberian Peninsula, South France and generally around the northern Mediterranean region (Cheddadi *et al.*, 1997). Annual precipitation minus evaporation (P–E), an indicator of moisture balance, was 50 to 200 mm greater than the present in South and West Europe (Cheddadi *et al.*, 1997) which means that the present open vegetation types (xerophytic woodland, shrubland or steppe) were replaced by deciduous forest (Prentice *et al.*, 1992). It is not clear if this increase in moisture and vegetation cover produced an increase or decrease in the runoff and sediment production.

It seems clear that, after 6 kyr BP, conditions became gradually drier towards the present time (Harrison *et al.*, 1991) with the exception of some superimposed short periods of cold/wet conditions during the Late Bronze Period (2,650 ^{14}C yr BP or 800 BC; Van Geel *et al.*, 1998) and the LIA (LIA; Le Roy Ladurie, 1967; Grove, 1988). There are relatively few regional references indicating alluviation in NE Spain at 6 to 5 ka, based on luminescence and radiocarbon datings (Fuller *et al.*, 1996; Macklin *et al.*, 1994). After 5 kyr BP, the slash-and-burn nomadic Neolithic agriculture was replaced by a progressive occupation of larger areas of land during the Bronze and Iron Ages (Butzer, 1980). Intense agriculture, grazing activities and deforestation probably caused progressive erosion of soils and higher sediment delivery to the rivers. However, the increase of human pressure (late Iron Age, Roman Period) was not always synchronous with acceleration in morphogenetic activity in the Mediterranean (Provansal, 1995b). Agricultural practices in South France during the late Neolithic (ca 5,000 BP) gave rise to increasing rates of sediment yield (between 2 and $20 \, \mathrm{Tn \, Ha^{-1} \, yr^{-1}}$), although the largest erosion rates (up to $70-100 \, \mathrm{Tn \, Ha^{-1} \, yr^{-1}}$) occurred during anomalously "cold/wet" climatic periods, such as the Late Bronze, or Late Medieval Age (Jorda *et al.*, 1991; Jorda and Provansal, 1996; Site 21).

General phases of aggradation in the northwestern Mediterranean (Figure 9.2), associated with high river discharge (including torrential floods) and sediment load, have been described during the Late Bronze Age (ca 700–500 BC), post-Roman times (ca AD 500–1000) and/or the beginning of Modern Times (ca AD 1500–1700) (Vita-Finzi, 1969; Veggiani, 1983; Astaras and Sotiriadis, 1988; Biondi and Coltorti, 1982; Coltorti, 1991; Provansal, 1992; 1995a; Cremaschi *et al.*, 1994; Abbott and Valastro, 1995; Fuller *et al.*, 1998; Brown, 1998; Schulte, 2002). In the eastern Mediterranean (Southern Greece; Site 32), aggradation phases occurred at 3,350–4,650, 2,000–2,300 and 250–1,550 $^{14}\mathrm{C \, yr \, BP}$ (Pope and Van Andel, 1984; Jameson *et al.*, 1994; Grove, 2001).

During these relatively cold periods, higher flood frequency and magnitude have been reported in Central Spain (Benito *et al.*, 1996; 2002; Sites 17 & 18); East Spain (Butzer *et al.*, 1983; Carmona González, 1995), in South France (Jorda, 1985; Provansal, 1992; 1995a; Sites 21 & 22), Italy (Coltorti, 1997; Sites 25 & 26) and Greece (Bintliff, 1982; Site 33). There appears to be a link between glacier advances in the Alps (Löbben Cold Phase at 1930–1250 BC; Göschener Cold Phase I at 960–290 BC; Göschener Cold Phase II at AD 500–800; and LIA advances at ca AD 1320, 1600, 1700 and 1810; Holzhauser, 1997) and flood frequency and alluviation phases in Mediterranean Europe (Grove, 2001).

Higher flood frequency during these periods together with high sediment influx may be the cause of channel metamorphosis from meandering to braiding, especially in the coastal plains of the Rhone, (Arnaud–Fassetta and Provansal, 1999), South France (Bravard, 1989), as well as in rivers draining into the Adriatic coast (Baldetti *et al.*, 1983; Coltorti, 1991; Coltorti, 1997). The main progradation stage at river deltas also occurred either after the Roman Period (e.g. Po and Medjerda river deltas) or, usually, during the sixteenth and seventeenth centuries (e.g. Rhone, Ebro and Po deltas; Briand and Maldonado, 1997).

In periods with lower fluvial activity, such as 400 BC to AD 500 and AD 1100 to 1400 (Figure 9.2), the Mediterranean palaeohydrological regime was characterised by a drought-dominated regime, with low flow conditions, lower frequency of torrential rainfall and a drier climate, which produced a general shift from alluvial deposition to river incision in many sites of South France (Provansal, 1995a; Jorda and Provansal, 1996), Spain (Alonso and Garzón, 1994; Cuenca Payá and Walker, 1995) and South Italy (Abbott and Valastro, 1995).

Since 1900, extensive downcutting of alluvial floodplains has been observed, in most cases accelerated by the construction of dykes, by river straightening and by gravel mining activities. During the last 50 years, these human activities have induced greater downcutting than occurred during the whole Holocene (Coltorti, 1997).

6 DISCUSSION AND CONCLUSIONS

The main features of the general atmospheric circulation patterns over the Mediterranean region were maintained at least during the last 5,000 years. The LGM and Tardiglacial circulation patterns over the Mediterranean region are less clear, and differences from the present Mediterranean climate, and therefore, in hydrological regime, can be expected. In the western Mediterranean, glacial periods produced colder and drier conditions but increased the seasonality of precipitation (Guiot *et al.*, 1993), with lower evapotranspiration, and lower total annual precipitation. Some of these atmospheric–climatic characteristics were probably repeated to some extent during the Late Pleistocene–Holocene "abrupt" climate changes such as the Younger Dryas interval or the second abrupt Holocene climate change (ca 8,200 cal yr BP; Alley *et al.*, 1997). In contrast, warmer climatic episodes are characterised by soil development, minimal river activity and landscape stability (Rose and Meng, 1999). "Transitional periods" from warm to cold could provide enough moisture influx and high seasonal changes in rainfall and runoff patterns to enhance river dynamics and fluvial aggradation.

The Mediterranean basin is poor in Late Pleistocene–Holocene palaeohydrological data. Present Mediterranean river hydrology includes a very complex rainfall–runoff pattern across the region, which makes it even more difficult to interpret the broad patterns of palaeohydrological change in terms of changing regional and seasonal climates. Besides these limitations, major fluvial activity and aggradation phases coincide at the regional scale and at some periods, there is a coincidence across the whole Mediterranean (Figure 9.3). In the northwestern Mediterranean, general fluvial activity during the Holocene occurred at least at ∼8.5–8 ^{14}C kyr BP (9.5–8.7 cal kyr BP), ∼7–6 ^{14}C kyr BP (7.8–6.8 cal kyr BP), ∼4.2–3.5 ^{14}C kyr BP (4.7–3.7 cal kyr BP), 2.8–2.5 ^{14}C kyr BP (700–500 BC), AD 500–1000 and/or AD 1600–1700. In the eastern Mediterranean, aggradational phases occurred at ∼7–6 ^{14}C kyr BP (7.8–6.8 cal kyr BP), ∼4.5–3.3 ^{14}C kyr BP (5.3–3.5 cal kyr BP), 2.3–2 ^{14}C kyr BP and at the LIA (AD 1600–1810). The lack of coincidence for some periods may reflect seesaw behaviour between the east and west Mediterranean, which is also observed in other present climatic variables such as precipitation, temperature and atmospheric pressure. For the Late Pleistocene, chronological determinations are not sufficiently precise to establish generalised aggradation phases. In the southern Mediterranean (Figures 9.2, 9.3), increased fluvial activity occurred at 8,300–5,000 ^{14}C yr BP (9–5.7 cal kyr BP), 1,300–300 ^{14}C yr BP, AD 200–400 and AD 900–1000. It is clear from fluvial and historical archives that hydrological regime and fluvial morphogenesis in the Mediterranean area is closely linked to the climate and rainfall seasonality. Since most of the moisture is provided during periods or seasons of concentrated rainfall, changes in flood frequency and magnitude were related to, or even led to, changes in the hydrological regime. Periods with increased flood magnitude and/or frequency are strongly associated with increased moisture influx.

Since Vita-Finzi's work on Mediterranean valleys (1969), chronological precision has improved our knowledge of fluvial activity and palaeohydrology during the Late Pleistocene and Holocene. However, this development had less impact in fluvial palaeohydrology than in other scientific disciplines devoted to the reconstruction of Late Quaternary environmental changes (e.g. pollen records or lake-level reconstructions). Lewin *et al.* (1995) identified a number of problems and research needs in Mediterranean Quaternary river environments that are also critical to an understanding of Late Pleistocene–Holocene palaeohydrological changes. The main problems relate to the scarcity of alluvial chronologies in sites with an appropriate understanding of palaeohydrology and climate process–response signals in the sedimentary record. Some achievements have been realised in reconstructing palaeoflood sequences during critical periods of environmental change, which have an unequivocal palaeohydrological interpretation. As we continue to understand more about patterns of hydrological change in the past in a very sensitive region such as the Mediterranean, we will be able

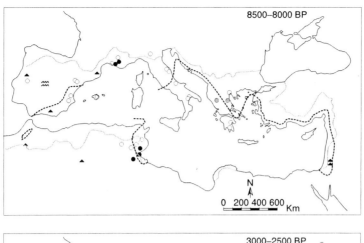

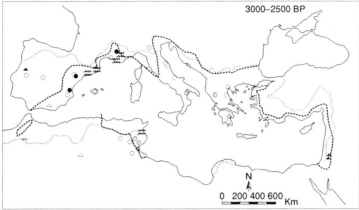

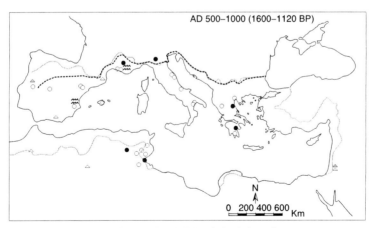

● ⊚ ○   Fluvial activity: High, Low (or incision), non data
▲ △ ⬠△   Lake level record: High, low, similar to present, non-data
〰〰   Frequent flooding
⸱‿⸱   Mediterranean climate during the indicated period (radiocarbon years)
‿   Present coastline
⸱‿⸱   Present Mediterranean climate

Figure 9.3   Palaeohydrological information on fluvial activity (mainly aggradation processes) and flooding in the Mediterranean at selected periods. The Mediterranean climate boundary is based on pollen ratios of summer-dry woodland across the region (Huntley and Birks, 1983; Roberts, 1998; Jalut *et al.*, 2000)

to understand the impact on hydrology and society of the different forecast scenarios of global warming.

ACKNOWLEDGEMENT

This research was supported by the Spanish Committee for Science and Technology (CICYT) grant HID99-0850, FEDER Project 1FD97-2110-CO2-02, REN-2001-1633 and by the European Commission (DG XII), through research contract number EVG1-CT-1999-00010 (Systematic, Palaeoflood and Historical data for the improvement of flood Risk Estimation, "SPHERE" Project). I am very grateful to Ken Gregory, Chris Gregory and Varyl Thorndycraft for the critical review of the original manuscript, and for his very useful comments and suggestions.

REFERENCES

Abbott, J. and Valastro, S., 1995. The Holocene alluvial records of the chorai of Metapontum, Basilicata and Croton, Calabria, Italy. In J. Lewin, M.G. Macklin and J.C. Woodward (eds), *Mediterranean Quaternary River Environments*. Balkema, Rotterdam, 195–205.

Allen, J.R., Huntley, B. and Watts, W.A., 1996. The vegetation and climate of north-west Iberia over the last 14000 yr. *Journal of Quaternary Science*, **11**, 125–147.

Alley, R.B., Mayewski, P.A., Sowers, T., Stuiver, M., Taylor, K.C. and Clark, P.U., 1997. Holocene climatic instability: a prominent, widespread event 8200 yr ago. *Geology*, **25**, 483–486.

Alonso, A. and Garzón, G., 1994. Quaternary evolution of a meandering gravel bed river in Central Spain. *TerraNova*, **6**, 456–475.

Arnaud-Fassetta, G., 2002. Geomorphological records of a 'flood-dominated' regime in the Rhône Delta (France) between the 1st century BC and the 2nd century AD. What correlations with the catchment paleohydrology? *Geodinamica Acta*, **15**, 79–92.

Arnaud-Fassetta, G. and Provansal, M., 1999. High frequency variations of water flux and sediment discharge during the Little Ice Age (1586–1725 AD) in the Rhône Delta (Mediterranean France). Relationship to the catchment basin. *Hydrobiologia*, **410**, 241–250.

Astaras, T.A. and Sotiriadis, L., 1988. The evolution of the Thessaloniki-Giannitsa plain in Northern Greece during the last 2500 years – from Alexander the Great era until today. In G. Lang and C. Schlüchter (eds), *Lake, Mire and River Environments*, Balkema, Rotterdam, 105–114.

Aura, J.E., Villaverde, V., González Morales, M., González Sainz, C., Zilhão, J. and Straus, L.G., 1998. The Pleistocene-Holocene transition in the Iberian Peninsula: continuity and change in human adaptations. *Quaternary International*, **49/50**, 87–103.

Baldetti, E., Grimaldi, F., Moroni, M., Compagnucci, M. and Natali, A., 1983. *Le basse valli del Musone e del Potenza nel Medioevo*. Archivio Storico Santa Casa, Loreto, 94.

Ballais, J.L. and Crambes, A., 1992. Morphogenèse holocène, géosystèmes et anthropisation sur la Montagne Sainte-Victoire. *Méditerranée*, **1–2**, 29–41.

Ballais, J.L. and Heddouche, A., 1991. Bas Sahara septentrional et Grand Erg Oriental. In *Paléomilieux et peuplements préhistoriques sahariens au Pléistocène supérieur*, Publication of IGCP Project 252, Solignac, 1–21.

Ballais, J.L., 1991. Évolution holocène de la Tunisie saharienne et présaharienne. *Méditerranée*, **4**, 31–38.

Ballais, J.L., 1995. Alluvial Holocene terraces in eastern Maghreb: climate and anthropogenical controls. In J. Lewin, M.G. Macklin and J.C. Woodward (eds), *Mediterranean Quaternary River Environments*, Balkema, Rotterdam, 183–194.

Barker, G.W. and Hunt, C.O., 1995. Quaternary valley floor erosion and alluviation in the Biferno Valley, Molise, Italy: the role of tectonics, climate, sea level change and human activity. In J. Lewin, M.G. Macklin and J.C. Woodward (eds), *Mediterranean Quaternary River Environments*, Balkema, Rotterdam, 145–157.

Bartov, Y., Stein, M., Enzel, Y., Agnon, A. and Reches, Z., 2002. Lake levels and sequence stratigraphy of Lake Lisan, the Late Pleistocene precursor of the Dead Sea. *Quaternary Research*, **57**, 9–21.

Beckinsale, R.P., 1969. River regimes. In R.J. Chorley (ed.), *Water, Earth and Man*, Methuen, London, 455–471.

Begin, Z.W., Broecker, W., Buchbinder, B., Druckman, Y., Kaufman, A., Magaritz, M. and Neev, D., 1985. Dead Sea and Lake Lisan levels in the last 30,000 years: a preliminary report. *Israel Geological Report*, **29/85**, 1–18.

Benito, G., Machado, M.J. and Pérez-González, A., 1996. Climate change and flood sensitivity in Spain. In J. Branson, A.G. Brown and K.J. Gregory (eds), *Global Continental Changes: The Context of Palaeohydrology*, Geological Society of London, Special Publication No. 115, London, 85–98.

Benito, G., Machado, M.J., Pérez-González, A. and Sopeña, A., 1998. Palaeoflood analysis of the Tagus River (Central Spain). In G. Benito, V.R. Baker and K.J. Gregory (eds), *Palaeohydrology and Environmental Change*. John Wiley, Chichester, 317–333.

Benito, G., Sopeña, A., Sánchez-Moya, Y., Machado, M.J. and Pérez-González, A., 2002. Palaeoflood magnitude and frequency in the context of the Late Pleistocene-Holocene climatic changes (Tagus River, Central Spain). In M.B. Ruiz Zapata (ed.), *Quaternary Climatic Changes and Environmental Crises in the Mediterranean Region*, Universidad de Alcalá de Henares, in press.

Bintliff, J.L., 1975. Mediterranean alluviation: new evidence from archaeology. *Proceedings of the Prehistoric Society*, **41**, 78–84.

Bintliff, J.L., 1982. Climatic change, archaeology and Quaternary science in the eastern Mediterranean region. In A. Harding (ed.), *Climatic Change in Later Pre-History*, Edinburgh University Press, UK 143–161.

Biondi, E. and Coltorti, M., 1982. The Ensino flood plain during the Holocene. *Abstr. 11th INQUA Congr*, Vol. 3, 45. Moscow.

Bond, G.C., Heinrich, H., Broecker, W., Labeyrie, L., McManus, J., Andrews, J., Huon, S., Jantschik, B., Clasen, S., Simet, C., Tedesco, K., Klas, K., Bonani, G. and Ivy, S., 1992. Evidence for massive discharges of icebergs into the North Atlantic ocean during the last glacial period. *Nature*, **360**, 245–249.

Bravard, J.P., 1989. La Metamorphose des rivieres des Alpes Francaises a la fin du moyen-age et a l'epoque moderne. *Bulletin de la Société Géographique de Liège*, **25**, 145–157.

Briand, F. and Maldonado, A. (eds), 1997. *Transformations and evolution of the Mediterranean coastline*. CIESM Science Series 3, Bulletin de l'Institut Océanographique, Monaco, Special No. 18, 249.

Broecker, W.S., 1998. Paleocean circulation during the last deglaciation: a bipolar seasaw? *Paleoceanography*, **13**, 119–121.

Brown, A.G., 1998. Fluvial evidence of the Medieval Warm Period and the Late Medieval climate deterioration in Europe. In G. Benito, V.R. Baker and K.J. Gregory (eds), *Palaeohydrology and Environmental Change*. John Wiley, Chichester, 43–52.

Brückner, H., 1986. Man's impact on the evolution of the physical environment in the Mediterranean Region in historical times. *GeoJournal*, **13**(1), 7–17.

Brun, A., 1989. Microflores et paléovégétations en Afrique du Nord depuis 30,000 ans. *Bulletin de la Société Géologique de France*. **8**, V, 1, 25–33.

Bruneton, H., Arnaud-Fassetta, G., Provansal, M. and Sistach, D., 2001. Geomorphological evidence for fluvial change during the Roman period in the lower Rhône valley (southern France). *Catena*, **45**, 287–312.

Burjachs-Casas, F., Rodó, X. and Comín, F.A., 1996. Gallocanta: ejemplo de secuencia palinológica en una laguna efímera. In B. Ruiz-Zapata (ed.), *Estudios Palinológicos*, XI Simposio de Palinología, Universidad de Alcalá, Alcalá de Henares, 25–29.

Burke, K.A., Durotoye, A.B. and Whiteman, A.J., 1971. A Dry Phase south of the Sahara 20,000 years ago. *West African Journal of Archaeology*, **1**, 1–8.

Butzer, K.W., Miralles, I. and y Mateu, J.F., 1983. Urban geo-archaeology in Medieval Alzira (Prov. Valencia, Spain). *Journal of Archaeological Science*, **10**, 333–349.

Butzer, K.W., 1974. Accelerated soil erosion. In I. Manners and M.W. Mikesell (eds), *Perspectives on Environment*, Association of American Geographers, Washington, DC, 57–78.

Butzer, K.W., 1975. Pleistocene littoral-sedimentary cycles of the Mediterranean Basin: a Mallorquin view. In K.W. Butzer and G.L. Isaac (eds), *After the Australopithecines*, Mouton, The Hague, 25–71.

Butzer, K.W., 1980. Holocene alluvial sequences: problems of dating and correlation. In R.A. Cullingford, D.A. Davidson and J. Lewin (eds), *Timescales in Geomorphology*, John Wiley, Chichester, 131–142.

Carmona González, P., 1995. Niveles morfogenéticos cuaternarios en los sistemas fluviales de la depresión valenciana. In *El Cuaternario del País Valenciano*, Universitat de Valencia AEQUA, Valencia, 97–104.

Cheddadi, R., Guiot, J. and Lamb, H.F., 1997. Holocene climatic change in Morocco: a quantitative reconstruction from pollen data. *Climate Dynamics*, **14**, 883–890.

Clark, J.D., 1980. Human populations and cultural adaptations in the Sahara and the Nile during prehistoric times In M.A.J. Williams and H. Faure (eds), *The Sahara and the Nile*, Balkema, Rotterdam, 527–585.

COHMAP Members, 1988. Climatic changes in the last 18,000 years: observations and model simulations. *Science*, **241**, 1043–1052.

Coltorti, 1991. Modificazioni morfologiche oloceniche melle piane alluvionali marchigiane: alcuni esempi nei fiumi Misa, Cesano e Musone. *Geografia Fisica e Dinamica Quaternaria*, **14**(1), 73–86.

Coltorti, M., 1997. Human impact in the Holocene fluvial and coastal evolution of the Marche region, Central Italy. *Catena*, **30**, 311–335.

Cremaschi, M., Marchetti, M. and Ravazzi, C., 1994. Geomorphological evidence for land surfaces cleared from forest in central Po plain (northern Italy) during the Roman Period. In B. Frenzel, L. Reisch and M.M. Weiß (eds), *Evaluation of Land Surfaces Cleared from Forest in the Mediterranean Region during the Time of the Roman Empire*, Fischer, Stuttgart, 119–132.

Cuenca Payá, A. and Walker, M.J., 1995. Terrazas fluviales en la zona bética de la Comunidad Valenciana. *El Cuaternario del País Valenciano*, Universitat de Valencia & AEQUA, Valencia, 105–114.

Dansgaard, W., Johnsen, S.J., Clausen, H.B., Dahl-Jensen, D., Gundestrup, N.S., Hammer, C.U., Hvidberg, C.S., Steffensen, J.P., Sveinbjörnsdottir, A.E., Jouzel, J. and Bond, G., 1993. Evidence for general instability of past climate from a 250-kyr ice-core record. *Nature*, **364**, 218–220.

Davis, B.A.S., 1994. *Paleolimnology and Holocene Environmental Change from Endorheic Lakes in the Ebro Basin, North-East Spain*. Ph.D. Thesis, University of Newcastle Upon Tyne, Tyne, 317.

Delibrias, G., Rognon, P. and Weisrock, A., 1976. Datation de plusieurs épisodes à 'limons roses' dans le quaternaire récent de l'Atlas Atlantique Marocain. *Comptes rendus de l'Académie des Sciences de Paris*, **282**, 593–596.

DeMenocal, P., Ortiz, J., Guilderson, T., Adkins, J., Sarnthein, M., Baker, L. and Yarusinski, M., 2000. Abrupt onset and termination of the African Humid Period: rapid climate responses to gradual insolation forcing. *Quaternary Science Reviews*, **19**, 347–361.

Demitrack, A., 1986. *The Late Quaternary Geologic History of the Larissa Plain, Thessaly, Greece: Tectonic, Climatic and Human Impact on the Landscape*. Unpublished Ph.D. Dissertation, Stanford University, Stanford.

Devereux, C.M., 1982. Climate speeds erosion of the Algarve's valleys. *The Geographical Magazine*, **54**, 10–17.

Farrand, W.R., Stearns, C.H. and Jackson, H.E., 1982. Environmental setting of Capsian and related occupations in the high plains of eastern Algeria. *Geological Society of America*, **14**(7), 487.

Frumkin, A., Magaritz, M., Carmi, I. and Zak, I., 1991. The Holocene climatic record of the salt caves of Mount Sedom, Israel. *The Holocene*, **1**(3), 191–200.

Fuller, I.C., Macklin, M.G., Lewin, J., Passmore, D.G. and Wintle, A.G., 1998. River response to high-frequency climate oscillations in southern Europe over the Past 200 k.y. *Geology*, **26**(3): 275–278.

Fuller, I.C., Macklin, M.G., Passmore, D.G., Brewer, P.A., Lewin, J. and Wintle, A.G., 1996. Geochronologies and environmental records of Quaternary fluvial sequences in the Guadalope basin, northeast Spain, based on luminescence dating. In J. Branson, A.G. Brown and K.J. Gregory (eds), *Global Continental Changes: The Context of Palaeohydrology*, Special Publication No. 115, Geological Society of London, London, 99–120.

Gasse, F. and Van Campo, E., 1994. Abrupt post glacial climatic events in West Asia and North Africa monsoon domains. *Earth and Planetary Science Letters*, **126**, 435–456.

Gasse, F., Téhet, R., Durand, A., Gibert, E. and Fontes, J.Ch., 1990. The arid-humid transition in the Sahara and the Sahel during the last deglaciation. *Nature*, **346**, 141–156.

Gat, J.R. and Magaritz, M., 1980. Climatic variation in the eastern Mediterranean Sea area. *Naturuissenschaften*, **67**, 80–87.

Geyh, M.A., 1994. The paleohydrology of the eastern Mediterranean. In O. Bar-Yosef and R.S. Kra (eds), *Late Quaternary Chronology and Paleoclimates of the Eastern Mediterranean*, Radiocarbon, Tuscon, 131–145.

Grimm, E.C., Jacobson, G.L., Watts, W.A., Hansen, B.C.S. and Maasch, K.A., 1993. A 50,000-year record of climate oscillations from Florida and its temporal correlation with the Heinrich Events. *Science*, **261**, 198–200.

Grove, A.T., 2001. The 'Little Ice Age' and its geomorphological consequences in Mediterranean Europe. *Climatic Change*, **48**, 121–136.

Grove, J.M., 1988. *The Little Ice Age*. Routledge, London, 1–498.

Guilcher, A., 1979. *Précis d'hydrologie marine et continental*. Masson, Paris, 344.

Guiot, J., Harrison, S.P. and Prentice, I.C., 1993. Reconstruction of Holocene precipitation patterns in Europe using pollen and lake-level data. *Quaternary Research*, **40**, 139–149.

Gutiérrez-Elorza, M. and Peña-Monné, J.L., 1998. Geomorphology and late Holocene climatic change in Northeastern Spain. *Geomorphology*, **23**, 205–217.

Harrison, S.P. Saarse, L. and Digerfeldt, G., 1991. Holocene changes in lake levels as climate proxy data in Europe. In B. Frenzel, A. Pons, and B. Gläser, (eds), *Evaluation of Climate Proxy Data in Rrelation to the European Holocene*, Gustard Fischer Verlag, Stuttgart, 159–169.

Heinrich, H., 1988. Origin and consequences of cyclic ice rafting in the northeast Atlantic Ocean during the past 130,000 years. *Quaternary Research*, **29**, 142–152.

Holzhauser, H., 1997. Fluctuations of the Gosser Aletsch Glacier and the Gorner Glacier during the last 3200 years: new results. In B. Frenzel, G.S. Boulton, G. Gläser and U. Huckriedge (eds), *Glacier Fluctuations during the Holocene*, Fischer, Stuttgart, 35–58.

Huntley, B.B. and Birks, H.J.B., 1983. *An Atlas of Past and Present Pollen Maps for Europe: 0–13,000 Years Ago*. Cambridge University Press, Cambridge, 1–667.

Jalut, G., Esteban Amat, A., Bonnet, L., Gauquelin, T. and Fontugne, M., 2000. Holocene climatic changes in the Western Mediterranean, from south-east France to south-east Spain. *Palaeogeography, Palaeoclimatology, Palaeoecology*, **160**, 255–290.

Jameson, M.H., Runnels, C.N. and van Andel, T.H., 1994. *A Greek Countryside: The Southern Argolid from Prehistory to the Present Day*. Stanford University Press, Stanford, 676.

Jorda, M., 1985. La torrentialité holocène des Alpes françaises du sud. Facteurs anthropiques et parameters naturels de son evolution. *Cahiers ligures de Préhistoire*, N.S., **2**, 49–70.

Jorda, M., 1992. Morphogenèse et fluctuations climatiques dans les Alpes françaises du sud de l'Age du Bronze au haut Moyen Age. *Les Nouvelles de l'Archaeolie*, **50**, 14–21.

Jorda, M., Parron, C., Provansal, M. and Roux, M., 1991. Erosion et détritisme holocènes en Basse Provence calcaire. L'impact de l'anthropisation. *Physio-Géo*, **22/23**, 37–47.

Jorda, M. and Provansal, M., 1990. Terrasses de culture et bilan érosifen región méditerranéene. Le bassin-versant du Vallat de Monsieur (Basse-Provence). *Méditerranée*, **3–4**, 55–61.

Jorda, M. and Provansal, M., 1996. Impacts de l'anthropisation et du climat sur le détritisme en France du sud-est (Alpes du sud et Provence). *Bulletin de la Societé Géologique de France*, **167**(1), 159–168.

Juliá, R., Guiralt, S., Burjachs, F., Roca, J.R. and Wansard, G., 1998. Short climate events in the Mediterranean Iberian Peninsula during the Lateglacial and the Early Holocene transition. *Terra Nostra*, **98**(6), 65–69.

Knox, J.C. 1984. Fluvial responses to small scale climate changes. In J.E. Costa and P.J. Fleisher (eds), *Developments and Applications of Geomorphology*, Springer-Verlag, Berlin, 318–342.

Lamb, H.F., Eicher, U. and Switsur, V.R., 1989. An 18,000-year record of vegetational, lake level and climate change from the Middle Atlas, Morocco. *Journal of Biogeography*, **16**, 65–74.

Lamb, H.H., 1971. Climates and circulation regimes developed over the northern hemisphere during and since the Last Ice Age. *Palaeogeography, Palaeoclimatology, Paleoecology*, **10**, 125–162.

Le Roy Ladurie, E., 1967. *Historie du climat depuis l'an mil*, Flammarion, Paris, 2nd ed. 1983.

Lewin, J., Macklin, M.G. and Woodward, J.C., 1991. Late Quaternary fluvial sedimentation in the Voidomatis Basin, Epirus, Northwest Greece. *Quaternary Research*, **35**, 103–115.

Lewin, J., Macklin, M.G. and Woodward, J.C., 1995. Mediterranean Quaternary river environments-some future research needs. In J. Lewin, M.G. Macklin and J.C. Woodward (eds), *Mediterranean Quaternary River Environments*, Balkema, Rotterdam, 283–284.

López-Avilés, A., Ashworth, P.J. and Macklin, M.G., 1998. Floods and Quaternary sedimentation style in a bedrock-controlled reach of the Bergantes River, Ebro Basin, Northeast Spain). In G. Benito, V.R. Baker and K.J. Gregory (eds), *Palaeohydrology and Environmental Change*. John Wiley, Chichester, 181–196.

Macklin, M.G., Fuller, I.C., Lewin, J., Maas, G.S., Passmore, D.G., Rose, J., Woodward, J.C., Black, S., Hamlin, R.H.B. and Rowan, J.S., 2002. Correlation of fluvial sequences in the Mediterranean basin over the last 200 ka and their relationship to climate change. *Quaternary Science Reviews*, **21**, 1633–1641.

Macklin, M.G., Lewin, J. and Woodward, J.C., 1995. Quaternary fluvial systems in the Mediterranean basin. In J. Lewin, M.G. Macklin and J.C. Woodward (eds), *Mediterranean Quaternary River Environments*, Balkema, Rotterdam, 1–25.

Macklin, M.G., Passmore, D.G., Stevenson, A.C., Davis, B.A. and Benavente, J.A., 1994. Responses of rivers and lakes to Holocene environmental change in the Alcañiz region, Teruel, north-east Spain. In A.C. Millington and K. Pye (eds), *Environmental Change in Drylands: Biogeographical and Geomorphological Perspectives*, John Wiley, Chichester, 113–130.

Masachs, V., 1950. Aportación al conocimiento del régimen fluvial mediterráneo. *Comptes Rendus du Congrès International de Geógraphie, Union Géographique Internationale. Lisbonne 1949*. Tome II. 359–390.

Neboit, R., 1984. Genèse des terrasses fluviatiles holocènes en Sicile et en Italie Méridionale. *Bulletin de l'Association française pour l'étude du Quaternaire*, **1-2-3**, 157–160.

Nesci, O. and Savelli, D., 1986. Cicli continentali tardo-quaternari lungo i tratti vallivi mediani delle Marche settentrionali. *Geografia Fisica e Dinamica Quaternaria*, **9**, 192–211.

Ouda, B., Zouari, K., Ouezdou, H.G., Chkir, N. and Causse, C., 1998. Nouvelles données paléo-environmentales pour le Quaternaire recent en Tunisie centrale (basin de Maknassy). *Comptes rendus de l'Académie des Sciences de Paris*, (Sciences de la terre et des planètes), **326**, 855–861.

Page, W.D., 1972. The geological setting of the archaeological site at oued el Akarit and the palaeoclimatic significance of gypsum soils (S-Tunisia) Unpublished Thesis. University of Colorado, CO.

Pardé, M., 1950. Sur les régimes fluviaux méditerranées. *Comptes Rendus du Congrès International de Geógraphie, Union Géographique Internationale. Lisbonne 1949*. Tome II. 391–420.

Pérez-Obiol, R. and Juliá, R., 1994. Climatic Change on the Iberian Peninsula recorded in a 30,000-yr pollen record from Lake Banyoles. *Quaternary Research*, **41**, 91–98.

Petit-Maire, N., Burollet, P.F., Ballais, J.L., Fontugne, M., Rosso, J.C. and Lazaar, A., 1991. Paléoclimats holocènes du Sahara septentrional. Dépôts lacustres et terrasses alluviales en bordure du Grand Erg Oriental à l'extreme-sud de la Tunesie. *Comptes rendus de l'Académie des Sciences de Paris*, **312**, série II, 1661–1666.

Pons, A. and Reille, M., 1988. The Holocene-and Upper Pleistocene pollen record from Padul (Granada, Spain): a new study. *Palaeogeography, Palaeoclimatology, Palaeoecology*, **66**, 243–263.

Pope, K.O. and van Andel, T.H., 1984. Late Quaternary alluviation and soil formation in the southern Argolid: its history, causes and archaeological implications. *Journal of Arqueological Sciences*, **11**, 281–306.

Prentice, I.C., Guiot, J. and Harrison, S.P., 1992. Mediterranean vegetation, lake levels and palaeoclimate at the last glacial maximum. *Nature*, **360**, 658–660.

Provansal, M. 1995a. Holocene sedimentary sequences in the Arc River delta and the Etang de Berre in Provence, southern France. In J. Lewin, M.G. Macklin and J.C. Woodward, (eds), *Mediterranean Quaternary River Environments*, Balkema, Rotterdam, 159–165.

Provansal, M. 1995b. The role of climate in landscape morphogenesis since the Bronze Age in Provence, southeastern France. *The Holocene*, **5**(3), 348–353.

Provansal, M., 1992. Le rôle du climat dans la morphogénèse à la fin de l'Age du Fer et dans l'Antiquité en Basse Provence. *Les nouvelles de l'Archéologie*, **50**, 21–26.

Ritchie, J.C., Eyles, C.H., Haynes, C.V., 1985. Sediment and pollen evidence for an early to mid Holocene humid period in the eastern Sahara. *Nature*, **314**, 352–355.

Roberts, N., 1998. The Holocene. *An Environmental History*. Blackwell, Oxford, 316.

Roberts, N., Lamb, H.F., El Hamouti, N. and Barker, P., 1994. Abrupt Holocene hydroclimatic events: palaeolimnological evidence from North-West Africa. In A.C. Millington and K. Pye (eds), *Environmental Change in Drylands. Biogeographical and Geomorphological Perspectives*, John Wiley, Chichester, 165–175.

Roca, J.R. and Juliá, R., 1997. Late Glacial and Holocene climatic changes and desertification expansion based on biota content in the Salines sequence, Southeastern Spain. *Geobios*, **30**, 823–830.

Rognon, P., 1984. Signification dynamique et climatique des formations et terrasses fluviatiles en Africa du Nord et au proche Orient. *Bulletin de l'Association française pour l'etude du Quaternaire*, **1-2-3**, 161–169.

Rognon, P., 1987. Aridification and abrupt climate events on the Saharan northern and southern margins, 20,000 years BP to present. In W.H. Berger and C.D. Labeyrie (eds), *Abrupt Climatic Change: Evidence and Implications*. Reidel, Dordrecht, 209–220.

Rose, J. and Meng, X. 1999. River activity in small catchments over the last 140 ka in North-east Mallorca. In T. Quine and A.J. Brown (eds), *Fluvial Processes and Environmental Change*, Wiley, Chichester, 91–102.

Schulte, L., 2002. Climatic and human influence on river systems and glacier fluctuations in southeast Spain since the Last Glacial Maximum. *Quaternary International*, **93–94**, 85–100.

Schumm, S.A., 1977. *The Fluvial System*. Wiley, New York, 338

Servant, M., 1973. *Séquences continentals et variations climatiques: Evolution du basin du Tchad au Cénozoique supérieur*. Thèse d'État, ORSTOM, Paris.

Shackleton, J.C. and van Andel, T.H., 1985. Late palaeolithic and mesolithic coastlines of the western Mediterranean. *Cahiers Ligures de Préhistoire et Protoprehistoire*, **2**, 7–19.

Starkel, L., 1999. 8500-8000 yrs BP Humid Phase- Global or Regional? Science Reports of Tohoku University, 7th Series *Geography*, **49**(2), 105–133.

Starkel, L., 2002. Change in the frequency of extreme events as the indicator of climatic change in the Holocene (in fluvial systems). *Quaternary International*, **91**, 25–32.

Stocker, T.F., 1999. Abrupt climate changes: from the past to the future- a review. *International Journal of Earth Sciences*, **88**, 365–374.

Street, F.A. and Grove, A.T., 1976. Environmental and climatic implications of late Quaternary lake-level fluctuations in Africa. *Nature*, **261**, 385–390.

Stuiver, M. and Reimer, P.J., 1993. Extended 14C database and revised CALIB radiocarbon. *Radiocarbon*, **35**, 215–230.

Swezey, C., Lancaster, N., Kocurek, G., Deynoux, M., Blum, M., Price, D. and Pion, J.C., 1999. Response of Aeolian systems to Holocene climatic and hydrologic changes on the northern margin of the Sahara: a high-resolution record from the Chott Rharsa basin, Tunisia. *The Holocene*, **9**(2), 141–147.

Valero-Garcés, B., Zeroual, E. and Kelts, K., 1998. Arid Phases in the western Mediterranean region during the Last Glacial Cycle reconstructed from lacustrine records. In G. Benito, V.R. Baker and K.J. Gregory, (eds), *Palaeohydrology and Environmental Change*, Wiley, London, 67–80.

Van Andel, Tj.H. and Zangger, W., 1990. Landscape stability and destabilization in the prehistory of Greece. In S. Bottema, G. Entjes-Nieborg and V. Van Zeist (eds), *Man's Role in the Shaping of the Eastern Mediterranean Landscape*. Balkema, Rotterdam, 139–157.

Van der Brink, L.M. and Janssen, C.R., 1985. The effects of human activities during cultural phases on the development of montane vegetation in the Serra de Estrela, Portugal. *Review of Paleobotany and Palynology*, **44**, 193–215.

Van der Knaap, W.O. and van Leeuwen, J.F.N., 1995. Holocene vegetation succession and degradation as responses to climatic change and human activity in the Serra de Estrela, Portugal. *Review of Paleobotany and Palynology*, **89**(3–4), 153–211.

Van Geel, B., Van der Plicht, J., Kilian, M.R., Klaver, E.R., Kouwenberg, J.H.M., Ressen, H., Reynaud-Farrera, I. and Waterbolk, H.T., 1998. The sharp rise of δ ^{14}C at ca. 800 cal BC. Possible causes, related climatic teleconnections and the impact on human environments. *Radiocarbon*, **40**, 335–350.

Veggiani, A., 1983. Degrado ambientale e dissesti idrogeologici indotti dal deterioramento climático nell'Alto medioevo in Italia: I casi riminesi. *Studi Romagnoli*, **34**, 123–146.

Vita-Finzi, C., 1969. *The Mediterranean Valleys*. Cambridge University Press, Cambridge, 140.

Weisrock, A., 1998. The upper Soltanian of S.W. Morocco. *Abstracts of the Third FLAG Conference*, September 15–19, Cheltengam, UK.

Weisrock, A., Delibrias, G., Rognon, P. and Coudé-Gaussen, G., 1985. Variations climatiques et morphogenèse au Maroc atlantique (30–33° N) à la limite Pléistocene-Holocene. *Bulletin de la Societé Géologique de France*, **8**(4), 565–569.

Wigley, T.M.L. and Farmer, G., 1982. Climate of the eastern Mediterranean and the Near East. In J.L. Bintliff and W. van Zeist (eds), *Palaeoclimates, Palaeoenviroments and Human Communities in the Eastern Mediterranean Region in Late Prehistory*, British Archaeological Reports (International Series) 133, Oxford, 3–37.

Woodward, J.C., Hamlin, R.H.B., Macklin, M.G., Karkanas, P. and Kotjabopoulou, P., 2001. Quantitative sourcing of slackwater deposits at Boila rockshelter: a record of late-glacial flooding and Palaeolithic settlements in the Pindus Mountains, Northern Greece. *Geoarchaeology*, **16**, 501–536.

Yechieli, Y., Magaritz, M., Levy, Y., Weber, U., Kafri, U., Woelfli, W. and Bonani, G., 1993. Late Quaternary geological history of the Dead Sea area, Israel. *Quaternary Research*, **39**, 59–67.

Zouari, K., 1988. *Géochimie et sédimentologie des dépôts continentaux d'origine aquatique du Quaternaire supérieur du Sud Tunisien: Interprétations paléohydrologiques et paléoclimatiques*, Thèse, Université Paris XI-Orsay, Paris, 216.

10 North American Palaeofloods and Future Floods: Responses to Climatic Change

J.C. KNOX
University of Wisconsin, Madison, Wisconsin, USA

1 INTRODUCTION

Because climate is the principal driving force in hydrologic systems, even modest climate changes have the potential to significantly influence changes in hydrologic activity, especially for extreme events. Palaeohydrologic information obtained from the geologic record extends the understanding of hydrologic responses to climate changes far beyond available instrumental records and thereby provides a framework for a better anticipation of hydrologic responses to possible direct and indirect effects of potential future climate changes. The relatively short durations of most instrumental records hamper their effectiveness for such endeavors. Although separating individual effects of climate change and land use associated with floods identifiable in the stratigraphic record can be difficult in some world regions, for much of North America exclusive of the southwest and Mexico, significant human influences on flood hydrology are restricted to the last 200 to 300 years (Knox, 2001; 2002). This chapter first considers evidence that ongoing and anticipated twenty-first century global climate change may be associated with significant changes in extreme hydrologic events, and then reviews palaeohydrologic evidence that illustrates how extreme hydrologic events responded to climatic changes during the Holocene.

The effects of even relatively minor changes in climate on flooding and stream flow are suggested but not always clear in North America's instrumental records. Ashmore and Church (2001) concluded that responses to climate changes associated with global warming in Canada would result in shifts in precipitation amounts, intensities of cyclonic storms, the proportion of precipitation falling as rain, shifts in glacier-mass balance, and reductions in the extent of permafrost. The Saint John River, New Brunswick, Canada, has experienced more frequent mid-winter jams and higher April flows, and it has been suggested that both these occurrences are because of global warming (Beltaos and Burrell, 2002). Climatic and hydrologic instrumental records for many areas of North America indicate that the mean and variance of climatic and hydrologic time series have changed during the late twentieth century, and these changes may be related to accelerated global warming (Angel and Huff, 1997; Brown and Braaten, 1998; Burn, 1994; Easterling *et al.*, 2000; Groisman *et al.*, 2001; Karl *et al.*, 1996; Karl and Knight, 1998; Knapp, 1994; Knox, 1983; 2000; Lins and Slack, 1999; Olsen *et al.*, 1999; Woodhouse and Overpeck, 1998). Given the relative short life of instrumental records for a basis of comparison, it is worth noting that a reconstructed

Palaeohydrology: Understanding Global Change. Edited by K.J. Gregory and G. Benito
© 2003 John Wiley & Sons, Ltd ISBN: 0-470-84739-5

record of the northern hemisphere mean temperature spanning six centuries shows that the magnitude of temperature rise since 1900 is unprecedented since at least AD 1400 (Mann et al., 1998). The dominant factor accounting for the anomalous temperature rise was attributed to rising levels of atmospheric greenhouse gases. Since hemispheric-scale patterns of storm tracks and frontal boundaries between unlike air masses are highly sensitive to hemispheric temperature gradients, the late twentieth century accelerated temperature rise presents a number of concerns about potential impacts on hydrologic systems. An excellent review of past research involving potential impacts of global warming on hydrologic processes, including extreme events, in North America and elsewhere is given in the Intergovernmental Panel on Climate Change report on regional impacts of climate change (Watson et al., 1998).

Climate model simulations suggest that global warming should produce an intensification of the hydrological cycle leading to more extremes of floods and droughts in the middle-latitude regions such as typified by North America (Houghton et al., 1996). A number of studies in the United States have detected changes in stream flow that tend to support the model predictions. For example, Lins and Michael (1994) concluded that stream flow has increased for much of the conterminous United States since the early 1940s, and Lettenmaier et al. (1994) showed that stream flow during the months of November to April had particularly increased in north-central United States. Lins and Slack (1999) concluded that various subregions of the United States have experienced statistically significant trends in flow that could be linked to hemispheric changes in ocean-atmosphere circulation, although the authors were uncertain whether more floods and droughts might be anticipated with further global warming expected in the coming decades. A related investigation in which the effects of spatial correlation among observational sites in the database was removed concluded that historical flood records in the United States do not exhibit significant upward or downward trends (Vogel et al., 2001). Such differences of conclusions reflect problems associated with limitations of short-duration instrumental records. Geologic records, on the other hand, offer long-term histories in which hydrologic responses to environmental change are more clearly defined (Jarrett and Tomlinson, 2000).

2 MODERN FLOOD RESPONSES TO LARGE-SCALE ATMOSPHERIC CIRCULATION CHANGES

Long, century-scale instrument records support the idea that extreme floods tend to be more sensitive to climatic change than high-frequency low-magnitude floods (Knox, 1983; 2000). The instrument record of annual maximum floods on the Upper Mississippi River (UMR), where an anomalous high frequency of large floods has occurred since about 1950, illustrates this relationship (Figure 10.1) (Knox, 1983; 1999; 2000; Knapp, 1994; Olsen et al., 1999). Independent evidence of changes in atmospheric circulation patterns was used to partition a 123-year-long annual flood series of the UMR at Winona, Minnesota into five climate episodes. The bracketing ages of 1895, 1920, 1950, and 1980 are the approximate times that have been associated with changes in hemispheric-scale ocean-atmosphere circulation regimes (Lamb, 1966; 1977; Kutzbach, 1970; Kalnicky, 1974; Knox et al., 1975 and McPhaden, 1999). The means and variances for annual maximum flood episodes were computed on logarithm-transformed data and then converted back for plotting on Figure 10.1. Analysis of variance and difference of mean tests for changes between episodes showed that only the episode between 1920 and 1949 was significantly different from the other groups. However,

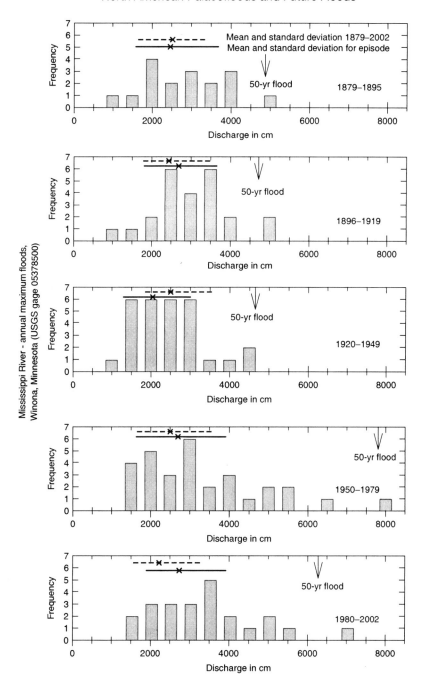

Figure 10.1 Magnitude and frequency of annual maximum floods on the Upper Mississippi River at Winona, Minnesota: 1879 to 2002. The temporal divisions are based on independent climate studies that suggest changes in hemispheric-scale patterns of atmospheric circulation that occurred about 1895, 1920, 1950, and about 1980 (Lamb, 1966; 1977; Kutzbach, 1970; Kalnicky, 1974; Knox *et al.*, 1975 and McPhaden, 1999). The magnitude of a flood of 50-year recurrence probability was computed for each episode using the Log–Pearson Type-III methodology (United States Water Resources Council, 1981. Data source for floods: US Geological Survey)

these standard statistical methods are not sufficiently sensitive to reflect the prominent differences in the numbers of extreme floods between episodes due to their overall low frequencies relative to the more common floods of low magnitude. A Log–Pearson Type-III flood-frequency methodology was used to compute the magnitudes of respective floods expected once in 50 years for the episodes (United States Water Resources Council, 1981). Magnitudes of the computed 50-year floods ranged between about 4,500 and 5,000 m^3s^{-1} prior to 1950, but have increased to 6,500 to 8,000 m^3s^{-1} since 1950 (Figure 10.1). Meanwhile, average flood magnitudes changed only moderately between climate episodes (Figure 10.1). During the late twentieth century, the Upper Mississippi Valley (UMV) recorded statistically significant increases in the incidence of extreme rainfalls that have been attributed in part to rapid global warming during the same period (Angel and Huff, 1997; Karl and Knight, 1998). The very large floods on the UMR since about 1950 are mainly the result of excesses of rainfall, snowmelt, or combined rainfall and snowmelt. Changes in land use and other human modifications of the floodplain environment have had only modest impacts on these large floods.

2.1 Examples of Regional Hydrologic Responses to Large-scale Atmospheric Circulation Patterns

Floods are closely linked to patterns of large-scale ocean-atmosphere circulation because they depend on the precipitable water delivered to a region by air masses. The potential water vapor loading capacity of air masses increases exponentially with air temperature at constant pressure, and changes in the seasonal source areas for air masses can have great impacts on regional precipitation, snowmelt, and flooding. Storm tracks and precipitation tend to concentrate along air mass boundaries, which in middle-latitude North America is closely connected to the general location of the middle-latitude jet stream. A change in the prevailing latitudinal position and/or the amplitude and frequency of waves in the jet stream can result in enormous changes in precipitation and runoff and explain in-phase versus out-of-phase hydrologic conditions between regions. Some example relations are shown in Figure 10.2.

The meridional (north–south) circulation pattern of Figure 10.2a represents a configuration that, in summer, normally would involve strong precipitation and flooding in the southwest, Midwest, and northeast, while the southeast would tend to remain hot, humid, and dry. Variations of this pattern characterized much of the spring and summer of 1993 when record flooding occurred on the UMR system. The circulation pattern represented by Figure 10.2b is also characterized by a strong meridional component, but with a configuration that, in summer, normally produces relatively warm or hot and dry conditions in the southwest, while at the same time leads to increased precipitation and flooding in the southeast. Meanwhile, the upper Midwest tends to be anomalously cool and dry with little flooding under the circulation regime of Figure 10.2b. Figure 10.2c is illustrative of a strongly zonal (west–east) westerly circulation. This zonal pattern is normally associated with excessive winter-season precipitation and flooding across the southern part of the continent and on the west coast where a strong onshore flow of moist air is often associated with large late-winter and spring floods that involve rain falling on a snow cover in the uplands. However, in the continental interior where a strong westerly circulation is associated with relatively warm/dry and/or cool/dry airflow that prevents penetration of the region by moist air from more tropical source regions, the climate tends to be relatively dry and floods tend to be fewer and of smaller magnitude. It is worth noting that the

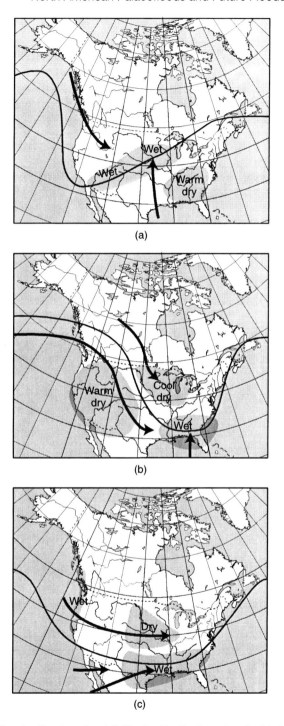

Figure 10.2 Regional climate of midlatitude North America is highly influenced by large-scale patterns in the troposphere's westerly circulation. Rossby waves in the jet stream (solid lines) denote the approximate boundaries of air masses from tropical and polar-source regions and signify locations of the main storm tracks that together strongly influence magnitudes and frequencies of floods

circulation pattern in Figure 10.2c tends to be associated with times of strong El Niño development in the tropical Pacific.

3   HOLOCENE PALAEOHYDROLOGY IN THE SOUTHWEST AND IN UPPER MISSISSIPPI VALLEY

Holocene palaeohydrologic information from North American sites indicates that floods and droughts have experienced episodic shifts involving large changes in magnitudes and recurrence frequencies. Proxy records of past hydrologic activity support a hypothesis that large-scale changes in the locations of seasonal air mass boundaries and storm tracks have greatly influenced regional magnitudes and recurrence frequencies of floods and droughts (Knox, 1983; Hirschboeck, 1987; Hirschboeck et al., 2000; Redmond et al., 2002). The episodic palaeohydrologic activity also implies that future climate changes associated with anticipated global warming are likely to be associated with significant changes in hydrologic activity. Evidence of former floods, the ages of which were estimated by radiocarbon dating, is available in many regions throughout much of North America. Because of space limitations in this volume, only a few examples can be provided to illustrate how palaeohydrologic data are useful for improving the understanding of hydrologic responses to climate change. The selected examples include two relatively long flood chronologies that represent the southwest United States (SW) and the UMV and a late Holocene record of extremely large floods recorded in the Gulf of Mexico marine sediments (Table 10.1). A few other palaeohydrologic data sets are included to examine their potential covariation with the SW, UMV, and the Gulf flood series (Table 10.1). Since all the palaeohydrologic data presented in Table 10.1 have been published separately elsewhere, only summary information will be presented here.

Palaeoflood data for the SW are largely derived from high-water marks that represent maximum elevations of slackwater sediments deposited during former large floods (e.g. Ely and Baker, 1985; Ely et al., 1993; Enzel, 1992; House et al., 2002; O'Connor et al., 1994; Webb et al., 1988 and Webb et al., 2002). Holocene palaeoflood data for the UMV are divided into three categories. Overbank floods on tributary watersheds smaller than about $200\,km^2$ represent the first category in which flood magnitudes were related to water depths competent to transport the largest cobbles or boulders found in Holocene flood deposits (Knox, 1993). Bankfull stage floods of one to two years recurrence probability represent the second category in which flood magnitudes have been estimated from the morphologic dimensions of palaeochannels preserved in the floodplain and terrace alluvium in tributary watersheds that were mostly smaller than about $200\,km^2$ (Knox, 1985; 2000). Overbank floods on the main channel UMR represent the third category in which vertical sedimentological changes in overbank floodplain and terrace alluvium (a form of slackwater deposition) have been used as a proxy for flood characteristics (Figure 10.3). Megafloods into the Gulf of Mexico were identified by vertical stratigraphic changes in sedimentology, planktonic faunal turnovers, and abrupt changes in $\delta\,^{13}C$ (Brown et al., 1999).

Perhaps the most striking phenomenon about the flood histories presented in Table 10.1 is the clustering of floods into discrete episodes that alternate between periods of very frequent recurrences of large floods and periods of very few or no large floods. Ely (1997) summarized many SW palaeoflood studies and concluded that floods were relatively frequent between about 5,800 and 4,200 years ago, but large floods were extremely rare or absent between about 4,200 and 2,400 years ago.

Table 10.1 Findings of palaeohydrology investigations from various North American study sites. All ages are expressed in calendar years BP (years before AD 1950)

Age Cal yr BP	Southwest floods Ely et al., 1993; Ely, 1997; O'Connor et al., 1994 Enzel et al., 1996	Upper Mississippi Valley tributary floods Knox, 1985; 1993, 2000	Upper Mississippi River main channel floods Knox, this paper	Major droughts – NE Gr. Plains and Minnesota (Laird et al., 1996; 1998a; 1998b; Fritz et al., 2000; Woodhouse and Overpeck, 1998; Dean, 1994; 1997; Dean et al., 1996; Valero-Garces et al., 1997; Clark, 1993)	Mississippi River megafloods to the Gulf of Mexico (Brown et al., 1999)	Gulf coast hurricanes (Liu and Fearn, 1993; 2000); southeast US bankfull floods (Leigh and Feeney, 1995)	West coast river flows (Goman and Wells, 2000; Ingram et al., 1996; Schimmelmann et al., 1998; Enzel, 1992; Chatters and Hoover, 1986)
0	600–0: frequent large floods		600–300: large floods	550–450:	300:	950–0: few large hurricanes	300–50: high flow; large floods 350 and 485 so. Calif.
500	850–650: fewer large floods	800–500: large overbank floods more frequent	700–600: small floods	950–750:			
1000	1600–1200: large flood on Colorado River	1000: bankfull floods 10–20% larger than present day	1000–750: moderately large floods	1250–1050:	1200:		930–560: Columbia River floods 3 times more frequent than present day
			1100–1000: small floods, but large flood at 1280				

(continued overleaf)

Table 10.1 *(continued)*

Age Cal yr BP	Southwest floods Ely et al., 1993; Ely, 1997; O'Connor et al., 1994	Upper Mississippi Valley tributary floods Knox, 1985; 1993, 2000	Upper Mississippi River main channel floods Knox, this paper	Major droughts – NE Gr. Plains and Minnesota (Laird et al., 1996; 1998a; 1998b; Fritz et al., 1998b; Woodhouse and Overpeck, 1998; Dean, 1994; Dean et al., 1996; Valero-Garces et al., 1997; Clark, 1993)	Mississippi River megafloods to the Gulf of Mexico (Brown et al., 1999)	Gulf coast hurricanes (Liu and Fearn, 1993; 2000); southeast US bankfull floods (Leigh and Feeney, 1995)	West coast river flows (Goman and Wells, 2000; Ingram et al., 1996; Schimmelmann et al., 1998; Enzel, 1992; Chatters and Hoover, 1986)
1500	2200–0: frequent large floods in general		1500–1300: small floods				1050–750: drought and low runoff from Sierra Nevada Mts.
2000		1800–1500: bankfull floods 5–10% smaller than present day	1800–1500: very large flood or floods; 2050–1950: small floods	1750–1580:	2000:	3600–950: large hurricanes relatively frequent	
2500		3300–1800: bankfull floods equivalent to present day; 3300–0: overbank flood occurrences equivalent to present-day 500-year flood	2500–2200: generally – large floods		2500:		3800–2000: relatively high flows to San Francisco Bay
3000		3000–2500: moderate to large floods			3000:		
3500					3500:		

3900: large flood – Mohave River, So. California

5800–3600: few large hurricanes

4700

4800–4300:

5600–5400:

6250–6000:

3800–2200: no evidence of large floods

5500–3300: overbank floods generally smaller than present-day 50-year flood

5000–3000: generally – small, but highly variable floods; very large flood at about 4700

5500–3300: bankfull floods 20–30% smaller than present-day

7000–5000: moderate floods; larger than 5000–3000, but smaller than 3000–1500

7000–6000: evidence for two extremely large overbank floods

4000

4500

5000
5500
6000

(continued overleaf)

Table 10.1 (continued)

Age Cal yr BP	Southwest floods Ely et al., 1993; Ely, 1997; O'Connor et al., 1994	Upper Mississippi Valley tributary floods Knox, 1985; 1993, 2000	Upper Mississippi River main channel floods Knox, this paper	Major droughts – NE Gr. Plains and Minnesota (Laird et al., 1996; 1998a; 1998b; Fritz et al., 2000; Woodhouse and Overpeck, 1998; Dean, 1994; 1997; Dean et al., 1996; Valero-Garces et al., 1997; Clark, 1993)	Mississippi River megafloods to the Gulf of Mexico (Brown et al., 1999)	Gulf coast hurricanes (Liu and Fearn, 1993; 2000); southeast US bankfull floods (Leigh and Feeney, 1995)	West coast river flows (Goman and Wells, 2000; Ingram et al., 1996; Schimmelmann et al., 1998; Enzel, 1992; Chatters and Hoover, 1986)
6500							
7000							
7500	7500–5500: bankfull floods 5–15% larger than present day			7500–7200:			
8000						9500–5000: bankfull flood 2 times larger than present day	
8500	9500–7500: bankfull floods 10–20% smaller than present day						
9000				9800–8500:			
9500							

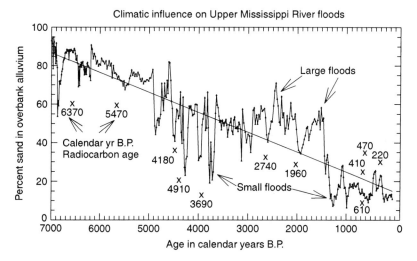

Figure 10.3 Percentage of sand in the total sediment fraction of overbank flood deposits on the Mississippi River Island 146, River Mile 663.4, Lansing Site 2. The 7,000-year sedimentary record represents 5.8 m of deposition and does not include 1.2 m of post AD 1850 sedimentation related to Euro-American agriculture. The trend line was fitted by least-squares regression and represents the normal fining-upward tendency for overbank floodplain vertical accretion. Large floods of high energy increase the amount of sand that is transported to the floodplain and are represented by positive (points plot above the regression line) departures from the average on the fining-upward trend line. Small floods of low energy are associated with relatively small amounts of sand transported to the floodplain and are represented by negative departures from the average on the fining-upward trend line. The proxy flood record indicates that floods of a given magnitude have occurred very episodically in response to climatic change. The calendar year BP age for respective data points was estimated from a least-squares regression equation of radiocarbon calendar age BP versus radiocarbon sample depth. The equation was significant at $p < 0.0001$, with a standard error of 340 years

SW floods, including large floods, became frequent about 2,400 years ago, except between about 800 and 600 years ago during the Medieval Warm Period when large floods decreased in frequency (Ely, 1997). Ely observed that periods of cool, wet climate in the western United States were associated with periods characterized by high frequencies of extreme floods, whereas extreme floods were relatively rare during periods of warm climate. Ely *et al.* (1993) and Ely (1997) report that large floods in the SW are typically produced either by winter storms or by storms related to dissipating tropical cyclones. Both systems are favored by a strongly developed meridional pattern in the middle-latitude upper-troposphere circulation that involves an anomalous trough of low pressure along or off the coast of the western United States similar to that shown in Figure 10.2a. Large floods from winter storms or tropical cyclones are also favored in the SW during multiple-year periods characterized by strong El Niño circulation regimes (Ely *et al.*, 1993). Figure 10.2c approximates a common winter-circulation regime typical of the strong El Niño years. On the other hand, the meridional pattern depicted by Figure 10.2b involves the development of a strong, regional anticyclonic (high-pressure) system that produces warm and dry climatic conditions in the SW and results in few large floods.

The UMV also shows episodic behavior of flooding characteristics that is strongly linked to shifts in climate (Table 10.1 and Figure 10.3). UMV floods tend to be larger and more frequent when the large-scale upper troposphere circulation involves a trough of low pressure over the western United States and a ridge of high pressure over southeastern United States, which together align the jet-stream axis and storm tracks across the region as indicated in Figure 10.2a. When prevailing large-scale troposphere circulation favors either a strong northwesterly or strong westerly airflow across the region as shown in Figures 10.2b and 10.2c, air masses of high water-vapor content and from tropical source regions are blocked from the region and large floods are rare. For the sample southwestern Wisconsin tributary watersheds that are mostly smaller than about $200 \, \text{km}^2$, the Holocene variations of large overbank floods and bankfull stage floods of one to two years' recurrence probability are broadly similar (Table 10.1). The similarity is unusual because the overbank floods largely result from heavy and/or long-duration summer rainstorms whereas the bankfull stage floods are mainly a result of spring snowmelt. The low resolution of radiocarbon-age control may account for some of the similarity in episode averaging, but persistence of similar flooding characteristics over long periods indicates that late Holocene seasonal climate changes primarily explain the parallel responses for large and small floods in these UMV tributaries.

The main channel UMR along Wisconsin's western border has a watershed that ranges from about 120,000 to $205,000 \, \text{km}^2$ between the upstream and downstream ends of the reach. Floods on this large river are dominated by snowmelt runoff, but excess spring and summer rainfall in association with the persistence of large-scale atmospheric circulation patterns similar to that shown in Figure 10.2a also produce relatively common large floods as illustrated by the large floods of 1993 and 2001. Table 10.1 shows that temporal variations in magnitudes of overbank floods on the main channel UMR since about 7,000 calendar years BP have shifted with climate in a way that is broadly similar to temporal variations for magnitudes of floods on the small tributaries. Thus, the general tendency for moderate to relatively large floods between about 7,000 and 5,500 years ago, followed by an episode of smaller floods between about 5,500 and 3,300 years ago, then returning to generally larger floods about 3,000 years ago is characteristic of both the small tributaries and the main channel of the UMR. These long-term flooding episodes closely reflect long-term shifts in the regional climate. For example, the shift to smaller magnitude floods between about 5,000 and 3,300 years ago was coincident with a shift to relatively warmer and drier climatic conditions in the region. Proxy-climate data presented by Bartlein et al. (1984), Winkler et al. (1986), and Baker et al. (1996) indicate that the climate shift was associated with an increase of 1.5°C in mean annual temperature above the preceding period and a reduction of the mean annual precipitation to a magnitude of about 15% less than the modern period. Denniston et al. (1999) used the isotopic analyses of $\delta^{13}C$ and $\delta^{18}O$ from the northeast Iowa cave stalagmites to infer that the dry climate episode between about 5,500 and 3,300 years ago was associated with a predominance of cool-weather precipitation. About 3,300 years ago, the UMV shifted to a cooler and wetter regional climate that has been punctuated by occasional droughts (Bartlein et al., 1984; Winkler et al., 1986; Baker et al., 1996; Laird et al., 1996; Fritz et al., 2000; Woodhouse and Overpeck, 1998; Dean, 1997; Clark, 1993). More frequent large floods on both the tributaries and the main channel UMR accompanied the shift to cooler and wetter climatic conditions (Table 10.1). Nearly all the largest floods of the approximately 70 palaeofloods documented for the tributaries in the UMV during the past 7,000 years were associated with flow depths that ranged between about 3.0 and 3.5 times the depth of their associated bankfull stage depths, which suggests for a given watershed size the existence of an upper limiting threshold flood magnitude determined by the region's climatology and physiography (Knox, 1993; 2000). The middle and late Holocene palaeoflood record for the SW also supports the idea that an upper limiting threshold flood magnitude exists for floods for a given watershed size in that region (Enzel et al., 1993).

The stratigraphic record of overbank floods on the UMR is of relatively high resolution and is well preserved because both the channel and the floodplain have been progressively aggrading throughout the Holocene. This high-resolution record, which was mainly determined from analyses of grain-size variability at 1-cm vertical increments, indicates a relation with climate change that is not apparent from the low-resolution tributary palaeoflood records (Table 10.1 and Figure 10.3). The results indicate that exceptionally large floods on the UMR have commonly accompanied the beginning of warm and dry climate episodes in the region. The recent high frequency of large floods on the UMR since the early 1990s may be a modern analogue because these floods have accompanied major hemispheric warming during the same period. Table 10.1 shows that Holocene proxy-climate evidence from areas marginal to the UMV headwaters and representative of north-central Minnesota and the northeastern Great Plains of North and South Dakota experienced major droughts that occurred about 4,800–4,300; 1,750–1,580; 1,250–1,050; 950–750, and 550–450 calendar years BP (before AD 1950). Ages for many of these droughts are the approximate times of the UMR large floods that were common about 4,700, 2,500–2,200, 1,800–1,500, 1280, 1,000–750, and 600–300 calendar years BP (Figure 10.3). Nonetheless, the general tendency for dry climate episodes to have smaller floods becomes more apparent as the duration of dry climate lengthens. For example, the relatively long dry episode in the UMV, between about 5,500 and 3,300 years ago, experienced a large flood or floods near its beginning, but overall was characterized by relatively smaller floods of high variability in the short term (Figure 10.3). The tendency for smaller and more variable floods to prevail during long dry episodes probably results because frontal zones at air mass boundaries, especially those involving strong cold fronts, are displaced poleward of the UMV. However, during the early phases of drought episodes, as warm air masses of high water vapor content do not fully dominate over the UMV region, cold, dry polar air masses can still occasionally intrude on the region to displace warm air masses and produce heavy precipitation across the UMV. Furthermore, warming during dry phases also increases the likelihood of spring rains falling on a melting snow cover to enhance flood magnitudes as what occurred during the spring of 2001, when the UMR experienced the second largest flood since the beginning of instrumentation in the late 1860s and 1870s.

4 ASSOCIATIONS OF SOUTHWEST AND UPPER MISSISSIPPI VALLEY PALAEOHYDROLOGY WITH PALAEOHYDROLOGY ELSEWHERE IN NORTH AMERICA

The associations between the SW and the UMV flood episodes and specific large-scale troposphere circulation patterns implies that palaeohydrologic activity in many, if not most, regions of North America probably has experienced nonstationary behavior, although the directions and magnitudes of change need not be the same as shown in Figure 10.2. The nonstationary behavior of floods in the palaeorecords shows that large changes in flood frequencies and/or magnitudes can occur abruptly and within relatively short time periods. An illustration is the observation of Chatters and Hoover

(1986) that the Columbia River floods in the Pacific northwest between about 1,000 and 600 years ago were three times more frequent than under present-day conditions. Elsewhere, during the latter half of this same period, large floods in the SW experienced reduced frequencies (Ely, 1997).

The out-of-phase relation between floods in the Pacific northwest and those of the SW between about 800 and 600 years ago reflects regional responses to the large-scale forcing effects of particular prevailing regimes of atmospheric circulation (Figure 10.2) and illustrates the importance of being cautious when applying local and regional palaeohydrology records to wider areas. Such regional differences are illustrated by proxy evidence for climate warming during the Medieval period 1,200 to 700 years ago when much of the Sierra Nevada region of California and the Canadian Rockies was relatively warm, while the climate of southeast United States apparently was a little different to that of later times (Hughes and Diaz, 1994; Luckman, 1994; Luckman et al., 1997; Stahle and Cleaveland, 1994). Earlier, when persistence of high flows characterized runoff to the San Francisco Bay area of northern California between about 3,800 and 2,000 years ago (Goman and Wells, 2000), large floods were relatively rare in the SW between about 4,200 and 2,400 years ago (Ely, 1997). When large floods in the SW were relatively frequent between about 5,800 and 4,200 years ago, they were relatively uncommon in the UMR tributaries during the same period (Table 10.1; Ely, 1997; Knox, 2000). Another example of regional differences to large-scale climate forcing is provided by the observation that the climate of the UMV is often out of phase with the climate in the southeast part of the continent. Bankfull stage floods reconstructed from relict stream channels in southeast Georgia indicate that flood magnitudes for the period between about 9,500 and 5,000 years ago were at least double those of modern streams in the region (Leigh and Feeney, 1995). However, the UMV tributary bankfull stage floods during much of the early Holocene before about 7,500 years ago were 10 to 20% smaller than their present-day equivalents (Knox, 1985; 2000). This sort of out-of-phase relation often occurs with strong zonal westerly troposphere circulation that concentrates the jet stream and the storm tracks over the southeast as what occurred during the 1930s when the mid-continent was in the middle of a severe drought. It is also likely that the storm tracks and the jet-stream axis were concentrated over the southeast during much of the early Holocene when remnants of the Laurentide Ice Sheet forced a strong zonal westerly to northwesterly circulation over the mid-continent (Knox, 1983).

Lamb (1966) showed that climatic epochs of many timescales, including the Holocene, tend to be associated with certain preferred large-scale atmospheric circulation patterns. The troposphere of the middle-latitude westerly circulation commonly develops regimes that are characterized by strongly meridional, high-amplitude waves during cool phases of climate that typically enhance short-term variability of both wetness and dryness in a region. Such meridional regimes are prone to migrate in the westerly circulation and involve slow-moving cyclones associated with persistent heavy rains followed by large floods that may affect wide regions throughout the middle latitudes. Such a phenomenon may explain the evidence of widespread flooding across much of middle-latitude North America about 400 years ago. Schimmelmann et al. (1998) identified a gray silt deposit in coastal California's Santa Barbara channel as representing deposition since about AD 1605 from an extreme flood that they attributed to a period of wet and cold climatic conditions in the region at that time. Schimmelmann et al. (1998) cite numerous studies that provide global-scale evidence of climate change in the first decade of the 1600s. They conclude that a southerly displacement of the circumpolar westerly circulation and jet stream brought intense precipitation to southern California during this decade. Schimmelmann *et al.* (1998) hypothesized that a southward displacement of the storm track over southern California with cool, wet conditions, and associated flooding resulted from a clustering of volcanic eruptions at that time. Their hypothesis is supported by dendrochronological evidence from the southeastern United States, where Stahle and Cleaveland (1994) identified the years between about AD 1596 and AD 1613 as including a cluster of extreme wet spring rainfalls that they hypothesized as being due to displacement of the Bermuda high-pressure system of that time. They reported that spring rainfall was above average from AD 1595 to AD 1614 in the southeast, but was also highly variable with more dry years than wet years.

4.1 "Megafloods" into the Gulf of Mexico

Several "megafloods" from the Mississippi River identified by Brown *et al.* (1999) in the Orca basin, northern Gulf of Mexico, occurred about 4,700, 3,500, 3,000, 2,500, 2,000, 1,200, and 300 calendar years ago. These floods led them to conclude that the late Holocene flooding into the Gulf was characterized by three distinct intervals of differing megaflood frequency and magnitude. It is noteworthy that the times of large floods into the Gulf of Mexico are in close general agreement with the episodes of large floods shown in Figure 10.3 for the UMR because the Holocene UMR contributes a relatively small proportion of the total watershed runoff of the combined Mississippi, Missouri, and Ohio River tributary drainages. The similarity suggests that, in this case, the hydrologic episodes represented by the two records probably signify a larger regional phenomenon characteristic of mid-continent North America (McQueen *et al.*, 1993). Brown *et al.* (1999) hypothesized that megafloods were favored during times when a more vigorous "Loop Current" possibly involved raised sea-surface temperatures and evaporation that in turn favored increased transport of moisture onto the mid-continent.

4.2 Large Coastal Floods Associated with the Gulf of Mexico Hurricanes

Flooding associated with hurricanes normally increases when Gulf and tropical Atlantic waters are anomalously warm, and hurricanes advect moisture and heavy rainfall over the lower Mississippi River tributaries and much of the southeast United States and/or coastal Mexico. Liu and Fearn (2000) analyzed a 7,000-year record of over-wash sand layers that were deposited during intense hurricanes that passed across a Florida coastal lake, and they found that the occurrences of catastrophic hurricanes has been episodic. They identified two relatively quiescent periods, between about 5,950 and 3,650 years ago and during the last 1,000 years, when few catastrophic hurricanes struck the northwestern Florida coast. During the intervening period, between about 3,650 and 1,000 years ago, the frequency of catastrophic hurricanes was relatively high. Farther west on the Gulf coast in Alabama, Liu and Fearn (1993) used similar methodology to identify major hurricane occurrences at about 3,450 to 3,250, 2,800, 2,250, 1,350, and 750 calendar years ago. Liu and Fearn (1993) attributed the millennial-scale variability in catastrophic hurricane landfalls to shifts in the positions of the jet stream and the semipermanent Bermuda high-pressure cell over the western Atlantic Ocean. The coastal palaeofloods of hurricanes show strong nonstationary behavior as observed for the palaeofloods of rivers in the SW and in the UMV. Temporal differences between the boundaries of the hurricane-flood episodes and boundaries for the flood episodes of the SW and the UMV probably reflect in part the time-transgressive effect between subtropical and mid-latitude climate systems.

4.3 Palaeohydrology in Mexico

The times of Holocene climate changes in Mexico show many similarities to the large-scale climatic discontinuities that were responsible for the flood episodes previously described for the United States. Although a palaeoflood investigation is under way in the Lower Panuco basin on eastern Mexico's Gulf Coast (Hudson, 2002), relatively little palaeoflood information is available, which provides detailed chronologies that can be compared with past climate changes. Metcalfe *et al.* (2000) provide an extensive review of presently existing evidence for Late Pleistocene and Holocene climatic change in Mexico. They note that the very early Holocene was probably quite dry, but that the environment became relatively wet over much of Mexico, except possibly the Basin of Mexico, by the early mid-Holocene about 7,000 years ago. They further suggest that modern arid climatic conditions were established in the Sonoran and Chihuahuan deserts about 4,500 years ago, while Yucatan remained relatively wet. A severe drought episode about 1,000 years ago occurred in Yucatan and central Mexico and may have been the driest period of the entire Holocene (Metcalfe *et al.*, 2000). Given that much of the Mexican landscape has shifted through semiarid and arid phases during the Holocene, it is likely that palaeohydrologic changes were relatively profound.

4.4 Palaeohydrology in Southern Canada

Published, long palaeoflood chronologies are not available for Canadian sites, although a number of investigations are under way. The great flood of 1997 on southern Manitoba's Red River served as a catalyst for several new palaeoflood investigations on that river system. St George and Nielsen (2000) used tree-ring analyses from living and fossil oak trees, along a 100-km reach of the Red River, to identify large floods that occurred between AD 1463 and AD 1999. Their results indicated that two extremely large floods had occurred on the Red River as recently as the early nineteenth century. Tree-ring analyses were also used to reconstruct lake levels over the past 200 years for two large lakes in subarctic northern Quebec (Begin, 2000), and for lake sites in western Quebec (Tardif and Bergeron, 1997). The western Quebec study concluded that major ice floods along the southern limit of the boreal forest have been increasing since the end of the Little Ice Age owing to greater penetration of the region by warm and humid air masses from the south. Farther west in the driest sector of southern Canada's prairie grasslands, known as the Palliser Triangle, several palaeohydrology-related studies have been undertaken to help anticipate the biophysical impacts of potential future climatic changes. For example, diatom-based proxies of past salinity levels in the lakes of the region imply that lake salinity was quite high between about 11,000 and 9,400 years ago, but by about 8,700 years ago they had become freshwater lakes (Wilson and Smol, 1999; Last *et al.*, 1998). The transition to freshwater toward the middle Holocene is consistent with results of a palaeohydrology study of a prairie pothole lake in southern Saskatchewan where, following a very dry period, water levels rose between about 8,500 and 6,700 years ago (Yansa, 1998; Vance *et al.*, 1992). It is worth noting that the period of rising water levels in southern Saskatchewan is very similar to the period in the UMV when bankfull stage floods on small tributary watersheds were averaging 10 to 20% larger than their modern counterparts. However, a number of other palaeohydrology studies suggest that about 5,000 years ago an onset of cooler and wetter climatic conditions occurred in the northern Great Plains of south-central Canada (e.g. Last and Sauchyn, 1993). The period between about 5,000 and 3,000 years ago in the UMV apparently involved a shift toward a warmer and drier environment. Palaeohydrology investigations in western Canada's central Alberta showed that shallow basins were empty during the early Holocene, but began flooding about 8,000 years ago and that most had filled between 6,500 and 4,500 years ago (Schweger and Hickman, 1989). Collaborative research that was part of the Palliser Triangle project used plant macrofossils, diatoms, ostracods, algal pigments, sedimentology, mineralogy, and stable isotope geochemistry as proxies of past climatic variability, and the results showed that the historic record of the last 100 years does not adequately capture the range of climatic variability for even the last 2,000 years (Lemmen *et al.*, 1997).

5 SUMMARY AND CONCLUSIONS

Climate is the principal force that drives hydrologic systems. Extraordinary global warming that accelerated during the late twentieth century and which may continue during the twenty-first century if increased loading of the atmosphere with greenhouse gases continues, implies that important changes might be anticipated in the hydrologic cycle. Global-circulation models support the idea that global warming enhances the intensity of the hydrologic cycle that, in turn, may contribute to increased frequencies of extreme hydrologic events, including floods and droughts. Such natural hazards are of particular concern because both greatly affect food resources and human life. The period of instrumented observations used for documenting hydrologic responses to climate change in North America, as in most World regions, is too short to adequately determine the range of possible shifts in extreme events with potential future climate changes. Palaeohydrologic evidence contained in the stratigraphic archives of river, lake, and marine deposits provides an alternative source of information that characterizes how climate changes affected hydrologic activity in the past and how climate changes might affect hydrologic activity in the future.

Palaeohydrologic data from the middle latitudes of North America demonstrate that magnitudes and recurrence frequencies of floods, droughts, and hurricanes have experienced large changes in response to relatively modest climate changes during the Holocene (post glacial). Early Holocene climates of North America were still responding to the forcing effects of the disintegrating Laurentide Ice Sheet in northern Canada, and therefore are not proper analogues of recent and near-future climates. Middle and late Holocene environments, on the other hand, involve forcing factors that represent climatic and hydrologic conditions that approximate modern climates. The palaeohydrologic data that have been reviewed here indicate a number of relationships that are useful for water-resources planning and policy decisions. Some of the more important observations from the North American palaeohydrology data are listed below:

- Relatively modest middle and late Holocene climate changes appear to be principally responsible for strong clustering in magnitudes and recurrence frequencies of floods, droughts, and other natural hazards such as hurricanes.
- Low-frequency, extreme hydrologic events are more sensitive and show greater responses to climate changes than do high-frequency, small-magnitude (mean to modal range) hydrologic events.
- Extreme floods and/or extreme droughts tend to either "turnon" or "turnoff" with shifts in the frequencies of certain patterns of large-scale ocean-atmosphere circulation.

- Long-term warm climate episodes tend to be characterized by smaller and more highly variable floods than cool-climate episodes, but frequent very large floods have often been associated with the early phases of warm episodes in the UMV.
- Strong teleconnections frequently exist between regional flood episodes in North America, and may reflect either in-phase or out-of-phase relations depending on prevailing hemispheric-scale ocean-atmosphere circulation patterns.
- The long-term Holocene palaeoflood records of the southwest and the UMV support the idea that for any given watershed size a threshold upper-maximum flood magnitude exists for each of the regions.

In summary, Holocene palaeohydrologic data from the North American sites show that the assumption of stationary mean and stationary variance, which underlies much of the methodology used in probability assessments of hydrologic events, is not adequately met. Short-term instrumental records often poorly reflect the range of hydrologic conditions that occurred during the middle and late Holocene and that might occur in the future. Knowledge of palaeohydrologic conditions during Holocene climate episodes and during Holocene climatic changes facilitates a more accurate projection of hydrologic responses that are possible under potential future climate changes. Experience gained from palaeohydrology can be used to assign weighted probabilities to specified hydrologic events based on given climate scenarios and result in improved planning and policy decisions that involve water resources and natural hazards.

ACKNOWLEDGEMENTS

The research on palaeofloods of UMV tributaries was supported by NSF grants EAR-8511280, EAR-9206854, and EAR-9409778. The research on palaeofloods of the UMR was supported by NSF grants EAR-9807715 and ATM-0112614. An Evjue–Bascom research award from the University of Wisconsin, Madison, provided additional support. I especially thank Mike Daniels, Lindsay Theis, and Travis Tennessen for their field and laboratory assistance related to palaeoflood research on the UMR and Richard Worthington for advice and assistance with graphical illustrations. I also thank Ken Gregory and Gerardo Benito for their patience and contributions that have led to this volume.

REFERENCES

Angel, J.R. and Huff, F.A., 1997. Changes in heavy rainfall in midwestern United States. *Journal of Water Resources Planning & Management – ASCE*, **123**, 246–249.

Ashmore, P. and Church, M., 2001. The impact of climate change on rivers and river processes in Canada. *Bulletin of the Geological Survey of Canada Bulletin*, **555**, 1–48.

Baker, R.G., Bettis III, E.A., Schwert, D.P., Horton, D.G., Chumbley, C.A., Gonzàlez, L.A. and Reagan, M.K., 1996. Holocene paleoenvironments of Northeast Iowa. *Ecological Monographs*, **66**, 203–234.

Bartlein, P.J., Webb, T.I. and Fleri, E., 1984. Holocene climatic change in the Northern Midwest: pollen-derived estimates. *Quaternary Research*, **22**, 361–374.

Begin, Y., 2000. Reconstruction of subarctic lake levels over past centuries using tree rings. *Journal of Cold Regions Engineering*, **14**(4), 192–212.

Beltaos, S. and Burrell, B.C., 2002. *Extreme Ice Jam floods Along the Saint John River*. International Association of Hydrological Sciences Publication 271, New Brunswick, Canada, 9–14.

Brown, R.D. and Braaten, R.O., 1998. Spatial and temporal variability of Canadian monthly snow depths, 1946–1995. *Atmosphere and Ocean*, **36**, 37–45.

Brown, P., Kennett, J.P. and Ingram, B.L., 1999. Marine evidence for episodic Holocene megafloods in North America and the northern Gulf of Mexico. *Paleooceanography*, **14**(4), 498–510.

Burn, D.R., 1994. Hydrologic effects of climatic change in west-central Canada. *Journal of Hydrology*, **160**, 53–70.

Chatters, J.C. and Hoover, K.A., 1986. Changing late Holocene flooding frequencies on the Columbia River, Washington. *Quaternary Research*, **26**(3), 309–320.

Clark, J.S., 1993. Fire, climate change, and forest processes during the past 2000 years. In J.P. Bradbury and W.E. Dean (eds), *Elk Lake, Minnesota: Evidence for Rapid Climate Change in the North-Central United States*. Special Paper 276, Geological Society of America, Boulder, 295–308.

Dean, J.S., 1994. The medieval warm period on the Southern Colorado plateau. *Climatic Change*, **26**(2–3), 225–241.

Dean, W.E., 1997. Rates, timing, and cyclicity of Holocene eolian activity in North-Central United States: evidence from varved lake sediments. *Geology*, **25**, 331–334.

Dean, W.E., Ahlbrandt, T.S., Anderson, R.Y. and Bradbury, J.P., 1996. Regional aridity in North America during the middle Holocene. *The Holocene*, **6**, 145–155.

Denniston, R.F., Gonzàlez, L.A., Asmerom, Y., Baker, R.G., Reagan, M.K. and Bettis III, E.A., 1999. Evidence for increased cool season moisture during the middle Holocene. *Geology*, **27**(9), 815–818.

Easterling, D.R., Meehl, G.A., Parmesan, C., Changnon, S.A., Karl, T.R. and Mearns, L.O., 2000. Climate extremes: observations, modeling and impacts. *Science*, **289**, 2068–2074.

Ely, L.L., 1997. Response of extreme floods in the Southwestern United States to climatic variations in the late Holocene. *Geomorphology*, **19**(3–4), 175–201.

Ely, L.L. and Baker, V.R., 1985. Reconstructing paleoflood hydrology with slackwater deposits: Verde River, Arizona. *Physical Geography*, **6**, 103–126.

Ely, L.L., Enzel, Y., Baker, V.R. and Cayan, D.R., 1993. A 5000-year record of extreme floods and climate change in the southwestern United States. *Science*, **262**, 410–412.

Enzel, Y., 1992. Flood frequency of the Mojave river and the formation of late Holocene playa lakes, Southern California, USA. *The Holocene*, **2**(1), 11–18.

Enzel, Y., Ely, L.L., House, P.K. and Baker, V.R., 1996. Magnitude and frequency of Holocene palaeofloods in the southwestern United States: a review and discussion of implications. In J. Branson, A.G. Brown and K.J. Gregory (eds), *Global Continental Changes: The Context of Palaeohydrology*. Special Publication 115, Geological Society, London, 121–137.

Enzel, Y., Webb, R.H., Ely, L.L., House, P.K. and Baker, V.R., 1993. Paleoflood evidence for a natural upper bound to flood magnitudes in the Colorado river basin. *Water Resources Research*, **29**(7), 2287–2297.

Fritz, S.C., Engstrom, D.R., Ito, E., Yu, Z. and Laird, K.R., 2000. Hydrologic variation in the Northern Great Plains during the last two millennia. *Quaternary Research*, **53**(2), 175–184.

Goman, M. and Wells, L., 2000. Trends in river flow affecting the northeastern reach of the San Francisco Bay estuary over the past 7000 years. *Quaternary Research*, **54**(2), 206–217.

Groisman, P.Y., Knight, R.W. and Karl, T.R., 2001. Heavy precipitation and high streamflow in the contiguous United States: trends in the twentieth century. *Bulletin of the American Meteorological Society*, **82**, 219–246.

Hirschboeck, K.K., 1987. Catastrophic flooding and atmospheric circulation anomalies, (USA). In: L. Mayer and D. Nash (eds), *Catastrophic Flooding. Binghamton Symposia in Geomorphology, International Series*, Allen & Unwin, Boston, Massachusetts, Vol. 18, 23–56.

Hirschboeck, K.K., Ely, L.L. and Maddox, R.A., 2000. Hydroclimatology of meteorologic floods. In E.E. Wohl (ed.), *Inland Flood Hazards*. Cambridge University Press, Cambridge, 39–72.

Houghton, J.T., Meira Filho, L.G., Callander, B.A., Harris, N., Kattenberg, A. and Maskell, K., 1996. *Climate Change 1995*. Cambridge University Press, Cambridge.

House, P.K., Pearthree, P.A. and Klawon, J.E., 2002. Historical flood and paleoflood chronology of the lower Verde River, Arizona: stratigraphic evidence and related uncertainties. In

P.K. House, R.H. Webb, V.R. Baker and D.R. Levish (eds), *Ancient Floods Modern Hazards: Principles and Applications of Paleoflood Hydrology*. American Geophysical Union, Washington, DC, 267–293.

Hudson, P.F., 2002. Variability of natural levee deposits in the Lower Panuco basin, Mexico. 98th Annual Meeting, Abstracts, Association of American Geographers, Los Angeles, CA.

Hughes, M.K. and Diaz, H.F., 1994. Was there a 'Medieval Warm Period', and if so, where and when?. *Climatic Change*, **26**, 109–142.

Ingram, B.L., Ingle, J.C. and Conrad, M.E., 1996. A 2000 yr record of Sacramento-San Joaquin river inflow to San Francisco Bay estuary, California. *Geology*, **24**(4), 331–334.

Jarrett, R.D. and Tomlinson, E.M., 2000. Regional interdisciplinary paleoflood approach to assess extreme flood potential. *Water Resources Research*, **36**(10), 2957–2984.

Kalnicky, R.A., 1974. Climatic change since 1950. *Annals of the Association of American Geographers*, **64**, 100–112.

Karl, T.R. and Knight, R.W., 1998. Secular trends of precipitation amount, frequency, and intensity in the United States. *Bulletin of the American Meteorological Society*, **79**, 231–241.

Karl, T.R., Knight, R.W., Easterling, D.R. and Quayle, R.G., 1996. Indices of climate change for the United States. *Bulletin of the American Meteorological Society*, **77**, 279–303.

Knapp, H.V., 1994. Hydrologic trends in the upper Mississippi river basin. *Water International*, **19**(4), 199–206.

Knox, J.C., 1983. Responses of river systems to Holocene climates. In H.E. Wright Jr. (ed.), *Late Quaternary Environments of the United States: Volume 2, The Holocene*. University of Minnesota Press, Minneapolis, MN, 26–41.

Knox, J.C., 1985. Responses of floods to Holocene climatic change in the upper Mississippi valley. *Quaternary Research*, **23**(3), 287–300.

Knox, J.C., 1993. Large increases in flood magnitude in response to modest changes in climate. *Nature*, **361**, 430–432.

Knox, J.C., 1999. Long-term episodic changes in magnitudes and frequencies of floods in the Upper Mississippi River Valley. In A.G. Brown and A.G. Quine (eds), *Fluvial Processes and Environmental Change*. John Wiley & Sons, Chichester, 255–282.

Knox, J.C., 2000. Sensitivity of modern and Holocene floods to climate change. *Quaternary Science Reviews*, **19**(1–5), 439–457.

Knox, J.C., 2001. Agricultural influence on landscape sensitivity in the upper Mississippi river valley. *Catena*, **42**(2–4), 193–224.

Knox, J.C., 2002. Agriculture, erosion, and sediment yields. In: A.R. Orme (ed), *The Physical Geography of North America*. Oxford University Press, Oxford, 482–500.

Knox, J.C., Bartlein, P.J., Hirschboeck, K.K. and Muckenhirn, R.J., 1975. The Response of Floods and Sediment Yields to Climate Variation and Land Use in the Upper Mississippi Valley. Report 52, University of Wisconsin, Madison, Institute for Environmental Studies, Madison, 1–76.

Kutzbach, J.E., 1970. Large-scale features of monthly mean northern hemisphere anomaly maps of sea-level pressure. *Monthly Weather Review*, **98**, 708–716.

Laird, K.R., Fritz, S.C. and Cumming, B.F., 1998a. A diatom-based reconstruction of drought intensity, duration, and frequency from Moon Lake, North Dakota: a sub-decadal record of the last 2300 years. *Journal of Paleolimnology*, **19**(2): 161–179.

Laird, K.R., Fritz, S.C., Cumming, B.F. and Grimm, E.C., 1998b. Early-Holocene limnological and climatic variability in the Northern Great Plains. *The Holocene*, **8**(3), 275–285.

Laird, K.R., Fritz, S.C., Maasch, K.A. and Cumming, B.F., 1996. Greater drought intensity and frequency before AD 1200 in the Northern Great Plains, USA. *Nature*, **384**, 552–554.

Lamb, H.H., 1966. On the nature of certain climatic epochs which differed from the modern (1930-39) normal. In H.H. Lamb (ed.), *The Changing Climate*. Methuen and Co., Ltd, London, 58–112.

Lamb, H.H., 1977. *Climate: Present, Past, and Future: Climatic History and the Future, Volume 2, Climatic History and the Future*. Metheun & Co. Ltd, London, 1–835.

Last, W.M. and Sauchyn, D.J., 1993. Mineralogy and lithostratigraphy of Harris lake, south-western Saskatchewan, Canada. *Journal of Paleolimnology*, **9**(1), 23–39.

Last, W.M., Vance, R.E., Wilson, S. and Smol, J.P., 1998. A multi-proxy limnologic record of rapid early-Holocene hydrologic change on the Northern Great Plains, Southwestern Saskatchewan, Canada. *The Holocene*, **8**(5), 503–520.

Leigh, D.S. and Feeney, T.P., 1995. Paleochannels indicating we climate and lack of response to lower sea level, Southeast Georgia. *Geology*, **23**, 687–690.

Lemmen, D.S., Vance, R.E., Wolfe, S.A. and Last, W.M., 1997. Impacts of future climate change on the southern Canadian Prairies: a paleoenvironmental perspective. *Geoscience Canada*, **24**(3), 121–133.

Lettenmaier, D.P., Wood, E.F. and Wallis, J.R., 1994. Hydro-climatological trends in the continental United States, 1948–1988. *Journal of Climate*, **7**, 586–607.

Lins, H.F. and Michael, P.J., 1994. Increasing U.S. streamflow linked to greenhouse forcing. *EOS Transactions*, **75**, 281–283.

Lins, H.F. and Slack, J.R., 1999. Streamflow trends in the United States. *Geophysical Research Letters*, **26**, 227–230.

Liu, K.B. and Fearn, M.L., 1993. Lake-sediment record of late Holocene hurricane activities from coastal Alabama. *Geology*, **21**, 793–796.

Liu, K.B. and Fearn, M.L., 2000. Reconstruction of prehistoric landfall frequencies of catastrophic hurricanes in Northwestern Florida from lake sediment records. *Quaternary Research*, **54**, 238–245.

Luckman, B.H., 1994. Evidence for climatic conditions between ca. 900–1300 AD in the Southern Canadian Rockies. *Climatic Change*, **26**, 171–182.

Luckman, B.H., Briffa, K.R., Dones, P.D. and Schweingruber, F.H., 1997. Tree-ring based reconstruction of summer temperatures at the Columbia Icefield, Alberta, Canada, AD 1073–1983. *The Holocene*, **7**(4), 375–389.

Mann, M.E., Bradley, R.S. and Hughes, M.K., 1998. Global-scale temperature patterns and climate forcing over the past six centuries. *Nature*, **392**, 779–787.

McPhaden, M.J., 1999. Genesis and evolution of the 1997-98 El Niño. *Science*, **283**, 950–954.

McQueen, K.C., Vitek, J.D. and Carter, B.J., 1993. Paleoflood analysis of an alluvial channel in the South-Central Great Plains: Black Bear Creek, Oklahoma. *Geomorphology*, **8**(2–3), 131–146.

Metcalfe, S.E., O'Hara, S.L., Caballero, M. and Davies, S.J., 2000. Records of Late Pleistocene-Holocene climatic change in Mexico – a review. *Quaternary Science Reviews*, **19**, 699–721.

O'Connor, J.E., Ely, L.L., Wohl, E.E., Stevens, L.E., Melis, T.S., Kale, V.S. and Baker, V.R., 1994. A 4500-year record of large floods on the Colorado river in the Grand Canyon, Arizona. *Journal of Geology*, **102**, 1–9.

Olsen, J.R., Stedinger, J.R., Matalas, N.C. and Stahkhiv, E.Z., 1999. Climate variability and flood frequency estimation for the upper Mississippi and lower Missouri rivers. *Journal of the American Water Resources Association*, **35**, 1509–1523.

Redmond, K.T., Enzel, Y., House, P.K. and Biondi, F., 2002. Climate variability and flood frequency at decadal to millennial time scales. In P.K. House, R.H. Webb, V.R. Baker and D.R. Levish (eds), *Ancient Floods, Modern Hazards: Principles and Applications of Paleoflood Hydrology*. American Geophysical Union, Washington, DC, 21–45.

Schimmelmann, A., Zhao, M., Harvey, C.C. and Lange, C.B., 1998. A large California flood and correlative global climatic events 400 years ago. *Quaternary Research*, **49**, 51–61.

Schweger, C.E. and Hickman, M., 1989. Holocene paleohydrology of central Alberta: testing the general-circulation-model climate simulations. *Canadian Journal of Earth Sciences*, **26**(9), 1826–1833.

St. George, S. and Nielsen, E., 2000. Signatures of high-magnitude 19th-century floods in *Quercus macrocarpa* tree rings along the Red River, Manitoba, Canada. *Geology*, **28**(10), 899–902.

Stahle, D.W. and Cleaveland, M.K., 1994. Tree-ring reconstructed rainfall over the southeastern U.S.A. during the medieval warm period and little ice age. *Climatic Change*, **26**, 199–212.

Tardif, J. and Bergeron, Y., 1997. Ice-flood history reconstructed with tree-rings from the southern boreal forest limit, Western Quebec. *The Holocene*, **7**(3), 291–300.

United States Water Resources Council. 1981. Guidelines for Determining Flood Flow Frequency. Bulletin 17B. U.S. Government, Washington, DC.

Valero-Garces, B.L., Laird, K.R., Fritz, S.C., Kelts, K., Ito, E. and Grimm, E.C., 1997. Holocene climate in the Northern Great Plains inferred from sediment stratigraphy, stable isotopes, carbonate geochemistry, diatoms, and pollen at Moon Lake, North Dakota. *Quaternary Research*, **48**, 359–369.

Vance, R.E., Matthewes, R.W. and Clague, J.J., 1992. 7000 year record of lake-level change on the Northern Great Plains: a high-resolution proxy of past climate. *Geology*, **20**(10), 879–882.

Vogel, R.M., Zafirakou-Koulouris, A. and Matalas, N.C., 2001. Frequency of record-breaking floods in the United States. *Water Resources Research*, **37**(6), 1723–1731.

Watson, R.T., Zinyowera, M.C. and Moss, R.H., 1998. *The Regional Impacts of Climate Change*. Cambridge University Press, Cambridge, 1–517.

Webb, R.H., Blainey, J.B. and Hyndman, D.W., 2002. Paleoflood hydrology of the Paria River, southern Utah and northern Arizona, USA. In P.K. House, R.H. Webb, V.R. Baker and D.R. Levish (eds), *Ancient Floods, Modern Hazards: Principles and Applications of Paleoflood Hydrology*. American Geophysical Union, Washington, DC, 295–310.

Webb, R.H., O'Connor, J.E. and Baker, V.R., 1988. Paleohydrologic reconstruction of flood frequency on the Escalante River. In V.R. Baker, R.C. Kochel and P.C. Patton (eds), *Flood Geomorphology*. John Wiley & Sons, New York, 403–418.

Wilson, S.E. and Smol, J.P., 1999. Diatom-based salinity reconstructions from Palliser triangle lakes: a summary of two Saskatchewan case studies. *Bulletin of the Geological Survey of Canada*, **534**, 67–79.

Winkler, M.G., Swain, A.M. and Kutzbach, J.E., 1986. Middle Holocene dry period in the Northern Midwestern United States: lake levels and pollen stratigraphy. *Quaternary Research*, **25**, 235–250.

Woodhouse, C.A. and Overpeck, J.T., 1998. 2000 years of drought variability in the Central United States. *Bulletin of the American Meteorological Society*, **79**, 2693–2714.

Yansa, C.H., 1998. Holocene paleovegetation and paleohydrology of a prairie pothole in southern Saskatchewan, Canada. *Journal of Paleolimnology*, **19**, 429–441.

Section III   Low-latitude Regions

11 Palaeohydrological Reconstructions for Tropical Africa since the Last Glacial Maximum – Evidence and Problems

M. F. THOMAS[1] AND M. B. THORP[2]
[1]*University of Stirling, Stirling, UK*
[2]*National University of Ireland, Dublin, Republic of Ireland*

1 INTRODUCTION

A comprehensive review of the ecological and hydrologic changes that occurred during the last 20,000 years in the humid tropics was provided by Kadomura in 1995, and in the same year Thomas and Thorp (1995) provided an interpretation of the geomorphic response to these changes. Since that time, much progress has been made with the study of Quaternary climate change, particularly with respect to high-resolution records provided by ice and marine cores. Recent modelling exercises have also become more realistic with respect to estimations of tropical Sea Surface Temperatures (SSTs) during the Last Glacial Maximum (LGM) and the possible temperature changes on land in tropical latitudes (see Broecker, 1995; Guilderson *et al.*, 1994; Jolly *et al.*, 1998; Kutzbach *et al.*, 1998). Model predictions now converge with figures derived from analyses of empirical data (e.g. Bonnefille and Chalié, 2000; Webb *et al.*, 1997) to indicate a temperature depression of $4 \pm 2°C$, typically averaging 3.7°C for ocean cooling and 5.3°C over land (Kutzbach *et al.*, 1998). Consideration of the new data and models has led to new reviews of the palaeohydrology of the African continent, notably by Gasse (2000). Much of the new understanding depends on the application of the ideas of Broecker and Denton (1990) concerning ocean–atmosphere coupling and reorganisation during the later Quaternary. In turn, this has led to increased interest in the complexity of regional palaeoclimates across the tropics.

Aspects of this situation applicable to Africa (Figure 11.1) include

1. regionalisation of climate change in response to fluctuations in the African and Indian monsoons and the movement and vigour of the Intertropical Convergence Zone (ITCZ) (see Figure 11.2);
2. diachronous progression of Late Pleistocene climate change between the Equator and the tropical deserts, in the Northern Hemisphere;
3. complexities of circulation related to past positions of the Jet Streams and associated Westerlies in the drier margins of the tropics of both hemispheres;
4. the influence of changes in Walker cell activity and the strength of convergence, possibly also related to El Niño Southern Oscillation (ENSO) events.

Palaeohydrology: Understanding Global Change.   Edited by K.J. Gregory and G. Benito
© 2003 John Wiley & Sons, Ltd   ISBN: 0-470-84739-5

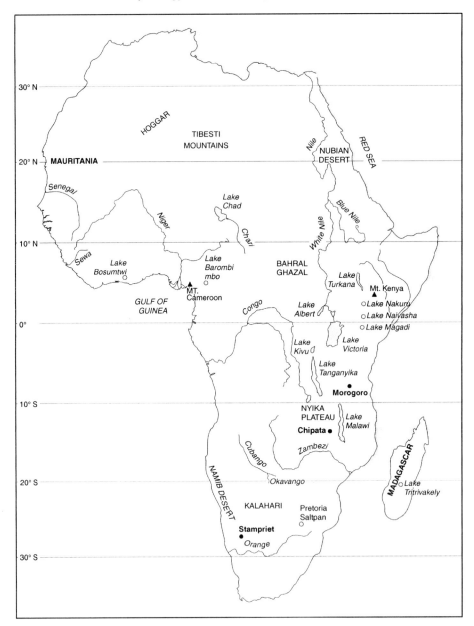

Figure 11.1   Locations referred to in the text

These considerations influence interpretations of empirical data. Postglacial warming, for example, was apparent earlier in the equatorial Congo basin (Preuss, 1990), and in Uganda, at around 17 cal kyr BP, and later in parts of West Africa, at ca 15 cal kyr BP; increased rainfall possibly not reaching the Sahel until ca 11 ka BP. There also appear to be differences between the northern and southern hemispheres, and beyond 10° S in East Africa a different rhythm may have been present. Nonetheless, the outline of climate and hydrologic change across Africa offered by Kadomura (1995) has remained robust.

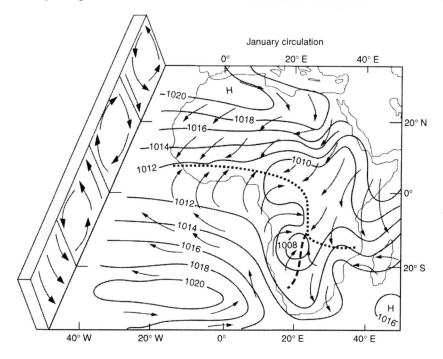

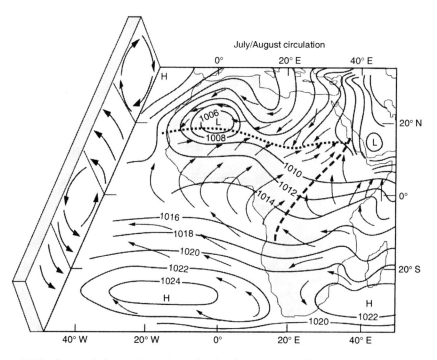

Figure 11.2 Present-day patterns of air circulation over Africa. Dotted lines indicate the position of the Intertropical Convergence Zone (ITCZ); broken lines the Congo air boundary. From Nicholson, S.E., 1996. A review of climate dynamics and climate variability in Eastern Africa. In T.C. Johnson and E.O. Odada (eds), *The Limnology, Climatology and Palaeoclimatology of the East African Lakes*. Gordon & Breach, Amsterdam, 25–56. Reproduced with permission from Taylor & Francis.

There is also debate concerning the precipitation changes required to match the empirical proxy data on the one hand and to satisfy the climate models and palaeocirculation patterns on the other. Allied to this is the problem of estimating the impact on surface hydrology of reduced evapotranspiration at times of lowered temperatures and weakened monsoons. Some writers have argued that only small changes in precipitation were required to lower some of the East African lakes by substantial amounts, while other data suggest more severe aridity, even within the inner humid tropics.

2   CLIMATIC CONDITIONS AT THE LAST GLACIAL MAXIMUM

2.1   Temperature Changes

Following fresh interpretations of ice core and ocean drilling records (Guilderson *et al.*, 1994; Thompson *et al.*, 1995; 1998; Broecker, 1995; Stute *et al.*, 1995; Ruddiman, 1997; Schultz *et al.*, 1998), there is now wide agreement that temperature depression in the humid tropics during the LGM was at least 5°C. The recent Community Climate Model, Version 1 (CCM1) output (Kutzbach *et al.*, 1998; Webb and Kutzbach, 1998) provides further support for this conclusion. Previous disagreement persisted for nearly two decades, following the acceptance of the CLIMAP models (1976; 1981) that limited change of ice age SSTs in tropical latitudes to ca 2°C. This view produced many conflicts with interpretations of empirical data (Rind and Peteet, 1985; Lautenschlager, 1991; Guilderson *et al.*, 1994).

Guilderson *et al.* (1994), show a probable 5°C cooling of the western tropical Atlantic at 18 to 19 ^{14}C kyr BP (22–21 cal kyr BP). This was followed by significant warming from 18 to 14.7 ^{14}C kyr BP, (21–17.5 cal kyr BP) by about 50% of the glacial–interglacial shift, but then by a rapid cooling by about 4°C around 13.7 ^{14}C kyr BP (16 cal kyr BP), as a result of rapid deglaciation and meltwater influx. However, by 11.4 ^{14}C kyr BP (13.5 cal kyr BP) temperatures had recovered to near present levels, after which variation has been within 2°C.

2.2   Precipitation Changes

There is less agreement concerning the fluctuation of precipitation amount, intensity and seasonality during this period. The recent modelling by Kutzbach *et al.* (1998) indicated that precipitation over much of the African tropics remained at 88% of present-day means at the LGM, a reduction of only 12%, although they note that the weakened summer monsoon may have led to a greater decline in SE Asia. Simulation for 6,000 cal yr BP predicted rainfall 5% above present. Jolly *et al.* (1998), relating pollen data to the CCM1 model outcomes, and comparing the vegetation simulation in BIOME1.2 for Africa, found good agreement with this model. In contrast, Dong *et al.* (1996) using the Universities Global Atmospheric Modelling Programme (UGAMP) GCM, indicate a regionally averaged decline in precipitation at 21 cal kyr BP of 52.6% for North Africa (10–30° N) and 65.3% for South Asia (20–35° N).

By 16 cal kyr BP or 13,500 BP (^{14}C), the CCM1 model shows that orbital forcing would have enhanced precipitation considerably in the low latitudes, but predicts that the greatest change to monsoon summer precipitation in North Africa and Asia came at 10 cal kyr BP (9 kyr ^{14}C BP). Predictions for 6 cal kyr BP are similar to 11,000 cal kyr BP. The UGAMP model shows an increase in rainfall above present-day means of 14.4% at 6 cal kyr BP in North Africa. When compared with predictions for 21 cal kyr BP, these figures suggest major fluctuations (>60%) in rainfall totals and, while these figures do not apply to the inner tropics, it is difficult to envisage that changes would have been negligible in neighbouring areas.

Pollen-inferred precipitation changes for equatorial mountain sites in East Africa have been re-examined by Bonnefille and Chalié (2000), who used 9 sites between 1,850- and 2,240-m altitude, and between 2 and 4° S latitude. More than 600 pollen spectra were examined, and 88 radiocarbon dates applied. They deduce that precipitation was reduced by an average 32% (20–44%) between 30 ^{14}C kyr BP and 15 ^{14}C kyr BP (18 cal kyr BP). In detail, several sharp swings of precipitation are documented, corresponding accurately with the records in Lakes Nakuru and Tanganyika. The greatest fall was around 22 to 21 cal kyr BP, when arboreal pollen fell markedly or was absent at several sites. The glacial–interglacial transition is poorly represented in these cores, but woody vegetation (montane forest) was re-established between 13 and 12 ^{14}C kyr BP (15.5–14 cal kyr BP). In the Holocene, forest cover was at its maximum between 10 and 8 ^{14}C kyr BP (11.5–9 cal kyr BP), and declined after 7 to 6 kyr BP (8–7 cal kyr BP), with marked forest decline at 5.8 and 3.8 ^{14}C kyr BP (6.6 and 4.2 cal kyr BP). Precipitation increases in the early Holocene (11–8 cal kyr BP) may have been 15 to 35% above present-day (Hastenrath and Kutzbach, 1983; Bonnefille and Chalié, 2000). However, these authors point out that the magnitude of increase was probably greater at 10° N than near the Equator.

Gasse (2000) indicates a negative shift in rainfall across southern Africa after 30 ka, with rainfalls 15 to 20% lower in Botswana, quoting evidence from the Pretoria Saltpan (Partridge et al., 1997). On the other hand, she takes the view that the rainfall depression averaging 30% during modern Sahelian droughts was far less severe in its impact than the persistent arid conditions at the LGM. Other estimates of precipitation reduction at the LGM, based on interpretations of proxy data from the humid tropics of Africa, Asia and South America, range from (25) 30 to 50 (65)% (Peters and Tetzlaff, 1990; Heaney, 1991; Van der Hammen and Absy, 1994; Verstappen, 1994).

On the question of Holocene aridity, around 4.5 to 4 cal kyr BP, Bonnefille and Chalié (2000) speculate that the data could be construed to imply a higher climatic variability at a century scale, rather than an overall aridity. This would fit the observations of periodically low flood levels in the Nile post 5 ka (Hassan, 1997; Bryson and Bryson, 1997). However, it is likely that "aridity" in the equatorial highlands of East Africa has never been severe, and that variations of temperature and CO_2, may have had a more significant effect on vegetation. Reviewing the evidence from Atlantic equatorial Africa, Vincens et al. (1998) concluded that the pollen evidence is consistent with a widespread arid event after 4 cal kyr BP. They also suggest that it may have been of long duration, until 1,300 BP at Lake Sinnda in Congo.

2.3 Vegetation Changes

Pollen spectra have been the most widely used proxy data source for palaeoclimatic interpretation, and inferences regarding shifts of regional rainfall means based on cores from lake and mire sites converge to show a decline of the rainforests over wide areas of the tropics. Servant et al. (1993) drew comparisons between Carajás in North Brazil, Salitre in SE Brazil, and Barombi Mbo in South Cameroun, and found that all three sites were forested until after 30 ^{14}C kyr BP. Regression or fragmentation of the rainforest appears to have commenced shortly after this at Lake Bosumtwi in South Ghana (Maley, 1987), but was delayed until after 23 ^{14}C kyr BP at Barombi Mbo. At these and many other sites in the humid tropics, the rainforest was not re-established until after 9.5 ^{14}C kyr BP (10.8 cal kyr BP). Maley (1987; 1996) locates forest survival at the LGM in Africa where the rainfall is high (generally 2,500–3,000+ mm a^{-1}) or where there is minimum seasonality, with the site at Lake Barombi Mbo situated close to the margin. Elenga *et al.* (1994) analysed pollen from a swamp on the southern Batéké Plateau in Congo (1–4° S, with rainfall 1,300–1,900 mm a^{-1}) and concluded that hydromorphous forest gave way to hygrophytic grassland after 24 ^{14}C kyr BP until around 13 ^{14}C kyr BP (15.5 cal kyr BP). Runge (2001a, Annex 1, Map), in his summary of palaeoenvironmental conditions in the Congo basin, shows that cooler, dry climates, with open grasslands or savannas and riparian or gallery woodlands became widespread after 40 ^{14}C kyr BP. Forested conditions persisted or were re-established after ca 5 kyr in the central and western parts of the basin, but the savannas and open woodlands returned by ca 25 ^{14}C kyr BP, and persisted until 15 to 11 cal kyr BP. For several millennia either side of the LGM, savannas characterised the whole of the Congo basin. Between 17 and 11 cal kyr BP, the forest returned to most areas and remained undisturbed for up to 6 kyr. A mid-Holocene (ca 4 cal kyr BP) dry interval lasting no more than 1 kyr is represented at most sites away from the centre of the basin, and forest recession also occurred widely after 2 to 3 cal kyr BP. However, it should be noted that we have hardly any data for the central Congo basin and that there are only 12 main sites currently representing around 1,000,000 km^2 of central Africa.

3   AFRICAN LAKE LEVELS

Lake-level data, particularly from East Africa, have always played a central role in inferences about past climates in the region, but their interpretation has been controversial. This is due to several reasons, which may be summarised as follows:

1. The rift-tectonic setting of the East African lakes introduces non-climatic factors to account for water-level changes: active rifting, possibly lava dams.
2. Comparison between the evidence from shallow lakes such as Lake Victoria and deep troughs such as Lake Tanganyika is difficult unless water-balance studies are incorporated because of differential influences of groundwater fluctuations and wind shear as a factor in evaporation.
3. Comparisons between lakes with local catchments such as Lake Bosumtwi in Ghana, which is an impact crater <11 km in diameter, and lakes with extensive catchments such as Lake Chad have to recognise that the latter has many different sources of surface water, including rivers draining from forested areas to the south.
4. Lake sedimentation can be complex and disturbance of fine sediment may have influenced some cores.
5. Lake shores attract human settlement and this may have affected the environment: by local deforestation and influence on sediment loads in streams entering the lake, thus potentially affecting the lake cores.

Nonetheless, lake-level data indicate major hydrological changes during the LateGlacial Stage (LGS), and the summaries provided by Street-Perrott and Grove (1979) and Street-Perrott *et al.* (1985) have provided a framework for both the status and trend of African lakes. The lowering of most lakes, and the drying out of many across the LGM, begins as far back as 23 kyr BP. A stepwise recovery after 13 ^{14}C kyr BP (15.5 cal kyr BP) is clearly evident and highest lake levels occurred after 11 ^{14}C kyr BP (12.5 cal kyr BP). Individual lakes, however, pose specific problems of interpretation (Figure 11.3).

Lake Victoria has occasioned some debate. Johnson *et al.* (1996), in agreement with Talbot and Livingstone (1989), show that the basin was almost dry after 17.3 ^{14}C kyr BP (20.5 cal kyr BP) and again between 14 and 13 ^{14}C kyr BP (17–15.5 cal kyr BP); its level reduced by 65 m against a maximum depth of 67 m. But the lake floor was

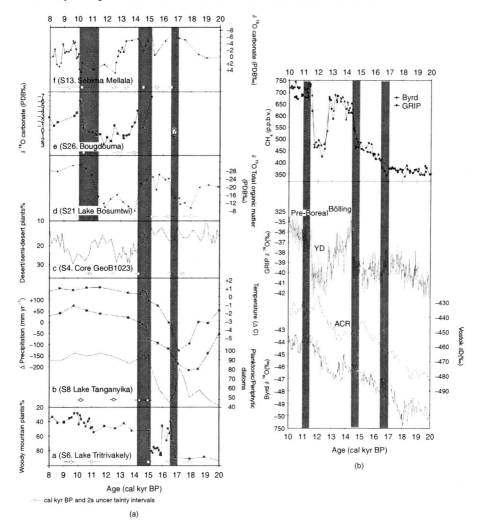

Figure 11.3 (a) Summary of some African lake and pollen records (20–8 cal kyr BP). (b) GRIP (Greenland), Vostok and Byrd (Antarctica) isotopic records. Vertical bands show the major warming/wetting phases in the African tropics. Reprinted from *Quaternary Science Reviews*, 19, Gasse, F., Hydrological changes in the African tropics since the last glacial maximum, 189–211, (2000), with permission from Elsevier Science

rapidly flooded around 12.7 ^{14}C kyr BP (15.2 cal kyr BP) (Talbot and Laerdal, 2000), and between 12.5 and 11.5 ^{14}C kyr BP (14.6–13.6 cal kyr BP) it briefly overflowed (Johnson *et al.*, 2000). Stager *et al.* (1997) examined diatoms from Lake Victoria that record abrupt changes since 11.4 ^{14}C kyr BP (13.2 cal kyr BP). These changes indicate a variably dry Younger Dryas, (13.2–11.5 cal kyr BP; 11.4–10 ^{14}C kyr BP), followed by a major outflow and establishment of open-basin conditions at 11.2 cal kyr BP (Johnson *et al.*, 2000). Increased stratification and reduced diatom productivity in the water column occurred between 8.8 and 6.8 ^{14}C kyr BP (9.8–7.5 cal kyr BP), crossing the important 8.2 cal kyr BP climatic event but lasting for much longer. Water column mixing increased again after 6 cal kyr BP. Seasonal climates after this lasted until 2,200 ^{14}C yr BP (2,160–2,300 Cal BP) and a dry "Little Ice Age" event, 600 to 200 ^{14}C yr BP

(560/620–160/280 Cal BP). A claim by Jolly *et al.* (1998) that Lake Victoria was at a high stand at the LGM appears unsubstantiated.

Beuning *et al.* (1997) have recently confirmed a similar pattern of behaviour for Lake Albert. Here two palaeosols, dated to 18 ^{14}C kyr BP (21.5 cal kyr BP) and 13 ^{14}C kyr BP (15.5 cal kyr BP), respectively, mark the drying out of the lake at the core sites to 54 m below the present level. Potentially, this would leave only ca 10 m of water in the lake at its lowest prolonged stand which ended with flooding after 13 ^{14}C kyr BP (15.5 cal kyr BP). Although Beuning *et al.* (1997) thought the overflow of Lake Victoria into the White Nile was delayed until the mid-Holocene, Talbot *et al.* (2000) have concluded that this link was established much earlier, around 11.5 ^{14}C kyr BP (13.4 cal kyr BP). For Lake Tanganyika, Gasse *et al.* (1989) used an analysis of diatomaceous sediments to estimate a lowering of water level by 350 to 400+ m during the LGS (their Stage II, 21.7–12.7 ^{14}C kyr BP = 14.3 cal kyr BP). However, Baltzer (1991) and Vincens *et al.* (1993) argue for a lower estimate of water-level reduction at the LGM of 250 to 300 m, which they claim would correspond with only a 15 to 23% (max) reduction in precipitation. Baltzer (1991), Vincens *et al.* (1993) and Gasse (2000), all converge on a reduction in precipitation at the LGM ranging from 14 to 24%. This is less than the average 32% (20–15 ka) estimated by Bonnefille and Chalié (2000), who also suggest that the greatest reductions in rainfall came at 19 ^{14}C kyr BP (22.5 cal kyr BP) (−45%) and 18 to 16 ^{14}C kyr BP (21.5–19 cal kyr BP) (−42%). Bergonzini *et al.* (1997) suggest a figure of 25%, considered a minimum by Tyson *et al.* (1997). It is also interesting to note that Cohen *et al.* (1997) found that the level of Lake Tanganyika has been remarkably stable over the last 2,800 years, an indication that short-term, minor precipitation changes have had little effect on the lake level. Lake Abhé in the Ethiopian Highlands was also low at the LGM (Gasse, 1977; Gasse and Street, 1978).

Across the continent in the rainforest zone of Ghana, Lake Bosumtwi displays complex behaviour but was generally low from ca 28 ^{14}C kyr BP until after 12.5 ^{14}C kyr BP (15 cal kyr BP), with dry episodes around 35.9, 21.9, 17.5 ^{14}C kyr BP. The reappearance of the rainforest occurred at ca 9,500 ^{14}C kyr BP (10.6 cal kyr BP) (Maley, 1991; 1996; Talbot and Johannessen, 1992; Talbot *et al.*, 1984). Although much has been made of the data from Lake Bosumtwi, Turner *et al.* (1996) have used a water-balance model based on historical records to suggest that the full range of lake levels observed in terrace deposits could have arisen from stochastic climatic variations similar to those observed this century! This emphasises the sensitivity of small lakes to short-term climate fluctuation. Situated across the Sahelian zone, Lake Chad responds to both groundwater recharge and inflows from the south (Rivers Logone and Chari). Edmunds *et al.* (1999) indicate that groundwater recharge ceased after 23.5 ^{14}C kyr BP, despite a temperature reduction of 6°C, and that the lake shrank to 7% of its modern extent.

Not all lakes in tropical East Africa show consistent behaviour. Lake Malawi (10–12° S latitude), for example, appears to have been relatively high at the LGM and prior to 40 ^{14}C kyr BP (Finney and Johnson, 1991; Scholtz and Finney, 1994), and was correspondingly low from 40 to 28 ^{14}C kyr BP, and from 10 to 6 ^{14}C kyr BP (11.5–6.8 cal kyr BP). Owen *et al.* (1990) also state that the lake was low prior to 25 ^{14}C kyr BP and at 10,740 ± 130 ^{14}C yr BP (12.6 cal kyr BP). According to De Busk (1998), two pollen cores from Lake Malawi record the vegetation history of the catchment, and show extremely dry conditions from 37.5 to 35.9 ^{14}C kyr BP, which were followed by a cold, moist climate supporting montane forests until 34 ^{14}C kyr BP. Warm-dry conditions led to the contraction of forest between 34 and 26.4 ^{14}C kyr BP, but the forest expanded during the LGM, indicating a lack of aridity in the catchment at this time. The Holocene record does not indicate that major fluctuations though drier conditions occurred between 8 and 6.1 ^{14}C kyr BP (9–7 cal kyr BP).

In neighbouring Zambia, Stager (1988) found that Lake Cheshi did not decline in level until 15 to 13 ^{14}C kyr BP (18–15.5 cal kyr BP). These records come from Sites 14 to 17° S latitude, and there is a possibility that the rainfall response to changes in the late Pleistocene circulation was reversed in this zone. This is supported by Optically Stimulated Luminescence (OSL) dating of regional dune building in Western Zambia, which, according to O'Connor and Thomas (1999) peaked at 32 to 27 ka, 16 to 13 ka, and 5 to 4 ka. Such observations emphasise the need to consider regional as well as global and zonal patterns of climate and hydrologic change. According to Jolly et al. (1998), there is a split between lakes east and west of the 34° E meridian, but this remains somewhat speculative.

The global cooling post 40 kyr BP was accompanied by changes in global atmospheric and oceanic circulations as well as in ocean–land heat transfers. Weakening of the Intertropical Convergence (ITCZ), and southward displacement of the northern Jetstream, for example, almost certainly reduced the penetration of the SW monsoon into the Sahel and might also have restricted the penetration of rain-bearing air masses from the Indian Ocean into East Africa (Figure 11.2). Gasse (2000) also points out that orbital precession is in antiphase between northern and southern hemispheres and this should predict a wet LGM in the southern hemisphere. However, in southern Africa, a dry LGM is documented from the evaporite geochemistry in the Pretoria Saltpan (Partridge et al., 1997), which suggests drying out post 30 ka, with a loss of precipitation of 15 to 20%. In Botswana, Holmgren et al. (1995) indicate warm and wet conditions from 51 to 43 ^{14}C kyr BP (Isotope Zone 3), and dry periods possibly corresponding with Heinrich events at 43, 26, 24, and 22 ^{14}C kyr BP. However, Gasse (2000) notes evidence for higher δ ^{18}O values from groundwater at Stampriet in Namibia, which are thought to indicate the northward extension of westerly winds to 25° S. Lowered LGM SSTs (by 2.5–3.5°C) are indicated by two ocean drilling sites, in the South Atlantic (GeoBio1023) and the Indian Ocean (MD79257) (Bard et al., 1997). In the nearby highlands of Madagascar, Lake Tritrivakely was ephemeral or dry at the LGM (Gasse and Van Campo, 1998). After this time the records become less clear, a reduction in diatoms at the time of the Younger Dryas suggests a dry episode but the pollen spectra for this site show little change (Figure 11.4). The apparent dryness of southern Africa at the LGM may be attributable to the impact of ocean circulation and lowered SSTs on vapour transport and monsoon circulation. On the other hand, any increase in precipitation from the northern penetration of the Westerlies is difficult to detect at these sites.

4 POSTGLACIAL CLIMATE CHANGES

Antarctic warming is evident at Vostok from 18.5 cal kyr BP (earlier at Byrd), while Arctic changes came later (14.6 cal kyr BP – Bölling) (Blunier et al., 1998; Gasse, 2000). A stepwise onset of postglacial conditions is deduced by Gasse (2000). In South Africa, warming began around 18 cal kyr BP at the Pretoria Saltpan (Partridge et al., 1997) and wetter conditions ensued after 17 cal kyr BP. In East Africa, most lakes started to refill around 15 cal kyr BP. In Lake Tanganyika, warming and precipitation increase took place 18 to 17 cal kyr BP and 15 to 14.5 cal kyr BP (Chalié, 1995). At Lake Victoria, dry conditions ended at 17.9 cal kyr BP but recurred, terminating around 15.3 cal kyr BP (Johnson et al., 1996). According to Beuning et al. (1997), the

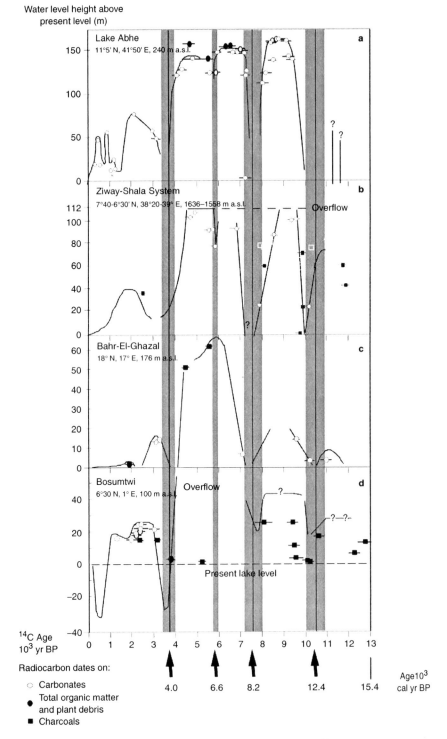

Water level height above
present level (m)

Figure 11.4 Changes in lake levels of the past 15.4 cal kyr (13 ^{14}C kyr BP). Vertical bands indicate weak monsoon (dry) episodes. Reprinted from *Quaternary Science Reviews*, 19, Gasse, F., Hydrological changes in the African tropics since the last glacial maximum, 189–211, (2000), with permission from Elsevier Science

behaviour of Lake Albert replicates these results. A cool-dry interval at the time of the Younger Dryas is evident in Lakes Kivu (Haberyan and Heckly, 1987) and Magadi (Williamson *et al.*, 1993; Roberts *et al.*, 1993).

At Lake Bosumtwi, the record is complex: a high lake level was recorded after 15.5 cal kyr BP, but dry excursions occurred around the Younger Dryas, before the lake rose to its highest level around 11.5 cal kyr BP, a time when Lake Barombi Mbo in Cameroun also experienced high water levels (Maley and Brenac, 1998) (Figures 11.4 and 11.5). In the Sahel, the reactivation of the monsoon appears to have occurred in two steps: 15 to 14.5 cal kyr BP and 11.5 to 11 cal kyr BP (Street-Perrott and Perrott, 1990) with a dry Younger Dryas episode intervening. In the Indian Ocean, Sirocko *et al.* (1993) recorded intensification at 16.0, 14.5, and 11.5 cal kyr BP. According to Gasse (2000), these changes dominate Africa north of 10° S. The earlier phase was weak in West Africa, while the 11.5 to 11 cal kyr BP period became very wet in "northern" Africa, so defined, but southern Africa became drier in the early Holocene.

5 HOLOCENE CLIMATE FLUCTUATIONS

Equatorial and northern Africa experienced a favourable precipitation–evapotranspiration (P-E) balance from 11 to 7.8 cal kyr BP, illustrated by the diatom record of Lake Victoria (Stager and Mayewski, 1997). Hastenrath and Kutzbach (1983) estimated precipitation 20 to 35% higher around 9 cal kyr BP, from Lakes Naivasha, Nakuru and Victoria. Lake Victoria overflowed into the White Nile at 11.5 ^{14}C kyr BP (13.5 cal kyr BP) (Talbot *et al.*, 2000; Williams *et al.*, 2000). Lake Kivu overflowed around 10.5 ^{14}C kyr BP (12.0 cal kyr BP) and Lake Tanganyika overflowed into the Congo around 11 ^{14}C kyr BP (13.0 cal kyr BP). A strong palaeoclimatic (palaeomonsoon) signal at 11.5 ^{14}C kyr BP (13.5 cal kyr BP) is evident from a lacustrine core at Sacred Lake, Mt Kenya, which also shows the impact of the 23,000-year precessional cycle (Olago *et al.*, 2000). There is also evidence for the former existence of lakes in the larger alluvial/lacustrine basins. The "Megachad" phases in the Lake Chad basin are thought to have occurred at ca 9 cal kyr BP and 6.5 cal kyr BP (Servant and Servant-Vildary, 1980), though Durand (1995) has disputed some of the evidence for elevated shorelines. In the Faiyum depression, Egypt, Hassan (1981) recorded evidence for high lake stands ca 11.5 cal kyr BP, from 8.5 to 7 ^{14}C kyr BP (9.5–7.8 cal kyr BP) and 6.5 to 5.1 ^{14}C kyr BP (7.3–5.8 cal yr BP). Other lakes may have occurred in the Bahr-el-Ghazal, Middle Niger, and even the Congo Basin (Peters and O'Brien, 2001).

In West Africa, Lake Bosumtwi contained freshwater from ca 15 to 10 cal kyr BP, but with a short dry excursion around the Younger Dryas (12.4 cal kyr BP). By ca 11.2 cal kyr BP, the lake was 185 m deep and may have overflowed. It remained high until a brief fall from 8.3 to 8 cal kyr BP, and again from 4.2 to 4 cal kyr BP, when it declined to −30 m. The claim by Gasse (2000) that the region experienced wet conditions more or less continuously from 8 to 7.5 cal kyr BP until 4 ka, becoming progressively drier until 2.2 cal kyr BP, may hide significant changes in precipitation during this period. However, the comparison offered between Lake Bosumtwi and the Bahr-el-Ghazal depression is striking (Figure 11.4) and shows clearly the short dry episodes at 12.4, 8.2, and 4.0 cal kyr BP. In Ethiopia, Lake Turkana experienced an abrupt change towards arid conditions at 4 cal kyr BP (Ricketts *et al.*, 1996), and Said (1993) notes that the main Nile flood was reduced after 4.2 cal kyr BP. The monsoon-supported Nubian lakes (Hoelzmann *et al.*, 2000; Pachuv *et al.*, 1997, 2000) disappeared and dry conditions across northern Africa at this time are well documented.

In the Sahel, moist conditions arrived later than in the humid tropics, but an early-mid-Holocene pluvial period became well established around 10 cal kyr BP and persisted until 6 cal kyr BP, but with a dry interval between 8.5 and 8 cal kyr BP (Cremaschi and Di Lernia, 1998). The northern limit of this "palaeosahel" reached to 22° N (Petit-Maire *et al.*, 1993). The hydrology of the lakes along the Saharan fringe of West Africa is complex, and influenced by local precipitation patterns, groundwater flows and convergent surface flows from neighbouring highlands. All lakes do not respond in an identical manner to climate change of given magnitude. A large palaeolake in NW Sudan (precipitation today 15 mm y^{-1}) appeared between 9.5 and 4 ^{14}C kyr BP (10.7–4.4 cal kyr BP), and according to Hoelzmann *et al.* (2000), this would have required a precipitation of 500 mm. Guo *et al.* (2000) present results from a statistical comparison of 560 radiocarbon dates from the Sahara and recognise drier episodes at 8 ka and either side of 6 ka, but the most severe drying out occurred after 5 ka, reaching its greatest severity by 4 ka, after which recovery was short-lived and hyperaridity persists (Figure 11.5). De Menocal *et al.* (2000) refer to the "African Humid Period", in connection with the terrigenous input to an ocean core site (658 C) off the Mauritanian coast, which they date from 14.8 to 5.5 cal kyr BP. Recent analyses of aeolian sediments corroborate these findings, showing sediment mobilisation prior to 11 cal kyr BP and after 5 cal kyr BP (Swezey, 2001). An arid phase in Sénégal at 4.5 cal kyr BP, followed by a brief wet episode at 3.5 cal kyr BP corresponds well

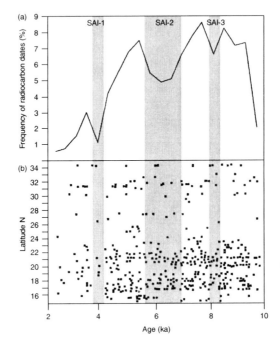

Figure 11.5 Temporal and latitudinal distribution of surface freshwater indicators from the Sahara. (a) Age frequency of the radiocarbon-dated markers, calculated at 400-year intervals and expressed as a percentage of the total number (560 dates). (b) Latitudinal distribution of the dated markers. Vertical bands indicate drier episodes. All dates are given as radiocarbon years BP (^{14}C kyr BP). Reproduced from Guo, Z., Petit-Maire, N. and Kropelein, S., 2000. Holocene non-orbital climatic events in present-day arid areas of northern Africa and China. *Global and Planetary Change*, **26**, 97–103, with permission from Global and Planetary Change

with the statistical findings (Guo *et al.*, 2000). On the other hand, the central Sahara experienced lacustrine conditions and fluvial activity in antiphase to the experience of the Sahel, and was moist during the LGM 20 to 15.5 cal kyr BP, and again from 15 to 12.5 cal kyr BP.

Gasse (2000) identifies three major Holocene climate shifts: 8.5 to 7.8, 7.0 to 6.6 and 4.5 to 3.5 cal kyr BP. The first of these shifts corresponds with a prominent event in the Greenland ice cores that led to cool, wet conditions in the North Atlantic region. Loss of rainfall and/or increased evaporation appears to have affected many sites in northern tropical Africa during this period, but the duration and intensity of the event is unclear. A later dry episode at 7.0 to 6.6 cal kyr BP is picked up at a number of northern sites but appears to have been of low amplitude. On the other hand, a dry event around 4.5 to 4.0 cal kyr BP appears to have been very widespread and of much greater intensity in many areas (Lamb *et al.*, 1995). This affected the Mediterranean and Mesopotamia, and possibly led to the collapse of their urban civilisations (Dalfes *et al.*, 1997). Gasse and Van Campo (1994) hold that weakened monsoons correspond with 8 ka and 4 ka events and also with reduced North Atlantic SST. An exception may be the equatorial Atlantic littoral, where Elenga *et al.* (2001) have found evidence of the continuity of dense swamp forests from 7 to 3 cal kyr BP, with a major regression after 3.0 cal kyr BP.

Early Holocene climates south of ca 10° S latitude have been found to be dry. Lake Malawi was 100 to 150 m lower in early Holocene (Finney and Johnson, 1991), a decrease in precipitation of at least 25%. In southern Africa, a site at Stampriet was dry from 10.9 to 9.3 cal kyr BP, but wet and warm from 6.3 to 4.8 cal kyr BP. It appears that the Holocene Optimum, at least in the higher southern latitudes was delayed until ca 8.8 cal kyr BP (Gingele, 1996). On the other hand, the northern Kalahari was dry between 4.8 and 3.3 cal kyr BP (Ning Shi *et al.*, 1998). Maximum aridity in Madagascar occurred around 4.2 cal kyr BP (Gasse and Van Campo, 1998). No thorough review has been made or is really possible concerning evidence for recent climate changes, but De Menocal *et al.* (2000) show that SSTs off West Africa fell by 3 to 4°C during the Little Ice Age, 1350 to 1850 AD.

6 LATE-QUATERNARY PALAEOHYDROLOGY AND FLUVIAL ACTIVITY

Climate cooling set in well before the LGM, possibly shortly after 40 ka BP, but climates appear to have remained moist for nearly 10 kyr, interrupted by short episodes of aridity. As we approach the LGM (post 23 kyr BP), the drying out of climate becomes apparent across tropical Africa north of 10° S. The situation in southern tropical Africa remains complex and to some extent uncertain, mainly because there is an absence of both highland peat bogs and rift-valley lakes, which have supplied the major understanding of climate change farther north. Several studies of Lake Malawi seem to confirm that it was out of phase with the northern lakes, but in phase with periods of dune building in western Zambia. In southern Africa, including Botswana and Namibia, cold–dry conditions prior to the LGM and increased wetness at that time are replicated at several sites (Dollar, 1998). It now seems possible, therefore, to partition Africa into the inner equatorial and northern tropical areas, and those beyond 10° S, though this parallel cannot be used as a boundary. East–west contrasts may also exist in southern Africa, particularly in the Kalahari, where differences of opinion persist (Thomas, 1994).

In central Africa, there is some evidence from the Congo of high rates of morphodynamic activity from >40 to ca 37 ^{14}C kyr BP (De Ploey, 1965; 1968; Preuss, 1986), but discussion of this and similar phenomena relating to Isotope Stage 3 lie outside the scope of this review. However, from 35 to 36 ^{14}C kyr BP until ca 29 to 28 ^{14}C kyr BP conditions were warm, humid and forested across central and western Africa, after which conditions became progressively cooler and drier. Forest regression at this time is documented from around Kinshasa, the Batéké Plateau near Brazzaville, in the central Congo basin (see Runge, 2001), and also in Cameroun and probably across West Africa. At the time of the LGM (21–18 cal kyr BP) cool-dry conditions predominated, with many rainforest areas reduced to forest–savanna mosaics or more open vegetation.

In the interior Congo Basin, Preuss (1990) has recorded deposition by braided, sand-bed streams from 23 to 7.2 ^{14}C kyr BP (23–18.5 cal kyr BP), followed by progressively higher humidity and a return to meandering rivers between 17.2 and 11.5 ^{14}C kyr BP (18.5–13 cal kyr BP). But this apparently very early return to humid conditions is not found in the East African lakes, nor is it apparent from the marine record of terrigenous sedimentation from the Congo River (Marret et al., 1999, Jansen et al., 1984), where arid conditions lasted until 14.5 cal kyr ^{14}C BP. Indications of strong morphodynamic activity in the eastern part of the Congo Basin were interpreted as indicators of more open, savanna conditions by Runge (1992; 1996), who obtained dates of 17 to 18 ^{14}C kyr BP (20.2–21.5 cal kyr BP) for unidentified trees buried in allochthonous sediments.

Drainage from the East African lakes into the White Nile has attracted much attention. According to Talbot et al. (2000), the overflow of Lakes Victoria and Albert into the Nile drainage network had occurred by 11.5 ^{14}C kyr BP (13.5 cal kyr BP), before the Younger Dryas dry interval, when Lake Victoria may have been closed. Evidence downstream in the lower White Nile (Williams, 1980; Williams et al., 2000) confirms late Pleistocene high water levels, followed by a marked regression around the time of the Younger Dryas (12.8–12.2 cal kyr BP). Wetter and warmer climates then prevailed around 9.5 ^{14}C kyr BP (11.0 cal kyr BP) and between 8.5 and 7.0 ^{14}C kyr BP (9.5–7.8 cal kyr BP). According to Adamson et al. (1980), the lower Nile was a wide braided stream from 20 to 12.5 ^{14}C kyr BP (22.0–14.5 cal kyr BP). The sapropel muds in the East Mediterranean were dated at 11.8 to 10.4 ^{14}C kyr BP (13.6–12.8 cal kyr BP) and 9.0 to 8.0 ^{14}C kyr BP (10.2–9.0 cal kyr BP) by Rossignol-Strick et al. (1982) and interpreted as marking periods of high discharges entering from the River Nile.

Kadomura (1986) pointed out the correspondence between the earlier sapropel date and the increased influxes of sediment to the submarine fans of the Congo (Giresse and Lanfranchi, 1984; Giresse et al., 1982) and Niger (Pastouret et al., 1978) rivers. A more recent core from the Congo fan (Marret et al., 2001) offers a high-resolution record of palaeodischarges since the LGM. Arid-cool conditions persisted after the LGM from 18.0 to 13.5 cal kyr BP, revealed in low, freshwater flux indicators and relatively high grass pollen. Although some variability occurs around 16.5 cal kyr BP, a rapid increase in forest pollen, and reduction of grass pollen is accompanied by high sedimentation rates and inferred palaeodischarges from 13.5 to 12.9 cal kyr BP (there is brief dry excursion 13.3–13.1 cal kyr BP). After 12.9 cal kyr BP, high discharges occur with evidence of widespread lowland rainforests.

A new core from the Niger fan has also been analysed (Zabel et al., 2001) but the low resolution of these cores reveals only the broad outlines of the river's Quaternary history. The authors consider the relationships between palaeodischarge, based mainly on an inverse relationship with marine productivity ($CaCO_3$ accumulation rate); chemical weathering indices (using elemental fluxes (especially K/Al), and estimates of suspended sediment (K, Ti, Zr accumulation rates). They conclude that periods of low discharge correspond with dry cool periods, high sediment load and low weathering rates (high K/Al from feldspar). Periods of high discharge correspondingly relate to low suspended sediment and increased weathering (low K/Al from clays). Aeolian contributions to the core are considered to be <15%. They attribute these relationships to the importance of vegetation and soil cover in controlling sediment supply to the river. While this reasoning agrees with theory and other observations, the core fails to record the marked sediment pulse post $11,500 \pm 650$ [14]C yr BP (ca 13.3 cal kyr BP) noted by Pastouret *et al.* (1978). The authors do not comment on this, but it emphasises the way in which single cores in submarine fans derived from very large river systems cannot be regarded as providing representative data.

The behaviour of Africa's three largest river systems during the postglacial transition can now be regarded as clearly revealed, but we still have very little to inform us about their Holocene histories. It can be argued that more information can be derived from small river systems with limited catchment areas, but there have been few such studies. In fact, it is difficult to point to any recent work that would test the conclusions drawn from rivers in Sierra Leone and Ghana (Hall *et al.*, 1985; Thorp and Thomas, 1992; Thomas and Thorp, 1980; 1995). Both these areas lie within the rainforest zone today, but might be considered marginal and sensitive to the effects of climate change. The valleys of headwater streams of the Rivers Moa and Sewa, in forest marginal areas of Sierra Leone (9° N latitude, rainfall 2,000+ mm a^{-1}), yielded no radiocarbon-dated (from >30 dates) sediments between ca 20.5 ^{14}C kyr BP (24 cal kyr BP) and 12.7 ^{14}C kyr BP (15.3 cal kyr BP), and this finding was confirmed by subsequent work on the River Birim in Ghana, within 50 km of Lake Bosumtwi (6° N. latitude, rainfall 1,750+ mm a^{-1}) (Thomas and Thorp, 1980; Hall *et al.*, 1985). Prior to ca 21.0 ^{14}C kyr BP (>24 kyr), the rivers had broad floodplains, represented today by a low terrace, with thick overbank, fine deposits and widespread swamps. During the hiatus of ca 7 kyr, there is no evidence for major sedimentation or entrainment of riparian forest trees due to bank erosion. Independent evidence for aridity supports the view that discharges were greatly reduced as a consequence of low rainfall. After 12.7 ^{14}C kyr BP (15.3 cal kyr BP), a different regime brought about episodes of strong scour and channel incision, accompanied by deposition of coarse gravels, presumably responding to high peak flows and possibly high rates of slope erosion during the terminal Pleistocene. There is also evidence of a Younger Dryas arid interval intervening before the main pluvial phase of the early Holocene that began after 10.0 ^{14}C kyr BP (11.5 cal kyr BP) and continued for around 3 kyr, marked by abundant sedimentation. Very thick overbank deposits spanning the pluvial are found in the middle Birim in Ghana and these may record the results of major palaeofloods of this period (Thomas, 1994; Thorp and Thomas, 1992) (Figure 11.6).

When the Holocene records from the sites in Sierra Leone and Ghana are re-examined, it is clear that a cluster of radiocarbon dates occurs between ca 11.0 and ca 8.0 to 7.8 ^{14}C kyr BP (13.0–8.6 cal kyr BP). After this, ages are scattered from 6.8 to 4.8 ^{14}C kyr BP (7.6–5.5 cal kyr BP), and then become clustered in larger numbers from 3.5 ^{14}C kyr BP (3.8 cal kyr BP) onwards. There are many dates of younger age but a specific rhythm or regional pattern is difficult to resolve. There are several reasons for this: the chances of preservation increase towards the present; the resolution of the data rises from 10^3 to 10^2 yr, revealing details unavailable in previous millennia;

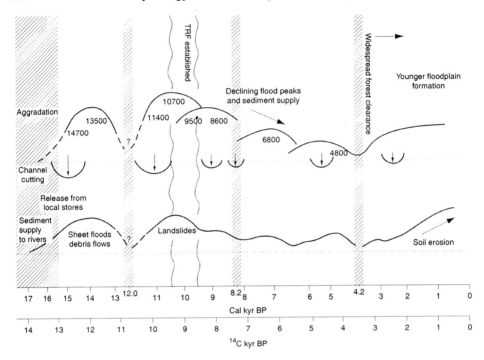

Figure 11.6 Late-Quaternary stream activity from humid tropical West Africa. Curve peaks indicate timing of major sedimentation events; troughs/arrows indicate probable periods of channel cutting. Vertical bands indicate periods of drier conditions. Adapted and revised from Thomas, M.F. and Thorp, M.B., 1995. Geomorphic response to rapid climatic and hydrologic change during the late Pleistocene and early Holocene in the humid and sub-humid tropics. *Quaternary Science Reviews*, **14**, 193–207, with permission from Quaternary Science Reviews

local cut-and-fill sequences are often preserved and probably represent internal readjustments within the fluvial system. Sampling also now picks up sediments of recent date within fingertip tributaries that often preserve no record of their earlier history.

The question of human occupation and the impact of forest clearance and agriculture on sediment yields may ultimately be shown to have affected the older records, but in the Sierra Leone sites, artefacts are associated only with the floodplain sediments younger than ca 4 ka.

Other comparisons remain few. Maley (2000) described thick fine-grained fluvial and lacustrine formations in the Tibesti and other Saharan upland areas with ^{14}C dates of 20.0 to 15.0 ^{14}C kyr BP (23.7–17.9 cal kyr BP) and 15.0 to ca 12.5 ^{14}C kyr BP (17.9–14.7 cal kyr BP), when surrounding lowlands were hyperarid. He concludes that these high altitude locations received low-intensity precipitation from depressions formed on the southern margins of the Westerlies, displaced southwards during the LGM when monsoon-derived precipitation was generally suppressed. Maley (2000) also suggests that it was runoff from the Red Sea Hills towards the lower Nile that accounts for terrace sediments in Nubia dating from this period, when the headwaters of the Nile system experienced dry climates and were detached from the lake system of East Africa. He notes the correspondence of these events with those in high north latitudes, marked by Heinrich events 1 and 2 (see also Leuschner *et al.*, 2000).

7   OTHER EVIDENCE OF MORPHODYNAMIC ACTIVITY

The understanding of catchment conditions also requires knowledge of slope activity and deposits, including landslides, alluvial fan deposits and more amorphous colluvium. Unfortunately, the dating of such deposits has until very recently proved difficult, because of the absence of accessible, buried organic remains. Most types of sediment can now be dated, using OSL techniques, providing they have been exposed to sunlight and rapidly buried.

There have been several accounts of widespread colluvium from central, eastern and southern Africa (Goudie and Bull, 1984; Meadows, 1983; 1985; Watson *et al.*, 1983; Wintle *et al.*, 1995; Runge, 1992; 1996; 2001b; Eriksson *et al.*, 1999; Sørensen *et al.*, 2001; Thomas, 1994; 1999; Thomas and Murray, 2001). But few recorded sequences have been securely dated. Eriksson *et al.* (1999) indicated two colluvial episodes in central Tanzania, one from 14.5 to 11.4 ka (OSL) and the other post 1 ka. Runge (1996) has dated allochthonous deposits from eastern Congo containing fallen trees to 17.0 to 18.0 ^{14}C kyr BP (20.0–21.5 cal kyr BP), around the LGM.

In E Brazil, Coelho-Netto (1999) recently summarised detailed work on hillslope colluvium and alluvium from the Bananal area, and this shows a remarkable concentration of radiocarbon-dated deposits around 10.0 ^{14}C kyr BP (11.5 cal kyr BP), with hardly any dates between 10 kyr BP and 1 kyr BP. But assumptions about the probable ages of other slope deposits are not justified. Thomas (1999) conjectured that prominent landslide deposits in East Zambia could date to the Pleistocene–Holocene transition, but subsequent dating using OSL determinations for intercalated sands returned ages > 180 to 300 ka (Thomas and Murray, 2001). Runge and Tchamié (2000) dated alluvial deposits from the Niantin river in North Togo; sediments that he hypothesised might relate to Lateglacial age aridity. But the Accelerator Mass Spectrometry (AMS) dates ranged between 50 to 250 yr BP, though possible contamination by modern carbon was thought possible.

Both Sørensen *et al.* (2001) and Thomas and Murray (2001) have analysed hillslope colluvium from east central Africa, which they found to be older than the expected Lateglacial age. In Zambia, one 4.8-m section through mainly sandy colluvium containing discontinuous thin beds of coarse material (pebbles, cobbles) ranged in OSL age from 65 ka (at −4.1 m) to Holocene (9 ka at −0.3 m) (Thomas and Murray, 2001). Similar results were obtained from a 12-m section in Morogoro District, Tanzania by Sørensen *et al.* (2001), but here the upper layers were late Holocene in age (4.5–3.5 ka). In Zambia, basal sediments in a neighbouring section through alluvial deposits could be at least 100 ka in age. However, in this area there is a second colluvium, which was not exposed at the time of sampling, but which is probably of younger age.

These few results suggest that different parts of the landscape have become destabilised at different times, and that in the case of the landslides in East Zambia, the slopes have not been reactivated for more than 200 ka, a remarkable fact in itself. The colluvial deposition apparent in this area may span most of the last glacial cycle, but it has not accumulated in a gradual manner with very low energy sand transport across piedmont slopes of 2 to 5 degrees. The occurrence of coarse, cobble-sized sediments at intervals through the section indicates pulses of higher energy. These were probably more than single storm events, and to transport coarse materials 0.5 to 1.0 km from the hillfoot in the absence of channelised flows, the piedmont slopes must have had sparse vegetation (they are under *brachystegia spp*, "miombo" woodland today).

8 CONCLUSIONS AND PROBLEMS

Major environmental changes occurred throughout Africa during the period for which we have data (mostly post 40 kyr BP, but this is changing). However, almost all estimates of rainfall change have been based on boundary values for reconstructed vegetation types, or on changes in palaeolake levels. The former usually arise from cores in moist depressions (mires and lakes) and may not characterise the vegetation of the wider landscape, while the latter can only be reliable when the total catchment, and groundwater contributions are considered alongside estimates of water balance.

In lacustrine and submarine cores, there is seldom evidence available for the direct study of landscape dynamics. Because the overwhelming majority of papers on the Quaternary is devoted to establishing and refining chronologies of climate change, there has been a neglect of the landscape systems (catchments) that yield and con-tribute sediment to sinks such as swamps, lakes and submarine fans. Evidence for morphodynamic activity, which would offer clues to surface water flows and slope instability, and possible climate change, is missing from most classic pollen sites and when gaps in the stratigraphic record are noted by palynologists, they are seldom made the subject of discussion. In fact, they also search for undisturbed sites, away from influxes of sediment.

Many undated deposits from the humid tropics, such as mudflows (recorded in southern Ghana: Thomas and Thorp, 1995), or massive colluvial fills and extensive fans around escarpments such as those of the Jos Plateau should be regarded as important indicators of climate change but disputed origins and an absence of dates have led to their neglect. Clast-supported "stonelines" and coarse colluvium also indicate surface processes that are not apparently active today in humid areas.

Fluvial sedimentation is also of wide significance in the context of climate change. The records that we have for river sediments have been discussed elsewhere (Thomas and Thorp, 1995; 1996; Thomas, 2000; Thorp and Thomas, 1992). A summary of conclusions applicable to the African record is shown in Table 1. This shows that humid conditions in northern tropical Africa persisted until ca 28/27 ^{14}C kyr BP in most areas and that the decline in precipitation and runoff took place after 23 ^{14}C kyr BP in West Africa. Increasing runoff may have occurred before 15 cal kyr BP in the inner equatorial regions, but later in West Africa. It was not until after this date that the re-integration of drainage in the Nile and Niger systems took place. There is evidence for a loss of activity in many rivers during the Younger Dryas, but the direct impact of any vegetation changes at this time is not known. Forest recovery was probably not achieved fully until after the Younger Dryas interval in most northern areas, only the inner Congo basin recovering earlier. The peak humidity that came after the Younger Dryas interval is reflected everywhere for which we have records and moist conditions persisted for several millennia. The importance of a decline in humidity after ca 8.2 to 7.8 cal kyr BP, in response to cooling of northern climates and oceans can be inferred from the West African record. Fewer recorded sedimentary units occur after this date until ca 7 cal kyr BP or later, and this can also be seen in the Nile terrace record. But this was not a period of intense aridity and it is thought that the vegetation and soil cover remained intact promoting low sediment yields. The importance of the mid-Holocene arid period after 4.5 cal kyr BP can also be inferred from the absence of alluvial sediments in the Moa–Sewa system in Sierra Leone. Information from the Birim and from the Nile does not indicate an immediate recovery from this event, which may have lasted more than one millennium. Renewed activity is evident in Sierra Leone from 3.6 cal kyr BP, but only later in the Birim, from ca 2 cal kyr BP.

Table 11.1 Observed fluvial responses in Africa to late quaternary environmental changes

---

1. Pre-27 ka (^{14}C yr BP)
-   Prolonged depositional phase during post 58-ka cooling marked by extensive floodplains (now low terraces), most dated post 40 ka, and by fan deposits; some cut-and-fill episodes as at ca 36 ka probably mark humid/arid fluctuations

2. Post 27/24–15 ka (cal kyr BP)

Larger rivers
-   braided river channels in Congo (after 28 ka BP) → ca 17 ka
-   return to meandering or anastomosing channels diachronous from 17–15 ka →
-   infill of incised channels with localised floodplain formation starting in the early Holocene (dambos formed in headwater valleys in plateau areas such as central Africa).

Smaller rivers
-   wide floodplains with backswamps and fine overbank sediments (>30 – >23^{14}C kyr BP)
-   reduced sedimentation with few or no recorded units during the LGS, but possible incision towards the end of the stage (23–15 ka)
-   channel incision in the late Pleistocene (post 15 ka), with deposition of coarse gravels and sands as braid bars in wide open channels with little overbank sedimentation.

3. Late Pleistocene – early Holocene fluctuations (post 15 ka)
-   cut-and-fill sequences during the Holocene, leading to both lateral and vertical accretion within mainly meandering channel-floodplain systems (generally declining flood peaks and mean discharges with small channels inset into former floodplains)
-   depositional units converge on certain periods but bracketing dates vary (mainly from Nile, Birim, Moa and Sewa systems):

    Unit 1   15 ka → 12.5 ka
    Unit 2   11 ka → 8.2 ka
    Unit 3   4.8 ka → 4 (3.5) ka
    Unit 4   2.0 (3) ka →

Many problems remain in the interpretation of African palaeohydrology, and there have been few studies of either the major river systems or of representative catchments across the continent. On the other hand, the study of palaeolakes has shown significant advances and high-resolution cores from marine sites have added further information on palaeoclimate. The interpretation of pollen spectra from lacustrine sediments and from swamps remains central to the reconstruction of palaeoenvironments, but the inferences made about landscape patterns, as opposed to local or regional pollen averages remain limited by the infrequency of sample sites.

The understanding of palaeohydrology requires direct evidence of former fluvial and slope activity, as well as information concerning palaeolake levels. The study of palaeofloods, which has made significant advances in other regions has made little progress in Africa.

The great exception to this generalisation is the unique record of the Nile floods kept at Cairo in Egypt, the so-called *Nilometer*. We have 900 years of flood records for the Nile. Episodes of high floods occurred in the early 800s AD (ca 1,300 kyr BP) and around 1100 AD (ca 1,000 kyr BP) during the Medieval Warm Period, but also between 1350 and 1470 AD (Hassan, 1981; Goodfriend and Stanley, 1999). According to Hassan (1981), these show correlation with water levels in Lake Chad, but periodicities within the data set are complex (De Putter *et al.*, 1998), and it is not possible to extrapolate such data to other river systems in Africa.

REFERENCES

Adamson, D.A., Gasse, F., Street, F.A. and Williams, M.A.J., 1980. Late Quaternary history of the Nile. *Nature*, **287**, 50–55.

Baltzer, F., 1991. Late Pleistocene and recent detrital sedimentation in the deep parts of northern Lake Tanganyika (East African Rift). *Special Publication International Association of Sedimentology*, **13**, 147–173.

Bard, E., Rostek, F. and Sonzogni, C., 1997. Interhemispheric synchrony of the last deglaciation inferred from alkenone palaeothermometry. *Nature*, **385**, 707–710.

Bergonzini, L., Chailié, F. and Gasse, F., 1997. Palaeoevaporation and palaeoprecipitation in the Tanganyika basin at 18,000 years BP, inferred from hydrologic and vegetation proxies. *Quaternary Research*, **47**, 295–305.

Beuning, K.R.M., Talbot, M.R. and Kelts, K., 1997. A revised 30,000 palaeoclimatic and palaeohydrologic history of Lake Albert, East Africa. *Palaeogeography, Palaeoclimatology, Palaeoecology*, **136**, 259–279.

Blunier, T., Chappellaz, J., Schwander, J., Dallenbach, A., Stauffer, B., Stocker, T.F., Raynaud, D., Jouzel, J., Claussen, H.B., Hammer, C.U. and Johnsen, S.J., 1998. Asynchrony of Antarctic and Greenland climate change during the last glacial period. *Nature*, **394**, 739–743.

Bonnefille, R. and Chailié, F., 2000. Pollen-inferred precipitation time-series from equatorial mountains, Africa, the last 40 kyr BP. *Global and Planetary Change*, **26**, 25–50.

Broecker, W.S., 1995. Cooling the tropics. *Nature*, **376**, 212, 213.

Broecker, W.S. and Denton, G.H., 1990. The role of ocean-atmosphere reorganisations in glacial cycles. *Quaternary Science Reviews*, **9**, 305–341.

Bryson, R.A. and Bryson, R.U., 1997. High resolution simulations of regional Holocene climate: North Africa and the Near East. In H.N. Dalfes, G. Kukla and H. Weiss (eds), *Third Millennium BC Climate Change and Old World Collapse*. NATO ASI Series 149, Springer-Verlag, Berlin, 565–593.

Chalié, F., 1995. Paléoclimats du bassin Tanganyika Sud au cours des 25 derniers mille ans: reconstitution quantitative par le traitement statistique des données polliniques. *Comptes Rendus de l'Académie des Sciences Paris*, **320**, 205–210.

CLIMAP Project Members, 1976. The surface of the ice-age earth. *Science* **191**, 1131–1138.

CLIMAP Project Members, 1981. Seasonal reconstruction of the earth's surface at the last glacial maximum. *Geological Society of America, Map and Chart Series*, **36**.

Coelho-Netto, A.L., 1999. Catastrophic landscape evolution in a humid region (SE Brazil): inheritances from tectonic, climatic and land use induced changes. *Geografia Fisica Dynamica Quaternaria* Supplement **111**, 21–48.

Cohen, A.S., Talbot, M.R., Awramik, S.M., Dettman, D.L. and Abell, P., 1997. Lake level and palaeoenvironmental history of Lake Tanganyika, Africa, as inferred from late Holocene and modern stromatolites. *Geological Society of America Bulletin*, **109**, 444–460.

Cremaschi, M. and Di Lernia, S., (eds), 1998. *Wadi Teshuinat Palaeoenvironment and Prehistory in South-Western Fezzan (Libyan, Sahara)*. All'Insegna del Giglio, CNR, Milano, pp. 332.

Dalfes, H.N., Kukla, G. and Weiss, H., (eds), 1997. *Third Millennium BC Climate Change and Old World Collapse*. NATO ASI Series 149, Springer-Verlag, Berlin, 565–593.

De Busk, G.H., 1998. A 37,500-year pollen record from Lake Malawi and implications for the biogeography of afromontane forests. *Journal of Biogeography*, **25**, 479–500.

De Menocal, P., Ortiz, J., Guilderson, T., Adkins, J., Sarnthein, M., Baker, L. and Yarusinski, M., 2000. Abrupt onset and termination of the African humid period: rapid climate responses to gradual insolation forcing. *Quaternary Science Reviews*, **19**, 347–361.

De Ploey, J., 1965. Position géomorphologique, génèse et chronologie de certains dépots superficiels au Congo occidentale. *Quaternaria*, **7**, 131–154.

De Ploey, J., 1968. Quaternary phenomena in the western Congo. *Means of Correlation of Quaternary Succession 8*. Proceedings VII Congress INQUA, University of Utah Press, 501–515.

De Putter, T., Loutre, M.F. and Wansgard, T., 1998. Decadal periodicities of Nile river historical discharge (A.D. 622–1470) and climatic implications. *Geophysical Research Letters*, **25**, 3193–3196.

Dollar, E.S.J., 1998. Palaeofluvial geomorphology in Southern Africa: a review. *Progress in Physical Geography*, **22**, 325–349.

Dong, B., Valdes, P.J. and Hall, N.M.J., 1996. The changes of monsoonal climates due to earth's orbital perturbations and ice age boundary conditions. *Palaeoclimates*, **1**, 203–240.

Durand, A., 1995. Geomorphological records of neotectonics in the Lake Chad basin: the changes in drainage pattern and the origin of the pseudo shoreline on Lake Megachad in the Kadzel region, Niger. *Comptes Rendus, Académie des Sciences, Paris*, **321**, 223–229.

Edmunds, W.M., Fellman, E. and Goni, I.B., 1999. Lakes groundwater and palaeohydrology in the Sahel of NE Nigeria – evidence from hydrogeochemistry. *Journal of the Geological Society*, **156**, 345–355.

Elenga, H., Schwartz, D. and Vincens, A., 1994. Pollen evidence of late Quaternary vegetation and inferred climate changes in Congo. *Palaeogeography, Palaeoclimatology, Palaeoecology*, **109**, 345–356.

Elenga, H., Vincens, A., Schwartz, D., Fabing, A., Bertaux, J., Wirrmann, D., Martin, L. and Servant, M., 2001. The Songolo estuarine swamp (South Congo) during the middle and late Holocene. *Bulletin de la Société Géologique de France*, **172**, 359–366.

Eriksson, M.G., Olley, J.O. and Payton, R.W., 1999. Late Pleistocene colluvial deposits in central Tanzania: erosional response to climatic change? *GFF*, **121**, 198–201.

Finney, B.P. and Johnson, T.C., 1991. Sedimentation in Lake Malawi (East Africa) during the past 10,000 years: a continuous palaeoclimatic record from the southern tropics. *Palaeogeography, Palaeoclimatology and Palaeoecology*, **85**, 351–356.

Gasse, F., 1977. Evolution of Lake Abhé (Ethiopia and TFAI) from 70,000 BP. *Nature*, **290**, 42–45.

Gasse, F., 2000. Hydrological changes in the African tropics since the last glacial maximum. *Quaternary Science Reviews*, **19**, 189–211.

Gasse, F. and Street, F.A., 1978. Late Quaternary lake-level fluctuations and environments of the northern rift valley and Afar region (Ethiopian and Djibouti). *Palaeogeography, Palaeoclimatology, Palaeoecology*, **25**, 145–150.

Gasse, F. and Van Campo, E., 1994. Abrupt post glacial climatic events in West Asia and North Africa monsoon domains. *Earth and Planetary Science Letters*, **126**, 435–456.

Gasse, F. and Van Campo, E., 1998. A 40,000 years pollen and diatom record from the southern tropics (Lake Tritrivakely, Madagascar Plateaux). *Quaternary Research*, **49**, 299–311.

Gasse, F., Lédée, V., Massault, M. and Fontes, J.C., 1989. Water level fluctuations of Lake Tanganyika in phase with oceanic changes during the last glaciation and deglaciation. *Nature*, **342**, 57–59.

Gasse, F., Téhet, R., Durand, A., Gibert, E. and Fontes, J.Ch., 1990. The arid-humid transition in the Sahara and Sahel during the last deglaciation. *Nature*, **346**, 141–156.

Gingele, F.X., 1996. Holocene climatic optimum in southwest Africa – evidence from the marine clay mineral record. *Palaeogeography, Palaeoclimatology, Palaeoecology*, **122**, 77–87.

Giresse, P., Bongo-Passi, G., Delibrias, G. and Du Plessy, J.C., 1982. La lithostratigraphie des sediments hemiplagiques du delta profond du fleuve Congo et ses indications sur les paléoclimats de la fin du Quaternaire. *Bulletin Société Géologie française, Série*, **24**, 803–815.

Giresse, P. and Lanfranchi, R., 1984. Les climats et les océans de la région congolaise pendant l'Holocène–Bilans selon les échelles et les méthodes de l'observation. *Palaeoecology of Africa*, **16**, 77–88.

Goodfriend, G.A. and Stanley, D.J., 1999. Rapid sand-plain accretion in the Northeastern Nile delta in the 9th Century AD, and the demise of the port of Pelusium. *Geology*, **27**, 147–150.

Goudie, A.S. and Bull, P.A., 1984. Slope processes and colluvium deposition in Swaziland: an SEM analysis. *Earth Surface Processes and Landforms*, **9**, 289–299.

Guilderson, T.P., Fairbanks, R.G. and Rubenstone, J.L., 1994. Tropical temperature variations since 20,000 years ago: modulating inter-hemispheric climate change. *Science*, **263**, 663–665.

Guo, Z., Petit-Maire, N. and Kropelein, S., 2000. Holocene non-orbital climatic events in present-day arid areas of northern Africa and China. *Global and Planetary Change*, **26**, 97–103.

Haberyan, K.A. and Heckly, R.E., 1987. The late Pleistocene and Holocene stratigraphy and palaeolimnology of Lakes Kivu and Tanganyika. *Palaeogeography, Palaeoclimatology, Palaeoecology*, **61**, 169–197.

Hall, A.M., Thomas, M.F. and Thorp, M.B., 1985. Late Quaternary alluvial placer development in the humid tropics: the case of the Birim Diamond Placer, Ghana. *Journal of the Geological Society*, **142**, 777–787.

Hastenrath, S. and Kutzbach, J.E., 1983. Palaeoclimatic estimates from water and energy budgets of East African lakes. *Quaternary Research*, **19**, 141–153.

Hassan, F.A., 1981. Historical Nile floods and their implications for climate change. *Science*, **212**, 1142–1145.

Hassan, F.A., 1997. Nile floods and political disorder in Early Egypt. In H.N. Dalfes, G. Kukla and H. Weiss (eds), *Third Millennium BC Climate Change and Old World Collapse*. NATO ASI Series 149, Springer-Verlag, Berlin, 565–593.

Heaney, L.R., 1991. A synopsis of climatic and vegetation change in Southeast Asia. *Climate Change*, **19**, 53–61 (Special Issue: Tropical Forests and Climate).

Hoelzmann, P., Kruse, H.J. and Rottinger, F., 2000. Precipitation estimates for the Eastern Saharan palaeomonsoon based on a water balance model of the West Nubian Palaeolake basin. *Global and Planetary Change*, **26**, 105–120.

Holmgren, K., Karlén, W. and Shaw, P., 1995. Palaeoclimatic significance of stable isotope composition and petrology of a late Pleistocene stalagmite from Botswana. *Quaternary Research*, **43**, 320–328.

Jansen, J.H.F., van Weering, T.C.E., Gieles, R. and van Iperen, J., 1984. Late Quaternary oceanography and climatology of the Zaire-Congo fan and the adjacent Eastern Angola basin. *Netherlands Journal of Sea Research*, **17**, 201–249.

Johnson, T.C., Kelts, K. and Odada, E., 2000. The Holocene history of Lake Victoria. *Ambio*, **29**, 2–11.

Johnson, T.C., Scholtz, C.A., Talbot, M.R., Kelts, K., Ricketts, R.D., Ngobi, G., Beuning, K., Ssemmanda, I. and McGill, J.W., 1996. Late Pleistocene desiccation of Lake Victoria and rapid evolution of Cichlid fishes. *Science*, **273**, 1091–1093.

Jolly, D., Harrison, S.P., Damnati, B. and Bonnefille, R., 1998. Simulated climate and biomes of Africa during the late Quaternary: comparison with pollen and lake status data. *Quaternary Science Reviews*, **17**, 629–657.

Kadomura, H., 1986. Late glacial-early Holocene environmental changes in tropical Africa: a comparative analysis with deglaciation history. *Geographical Reports, Tokyo Metropolitan University*, **21**, 1–21.

Kadomura, H., 1995. Palaeoecological and palaeohydrological changes in the humid tropics during the last 20,000 years, with reference to equatorial Africa. In K.J. Gregory, L. Starkel and V.R. Baker (eds), *Global Continental Palaeohydrology*. John Wiley, Chichester, 177–202.

Kutzbach, J., Gallimore, R., Harrison, S., Behling, P., Selin, R. and Laarif, F., 1998. Climate and biome simulations for the past 21,000 years. *Quaternary Science Reviews*, **17**, 473–506.

Lamb, H.F., Gasse, F., Ben Kaddour, A., El Hamouti, N., Van der Kaars, S., Perkins, T., Pearce, N.J. and Roberts, C.N., 1995. Relation between century-scale Holocene intervals in tropical and temperate zone. *Nature*, **373**, 134–137.

Lautenschlager, M., 1991. Simulation of the ice age atmosphere – January and July means. *Geologische Rundschau*, **80/3**, 513–534.

Leuschner, D.C. and Sirocko, F., 2000. The low-latitude monsoon climate during Dansgaard-Oeschger cycles and Heinrich events. *Quaternary Science Reviews*, **19**, 243–254.

Maley, J., 1987. Fragmentation de la forêt dense humide Africaine et extension des biotopes montagnards au Quaternaire récent: nouvelles données polliniques et chronologiques. Implications paleoclimatiques et biogéographiques. *Palaeoecology of Africa*, **18**, 307–334.

Maley, J., 1991. The African rainforest vegetation and palaeoenvironments during the Quaternary. *Climatic Change*, **19**, 79–98.

Maley, J., 1996. The African rain forest – main characteristics and changes in vegetation and climate from the Upper Cretaceous to the Quaternary. In I.J. Alexander, M.D. Swaine and R. Watling (eds.). *Essays on the Ecology of the Guinea-Congo Rain Forest*. Proceedings of the Royal Society of Edinburgh (Section B), **104**, 31–73.

Maley, J., 2000. Last glacial maximum lacustrine and fluviatile formations in the Tibesti and other Saharan mountains, and large-scale climatic teleconnections linked to the activity of the Subtropical Jet stream. *Global and Planetary Change*, **26**, 121–136.

Maley, J. and Brenac, P., 1998. Vegetation dynamics, palaeoenvironments and climatic changes in the forests of Western Cameroun during the last 28,000 years B.P. *Review of Palaeobotany and Palynology*, **99**, 157–187.

Marret, F., Scourse, J., Jansen, J.H.F. and Schneider, R., 1999. Changements climatiques et paleoceanographiques en Afrique centrale Atlantique au cours de la derniere deglaciation: contribution palynologique. *Comptes Rendus de l'Academie des Sciences, Serie IIa*, **329**, 721–726.

Meadows, M.E., 1983. Past and present environments of the Nyika Plateau, Malawi. *Palaeoecology of Africa*, **16**, 353–390.

Meadows M.E., 1985. Dambos and environmental change in Malawi, central Africa. In M.F. Thomas and A.S. Goudie A.S. (eds), Dambos: small channelless valleys in the tropics. *Zeitschrift für Geomorphologie* Supplementband **52**, 147–169.

Nicholson, S.E., 1996. A review of climate dynamics and climate variability in Eastern Africa. In T.C. Johnson and E.O. Odada (eds), *The Limnology, Climatology and Palaeoclimatology of the East African Lakes*. Gordon & Breach, Amsterdam, 25–56.

Ning, Shi, Dupont, L.M., Beug, H.J. and Schneider, R., 1998. Vegetation and climate changes during the last 21,000 years in SW Africa based on a marine pollen record. *Vegetation History and Arachaeobotany*, **7**, 127–140.

O'Connor, P.W. and Thomas, D.S.G., 1999. The timing and environmental significance of late Quaternary linear dune development in western Zambia. *Quaternary Research*, **52**, 44–55.

Olago, D.O., Street-Perrott, F.A., Perrott, R.A., Ivanovich, D., Harkness, D.D. and Odada, E.O., 2000. Long-term temporal characteristics of palaeomonsoon dynamics in equatorial Africa. *Global and Planetary Change*, **26**, 159–171.

Owen, R.B., Crossley, R., Johnson, T.C., Tweddle, D., Kornfield, I., Davison, S., Eccles, D.H. and Engstrom, D.E., 1990. Major low levels of Lake Malawi and their implications for speciation of cichlid fishes. *Proceedings of the Royal Society of London*, **240**, 519–553.

Pachur, H.J. and Rottinger, F., 1997. Evidence for a large extended palaeolake in the Eastern Sahara as revealed by spaceborne radar lab images. *Remote Sensing of Environment*, **61**, 437–440.

Pachur, H.J. and Hoelzmann, P., 2000. Late Quaternary palaeoecology and palaeoclimates of the eastern Sahara. *Journal of African Earth Sciences*, **30**, 929–939.

Partridge, T.C., De Menocal, P.B., Lorentz, S.A., Paiker, M.J. and Vogel, J.C., 1997. Orbital forcing of climate change over South Africa: a 200,000 year rainfall record from the Pretoria Saltpan. *Quaternary Science Reviews*, **16**, 1125–1133.

Pastouret, L., Chamley, H., Delibrias, G., Duplessy, J.C. and Theide, J., 1978. Late Quaternary climatic changes in Western Tropical Africa deduced from deep-sea sedimentation off the Niger Delta. *Oceanologica Acta*, **1**, 217–232.

Peters, C.R. and O'Brien, E.M., 2001. Palaeo-lake Congo: implications for Africa's late Cenozoic climate – some unanswered questions. *Palaeoecology of Africa*, **27**, 11–18.

Peters, M. and Tetzlaff, G., 1990. West African palaeosynoptic patterns at the last glacial maximum. *Theoretical and Applied Climatology*, **42**, 67–79.

Petit-Maire, N., Sanlaville, P. and Zan, Z.W., 1995. Oscillations de la limite nord du domaine des moussons africaine, indienne, et asiatique, au cours de dernier cycle climatique. *Bulletin Société Géologique de France*, **166**, 213–220.

Preuss, J., 1986. Jungpleistozäne Klimaänderungen im Congo-Zaire-Becken. *Geowissenschaften in unserer Zeit.*, **4**, 177–187.

Preuss, J., 1990. L'évolution des paysages du bassin intérieur du Zäire pendentes les quarantes derniers millénaires. In R. Lanfranchi and D. Sccwartz (eds), *Paysages Quaternaires de lAfrique centrale Atlantique*. ORSTOM, Paris, 260–270.

Rind, D. and Peteet, D., 1985. Terrestrial conditions at the last glacial maximum and CLIMAP sea-surface temperature estimates: are they consistent?. *Quaternary Research*, **24**, 1–22.

Ricketts, R.D. and Johnson, T.C., 1996. Climate change in the Turkana basin as deduced from a 4000 year long δO^{18} record. *Earth and Planetary Science Letters*, **142**, 7–17.

Roberts, N., Taieb, M., Barker, P., Danati, B., Icole, M. and Williamson, D., 1993. Timing of the Younger Dryas event in East Africa from lake-level changes. *Nature*, **366**, 146–148.

Rossignol-Strick, M., Nesteroff, W., Olive, P. and Vergnaud-Grazzini, C., 1982. After the deluge: Mediterranean stagnation and sapropel formation. *Nature*, **295**, 105–110.

Ruddiman, W.F., 1997. Tropical Atlantic terrigenous fluxes since 25,000 yrs BP. *Marine Geology*, **136**, 189–207.

Runge, J., 1992. Geomorphological observations concerning palaeoenvironmental conditions in eastern Zaire. *Zeitschrift für Geomorphologie, N.F.*, Supplementband **91**, 109–122.

Runge, J., 1996. Palaeoenvironmental interpretation of geomorphological and pedological studies in the rain forest "core areas" of Eastern Zaire (Central Africa). *South African Geographical Journal*, **78**, 91–97.

Runge, J., 2001a. Central African palaeoclimates and palaeoenvironments since 40 ka – an overview. *Palaeoecology of Africa*, **27**, 1–10.

Runge, J., 2001b. On the age of stone-lines and hillwash sediments in the eastern Congo basin – palaeoenvironmental implications. *Palaeoecology of Africa*, **27**, 1–10.

Runge, J. and Tchamié, T., 2000. Inselberge, Rumpfflächen und Sedimente kleiner Flusseinzugsgebiete in Nord-Togo: Altersstellung und morphodynamische Landschaftsgeschichte. *Zentralblatt für Geologie and Palaeontologie*, **1**, 497–508.

Said, R., 1993. *The River Nile: Geology Hydrology and Utilisation*. Pergamon, Oxford, 320.

Scholtz, C.A. and Finney, B.P., 1994. Late Quaternary sequence stratigraphy of Lake Malawi (Nyasa), Africa. *Sedimentology*, **41**, 163–179.

Schultz, H., von Rad, U. and Erlenkeuser, H., 1998. Correlations between Arabian Sea and Greenland climate oscillations of the past 110,000 years. *Nature*, **393**, 54–57.

Servant, M. and Servant-Vildary, S., 1980. L'environment Quaternaire du bassin du Tchad au Cénozoic Supérieur. In M.A.J. Williams and H. Faure (eds), *The Sahara and the Nile*. Balkema, Rotterdam, 133–162.

Servant, M., Maley, J., Turcq, B., Absy, M.L., Brenac, P., Fournier, J. and Ledru, M.P., 1993. Tropical forest changes during the Late Quaternary in African and South American lowlands. *Global and Planetary Change*, **7**, 25–40.

Sirocko, F., Sarnthein, M., Erlenkeuser, H., Lange, H., Arnold, M. and Duplessy, J.C., 1993. Century-scale events in monsoonal climate over the past 24,000 years. *Nature*, **364**, 322–324.

Sørensen, R., Murray, A.S., Kaaya, A.K. and Kilasara, M., 2001. Stratigraphy and formation of late Pleistocene colluvial apron in Morogoro district, Central Tanzania. *Palaeoecology of Africa*, **27**, 95–116.

Stager, J.C., 1988. Environmental changes at Lake Cheshi, Zambia since 40,000 years BP. *Quaternary Research*, **29**, 54–65.

Stager, J.C., Cumming, B. and Meeker, L., 1997. A high resolution 11,400 diatom record from Lake Victoria, East Africa. *Quaternary Research*, **47**, 81–89.

Stager, J.C. and Mayewski, P.A., 1997. Abrupt early to Mid-Holocene climatic transition registered at the equator and the poles. *Science*, **276**, 1834, 1835.

Street-Perrott, F.A. and Grove, A.T., 1979. Global maps of lake fluctuations since 30,000 BP. *Quaternary Research*, **261**, 385–390.

Street-Perrott, F.A. and Perrott, R.A., 1990. Abrupt climatic fluctuations in the tropics: the influence of Atlantic Ocean circulation. *Nature*, **343**, 607–612.

Street-Perrott, F.A., Roberts, N. and Metcalfe, S., 1985. Geomorphic implications of late Quaternary hydrological and climatic changes in the northern hemisphere tropics. In I. Douglas and T. Spencer (eds), *Environmental Change and Tropical Geomorphology*. Allen & Unwin, London, 164–183.

Stute, M., Forster, M., Frischkorn, H., Serejo, A., Clark, J.F., Schlosser, P., Broecker, W.S. and Bonani, G., 1995. Cooling of tropical Brazil (5°C) during the last glacial maximum. *Science*, **269**, 379–383.

Swezey, C., 2001. Eolian sediment responses to late Quaternary climate changes: temporal and spatial patterns in the Sahara. *Palaeogeography, Palaeoclimatology, Palaeoecology*, **167**, 119–155.

Talbot, M.R. and Johannessen, T., 1992. A high resolution palaeoclimatic record of the last 27,500 years in tropical West Africa from the carbon and nitrogen isotopic composition of lacustrine organic matter. *Earth and Planetary Science Letters*, **110**, 23–37.

Talbot, M.R. and Laerdal, T., 2000. The late Pleistocene-Holocene palaeolimnology of Lake Victoria, East Africa based upon elemental and isotopic analyses of sedimentary organic matter. *Journal of Limnology*, **23**, 141–164.

Talbot, M.R. and Livingstone, D.A., 1989. Hydrogen index and carbon isotopes of lacustrine organic matter as lake-level indicators. *Palaeogeography, Palaeoclimatology, Palaeoecology*, **70**, 121–137.

Talbot, M.R., Livingstone, D.A., Palmer, P.G., Maley, J., Melack, J.M., Delibrias, G. and Gulliksen, S., 1984. Preliminary results from sediment cores from lake Bosumtwi, Ghana. *Palaeoecology of Africa*, **16**, 173–192.

Talbot, M.R., Williams, M.A.J. and Adamson, D.A., 2000. Strontium isotope evidence for late Pleistocene re-establishment of an integrated Nile drainage network. *Geology*, **28**, 343–346.

Thomas, M.F., 1994. *Geomorphology in the Tropics*. John Wiley, Chichester, 460.

Thomas, M.F., 1999. Evidence for high energy landforming events on the central African plateau: Eastern province, Zambia. *Zeitschrift für Geomorphologie*, N.F., **43**, 273–297.

Thomas, M.F., 2000. Late Quaternary environmental changes and the alluvial record in humid tropical environments. *Quaternary International*, **72**, 23–36.

Thomas, M.F. and Thorp, M.B., 1980. Some aspects of the geomorphological interpretation of Quaternary alluvial sediments in Sierra Leone. *Zeitschrift für Geomorphologie, N.F.*, Supplementband **36**, 140–161.

Thomas, M.F. and Thorp, M.B., 1995. Geomorphic response to rapid climatic and hydrologic change during the late Pleistocene and early Holocene in the humid and sub-humid tropics. *Quaternary Science Reviews*, **14**, 193–207.

Thomas, M.F. and Thorp, M.B., 1996. The response of geomorphic systems to climatic and hydrological change during the Later Glacial and early Holocene in the humid and sub-humid tropics. In J. Branson, A. Brown and K.J. Gregory (eds), *Global Continental Changes: The Context of Palaeohydrology*. Geological Society Special Publication, **155**, The Geological Society, London, 139–153.

Thomas, M.F. and Murray, A.S., 2001. On the age and significance of Quaternary colluvium in Eastern Zambia. *Palaeoecology of Africa*, **27**, 117–133.

Thompson, L.G., Mosley-Thompson, E., Davis, M.E., Lin, P.N., Henderson, K.A., Cole-Dai, J., Bolzan, J.F. and Liu, K.-B., 1995. Late glacial stage and Holocene tropical ice core records from Huascarán, Peru. *Science*, **269**, 46–50.

Thorp, M.B. and Thomas, M.F., 1992. The timing of alluvial sedimentation and floodplain formation in the lowland humid tropics of Ghana, Sierra Leone, and Western Kalimantan (Indonesian Borneo). *Geomorphology*, **4**, 409–422.

Turner, B.F., Gardner, L.R. and Sharp, W.E., 1996. The hydrology of Lake Bosumtwi, a climate-sensitive lake in Ghana, West Africa. *Journal of Hydrology*, **183**, 243–261.

Tyson, P.D., Gasse, F., Bergonzioni, L. and D'Abreton, P., 1997. Aerosols, atmospheric transmissivity and hydrological modelling of climate change over Africa S of the equator. *International Journal of Climatology*, **39**, 583–601.

Van der Hammen, T. and Absy, M.C., 1994. Amazonia during the last glacial. *Palaeogeography, Palaeoclimatology, Palaeoecology*, **109**, 247–261.

Verstappen, H.Th., 1994. Climatic change and geomorphology in south and South-East Asia. *Geo-Eco-Trop*, **16**, 101–147.

Vincens, A., Chalié, F., Bonnefille, R., Guiot, J. and Tiercelin, J.J., 1993. Pollen-derived rainfall and temperature estimates from Lake Tanganyika and their implication for late Pleistocene water levels. *Quaternary Research*, **40**, 343–350.

Vincens, A., Schwartz, D., Bertaux, J., Elenga, H. and de Namur, C., 1998. Late Holocene climatic changes in western equatorial Africa inferred from pollen from Lake Sinnda, Southern Congo. *Quaternary Research*, **50**, 34–45.

Watson, A., Price-Williams, D. and Goudie, A.S., 1983. Palaeoenvironmental interpretation of colluvial sediments and palaeosols of the Late Pleistocene hypothermal in Southern Africa. *Palaeogeography, Palaeoclimatology and Paleoecology*, **45**, 225–250.

Webb III, T. and Kutzbach, J.E., 1998. An introduction to 'Late Quaternary Climates: data synthesis and model experiments'. *Quaternary Science Reviews*, **17**, 465–471.

Webb, R.S., Rind, D.H., Lehman, S.J., Healy, R.J. and Sigman, D., 1997. Influence of ocean heat transport on the climate of the last glacial maximum. *Nature*, **385**, 695–699.

Williams, M.A.J., 1980. Late Quaternary depositional history of the Blue and White Nile rivers in Central Sudan. In M.A.J. Williams and H. Faure (eds), *The Sahara and the Nile*. Balkema, Rotterdam, 37–62.

Williams, M.A.J., Adamson, D., Cock, B. and McEvedy, R., 2000. Late Quaternary environments of the White Nile region, Sudan. *Global and Planetary Change*, **26**, 305–316.

Williamson, D., Taieb, M., Damnati, B., Icole, M. and Thouveny, N., 1993. Equatorial extension of the Younger Dryas event: rock magnetic evidence from Lake Magadi (Kenya). *Global and Planetary Change*, **7**, 235–242.

Wintle, A.G., Botha, G.A., Li, S.H. and Vogel, J.C., 1995. A chronological framework for colluviation during the last 110 kyr in KwaZulu/Natal. *South African Journal of Science*, **91**, 134–139.

Zabel, M., Schneider, R.R., Wagner, T., Adegbie, A.T., de Vreis, U. and Kolonic, S., 2001. Late Quaternary climate changes in Central Africa as inferred from terrigenous input to the Niger fan. *Quaternary Research*, **56**, 207–217.

12 The Late-Quaternary Palaeohydrology of Large South American Fluvial Systems

E.M. LATRUBESSE

Federal University of Goiás, Goiânia, Brazil

1 INTRODUCTION

South America has some of the largest rivers in the world in terms of water discharge, including the Amazon River and its tributaries, the Negro and Madeira Rivers, the Orinoco, Paraná, Tocantins and other large rivers such as the São Francisco, the Magdalena and the Uruguay River (Figure 12.1). The rivers listed above drain an area of ~11,800,000 km^2 or approximately 66% of South America. The large South American rivers discharge ca 41% of the total global river water to the oceans. South America has not only the largest rainforest of the earth, the Amazon, but also has the Cerrado savanna and large tropical plains such as the Llanos of Colombia and Venezuela, the Chaco and the Pampa. Despite the enormous potential for the study of fluvial systems, research on hydrology, geomorphology and palaeohydrology of these major basins is comparatively scarce. This chapter focuses on the general geomorphologic and hydrologic background of the main basins of South America and analyzes the impact of global change on the fluvial systems during the Last Glacial and the Holocene. Recent research on fluvial deposits, vertebrate palaeontology, and palynology, together with geomorphological studies, demonstrates that the tropics of South America experienced dramatic climatic and palaeogeographic changes during the Late Quaternary, specifically during the Middle Pleniglacial and Upper Pleniglacial, followed by minor but significant changes during the Holocene.

Nevertheless, almost three quarters of South America is situated between the tropics, making it basically a tropical as well as a fluvial continent, where the responses of the tropical regions and its giant fluvial systems to Quaternary climate changes are not well understood. Little is known about the climatic changes that occurred during the last glaciation, more specifically, since the Middle Pleniglacial, and this chapter considers both existing and new data collected from the large fluvial basins of South America.

2 THE AMAZON BASIN

The Amazon is the largest fluvial basin of the world, with a drainage area of more than 6,000,000 km^2. To the west and to the southwest, it is limited by the Andes mountain chain, and to the north and the south by the Brazilian and the Guyana shields. With an average discharge of more than 209,000 m^3 s^{-1} and a sediment load

Palaeohydrology: Understanding Global Change. Edited by K.J. Gregory and G. Benito
© 2003 John Wiley & Sons, Ltd ISBN: 0-470-84739-5

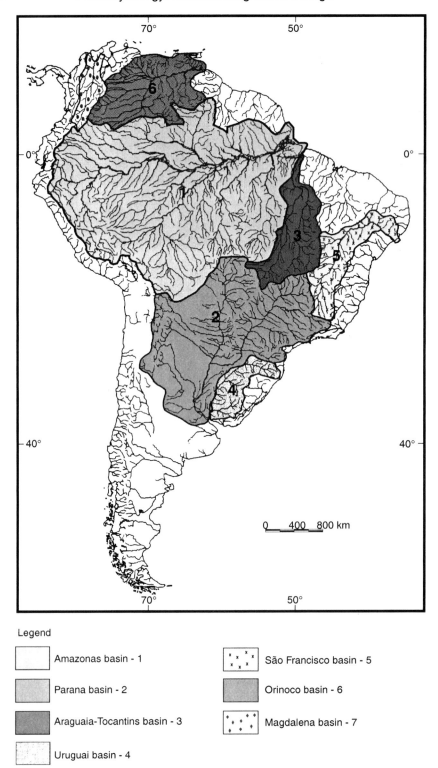

Figure 12.1   Major fluvial basins of South America

of more than 1.2 billions of tons per year of suspended load (Filizola, 1999; Meade *et al.*, 1983), the Amazon river is the largest and most unique fluvial system of the world in water discharge. The Amazon is the collecting system that receives water and sediments from a large variety of tributaries. Large tributary systems such as the Madeira, Negro, Japurá, Purus and others are ranked among the 20 largest rivers of the planet (Figure 12.2, Table 12.1).

The Amazon Rivers can be classified into three groups:

1. Fluvial systems with headwaters in the Andes chain (Ucayali, Marañón, Madre de Dios, Caquetá-Japurá, Putumayo-Iça, Pastaza).
2. Fluvial systems with headwaters in sedimentary lowlands (Purus Juruá, Javarí).
3. Fluvial systems with headwaters in the cratonic areas (Xingú, Negro, Tapajos, Trombetas).

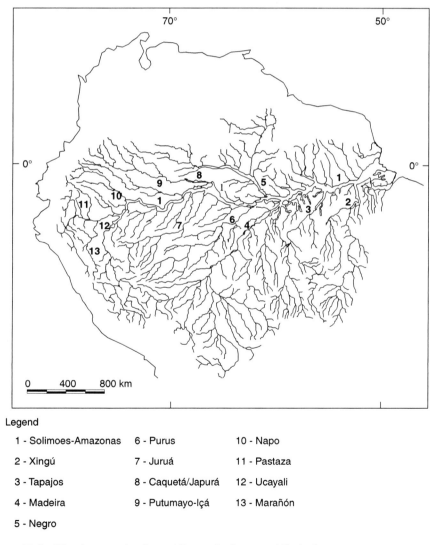

Legend

1 - Solimoes-Amazonas	6 - Purus	10 - Napo
2 - Xingú	7 - Juruá	11 - Pastaza
3 - Tapajos	8 - Caquetá/Japurá	12 - Ucayali
4 - Madeira	9 - Putumayo-Içá	13 - Marañón
5 - Negro		

Figure 12.2 The Amazon basin and the main Amazon tributaries

Table 12.1 Drainage area, average water discharge and sedimentary load for the main fluvial systems of South America

Basin	River	Drainage area (km²)	Mean annual discharge (m³ s⁻¹)	Sediment load (tons yr⁻¹)
Amazonas	Amazon	6,000,000	[1]209,000	[2]1,000 × 10⁶
	Madeira	1,360,000	[1]32,000	[3]~400/500 × 10⁶
	Negro	696,000	[1]~28,400	[3]6 × 10⁶
	Japurá/Caquetá	248,000	[1]18,620	[3]23 × 10⁶
	Purus	370,000	[1]11,000	[3]29 × 10⁶
	Juruá	185,000	[1]8,440	[3]26 × 10⁶
Orinoco	Orinoco	1,100,000	[2]35,000	[2]150 × 10⁶
Paraná	Paraná	2,600,000	[4]17,000	[5]112 × 10⁶
Tocantins	Tocantins	800,000	[6]~13,000	[7]
	Araguaia	360,000	[6]6,400	[7]18 × 10⁶
Magdalena	Magdalena	257,438	[8]7,200	[8]144 × 10⁶
São Francisco	São Francisco	650,000	[7]3,800	[7]16 × 10⁶
Uruguay	Uruguay	365,000	[9]4,660	[10]~6 × 10⁶

Source: [1]Filizola, 1999; [2]Meade, 1994; [3]Martinelli *et al.*, 1988; [4]Giacosa *et al.*, 2000; [5]Amsler and Prendes, 2000; [6]Data estimated from the Brazilian National Agency of Water-ANA; [7]ELETROBRAS, 1992; [8]Restrepo and Kjerfve, 2000; [9]Paoli *et al.*, 2000; [10]Amsler and Prendes, personal communication.

The Amazon rainforest is the largest tropical rainforest of the world and occupies more than 6 million km², spread mostly over the Amazon fluvial basin. Although the rainforest extends into various countries, 60% of the total area is located in Brazil (Capobianco, 2001). In reality, the Amazon region is a complex mosaic of different vegetation types, including closed rainforest, flooded forests, campina forests, savannas, bamboo forests, mangroves, montane and sub-montane forests and liana forests (Murça–Pires, 1984; Nelson and Oliveira, 2001). A humid and tropical climate prevails. Rainfall, averaging 2,000 mm yr⁻¹, increases in the northwest and in some parts of the Sub-Andean zone it can reach more than 4,000 mm yr⁻¹.

2.1 The Pleistocene Fluvial Record

Rivers of the Amazon basin have well-developed alluvial plains and terraces formed by Late Quaternary sediments. In several rivers, the position of the channel and the morphology and size of the alluvial plains have been related to neotectonic activity during the Late Quaternary (Sternberg, 1950; Tricart, 1977; Iriondo and Suguio, 1981; Dumont and Fournier, 1994; Latrubesse and Rancy, 2000; Latrubesse and Franzinelli, 2002). However, this chapter focuses its analysis mainly on the fluvial sedimentary and geomorphologic record as sources of palaeohydrological signals. Sandy and conglomerate sediments were deposited in Middle Pleniglacial times during the Late Pleistocene (ca 65,000–24,000 yr BP) in the Ucayali, Madre de Dios, Caquetá, Purus, Jurua and Negro fluvial systems. In the Peruvian Amazon, coarser sediments were deposited between 32 and >40 kyr BP. Dumont *et al.* (1992) found alluvial gravels that were up to 10 times coarser than the present sandy bed load in the Ucayali River, and coarse sediments in the Madre de Dios River with wood dated around 36 to 38 kyr BP.

Alluvial sedimentation was also recorded in rivers draining the northern Andes such as the Caquetá River, in the Colombian Amazon (named Japurá River in Brazilian territory) and the Pastaza River. In the Caquetá, sandy and gravel deposits with some layers of clay and peaty material are recorded in a lower terrace. Radiocarbon analyzes ranged between 30 kyr BP and infinite age (Van der Hammen *et al.*, 1992a). A large fan that extended over an area of approximately 60,000 km^2 was formed by the Pastaza River (Rasanen, 1993; Iriondo, 1994). The age of the sediments estimated by Rasanen (1993) indicates episodes of deposition in the Pastaza fan, occurring at least between 33 kyr BP and 7 kyr BP.

Alluvial deposits with characteristic Quaternary conglomerates were found in a lower terrace of some southwestern Amazon rivers including the Acre, Jurua, Purus and Madeira rivers (Latrubesse, 2000; Latrubesse and Rancy, 1998; Latrubesse and Kalicki, 2002). The most precise Quaternary data were obtained from the upper Jurua and Purus basins where more than 25 radiocarbon dates of wood and leaves in sediments were obtained. The coarse sediments were deposited during the Middle Pleniglacial and the early stages of the Upper Pleniglacial. In some rivers such as the Jurua and Madeira, the conglomerates have a rich fossiliferous level (Simpson and Paula Couto, 1981; Latrubesse and Rancy, 1998) containing Pleistocene fossil mammals. This "bone-bearing conglomerate" is found in the lower terrace, and facially represents a lag facies or a short period of strong morphogenetic channel activity with the formation of gravel channel bars. Sediments attributable to the Middle Pleniglacial were also found in the Moa River, an affluent of the Juruá River (Latrubesse and Rancy, 2000).

On rivers draining cratonic areas such as the Negro, Xingú, Tapajos and others, the Quaternary alluvial sedimentary record is not yet available in detail. There are some data from the upper Negro River basin (Latrubesse and Franzinelli, 1998). In the Tiquié, Vaupés and Curicuriarí rivers, Late Pleistocene alluvial sandy to coarser sediments frequently form a terrace level of about 14-m thickness produced by sandy to gravel deposits that include plant fragments (stems and leaves), organic matter and impregnation of iron oxide. Radiocarbon dating of trunks and organic matter gives results ranging from 27 kyr to more than 40 kyr BP (infinite age). This episode of sedimentation was provisionally correlated with the Middle Pleniglacial (Latrubesse and Franzinelli, 1998). The palaeohydrological regime of the basin was similar to that of the present producing supermature quartz sands, but the rivers were morphogenetically more active, moving sediments coarser than those of today and aggrading the fluvial system. This indicates more discharge variability and flood energy than at present. In the upper Rio Negro basin, the sedimentological record is similar with palynological results obtained on the watershed between the Caquetá and Negro rivers, where drier and more open vegetation covered a larger area than it does today, but probably with forest-like dominant type of vegetation (Van der Hammen *et al.*, 1992a).

Two large fans, occupying an area of thousands of square kilometers, were recorded in the Middle Amazon (Latrubesse, 2002). The fans were formed by two tributaries of the Madeira River: the Jiparana and the Aripuanã rivers, which drain the Brazilian Shield. The Aripuanã system is the larger one and extends over an area of more than 29,000 km^2. Fluvial inactive belts of up to 2.5 km in width and nearly 200 km in length were recognized in this large system. Typically the fluvial belts are relicts and fragmentary and not more than 10 to 20 km in length. Thermoluminescence (TL) dating of 20 ka BP suggests a Late Pleistocene age, indicating deposition during the Last Glacial Maximum (LGM). Palaeohydrological interpretation of the fans indicates past aridity in the area, replacement of the forest by savanna and avulsion processes in the rivers. Palynological results by Van der Hammen and Absy (1994) obtained at the Catira site in the Rôndonia State (ca 9° S, 63° W) close to the Jiparana River suggest arid conditions in the area with replacement of forest by savanna during part of the Middle and Upper Pleniglacial, and rainfall estimated to be below 1,000 to 500 mm during the LGM. The generation of the fans agrees with the vegetation and climatic scenario presented by Van der Hammen and Absy (1994) for this part of Amazon.

A younger late Pleistocene phase of sedimentation correlating with the Lateglacial was described in some rivers of the Amazon. In the Peruvian Amazon, Dumont et al. (1992) concluded that at 13 kyr BP, the discharge of the Ucayali River was 7 to 10 times smaller than at present. Sedimentation was recorded in the middle Caquetá (Japurá) River (<14 kyr BP) (van der Hammen et al., 1992b) and also in the cratonic basins of the Rio Negro and the tributaries of the Madeira River, on the Jiparana and Aripuanâ fans (Latrubesse and Franzinelli, 1998; Latrubesse, 2002). Holocene data showed that the Amazon suffered the effect of climatic oscillations, which affected the fluvial systems, but to a lesser extent than during the Late Pleistocene. Lower to Middle Holocene deposits are recorded in the Juruá, Purus, Caquetá Rivers, while Upper Holocene sediments younger than 3000 yr BP are also recorded in the Amazon River. Sedimentation during the early-middle Holocene is probably associated with a progressive decrease of precipitation reaching a peak of aridity during the Hypsithermal (7,000–4,500 yr BP).

No important climatic changes have been registered in the Amazon basin during the Late Holocene, although important and drastic changes in river behavior have been registered at that time in some basins including the Negro and Amazon rivers (Latrubesse and Franzinelli, 1998; Latrubesse and Franzinelli, 2002). A major part of the recent alluvial sediments of the Amazon/Solimões (Latrubesse and Franzinelli, 2002) and of the large and abundant islands of the Negro River were deposited in the Late Holocene and were thus more recent than the Middle Holocene climatic change. It is proposed that the Amazon and Negro rivers had a complex and delayed response to variations in the hydrological conditions that were provoked by changes in the Middle Holocene regional climate, approximately during the last 4,000 to 3,500 yr BP. The Negro changed from a river carrying relatively abundant suspended load and with dominant vertical accretion to a river with less suspended load and "black water", while the Amazon River evolved from a river with high rates of vertical accretion to a river with more lateral activity. Late Holocene climate oscillations have been a relatively minor influence on the river channels and on floodplain development, which were affected by continuing autogenic processes (Latrubesse and Franzinelli, 2002).

2.2 Aeolian Sediments in Amazonia

Large fields of aeolian deposits have recently been identified in the Amazon. Sand fields are currently covered by a "campina" vegetational association, composed of low, sparse trees and abundant tall grass. Parabolic dunes with ENE–WSW and NE–SW orientation, probably of the Holocene and the recent age, were identified in the "Pantanal Setentrional". This is the largest area with sandy deposits in the Amazon, and covers nearly 100,000 km^2 between the middle Negro and Branco rivers (Carneiro Filho, 1992; Nelson, 1994; Santos, 1992; Santos et al., 1993). Smaller sand fields between Manaus and the Atlantic coast were described by Iriondo and Latrubesse (1994).

To the north, in the state of Roraima (Brazil) and in western Guyana, aeolian silty and sandy deposits nearly 2 m thick are spread over more than 10,000 km^2 and longitudinal and parabolic dunes with NE–SW orientation are distinguishable to the north of the Tucano range area and in the Cauame area. Silty sediments were deposited peripherally to the west of the aeolian system and also in intramontane valleys. These sediments were interpreted as deposits of an aeolian system during the Late Pleistocene (Latrubesse and Nelson, 2001). A more recent drier period, probably of early-middle Holocene age, is indicated by the existence of pans or blowout hollows. Recently, important results were obtained (Carneiro Filho *et al.*, 2002) on the chronology of the dune fields in the Amazon basin, with TL dating in the palaeodune fields of Catrimani, Aracá, Temedauí, Tucano and Anauã, indicating periods of aeolian activity from the end of the Middle Pleniglacial to the early Holocene (32 ka BP to 8 ka BP). In the area between Manaus and Parintins, in middle Amazonia, silty loams, clayey loams and clayey sands cover the landscape as an irregular mantle, with the upper sections formed by massive loam interpreted to be primarily of aeolian origin (Iriondo and Latrubesse, 1994).

3 THE PARANA BASIN

The Parana is the second largest basin of South America considering its drainage area of 2.6 million km^2, and extends into Argentina, Brazil, Paraguay and Bolivia. The Parana River ranks sixth in the world according to water discharge with a mean discharge of ca 17,000 m^3 s^{-1} and a sedimentary load of approximately 112 million tons a year (Figure 12.1, Table 12.1). The basin drains a variety of geological environments. The Parana sedimentary basin extends over more than 29% of the basin, 29.8% of the Chaco-Pampean region, while the Precambrian Brazilian shield and the Andean chain contribute 7.4% each, approximately. Other units such as the Argentinean Mesopotamia, the Pantanal basin and the Carboniferous sediments of the Upper Parana and other minor units comprise the remaining 26.4% (Paoli *et al.*, 2000). The climate also shows a large variety with a tropical humid climate in the Brazilian region, tropical with a dry season in the Chaco region, temperate in the Pampean region and humid mountainous to semiarid in the Andes. The Chaco tributaries contribute comparatively little water discharge but are the tributaries that are the main sources of suspended sediments. The Bermejo River, a relatively small river draining from the Andes, which crosses the Chaco, introduces just 5% of the water discharge of the Parana but more than 56% of the suspended sediment (more than 50 million tons per year) of the system (Amsler and Prendes, 2000). The water discharge is derived mainly from the Brazilian territory along the upper-Brazilian tributaries and from its main tributary, the Paraguay River. Up to the confluence with the Paraguay, the river has an average discharge of 12,400 m^3 s^{-1}, and the contribution of the Paraguay is more than 3,800 m^3 s^{-1}.

The Quaternary record of the basin is analyzed by differentiating the basin into three main geomorphologic/sedimentary domains: the Parana River, the Pantanal/Paraguay basin and the Chaco system.

3.1 The Parana River

Significant advances have been made on the Quaternary history of the Upper Parana during recent years. Four main climatic events were inferred in the area during the Late Pleistocene and Holocene (Stevaux, 1994; 2000; Stevaux and Santos, 1998). The first event (40–8.5 kyr BP) is characterized by a braided channel with gravelly and sandy deposits. At that time the climate was drier than at present and the savanna vegetation was replaced by subtropical forest. Aeolian activity forming dunes and "pans" was dated by TL with an age of $23{,}540 \pm 2{,}240\,BP$ (Upper Pleniglacial) (Stevaux, 2000). The second change occurred during the Holocene, around 8.5 ka BP when the climate was wetter and the alluvial plain was excavated. An anastomosing pattern developed, construction of the Holocene floodplain began and a pluvial subtropical forest was installed in the area (Klein, 1975). A shorter drier episode between 3,500 and 2,500 years ago altered the fluvial system. Secondary channels were abandoned and backswamp areas along the alluvial plain were generated. From 2.5 ka to the present, down cutting of the channel occurred into the older units, migration occurred to the left margin, the river continued an anastomosing-braided pattern, and the present pluvial subtropical forest reached its maximum development (Klein, 1975). No significant data about fluvial deposits exist for the middle Parana area.

3.2   The Upper Paraguay/Pantanal Sub-basin

Interesting features are recorded in the Upper Paraguay basin, in the area known as the Pantanal, that is, a large flooded area of more than $100{,}000\,km^2$ located on the border of Brazil, Bolivia and Paraguay. The Upper Paraguay is a large sedimentary basin occupied by large coalescing fans formed by the Cuiabá, Taquarí, São Lourenço and other minor rivers draining into the Paraguay River (Tricart et al., 1984). With a semicircular shape and a diameter of approximately 250 km, the Taquari fan is the largest aggradational feature of the Pantanal. The fans were probably formed during the Late Quaternary, although some of them, such as the Taquari, continue to be partially active at present. Aeolian activity formed "pans" or deflation hollows during an arid period of the Holocene (Tricart et al., 1984; Klammer, 1982), probably during the late Holocene (3,500–1,000 years BP). This can be inferred very provisionally from a dating of molluscs in a shallow flooded "pan" of $3{,}820 \pm 70$ years BP (Assine et al., 1997).

3.3   The Chaco System

The Chaco is a large plain of approximately $840{,}000\,km^2$, which is spread between Bolivia, Paraguay and Argentina. It is situated on both sides of the Tropic of Capricorn, between the Amazon and the Pampa, and borders the Sub-Andean chain and the highlands of the Brazilian Planalto to the west. The climate of the Chaco ranges from semiarid tropical to humid tropical. Annual precipitation varies from west to east from 1,000 to 2,500 mm in the Sub-Andean zone, 400 mm in the Occidental Chaco and 1,200 mm in the Oriental Chaco (Schmieder, 1980). The vegetation is forest, savanna and swamp vegetation (Lopez Gorostiaga, 1984). The fluvial systems of the Chaco are mainly tributaries of the Parana basin such as the Salado, Pilcomayo and Bermejo. During some phases of the late Quaternary, the Parapetí River drained into the Amazon basin and into the Parana basin, whereas at present it flows to the Izozog swamps and connects with the Mamoré River, a tributary of the Amazon basin. The Grande River drains in the direction of the Amazon basin as a tributary of the Mamoré River. The rivers have their headwaters in the Andean and Sub-Andean zones and carry abundant suspended load during peak discharges. There is a scarcity of Quaternary data for the Chaco region although this is one of the more spectacular areas for the study of fluvial sedimentary records, geomorphology and palaeohydrology with the main synthesis of Chaco geomorphology provided by Iriondo (1993).

   A conspicuous geomorphologic characteristic of the Chaco is the formation of large Quaternary alluvial fans by the rivers mentioned above. During the Late Pleistocene, they formed complex sedimentary mega-depositional systems, which represent the largest system of coalescing fluvial fans in the world, each mega-fan having special characteristics. In general, they are formed by well-delimited alluvial belts, produced during humid periods of the Late Quaternary, and smaller and lesser stable palaeochannels during dry climates. Swamps and swampy areas are typical and Quaternary paludal deposits and present day swamps cover an area of 125,000 km^2 in the Bermejo and Pilcomayo regions. The fans were affected by two arid episodes in the Late Quaternary (Iriondo, 1993). The first one could have occurred during the LGM and the second during the late Holocene. However, dry episodes with dune reactivation were also recorded during the Middle Holocene in the northernmost fans such as the Parapetí (Kruck, 1996). During the LGM, the fans were formed by small shifting ephemeral channels, spill outs, and experienced contemporaneous aeolian activity that produced the formation of dune fields and loess deposition. TL and optically stimulated luminescence (OSL) dating by Kruck (1996) for the Pilcomayo fan deposits gave recorded ages that oscillated between 12 ka BP and 9 ka (Lateglacial-early Holocene). During the humid episodes of the Holocene, the fans were characterized by the development of well defined but unstable alluvial belts, which were abandoned by avulsion. TL and OSL dates on sediments of palaeochannels indicate avulsion of fluvial belts between 8 ka BP and approximately 3.5 to 3 ka BP (Kruck, 1996). Aeolian remobilization occurred in the Parapetí fan between approximately 6 ka BP and 3 ka BP.

During the late Holocene, a dry period was recorded between ~3,5 and 1 kyr BP and the widespread system of small ephemeral migrating channels was reactivated. Local dune fields, pans and loess-like sediments were deposited. In the Grande River, the last dunes were formed up to approximately 1.4 kyr BP. At present some of the giant fans continue to evolve. The Pilcomayo fan, for example, is the largest active fluvial fan in the world, which since the Late Quaternary has spread over an area of sedimentation of 200,000 km^2.

4 THE URUGUAY BASIN

The Uruguay drains an area of approximately 365,000 km^2 from the Brazilian territory, to the La Plata River (Paoli et al., 2000). The Uruguay and the Parana being the main former rivers of the La Plata basin meet in a large estuary. Its mean annual discharge in the lower course is approximately 4,660 m^3 s^{-1} but a peak discharge of 36,000 m^3 s^{-1} and low flows of 95 m^3 s^{-1} have been recorded (Iriondo, 1999) (Figure 12.1, Table 12.1).

Several units of Quaternary sediments and a good record of fossil mammals are found in the Uruguay fluvial basin. Related to the last glaciation, the fluvial sandy-to-sandy silty sediments of the Arroyo Feliciano Formation were deposited in large palaeochannels during the Middle Pleniglacial (Iriondo, 1999). During the Upper Pleniglacial, colluvial deposits spread in the minor valleys of the basin and loess deposits covered the fluvial landforms in part. A lower terrace is formed by alluvial deposits (conglomerates, sands and fine deposits with carbonate crusts and concretions), attributed to the Sopas Formation in Uruguay (Antón, 1975) and to the Touro Passo Formation in Brazil (Bombim, 1976). The deposits are indicators of semiarid conditions in the basin. Regarding the alluvial sediments of the Sopas Formation that are rich in vertebrates of Lujanean mammal stage, controversies exist about the age of these deposits. Iriondo (1999) suggests an age between 13 and 8 kyr BP and Anton (1975) suggests an age no greater than 30 kyr BP for these rich fossiliferous deposits. However, radiocarbon-dating by Ubilla and Perea (1999) on freshwater mollusc shells and wood, indicates ages of >43 kyr BP and >45 kyr BP respectively, which could indicate that the Sopas formation is older than previously suggested. However, correlation of the Sopas Formation with the last interglacial (120–140 ka BP), as indicated by Ubilla and Perea (1999), is very difficult to accept because the evidence is very weak. This is controversial when we consider that the Palmar Formation, another unit of alluvial deposits rich in palaeomammals, was positioned by Iriondo (1999) in the last Interglacial. That unit was recorded on the Argentinean banks of the rivers and was not recorded in Uruguayan territory, on the right bank. If this is correct, the Sopas Formation can probably be positioned provisionally in the Middle Pleniglacial.

The Dolores Formation is young and composed of alluvial deposits such as silts, clays, sandy to gravelly fine sediments and sands, related to both arid and cold conditions in the basin (Preciozzi *et al.*, 1985). Radiocarbon ages indicate an age of 11 ka BP (Ubilla, 1996). The Holocene record in the Uruguay basin is similar to that of the Parana system, with a warmer and humid climate during the Hypsithermal and a dry period during the Late Holocene (3,5–1,5 kyr BP) (Iriondo, 1999).

5   THE ORINOCO BASIN

The Orinoco River drains an area of about 830,000 km^2 in Venezuela and Colombia with the average water discharge being about 36,000 m^3 s^{-1} and the suspended-sediment discharge to the delta being more than 200 million tons (Nordin and Perez-Hernández, 1989). The Orinoco is the third largest river of the world and probably the eighth largest in terms of sediment discharge (Meade *et al.*, 1983) (Figure 12.1, Table 12.1). The most important sediment-carrying tributaries are the Guaviare, Meta and Apure rivers, which come from the Andes, cross the Llanos, enter the Orinoco by the left bank, and, like rivers draining the Guyana Shield in the Amazon basin, carry low concentrations of suspended sediment. However, they contribute large discharges and probably have substantial amounts of bed load (Nordin and Perez Hernandez, 1989). The Orinoco is characterized by high bedrock control and rapids at node points between which broad areas of floodplains develop. The Orinoco River floodplain can be divided into fringing floodplain and internal deltas (Hamilton and Lewis, 1990). The fringing floodplain of the Orinoco main channel extends from the Meta River to the delta and covers approximately 7,000 km^2 (Warne *et al.*, 2002). However, in the Apure, Arauca and Meta flooded areas, sediments are stored in an area of 70,000 km^2 (Welcomme, 1979). The tributaries form fanlike features in a swampy area, which is affected by the hydrologic dynamics of the Orinoco. Meade *et al.* (1983) estimated that one half of the amount of sediment transported by the Meta River is temporarily stored in fanlike or internal deltalike deposits.

The hydrologic regime is dominated by a tropical wet–dry climate with a high ratio of maximum to minimum discharges so that discharge average variability is about 26 but can oscillate between 54 and 8.1 (Nordin and Perez-Hernandez, 1989). During the low-flow season, deflation affects a large area of fluvial sediment that can be transported upstream by the wind each year.

Despite the enormous potential for palaeohydrological research, the Late Quaternary history of the Orinoco fluvial system is practically unknown although major hydrological changes affecting the Orinoco can be inferred from records of areas in Venezuela and Colombia. During the Late Pleistocene, the Orinoco basin had a climate drier than that at present and in the northwest of the Orinoco delta and in the lower Apure region, large aeolian dune fields were formed (Khobzy, 1981; Carbon and Schubert, 1994;

Iriondo, 1997). Inland, between the Meta and Orinoco Rivers, a loess or loess-like formation covered the Colombian Llanos (Gossen, 1971; Iriondo, 1997). It is inferred that the Orinoco had less discharge than today because at that time the northeast trade winds of the northern hemisphere were more intense and persistent in the area. Dunes were active in the Orinoco Llanos during part of the Middle Pleniglacial and the Upper Pleniglacial, with TL dating (Vaz and Garcia, 1989) indicating dune formation at $36,000 \pm 5,000$ years BP. Palaeosoils developed under the aeolian sediments dated back to $11,100 \pm 450$ years BP (Roa, 1979). The pollen record in the Colombian Llanos also agrees with the interpretation of this humid climatic condition during part of the Lateglacial (Behling and Hooghiemstra, 1998).

An arid phase towards the end of the Lateglacial and early Holocene ($11,600 \pm 1,600$ years BP) produced reactivation of aeolian activity and dune formation (Vaz and Garcia, 1989), although that dry period was less intense than that of the LGM. The pollen record in the Colombian Llanos indicates drier conditions between ca 10,000 and 5,000 yr BP (Behling and Hooghiemstra, 1998).

6 THE MAGDALENA RIVER

With an average discharge of $7,200 \, \text{m}^3 \, \text{s}^{-1}$ and a drainage area of $257,438 \, \text{km}^2$ the Magdalena River is the main system draining the Andes of Colombia to the Caribbean Sea (Restrepo and Kjerfve, 2000) (Figure 12.1, Table 12.1), with headwaters at an elevation of more than 3,000 m, and the Cauca, Sogamoso, San Jorge and Cesar Rivers as the main tributaries. The mean rainfall for the drainage basin is $2,050 \, \text{mm yr}^{-1}$ but some areas receive more than $3,000 \, \text{mm yr}^{-1}$. Despite having relatively small catchments, the Magdalena carries more sediment to the ocean than the Orinoco or the Parana, which have the largest basins and the highest water discharges. Milliman and Meade (1983) estimated an annual sediment load of $220 \times 10^6 \, \text{t yr}^{-1}$ while Restrepo and Kjerfve (2000) estimated a mean sediment load of $144 \times 10^6 \, \text{t yr}^{-1}$. From a geomorphic and sedimentary point of view, the Magdalena was considered an anastomosing river situated in a tectonically active foreland basin with wetlands, which maintained an anastomosing pattern, with average rates of vertical accretion of $3.8 \, \text{mm yr}^{-1}$ in response to the subsiding basin, at least during the last $8,000 \, \text{yr BP}$ (Smith, 1986). Van der Hammen (1986) proposed the existence of fluctuations in the flood stage of the Magdalena during the Holocene. On the basis of an analysis of borehole evidence, he proposed dry periods in ca 7,000 yr BP, 5,500 yr BP, 4,700 yr BP, 4,000 yr BP, 2,500 to 2,100 yr BP, 1,400 yr BP and 700 yr BP, although, as the method used does not include morpho-sedimentary analysis with a detailed study of facies, the results need to be considered with caution. Anastomosing rivers in a subsiding basin tend to develop changes in the areas flooded, swamps, channel-crevasse splays and channel avulsion. As both facies and sedimentary environments change along the alluvial plain, different types of sediments and differential rates of accumulation occur in the fluvial record, even without the direct influence of change in the hydrological regime.

7 CENTRAL BRAZIL: THE TOCANTINS AND THE SÃO FRANCISCO BASINS

Two main basins drain Central Brazil: the Tocantins–Araguaia River and the São Francisco River. There is little information about palaeohydrology and palaeoclimates of these basins.

The Tocantins–Araguaia basin is practically ignored in international literature on large rivers. With an area close to $800,000 \, km^2$, the Tocantins ranks in the twentieth or twenty-first position among world rivers, with drainage basin sizes close to that of the Mekong and Danube, although, only having a mean annual water discharge ca $12,000 \, m^3 \, s^{-1}$. The Araguaia–Tocantins River could be ranked tenth or eleventh according to water discharge. The Tocantins basin is located on the central highlands of Brazil at altitudes lower than 1,000 m, the geological setting comprises Precambrian rocks of the Brazilian Shield, Palaeozoic and Mesozoic rocks of the Paraná basin, the tertiary terrigenous sequence and widespread Quaternary deposits. The area is covered mainly by the Brazilian savannas ("Cerrado"), and by the Amazon rainforest in the lower course of the Tocantins. Annual rainfall varies between 1,200 mm in the upper basin and more than 2,000 mm in the lower basin. Two mistakes are frequently made in the fluvial literature concerning the Araguaia–Tocantins River. The first is that, although considered a tributary of the Amazon Basin, the water of the Araguaia–Tocantins River goes to the Atlantic Ocean, along the Para River, a channel situated to the south of the Marajo Island, while the Amazon discharge is diverted to the north of the Marajo island. The second mistake concerns the denomination of this river because, although the name Tocantins is traditionally used, the Araguaia is, from a geomorphological and hydrological point of view, the main drainage system. These rivers have few developed alluvial plains because they flow on bedrock with structural control by the Precambrian Shield rocks and faults. However, a large Quaternary sedimentary basin, the Bananal plain, is found in the middle Araguaia River covering more than $90,000 \, km^2$ and occupying approximately 12% of the fluvial basin. The Quaternary sediments that reach more than 100 km in width and extend more or less continuously in the north-south direction for about 700 km are almost unknown. Many inactive alluvial palaeochannels and other fluvial and swampy/lacustrine features occur in the Bananal plain including a number of underfit rivers that occupy the palaeoalluvial belts, and a drainage system on the Araguaia formation that is underdeveloped. This extensive plain is temporarily flooded during the rainy season as a result of both the local rainfall and a saturated water table. The Araguaia River has a well-developed alluvial belt in the middle course and stretches for more than 1,100 km. The middle Araguaia alluvial plain is a complex mosaic of Quaternary morpho-sedimentary units (Latrubesse and Stevaux, 2002) with Late Quaternary and Holocene sediments forming the floodplain. The older sediments (Late Pleistocene?) are composed of ferruginous coarse sands to conglomeratic sands indurated by iron oxides.

In the Holocene alluvial plain, three main units can be identified: a plain of accreted bars/islands and scrolls in the more sinuous reaches accompanying the channel; a palaeomeander-dominated unit; and an impeded floodplain, both of them situated marginally in relation to the channel. The impeded floodplain, formed by a discontinuous backswamp, is relatively the oldest unit of the floodplain. The palaeomeander-dominated unit is characterized by the presence of large oxbow lakes and scroll features. In general, this unit occupies an intermediate position between the accreted bar/islands unit and the impeded floodplain and it is assumed that this unit was formed earlier than the plain of accreted banks/islands. The accreted bars/islands unit is being generated by the channel activity at present and forms an irregular belt along both sides of the channel. Although the age of the units was not calibrated by absolute dating (very recent radiocarbon ages were obtained in the accreted bars/islands units), the existence of palaeomeanders and the impeded floodplain indicate hydrological changes in the system that occurred during the Holocene.

The São Francisco River is the main fluvial system draining the Brazilian semiarid northeastern region, with a drainage area of 650,000 km^2, and drains mainly the Pre-cambrian rocks of Central Brazil from the south to the north through the Cerrado and the Caatinga vegetation realms. Its mean water discharge is 3,800 m^3 s^{-1} and the sediment discharge 6 × 10^6 t yr^{-1}.

As in other South American rivers, information about the Quaternary fluvial record of the Araguaia and São Francisco rivers is almost nonexistent. Nevertheless there are some inferences, albeit controversial, about climatic changes affecting the Cerrado (savannas to dry forest) of Central Brazil. During the end of the Middle Pleniglacial and the LGM, arid and cooler conditions with grassland expansion and reduction of montane and cloud forests in south and southeast Brazil (Behling, 2002) were recorded. However, a relatively humid and warm climate between >32,400 to ca 32,000 yr BP and a cold and humid climate between 30,000 and 26,000 yr BP has also been reported (Salgado-Labouriau et al., 1998). For the LGM, both approaches agree with cold and dry conditions in Central Brazil extending up to the early Holocene. The interpretation of the middle-late Holocene is also controversial among palynologists for Central Brazil. While some suggest arid conditions from the Late Pleistocene to the late Holocene (Behling, 2002), others suggest better climatic conditions in the Holocene since 6,500 yr BP.

The main palaeoclimatic indicators in the basins may be due to the existence of large dune fields in the middle São Francisco River where Tricart (1974) described large longitudinal dune fields produced by deflation of E to ESE winds. The dunes formed during arid conditions of the LGM were stabilized during the Lateglacial. Humid climatic conditions were suggested in the area during ca 11,000 to 10,000 yr BP (Oliveira et al., 1999), with progressive warming and high humid levels between ca 10,000 and 6,500 yr BP. The return to "Caatinga/Cerrado" arid conditions could have started at 6,500 yr BP with increasing aridity from ca 4,500 yr BP until today.

8 DISCUSSION

As observed throughout this chapter, the palaeohydrological record of the large flu-vial basins of South America is scarce and, in some basins, very fragmentary. Some basins are relatively well studied, whereas others have no systematic fluvial studies. The best information that we have about Pleistocene chronology, palaeoecology and geology is related to the period before 24 kyr BP, possibly corresponding to the Mid-dle Pleniglacial (ca 65–24 kyr BP) (Figure 12.3). In the Amazon basin, sedimentation in the fluvial belts occurred during this period in rivers with their headwaters in the Andes and also in the lowlands of the southwestern Amazon (Dumont et al., 1992; Latrubesse and Franzinelli, 1998; Latrubesse and Ramonell, 1994; Rasanen et al., 1990; 1992; van der Hammen et al., 1992a). For some authors (Dumont et al., 1992; van der Hammen et al., 1992a), the alluvial sedimentation could have been directly associated with glacial advances and strong rains in the central and northern Andes where the rivers deposited sand and gravel. However, the rivers with headwaters in the lowlands of the southwestern Amazon (Purus and Jurua basins, for example) and those with headwaters in cratonic areas such as the Negro River also carried abun-dant sediment loads. In the lowlands, the occurrence of coarse sand and pebbles and fossil mammals found in the conglomerate deposits, clearly indicate the magnitude of hydrological changes. The Middle Pleniglacial would have been characterized by high precipitation in the Andes and a continuing change towards dry conditions in the

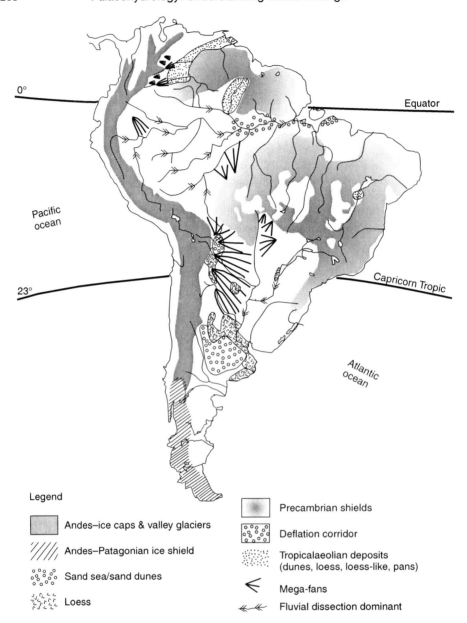

0°

Equator

Pacific
ocean

23°

Capricorn Tropic

Atlantic
ocean

Legend

Andes–ice caps & valley glaciers

Andes–Patagonian ice shield

Sand sea/sand dunes

Loess

Precambrian shields

Deflation corridor

Tropicalaeolian deposits
(dunes, loess, loess-like, pans)

Mega-fans

Fluvial dissection dominant

Figure 12.3 The palaeohydrological and palaeoenvironmental situation of South America at the end of the Middle Pleniglacial and the Last Glacial Maximum (LGM). Dissection (as indicated on the map) was dominant in the fluvial systems during the LGM, however, the end of the Middle Pleniglacial was characterized by sedimentation in the fluvial belts

lowlands, including the cratonic areas (Latrubesse, 2000). Recent results on dune-field chronologies (Carneiro Filho *et al.*, 2002), confirmed the claim of Latrubesse (2000) that the aridity in the lowlands of Amazonia began during the Middle Pleniglacial or early stages of the Upper Pleniglacial.

The Middle Pleniglacial–early Upper Pleniglacial episode of fluvial sedimentation is also well recorded in the Upper Parana basin. Coarse sediments were deposited at this time for the Parana River indicating a more arid period than at present (Stevaux, 1994; 2000; Stevaux and Santos, 1998).

The aridity in South America reaches its climax during the Upper Pleniglacial, when the aeolian sedimentation extended along the Venezuelan and Colombian Llanos and over parts of central and northern Amazonia. At that time, savanna vegetation reached its maximum extension in the Amazon, the Llanos had an arid climate and aeolian remobilization reached the core of the Amazon region. The Upper Pleniglacial was also a time of extensive aeolian activity as a response to increasing aridity also in Central Brazil, Chaco and Pampa. During the LGM, a large sand sea with large dunes was formed in the Pampean region (Argentina) and a mantle of loessic sediments was deposited in the Chaco–Pampean plain. At this time the Parana had a smaller discharge than it does today and dune fields formed on the older terraces of the Upper Parana River and deposition occurred in the Uruguay system during the Lateglacial. The Chaco mega-fans developed a large number of ephemeral and small channels and simultaneously suffered active deflation of the oldest alluvial belts. Dune fields formed in the Grande and Parapetí fans, and loessic sediments were deposited in the Chaco along the Sub-Andean foothills (Iriondo, 1993). The large basins of Central Brazil, such as the São Francisco and Tocantins–Araguaia, have also experienced climatic deterioration since the Middle Pleniglacial. During the LGM, the sand dunes of the Middle São Francisco River were active and the fluvial sediments suffered deflation.

Wind-circulation models were proposed for the Amazon and the Llanos area for times close to the late Pleistocene (ca 40–14 kyr BP), (Iriondo, 1997; Iriondo and Latrubesse, 1994; Latrubesse and Ramonell, 1994; Latrubesse, 2000). In the middle Amazon region, second-order changes in regional climate dynamics would have been sufficient to instigate a dry climate phase (Iriondo and Latrubesse, 1994). Trade winds would have been stronger and drier than at present, producing extensive deflation corridors and aeolian sedimentation. Northeastern trade winds were dominant approximately north of $0°$, removing sand and silt in the Llanos, "Pantanal Setentrional" and in Roraima. Southeastern trade winds that originated in the anticyclonic circulation of the southern hemisphere would have dominated south of $0°$ reaching as far as the western Chaco along the Sub-Andean foot-slopes, forming dunes and loess deposits. Considering the data mentioned, the palynological record in Carajas and Rondonia, the palaeomammals and the palaeohydrologic record of the southwestern Amazon show that it was characterized by a dominant savanna environment during the Upper Pleniglacial, with a dry season that was more pronounced and prolonged than that of the present. Aridity was also widespread in central and northeast Brazil producing the formation of sand dunes in the middle São Francisco.

Cold air masses of the South Pacific Anticyclonic (SPA) circulation were dominant in the Argentinean Pampas and part of the Eastern Chaco where the climate was dry and cold (Iriondo and García, 1993; Ramonell and Latrubesse, 1991), penetrating up to the Upper Parana and upper Araguaia basin and moving into the south and southwestern Amazon, causing clear falls in winter temperatures (Latrubesse and Ramonell, 1994). The SPA winds were strengthened by katabatic winds coming from the ice field of the Patagonian Andes (Iriondo and García, 1993). After 14 kyr BP, sedimentation in the fluvial belts occurred in response to climatic changes associated with the last deglaciation, the gradual recuperation of the rainforest occurred, and this sedimentation phase probably culminated with the marine transgression of the Middle Holocene.

In the early to middle Holocene, the plains of northern South America were dry. A climate drier than that of the present was also suggested for the Amazon basin during the Hypsithermal (Martin *et al.*, 1993). The Hypsithermal was characterized by humid and temperate conditions in the Chaco and Pampa plain as well as in the upper Parana and Uruguay basins (Iriondo and García, 1993; Stevaux and Santos, 1998) and dry conditions dominated that large region between 3,5 and 1 kyr BP.

The Late Holocene, mainly the last 1,000 years, seems to be a period of rapid recuperation of the fluvial systems accompanied by high rates of sedimentation. Many radiocarbon dates for the Amazon floodplain and some tributaries are very young (Late Holocene, <3,000 yr BP) and many of them indicate ages of less than 1,000 years (Absy, 1979; Latrubesse and Franzinelli, 2002; Sternberg, 1960; Vital and Stattegger, 2000).

9   CONCLUDING REMARKS

Knowledge of the Quaternary fluvial record of large South American rivers has advanced significantly in recent years. However, the richness, complexity and size of these rivers mean that the results reviewed in this chapter must be regarded as incipient and pioneering. Discovery of extensive and widespread aeolian fields in the Amazon is an advance that opens a new horizon in Quaternary research of tropical South America. As expressed by Latrubesse (2000), aeolian sediments and landforms can be the future key to the reconstruction of wind patterns and for the determination of different times of maximum aridity in the Amazon and, more generally, for the whole of tropical South America. Palaeogeographic reconstructions and estimates of temperature variability based exclusively on palynological studies should be made with caution, especially when they are not correlated with a regional scenario that considers well-documented sedimentologic and geomorphologic records and mapping.

When compared with the reality of other large fluvial basins of the world, which were or are being affected directly by intensive human use, the large South American rivers have particular potential for the study of the Quaternary palaeohydrology and for comparison of the palaeohydrological record with present-day hydrology and fluvial morpho-dynamics. Large parts of the Amazon fluvial system remain untouched by direct human intervention along the channels, while others, such as the Chaco's rivers, offer a large potential for understanding palaeohydrological river adjustment to climatic changes, metamorphosis (Baker, 1978) or understanding autogenic processes during the Late Quaternary, together with one of the more spectacular environments of alluvial sedimentation of the planet, the large and complex mega-fan features. As expressed by Baker (2000): *discoveries from field work in South America afford a spectacular opportunity to develop a synthesized understanding of realized hydrological change. This understanding can only be achieved by interpreting the indicators (signs) of past hydrological processes, that is, via... paleohydrology.*

Understanding past hydrology has become a necessary investigation in South America for at least two fundamental reasons. First, considering the global influence of tropical South America on the interaction of the water movement from the land surface to the atmosphere and hence the importance of South America in the global hydrologic and palaeohydrologic context (Baker, 2000); secondly, as a scientific priority tool reconciling environmental and political decisions about the role of large rivers for the future development of South American countries, in the light of the contrasting policies of use and conservation that are currently being proposed for these large basins.

REFERENCES

Absy, M.L., 1979. *A Palynological Study of Holocene Sediments in the Amazon basin*. Unpublished Ph.D. Thesis, University of Amsterdam, Amsterdam, 76.

Anton, D., 1975. *Evolución Geomorfológica del norte de Uruguay*. Dirección de suelos y fertilizantes, Ministerio de Agricultura y Pesca, p. 22.

Amsler, M. and Prendes, H.H., 2000. Transporte de sedimentos y procesos fluviales asociados. In C. Paoli and M. Schreider (eds), *El Río Paraná en su tramo medio. Centro de Publicaciones*. Universidad Nacional del Litoral-UNL, Santa Fé, 235–306.

Assine, L.M., Soares, P.C. and Angulo, R., 1997. Construção e abandono de lobos na evolução do leque do Rio Taquarí, Pantanal Mato-Grossense. *VI Congresso da Associação Brasileira de Estudos do Quaternário*. Extended Abstracts, ABEQUA, Curitiba, 431–433.

Baker, V., 1978. Adjustment of fluvial systems to climate and source terrain in tropical and subtropical environments. In A.D. Miall (ed.), *Fluvial Sedimentology*. Memoir 5, Canadian Society of Petroleum Geologists, Calgary, Canada 211–230.

Baker, V., 2000. South American paleohydrology: future prospects and global perspective. *Quaternary International*, **72**, 3–5.

Behling, H., 2002. South and Southeast Brazilian grasslands during late Quaternary times: a synthesis. *Palaeogeography, Palaeoclimatology, Palaeoecology*, **177**(1–2), 19–27.

Behling, H. and Hooghiemstra, H., 1998. Late quaternary palaeoecology and palaeoclimatology from pollen records of the savannas of the Llanos Orientales in Colombia. *Palaeogeography, Palaeoclimatology, Palaeoecology*, **139**, 251–267.

Bombim, M., 1976. Modelo paleoecológico evolutivo para o Neocuaternario da região da campanha oeste de Rio Grande do Sul (Brasil). *Comunicações Museum de Ciências. Pontificia Universidade Católica de Rio Grande do Sul*, **9**, 1–28.

Capobianco, J.P., 2001. Introdução. In J.P. Capobianco (ed.), *Biodiversidade na Amazônia Brasileira*. Instituto Sócio-Ambiental-ISA, São Paulo, 13–15.

Carbon, J. and Schubert, C., 1994. Late Cenozoic history of the Eastern Llanos of Venezuela: geomorphology and stratigraphy of the Mesa formation. *Quaternary International*, **21**, 91–100.

Carneiro Filho A., 1992. Observaçiones preliminares das dunas do Río Negro. In E. Franzinelli and E. Latrubesse (eds), *Resumos e Contribuiçoes Científicas. Simpósio. Internacional do Quaternário da Amazonia y 4 Reunión PICG 281*, UFAM, Manaus, p. 166.

Carneiro Filho, A., Schwartz, D., Tatumi, H. and Rosique, T., 2002. Amazonian paleodunes evidencing a drier climate during late Pleistocene-Holocene. *Quaternary Research*, **58**, 205–209.

Dumont, J.F. and Fournier, M., 1994. Geodynamic environment of Quaternary morphostructures of the Subandean foreland basins of Peru and Bolivia: characteristics and study methods. *Quaternary International*, **21**, 129–142.

Dumont J.F., Garcia. F. and Fournier, M., 1992. Registros de cambios climáticos para los depósitos y morfologías fluviales en la Amazonia Occidental. In: L. Ortlieb, J. Macharé (eds), *Paleo ENSO Records. International Symposium*. Extended Abstracts, ORSTOM, Lima, 87–92.

ELETROBRAS. 1992. *Diagnóstico das condições sedimentológicas dos principais rios brasileiros*. IPH-UFRGS/ELETROBRAS, p. 99.

Filizola, N.P., 1999. *O fluxo de sedimentos em suspensão nos rios da bacia Amazônica Brasileira*. ANEEL, Brasilia.

Giacosa, R., Paoli, C. and Cacik, P., 2000. Conocimiento del Regimen hidrológico. In C. Paoli and M. Schreider (eds), *El Río Paraná en su tramo medio*. Centro de Publicaciones, Universidad Nacional del Litoral-UNL, Santa Fé, 69–104.

Gossen, I., 1971. *Physiography and Soils of the Llanos Orientales, Colombia*. Publication of the ITC Enschede, Ser. B. 64, p. 198.

Hamilton, S.K. and Lewis Jr., W.M., 1990. Basin morphology in relation to chemical and ecological characteristics on lakes on the Orinoco river floodplain, Venezuela. *Archiv Fuer Hydrobiologie*, **119**(4), 393–425.

Iriondo, M., 1993. Late Quaternary and geomorphology of the Chaco. *Geomorphology*, **7**, 289–305.

Iriondo, M., 1994. The Quaternary of Ecuador. *Quaternary International*, **21**, 101–112.

Iriondo, M., 1997. Models of deposition of loess and loessoids in the upper Quaternary of South America. *Journal of South American Earth Sciences*, **10**(1), 71–79.

Iriondo, M., 1999. Estratigrafia del Cuaternario de la cuenca del Rio Uruguay. *XIII Congreso Geológico Argentino y III Congreso de Exploración de hidrocarburos, Actas IV*, Buenos Aires, 15–25.

Iriondo, M. and García, N.O., 1993. Climatic variations in the Argentine plains during the last 18,000 years. *Paleogeography, Palaeoclimatology, Palaeoecology*, **101**, 209–220.

Iriondo, M. and Latrubesse, E., 1994. A probable scenario for a dry climate in Central Amazonia during the late Quaternary. *Quaternary International*, **21**, 121–128.

Iriondo, M. and Suguio, K., 1981. Neotectonics of the Amazon plain. *INQUA Neotectonic, Bulletin*, **4**, 72–78.

Khobzy, J., 1981. Los campos de dunas del norte de Colombia y de los Llanos de la Orinoquia (Colombia y Venezuela). *Revista CIAF*, **6**(1–3), 257–292.

Klammer, G., 1982. Die palaeowuste des Pantanal von Mato Grosso und die pleistozane Kçi-mageschuchte des Brasilianischen randtropen. *Zeitschrift fur Geomorphologie*, **26**, 393–416.

Klein, R.M., 1975. Southern Brazilian phytogeographic features and probable influence of upper Quaternary climatic changes in the floristic distribution. *Boletim Paranaense de Geociências*, **33**, 67–88.

Kruck, W., 1996. Pleistoceno Superior y Holoceno del Chaco Paraguayo. *Memorias del XII Congreso Geológico de Bolivia*, Tarija Bolivia, 1217–1220.

Latrubesse, E., 2000. The late Pleistocene in Amazonia: a palaeoclimatic approach. In P. Smolka and W. Volkheimer (eds), *Southern Hemisphere Paleo and Neoclimates*. Springer-Verlag, Berlin, Germany 209–222.

Latrubesse, E., 2002. Evidence of Quaternary paleohydrological changes in middle Amazonia: the Aripuanã and Jiparaná fans like systems. *Zeitschrift fur Geomorphologie*, **129**, 61–72.

Latrubesse, E. and Franzinelli, E., 1998. Late Quaternary alluvial sedimentation in the Upper Rio Negro basin, Amazonia, Brazil: palaeohydrological implications. In G. Benito, V. Baker and K.J. Gregory (eds), *Palaeohydrology and Environmental Change*. Wiley, Chichester, 259–271.

Latrubesse, E. and Franzinelli, E., 2002. The Holocene alluvial plain of the middle Amazon river, Brazil. *Geomorphology*, **44**(3–4), 241–257.

Latrubesse, E. and Kalicki, T., 2002. Late Quaternary palaeohydrological changes in the Upper Purus basin, Southwestern Amazonia, Brazil. *Zeitschrift fur Geomorphologie*, **129**, 41–59.

Latrubesse, E. and Nelson, B., 2001. Evidence for late Quaternary aeolian activity in the Roraima-Guyana region. *Catena*, **43**, 63–80.

Latrubesse, E. and Ramonell, C., 1994. A climatic model for Southwestern Amazonia at last glacial times. *Quaternary International*, **21**, 163–169.

Latrubesse, E. and Rancy, A., 1998. The late Quaternary of the Upper Jurua River, Southwestern Amazonia, Brazil: geology and vertebrate paleontology. *Quaternary of South America and Antarctic Peninsula*, **11**, 27–46.

Latrubesse, E. and Rancy, A., 2000. Neotectonic influence on tropical rivers of Southwestern Amazon during the late Quaternary: the Moa and Ipixuna river basins, Brazil. *Quaternary International*, **72**, 67–72.

Latrubesse, E. and Stevaux, J.C., 2002. Geomorphology and environmental aspects of the Araguaia fluvial basin, Brazil. *Zeitschrift fur Geomorphologie*, **129**, 109–127.

Lopez Gorostiaga, O., 1984. *Formaciones vegetales del chaco Paraguayo*. Comisión Nacional de Desarrollo del chaco, Asunción, O.E.A. Série información básica No. 2, 34.

Martin, L., Fournier, M., Mourguiart, P., Sifeddine, A., Turq, B. and Flexor, J.M., 1993. Southern oscillation signal in South American palaeoclimatic data of the last 7,000 years. *Quaternary Research*, **33**, 1749–1762.

Martinelli, L.A., Forsberg, B.R., Meade, R.H. and Richey, J.E., 1988. Sediment delivery rates for the Amazon river and its principal Brazilian tributaries. *Chapman Conference*, AGU, Charleston, S.C. Comp. of Ext. Abstract, 77–81.

Meade, R.H., Nordin Jr., C.F., Hernandez, D.P., Mejía, B.A. and Godoy, J.M., 1983. Sediment and water discharge in rio Orinoco, Venezuela and Colombia. *Proceedings of the Second International Symposium on River Sedimentation: Water Resources and Electric Power Press*, Nanjing, China, 1134–1144.

Milliman, J.D. and Meade, R., 1983. World-wide delivery of river sediments to the oceans. *Journal of Geology*, **91**, 1–22.

Murça–Pires, J., 1984. The Amazonian forest. In H. Sioli (ed.), *The Amazon*. Dr. W. Junk Publishers, Dordtrecht, 581–601.

Nelson, B., 1994. Natural forest disturbance and change in the Brazilian Amazon. *Remote Sensing Review*, **10**, 105–125.

Nelson, B. and Oliveira, A.A., 2001. Área Botânica. In J.P. Capobianco (ed.), *Biodiversidade na Amazônia Brasileira*. Instituto Sócio-Ambiental-ISA, Sâo Paulo, 132–176.

Nordin Jr., C.F. and Perez-Hernandez, D., 1989. *Sand Waves, Bars and Wind-blown Sands of the Rio Orinoco, Venezuela and Colombia*. Water Supply Paper 2326A, *U.S. Geological Survey*, 74.

Oliveira, P.de, Barreto, A.M.F. and Suguio, K., 1999. Late Pleistocene/Holocene climatic and vegetational history of the Brazilian caatinga: the fossil dunes of the middle São Francisco river. *Palaeogeography, Palaeoclimatology, Palaeoecology*, **152**(3–4), 319–337.

Paoli, C., Iriondo, M. and Garcia, N., 2000. Características de las cuencas de aporte. In C. Paoli and M. Schreider (eds), *El Río Paraná en su tramo medio*. Centro de Publicaciones, Universidad Nacional del Litoral-UNL, Santa Fé, 29–68.

Preciozzi, F., Spoturno, J., Heinzen, W. and Rossi, P., 1985. *Carta Geológica del Uruguay a escala 1:500.000*, 1–90, map.Montevideo: Dirección Nacional de mineria y Geologia.

Ramonell, C. and Latrubesse, E., 1991. El loess de la Formación Barranquita: comportamiento del Sistema Eólico Pampeano en la provincia de San Luis, Argentina. *Third Meeting, IGCP 281, "Quaternary Climates of South América"*, Spec. Publ., Lima, No. 3, 69–81.

Rasanen, M., 1993. Geologia: la Geohistoria de la Amazonia Peruana. In R. Kalliola, M. Puhakka and W. Danjoy (eds), *Amazonia Peruana*. ONERN Universidad de Turku, Turku, Finland, 43–68.

Rasanen, M., Neller, R., Salo, J. and Jungens, H., 1992. Recent and ancient fluvial deposition system in the Amazonian foreland basin Peru. *Geological Magazine*, **129**(3), 293–306.

Rasanen, M., Salo, J.S., Jungnert, H. and Romero Pitman, L., 1990. Evolution of Western Amazon lowland relief: impact of Andean Foreland dynamics. *Terra Nova*, **2**, 320–332.

Restrepo, J.D. and Kjerfve, B., 2000. Magdalena river: interannual variability (1975–1995) and revised water discharge and sediment load estimates. *Journal of Hydrology*, **235**, 137–149.

Roa, P., 1979. Estudio de los médanos de los Llanos Centrales de Venezuela: evidencias de un clima desértico. *Acta Biológica Venezolana*, **1**, 19–49.

Salgado-Labouriau, M.L., Barberi, M., Ferraz-Vicentini, K.R. and Parizzi, M.G., 1998. A dry climatic event during the late Quaternary of tropical Brazil. *Reviews of Palaeobotany and Palynology*, **9**(2), 115–129.

Santos, J.O.S., 1992. O Pantanal setentrional e os campos de dunas da Amazonia. In: E. Franzinelli, E. Latrubesse (eds), *Intern. Symp. on the Quat. of Amazonia*, UFAM, Manaus, Abstracts and Sci. Contr., 110.

Santos, O., Nelson, B. and Giovannini, C.A., 1993. Campos de dunas: Corpos de areia sob leitos abandonados de grandes ríos. *Ciencia Hoje*, **16**(93), 22–25.

Schmieder, O., 1980. *Geografia de América Latina*. Fondo de Cultura Económica, México.

Simpson, G. and Paula Couto, C., 1981. Fossil mammals from the Cenozoic of Acre, Brazil III; Pleistocene Edentata, Pilosa, Proboscidea, sirenia, Perissodactyla and Artiodactyla. *Iheringia: Série Geológica*, **6**, 11–73.

Smith, D.G., 1986. Anastomosing river deposits sedimentation rates and basin subsidence, Magdalena river, Northwestern Colombia, South America. *Sedimentary Geology*, **46**, 177–196.

Sternberg, H., 1950. Vales tectônicos na planície Amazônica?. *Revista Brasileira de Geografia*, **12**(4), 3–26.

Sternberg, H., 1960. Radiocarbon dating as applied to a problem of Amazonian Morphology. *Comptes Rendus du XVIII Congrés International de Géographie*, 399–424.

Stevaux, J.C., 1994. The Upper Paraná river (Brazil): geomorphology, sedimentology and paleoclimatology. *Quaternary International*, **21**, 143–161.

Stevaux, J.C., 2000. Climatic events during the late Pleistocene and Holocene in the Upper Parana River: correlation with NE Argentina and South-Central Brazil. *Quaternary International*, **72**, 73–85.

Stevaux, J.C. and Santos, M. dos, 1998. Palaeohydrological changes in the Upper Paraná river, Brazil, during the late Quaternary: a facies approach. In G. Benito, V. Baker and K.J. Gregory (eds), *Palaeohydrology and Environmental Change*. Wiley, Chichester, 273–285.

Tricart, J., 1974. *Existence de périodes sèches au quaternaire en Amazonie et dans les regions voisines*, 145–158.

Tricart, J., 1977. Types de lits fluviaux en Amazonie Bresilienne. *Annales de Geographie*, **437**, 1–54.

Tricart, J., Pagney, P. and Frecaut, R., 1984. Le Pantanal (Bresil) Etude Ecogeographique. Geomorphologie. *Travaux et Documents de Geographie Tropicale*. CEGET, **52**: 7–92.

Ubilla, M., 1996. *Paleozoologia del Cuaternario continental de la cuenca norte del Uruguay: biogeografia, cronologia y aspectos climáticos-ambientales*. Unpublished PhD these, FC-Pedeciba, Uruguay, 232.

Ubilla, M. and Perea, D., 1999. Quaternary vertebrates of Uruguay: a biostratigraphic, biogeographic and climatic overview. *Quaternary of South America and Antarctic Peninsula*, **12**(1996–1997), 75–90.

van der Hammen, T., 1986. Fluctuaciones Holocénicas del nivel de inundaciones en la cuenca del bajo Magdalena-Cauca-San Jorge (Colombia). *Geologia Norandina*, **10**, 12–18.

van der Hammen, T. and Absy, M.L., 1994. Amazonia during the last glacial. *Palaeogeography, Palaeoclimatology, Palaeoecology*, **109**, 247–261.

van der Hammen, T., Duivenvoorden, J.F., Lips, J.M., Espejo, N. and Urrego, L., 1992a. The late Quaternary of the middle Caquetá river area (Colombian Amazonia). *Journal of Quaternary Science*, **7**, 45–55.

van der Hammen, T., Urrego, L., Espejo, N., Duivenvoorden, J. and Lips, J., 1992b. Late glacial and Holocene sedimentation and fluctuation of river water level in the Caquetá river area (Colombia, Amazonia). *Journal of Quaternary Science*, **7**(1), 57–67.

Vaz, J.E. and Garcia, M.J., 1989. Thermoluminescence dating of fossil sand dunes in Apure, Venezuela. *Acta Científica Venezolana*, **40**, 81, 82.

Vital, H. and Stattegger, K., 2000. Lowermost Amazon river: evidences of late Quaternary sea level fluctuations in a complex hydrodynamic system. *Quaternary International*, **72**, 53–60.

Warne, A., Meade, R., White, W.A., Guevara, E.H., Gibeaut, J., Smyth Aslan, A. and Tremblay, T., 2002. Regional controls on geomorphology, hydrology, and ecosystem integrity in the Orinoco delta, Venezuela. *Geomorphology*, **44**(3–4), 273–307.

Welcomme, R.L., 1979. *Fisheries Ecology of Floodplain Rivers*. Longman, London.

13 Late Pleistocene–Holocene Palaeohydrology of Monsoon Asia

V.S. KALE,[1] A. GUPTA[2] AND A.K. SINGHVI[3]
[1]University of Pune, Pune, India
[2]University of Leeds, Leeds, UK
[3]Physical Research Laboratory, Ahmedabad, India

1 INTRODUCTION

The monsoon system is a thermodynamic atmospheric circulation, characterised by strong seasonality of wind direction, temperature and precipitation (Ramage, 1971). The largest monsoon-dominated region in the world is in Asia (Figure 13.1). The Asian monsoon comprises the southwest (Indian) monsoon and the southeast (east Asian) monsoon. The former is the major source of precipitation over the Indian subcontinent and the western part of southeast Asia and the latter is the dominant influence over the eastern part of southeast Asia and east Asia. The two systems, although largely independent, interact and play a significant role in the global hydrologic cycle (An *et al.*, 2000; Kudrass *et al.*, 2001). Multi-proxy records from China and the Indian and North Pacific Oceans denote that the Indian as well as the east-Asian monsoon systems were established about 8 million years ago (An *et al.*, 2001). There is evidence of a stronger monsoon system at 3.5 million years and 2.6 million years (Qiang *et al.*, 2001). Continental and marine palaeoclimatic records further indicate that since the onset of glaciation in the northern hemisphere around 2.5 million years ago (Shackleton *et al.*, 1984), the strength of the Asian monsoon has varied on both long and short timescales (Overpeck *et al.*, 1996; Liu *et al.*, 1999; Lu *et al.*, 1999; Qiang *et al.*, 2001). These changes in the strength of the monsoons were linked to global processes (Schulz *et al.*, 1998; Wang *et al.*, 1999; Kudrass *et al.*, 2001), and had a profound impact on the palaeohydrology, palaeogeography and geomorphology of monsoon Asia.

This chapter attempts to reconstruct a framework for the late-Quaternary palaeo-climatic and palaeohydrological changes in monsoon Asia from the large number of investigations published so far (Table 13.1), from the Last Glacial Maximum (LGM) to the present, that is covering the last 18 ^{14}C kyr BP (radiocarbon) or 21.5 cal kyr BP (calibrated).

2 PALAEOHYDROLOGY DURING THE LAST GLACIAL MAXIMUM

During the last glacial period, only the highly elevated areas of monsoon Asia, such as the Tibetan Plateau and the Himalayan Mountains, provided favourable conditions for the formation of large glaciers (Seltzer, 2001) and ice caps. Three ice caps on the Tibet Plateau, namely, Dunde, Guliya and Dasuopu (Figure 13.1) have provided evidence

Palaeohydrology: Understanding Global Change. Edited by K.J. Gregory and G. Benito
© 2003 John Wiley & Sons, Ltd ISBN: 0-470-84739-5

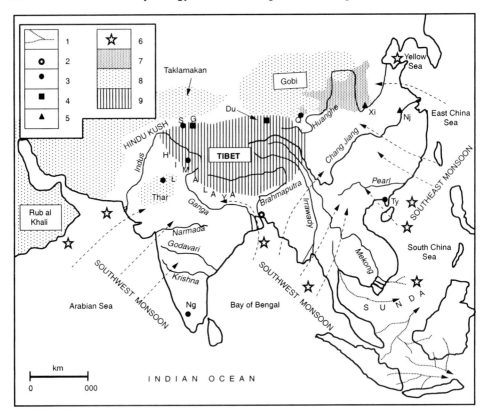

Figure 13.1 Map showing some important sites with climate proxy records mentioned in the text. The dashed lines with arrows show the wind patterns for the Asian summer monsoon (June to September). Key: 1 – extinct drainage of Sundaland; 2 – borehole sites; 3 – lake/peat sites; 4 – ice core sites; 5 – fluvial/flood sites; 6 – deep-sea core sites; 7 – Loess Plateau; 8 – deserts; 9 – Qinghai–Tibetan Plateau; Du – Dunde ice cap; G – Guliya ice cap; L – Lunkaransar Lake; Ng – Nilgiri; Nj – Nanjing flood site; Q – Qinghai Lake; S–Sumxi Co; Ty – Tianyang Lake; Xi – Xiaolangdi palaeofloods site

(Thompson *et al.*, 1989; 1997; 2000) of – (1) a pronounced climatic instability in the tropics; and (2) century-to-millennium scale fluctuations in the monsoon hydrologic cycle since the last glacial stage and even earlier.

Ice-core records from Tibet suggest that the climate during the LGM was cool, dry and variable (Thompson *et al.*, 1997). Evidence indicates the presence of large glaciers, depression of the glacier snowline by several hundred meters and atmospheric cooling of 5 to 7°C during the Last glacial (Lehmkuhl *et al.*, 1999; Seltzer, 2001). On the basis of pollen records from the Tianyang Lake (Figure 13.1), Zheng and Lei (1999) have inferred that the drop in temperature and precipitation during the LGM was much greater than during the three previous glacial periods (Oxygen Isotope Stages 6, 8, 10). Consequently, it is reasonable to suggest that during the LGM, the hydrological regime was notably different from that of the present.

A coeval weakening of the summer monsoon over Asia has been indicated by–(1) past lake levels and lake sediment/peat in China (Fang, 1991; Gasse *et al.*, 1991; Zheng

Table 13.1 Proxy palaeoclimatic records from monsoon Asia

Proxy records	Area	References
Lakes/peat	China	Fang (1991; 1993), Gasse *et al.* (1991), Lister *et al.* (1991), Zheng and Lei (1999), An *et al.* (2000)
	Thailand	Kealhofer and Penny (1998)
Lakes/peat	India	Bhattacharyya (1989), Singh *et al.* (1990), Sukumar *et al.* (1993), Mazari *et al.* (1996), Kusumgar *et al.* (1995), Prasad *et al.* (1997), Enzel *et al.* (1999), Phadtare (2000)
Loess	China	Kukla and An (1989), An *et al.* (1991), An *et al.* (1993), Maher *et al.* (1994), Porter and An (1995), Zhou *et al.* (1996), Chen *et al.* (1997)
Boreholes	Bangladesh	Goodbred and Kuehl (2000)
	China	Hori *et al.* (2001)
Speleothems, calc tufa and groundwater	Nepal	Denniston *et al.* (2000)
	India	Pawar *et al.* (1988), Yadava and Ramesh (1999), Sukhija *et al.* (1998)
Ice cores	Tibet–China	Thompson *et al.* (1989), Thompson *et al.* (1997), Thompson *et al.* (2000)
Deep-sea cores	Bay of Bengal/ Andaman Sea	Colin *et al.* (1998), Ahmed *et al.* (2000), Sangode *et al.* (2001)
	Arabian Sea	Van Compo *et al.* (1982), Duplessy (1982), Sirocko *et al.* (1993), Caratini *et al.* (1994), Overpeck *et al.* (1996), von Rad *et al.* (1999), Sarkar *et al.* (2000), Thamban *et al.* (2001)
	South China Sea	Huang *et al.* (1997), Wei *et al.* (1998), Wang *et al.* (1999), Pelejero *et al.* (1999)

and Lei, 1999; An *et al.*, 2000) and India (Bryson and Swain, 1981; Singh *et al.*, 1990; Sukumar *et al.*, 1993; Phadtare, 2000); and (2) isotopic, microfaunal and geochemical data from the Arabian Sea (Van Compo *et al.*, 1982; Sirocko *et al.*, 1993) and the South China Sea (Huang *et al.*, 1997; Wang *et al.*, 1999). A contrasting strengthening of the winter monsoon during the Last glacial stage and evidence of aridity and humidity associated with global cooling and warming, respectively, up to the millennium scale have also been deduced from land and oceanic records (Duplessy, 1982; Singh *et al.*, 1990; Porter and An, 1995; Colin *et al.*, 1998; Schulz *et al.*, 1998; Wang *et al.*, 1999).

The Asian Highlands (Himalaya, Tibet and Myanmar Ranges) are the source area of many large rivers including the Ganga, Brahmaputra, Indus, Mekong and Changjiang. The headwaters of these rivers are sustained at present by both glacial melt water and orographic monsoon rainfall. During the LGM, cold and dry conditions prevailed over much of the area, the southwest monsoon was weak and less moisture reached the headwaters of these large rivers (Duplessy, 1982). It is likely that even the springtime melt water was reduced (Emeis *et al.*, 1995). Consequently, the rivers must have experienced a significant reduction in water and sediment discharge, such that many rivers experienced only highly seasonal or ephemeral flows. River discharges to the Bay of Bengal and the Arabian Sea were markedly reduced (Cullen, 1981; Duplessy, 1982; Cayre and Bard, 1999).

Isotopic and geochemical analyses of deep confined groundwater in south India show that the period ca 18 ^{14}C kyr BP was marked by aridity (Sukhija *et al.*, 1998). An arid phase between ca 20 and 16 ^{14}C kyr BP is also indicated by peat in Nilgiri, south India (Sukumar *et al.*, 1993); and in southeast Asia a drop in rainfall by more than half has been suggested (Flenley, 1979; Verstappen, 1980). Planktonic foraminiferal assemblages from the South China Sea indicate a much higher temperature contrast between summer and winter (ca 9°C) during the Last glacial than that during the Holocene (Wei *et al.*, 1998), suggesting a greater seasonality. Increased aridity and seasonality during the LGM corresponds to reduced streamflow and flow magnitudes in the rivers of south and southeast Asia (Kale and Rajaguru, 1987; Verstappen, 1997).

At present, tropical cyclones forming over the adjoining seas are common components of the monsoon and are an important source of moisture. Pronounced continentality and cooler sea surfaces during the glacial times (Ruddiman, 1984) should have caused a diminution in the intensity and frequency of tropical storms. Oceanic reconstructions indicate fewer tropical cyclones in the South China Sea during the glacial period (Wang *et al.*, 1999). A decrease in the incidence of typhoons over Japan has been inferred by Sugai (1993). By implication, similar conditions should have prevailed in the seas adjoining the Indian subcontinent. The reduction in the flow magnitude and the increased seasonality had a striking effect on the fluvial systems originating in the Asian highlands.

Estimates of palaeomonsoon precipitation across the Chinese Loess Plateau indicate great decreases in rainfall in central China during the glacial periods (Maher *et al.*, 1994). In south China, the water discharge of the Changjiang River decreased remarkably, and the river occasionally became dry, exposing a 3- to 10-km-wide sandy channel bed. The dry sand bed provided the material for sand dunes on the southern bank of the river. Several phases of sand dune accretion indicate that the river was flowing only intermittently (Liu *et al.*, 1997). Studies in Thailand by Loeffler *et al.* (1984) also suggest greatly reduced river flows and sand dune activity in the seasonal rivers.

Apart from the decrease in precipitation and runoff, the cool and dry conditions during the LGM imply reduced weathering, affecting the sediment load of the Asian rivers. This is suggested by the clay mineralogical studies of sediments in the Bay of Bengal and the Andaman Sea deposited by the Ganga–Brahmaputra and the Irrawaddy rivers, respectively (Colin *et al.*, 1999). Sangode *et al.* (2001) suggest that the sediment supply to the Bay of Bengal during the LGM increased proportionally from rivers of peninsular India as the southwest monsoon weakened over the Himalaya.

The LGM period was also a period of eustatic low sea level (Fairbanks, 1989), when a large currently offshore area of southeast Asia and south Asia was exposed (Emmel and Curray, 1982; Pelejero *et al.*, 1999). The exposed area on the Sunda shelf off southeast Asia was a coastal lowland drained by several major river systems (Gupta *et al.*, 1987). Large rivers of monsoon Asia must have extended their courses across the exposed sea floor (Figure 13.1). Evidence of such courses is seen in submerged canyons (Liu *et al.*, 1992; von Rad and Tahir, 1997; Goodbred and Kuehl, 2000) or deltas (Emmel and Curray, 1982; Wang *et al.*, 1999).

Evidence of increased continentality has been found in many parts of monsoon Asia (Kale and Rajaguru, 1987; Prins and Postma, 2000). Reduced discharges in the rivers of peninsular India have been related to both the weakening of the summer monsoon and the seaward shifting of the shoreline by more than 200 km along the west coast of India (Kale and Rajaguru, 1987). Similar conditions should be expected for other rivers of monsoon Asia that were affected by a pronounced shoreline shift (ca 300–600 km).

Sedimentary response to such a drastic decrease in precipitation, water discharge and a falling sea level during the LGM is covered by relatively few studies. Williams and Clarke (1984), Kale and Rajaguru (1987) and Joshi and Kale (1997) have noted periods of aggradation in the interior of the Indian Peninsula during the LGM, Kale and Rajaguru (1987) recorded deposition by non-meandering and bedload-dominant streams during the glacial period in northwest Deccan, and Tandon *et al.* (1997), Juyal *et al.* (2000) and Srivastava *et al.* (2001) found evidence of disruption of drainage systems in western India. In the Indian Desert, drainage was seriously affected and the fluvial processes were largely dormant because of increased aridity (Kar *et al.*, 2001). Segmentation of streams by aeolian activity was a feature of this period (Kar, 1990).

In western Nepal, Monecke *et al.* (2001) found deposits of highly mobile, braided rivers produced under glacial conditions. Gupta *et al.* (1987) are of the view that the Old Alluvium of Singapore was laid down by braided rivers that were characterised by seasonal flows and large floods. The presence of similar deposits in southeast Asia suggests the widespread occurrence of seasonal rivers (Loeffler *et al.*, 1984).

In central China, Porter *et al.* (1992) found a strong association between cold phases and stream aggradation. The upper-middle reaches of the Changjiang responded to the reduced precipitation by aggradation. In the lower reaches, incision by rivers, graded to a lower sea level of the glacial stage appears to be a common phenomenon. This is suggested by the presence of incised valleys of the Ganga–Brahmaputra and Changjiang Rivers below early Holocene sediments (Goodbred and Kuehl, 2000; Hori *et al.*, 2001).

All the climate proxies of this time, thus, indicate a cooler and drier period in monsoon Asia in contrast to warm and wet conditions of the succeeding Holocene (Figure 13.2). During the LGM a low sea level, increased seasonality, decreased rainfall and reduced frequency of tropical cyclones is indicated in a number of studies. Fluvial activity was highly variable and generally subdued.

3   THE DEGLACIAL–EARLY HOLOCENE PERIOD

Sufficient oceanic evidence now indicates that the drier LGM climate ameliorated after about 13 to 12.5 ^{14}C kyr BP (15.3 to 14.7 cal kyr BP) in response to increased insolation and global warming (Sirocko *et al.*, 1993). This conclusion has been derived mainly from Arabian Sea and South China Sea multi-proxy records (Sirocko *et al.*, 1993; Overpeck *et al.*, 1996; Huang *et al.*, 1997; Wang *et al.*, 1999); Chinese and Indian lake sediment (Singh *et al.*, 1990; Fang, 1991; Lister *et al.*, 1991; An *et al.*, 2000) and Tibetan ice cores (Thompson *et al.*, 1997). Computer modelling (CCM1) also shows that the summer monsoon strengthened significantly after 13.5 ^{14}C kyr BP (16 cal kyr BP) (Kutzbach *et al.*, 1998). Reconstruction of Sea Surface Temperatures

(SSTs) using oxygen isotopes of planktonic foraminifera from the eastern Arabian Sea indicates a 1.5 to 2.5°C deglacial warming (Cayre and Bard, 1999). Marine records from the South China Sea also suggest increase in SSTs from the LGM to the Holocene (Wie et al., 1998; Steinke et al., 2001).

Intensification of the southwest and also the southeast monsoon since the last deglaciation is currently perceived as stepwise and dramatic (Marcontonio et al., 2001). Whilst Sirocko et al. (1993) have inferred that the monsoon intensified episodically between 14.3 and 8.7 ^{14}C kyr BP in four steps, Overpeck et al. (1996) concluded that the monsoon strength increased suddenly in two steps, 13 to 12.5 ^{14}C kyr and 10 to 9.5 ^{14}C kyr BP (ca 15.3–14.7 and 11.5–10.8 cal kyr BP). High-resolution SST records from the South China Sea show an abrupt warming (ca 1°C <200 yr) at the end of the last glaciation, approximately synchronous with the Bølling–Allerød transition (Steinke et al., 2001). A major intensification of the southwest monsoon at 11.5 cal kyr BP, coinciding with a major climatic transition in the Greenland record, has been identified by Sirocko et al. (1996).

Strengthening of the summer monsoon and related high precipitation occurred between 9.5 and 5.5 ^{14}C kyr BP (Sirocko et al., 1993; Overpeck et al., 1996). Glaciers on the Nanga Parbat expanded in the early Holocene, as a result of enhanced moisture levels (Phillips et al., 2000). In peninsular India, evidence of enhanced precipitation, discharge and groundwater levels is provided by the building of waterfall tufas between 9.8 and 8.1 ka (U/Th) in the rainshadow area of the Western Ghat (Pawar et al., 1988). Reconstruction of palaeorainfall over the Loess Plateau in China indicates much higher rainfall between 9 and 5 ^{14}C kyr BP (Maher et al., 1994). Fang (1991), on the basis of data from over 70 lakes distributed across China, found high lake levels and lake expansions between ca 9.5 and 3.5 cal kyr BP.

However, a number of recent investigations indicate that the timing of the Holocene climatic optimum was not synchronous over monsoon Asia (Fang, 1991; Shi et al., 1993; Lehmkuhl, 1997; An et al., 2000). It reached a maximum in northeastern and central China ca 10,000 to 7,000 years ago, in the middle and lower Changjiang basin around 7,000 to 5,000 years ago, in southern China around 3,000 years ago (An et al., 2000), and in western China between 7,500 and 3,500 years BP (Zhou et al., 1991). Pollen data from alpine peat suggests that in the central Himalaya, the highest monsoon intensity occurred between 6.0 and 4.5 cal kyr (Phadtare, 2000).

The abrupt deglacial intensification of the monsoon had a catastrophic impact on the drainage systems of monsoon Asia. In many Asian rivers, the glacial–interglacial transition was marked by an abrupt and immense increase in the wet monsoon flow. Widespread evidence now exists to suggest shrinkage of deserts, revival of fluvial activity in the Indian subcontinent and Tibet, and increased weathering and fluvial erosion in many parts during early Holocene (Colin et al., 1999; Lehmkuhl et al., 1999; Goodbred and Kuehl, 2000). The presence of river-transported silts at ca 11.8 ^{14}C kyr BP in the Qinghai Lake record (Lister et al., 1991) indicates the revival of fluvial activity on the Tibetan Plateau. On the Deccan Plateau, evidence of widespread overbank flooding by suspended-load dominant meandering streams from ca 17 to 10 ^{14}C kyr BP has been reported by Kale and Rajaguru (1987).

The increase in water discharge in many rivers at this time can also be attributed to an increase in the frequency of tropical cyclones (Huang et al., 1997; Wang et al., 1999) and reduced continentality as a result of high sea levels (Kale and Rajaguru, 1987). Appearance of storm-related early Holocene clay in the Taiwan lake sediments (Huang et al., 1997) suggests that the strengthening of the summer monsoon was accompanied by an increased number of tropical cyclones. Similarly, the early Holocene increase in the flood magnitude in the Ara River in Japan has been attributed to increased typhoon frequency (Grossman, 2001).

Hydrological characteristics of the fluvial systems were also significantly affected in the early Holocene by the impact of a strong monsoon rainfall against the mountain chains, extensive slope failures and consequently enhanced sediment supply. Unusually high sediment output and water discharge under conditions of an intensified early Holocene monsoon in the Himalayan rivers is revealed by chronostratigraphic data from the deltaic deposits of Ganga–Brahmaputra (Weber et al., 1997; Goodbred and Kuehl, 2000). Borehole data and volume calculations indicate that an enormous amount of sediment was deposited between 11 and 7 cal kyr BP (Goodbred and Kuehl, 2000). It is most likely that the enormous deposition of sediment in the Ganga–Brahmaputra delta was preceded by, or coincided with, equally phenomenal erosion in the Himalayan Mountains and sediment accumulation at the mountain–plains interface. The deltaic sedimentation could have been penecontemporaneous with the building of mega-fans (of rivers such as Kosi, Gandak, etc.) at the foot of the Himalaya.

Slope failure is a dominant denudational process in the Himalayan Ranges (Shroder, 1998), characterised by rapid uplift and high incision rates (Burbank et al., 1996). Landslides and debris flows that follow earthquakes and/or intense rainfall/snowmelt can block stream courses and create large lakes/dams. Later, sudden breaching of these dams generates catastrophic floods downstream. In addition, the advance of tributary glaciers into main valleys and the moraine dams created by retreating glaciers may also produce similar flood events. Numerous examples of floods, consequent to such dam failures have been reported from the Himalaya (Coxon et al., 1996; Hewitt, 1998; Shroder, 1998; Wohl and Cenderelli, 1998; Cornwell, 1998). The capacity and competence of such floods to transport sediment flux exceeds other denudational processes (Shroder, 1998). Evidence of postglacial enhanced temperatures and precipitation levels over the Himalaya suggests that landslide-dam, ice-dam and moraine-dam failure floods were a feature of the early Holocene humid period. Advance of glaciers, such as in the Nanga Parbat area, was favoured by increased moisture (Phillips et al., 2000). Sediment accumulation and later removal probably started to occur in the Himalaya early in the Holocene. In western Nepal, there is evidence of mobilisation and redeposition of morainic material by enormous debris flows. These debris flows were triggered by outburst floods from glacial lakes, or by strong monsoonal rains and earthquakes (Monecke et al., 2001).

Most climatic proxies suggest that the stronger monsoon in the early Holocene was associated with high lake levels, increased flow discharges, floods and scouring (Figure 13.2). Evidence of huge floods and scouring in the Changjiang River is provided by a large number of uprooted trees near Ichung–Nanjing (Figure 13.1) that have yielded ages between 6 and 4.5 ^{14}C kyr BP (Yang, 1991b). Analysis of slackwater flood deposits indicates that about five large floods ($>27,000\,\mathrm{m^3s^{-1}}$) had occurred on the Huanghe River at Xiaolangdi (Figure 13.1) between 8 and 6 ^{14}C kyr BP (Yang et al., 2000b). The largest known flood at 7,362 years BP on the river, with a magnitude of $42,900\,\mathrm{m^3s^{-1}}$, was comparable to the catastrophic flood in 1843 AD (Yang et al., 2000b). Similarly, the Pearl River, the second largest river in China in terms of discharge, also experienced a significant increase in flow (Wang et al., 1999). In Japan, Grossman (2001) deduced that large floods occurred on the Ara River in the early Holocene.

Proxy records of the Holocene palaeohydrology in southeast Asia are somewhat sparse, but there are sufficient indications of increased precipitation and stream discharge in the early Holocene (Verstappen, 1997). Estimation of discharge by Bishop and Godley (1994) suggests greater bankfull discharges in the Yom River (Thailand) during the early mid-Holocene humid phase. By implication, other river valleys of the region would also have experienced similar changes.

At present, high-magnitude floods, spaced over periods of a few years to decades, govern the channel morphology of many monsoonal rivers and the channel size increases with flood magnitude (Gupta, 1995). In the early Holocene, it is likely that the rivers responded by deepening and enlarging their channels to accommodate increased water discharges and to efficiently transport a large supply of sediments from upstream. Evidence from central China (Porter et al., 1992) and peninsular India is consistent with this notion. Widespread erosion and incision in the peninsular rivers was associated with enhanced monsoon precipitation. Rivers, such as the Son (Williams and Clarke, 1984), Sabarmati (Srivastava et al., 2001), Narmada (Gupta et al., 1999), Krishna and Godavari (Kale and Rajaguru, 1987) all responded to these changes by deepening and lowering their channels, giving rise to terraces. The early Holocene age of the soils of areally extensive interfluves in the upper Ganga Plains (Kumar et al., 1996) also implies that the Himalayan rivers were no longer aggrading. Similarly, the fluvio-sedimentary response of the Changjiang to the higher runoff in its upper-middle reaches was channel deepening (Yang, 1991a). By implication, other comparable rivers (Brahmaputra, Indus, Irrawaddy, Mekong, etc.) probably responded in a similar fashion, although neotectonic activities would have additionally accentuated incision and gorge formation (Hurtado et al., 2001).

Deglaciation and a rapid rise in sea level during the early Holocene has been documented in a variety of locations, including the Indus delta (von Rad and Tahir, 1997; Prins and Postma, 2000); the Ganga–Brahmaputra delta (Goodbred and Kuehl, 2000; Banerjee, 2000); South China Sea (Emmel and Curray, 1982; Pelejero et al., 1999) and southeastern China (Chen, 1999; Zheng and Li, 2000; Hori et al., 2001). The rapid and sometimes accelerated rise (such as between 14.6 and 14.3 cal kyr; Hanebuth et al., 2000) in sea level had a significant impact on the drainage systems. The most striking evidence is from southeast Asia where the distal parts of large drainage systems, which were draining the subaerially exposed Sunda shelf during the late Pleistocene, were submerged (Gupta et al., 1987; Pelejero et al., 1999). Consequently, southeast Asia witnessed a large-scale disruption and disintegration of the drainage network. Elsewhere, the transgression was associated with rapid delta growth and delta progradation due to enormous sediment discharge from upstream (von Rad and Tahir, 1997; Goodbred and Kuehl, 2000; Hori et al., 2001; Saito et al., 2001).

Evidence of cooling and strong seasonality associated with the Younger Dryas (YD) event have been found in the oxygen isotope records from the Arabian Sea (Cayre and Bard, 1999; Marcontonio et al., 2001), planktonic foraminiferal assemblages from the South China Sea (Wei et al., 1998; Steinke et al., 2001), central China loess (An et al., 1993), peat and aeolian-palaeosol sequence from east Asia (Zhou et al., 1996) and in lacustrine sediment from southeast Asia (Maloney, 1995). There is, however, very little direct evidence of fluvio-sedimentary response to this near-glacial event. This is not surprising as fluvial records on land lack the time-resolution and continuity required to understand the effects of shorter climatic events. A possibility of reduced discharge from the Irrawaddy and Salween Rivers into the Andaman Sea coinciding with the YD, has been suggested by Ahmed et al. (2000). There is also evidence of cessation of river-transported silt flux into the Qinghai Lake (China) in response to increased aridity around 10.8 ^{14}C kyr BP (Lister *et al.*, 1991). However, there is no evidence of a YD age glacier advance from Tibet (Lehmkuhl, 1997), which, as suggested by Phillips *et al.* (2000), may be attributed to drier conditions during this period.

After the early Holocene monsoon optimum, a progressively weakened monsoon and increasing aridity generally characterised the mid-Holocene (Steig, 1999). This change, after about 6 to 5 ^{14}C kyr BP, was much more gradual in comparison with the glacial–deglacial transition (Overpeck *et al.*, 1996). However, because of variations in latitude, altitude and distance from the sea, the termination of the early Holocene humid phase in various basins was not synchronous. A trend towards aridity after 5.5 cal kyr BP (4.8 ^{14}C kyr BP) is indicated by marine records (Sirocko *et al.*, 1993; Overpeck *et al.*, 1996); lake levels in China and northwest India (Singh *et al.*, 1990; Fang, 1991; Shi *et al.*, 1993; Enzel *et al.*, 1999); minor advance of glaciers in Lahul Himalaya (Owen *et al.*, 1997) and alpine peat from the higher Himalaya (Phadtare, 2000). Borehole data from the Ganga–Brahmaputra delta also indicates a drop in sediment discharge in the Himalayan rivers after about 7 cal kyr BP (Goodbred and Kuehl, 2000). Enzel *et al.* (1999) have reported the desiccation of Lunkaransar Lake after about 4,800 ^{14}C kyr BP in Rajasthan. A short phase of aggradation in the Son River (Williams and Clarke, 1984) and in the upper Godavari and Krishna Basins (Kale and Rajaguru, 1987) also indicates reduction in rainfall and discharge over parts of the Indian Peninsula. The deposits associated with the T_1 terrace in the Ganga Plains (Singh, 1996) were perhaps laid down during this phase. In Thailand, Bishop and Godley (1994) found evidence of lower discharges during the mid-late Holocene. Grossman (2001) noted a drop in flood magnitude on the Ara River, Japan around 5 ^{14}C kyr BP. Thus, the pattern of enhanced discharge, accelerated erosion and transported sediment of the early Holocene appears to have been replaced by a cooler, drier period (Steig, 1999) (Figure 13.2), which was geomorphologically less active.

4   LATE HOLOCENE PALAEOHYDROLOGY

During the late Holocene, temperature and precipitation conditions shifted in response to global climatic changes (Liu *et al.*, 1998). Although the shift was neither abrupt nor synchronous, a conspicuous change around ca 3.5 ^{14}C kyr BP is indicated by marine, peat and lake records (Table 13.2). This was also the time of decrease in the input of fluvial mud into the Indus (von Rad and Tahir, 1997), stabilisation of the present configuration of the Ganga–Brahmaputra delta (Goodbred and Kuehl, 2000), and reduced discharge in the Yom River, Thailand (Bishop and Godley, 1994).

Very few fine-resolution palaeoclimatic records exist to reconstruct millennial, century and decadal fluctuations in precipitation and runoff. Pollen from ice cores from the Dunde ice cap show evidence of relatively humid periods at 2.7 to 2.0, 1.5 to 0.8 and 0.6 to 0.8 cal kyr BP (Liu *et al.*, 1998). The records also indicate prominent changes during the Medieval Warm (MW) period (790–620 yrs BP) and the Little Ice Age (LIA) (330–80 yrs BP) (Liu *et al.*, 1998). This and other studies based on multi-proxy data (Zhou *et al.*, 1991; Sukumar *et al.*, 1993; Mazari *et al.*, 1996; Lehmkuhl, 1997; von Rad *et al.*, 1999; Denniston *et al.*, 2000) suggest – (1) relatively widespread drier periods from ca 2 to 1.5 ^{14}C kyr BP and around 1 ^{14}C kyr BP; (2) a significant warming during the MW period; and (3) a cooling during the LIA (Figure 13.3).

Fluvial records lack comparable temporal resolution and the corresponding fluctuations in flow and sediment pattern have been poorly understood, although there are a

Table 13.2 Timing of late Holocene change as indicated by palaeoclimate proxies

Climatic proxy/area	Date	Remark	Reference
Glacier – Tibet, China	3 ^{14}C kyr BP	Glacial advance	Lehmkuhl (1997)
Lakes – China	1000 BC	Dramatic drop in lake levels	Fang (1993)
Palaeoceanographic record – South China Sea	4 ^{14}C kyr BP	Cooling event	Wei *et al.* (1998)
High Himalaya	4–3.5 cal. yrs BP	Sharp decrease in temperature and rainfall	Phadtare (2000)
Ice core and marine core – Tibet and Arabian Sea data	3.4 ^{14}C kyr BP	Significant increase in the biogenic and lithogenic components	Rangarajan and Sant (2000)
Shallow water cores – Arabian Sea	3.5 ^{14}C kyr BP	General weakening of the monsoon	Nigam (1993)
Upwelling indices – Arabian Sea	3.5 ^{14}C kyr BP	Intensity of monsoon reduced	Naidu (1996)
Marine core – Arabian Sea	3.5–4 ^{14}C kyr BP	Decreased precipitation	von Rad *et al.* (1999)
Groundwater – south India	4 ^{14}C kyr BP	Onset of unstable climate	Sukhija *et al.* (1998)
Peat – Nilgiri, south India	3.5 ^{14}C kyr	Onset of aridity	Sukumar *et al.* (1993)
Loess and wind deposit – Korat Plateau, Thailand	3.5 ka	Drier conditions and loess deposition	Nutalaya *et al.* (1989a)

few geological and historical records. Caratini *et al.* (1994) have inferred a reduction in the discharge of the Kali River into the Arabian Sea since about 2.2 ^{14}C kyr BP, due to reduction in rainfall over the Western Ghat. Slackwater palaeoflood hydrology indicates distinct periods of large and moderate floods in central and western India during the last 2 ^{14}C kyr BP (Ely *et al.*, 1996, Kale *et al.*, 2000). Sequences of extreme floods occurred twice: between ca 400 to 1000 AD and after the 1950s (Kale, 1999). A significantly reduced frequency of large floods between ca AD 1500 and the late 1800s may reflect the regional influence of the LIA (Kale, 1999). Nigam and Khare (1992) have identified large floods in 2000 and 1500 BC in peninsular India from oceanographic records. Rao *et al.* (1963) also reported archaeological evidence of these two floods from western India (Gujarat). A period of major flood activity ca 1.7 to 1.8 ^{14}C kyr BP has been inferred in northeast Thailand (Nutalaya *et al.*, 1989b; Bishop and Godley, 1994).

Historical records are a valuable source of information on short-term climatic variability. In India, information on droughts is more extensive than that on floods. Information exists about major floods and cyclones in various river basins of India, but the records are mostly neither continuous nor uniform (Kale, 1998). In contrast, the Chinese historical documents provide a near-continuous record of thermal and

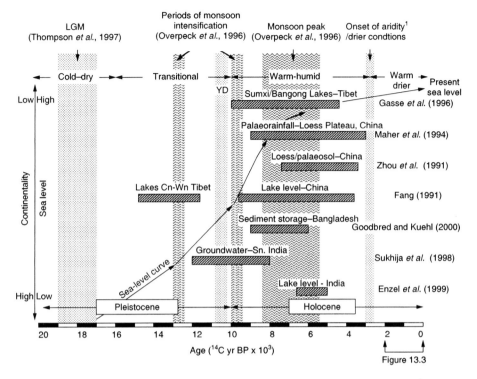

Figure 13.2 Periods of monsoon intensification during the last glacial/postglacial period indicated by proxy records. YD – Younger Dryas. [1] – After Singh, G., Wasson, R.J. and Agrawal, D.P., 1990. Vegetation and seasonal climatic changes since the last full glacial in the Thar Desert, Northwest India. *Review of Palaeobotany and Palynology*, **64**, 351–358; Fang, J.Q., 1991. Lake evolution during the past 30,000 years in China and its implications for environmental change. *Quaternary Research*, **36**, 37–60; Overpeck, J., Anderson, D., Trumbore, S. and Prell, W., 1996. The Southwest Indian monsoon over the last 18000 years. *Climate Dynamics*, **12**, 213–225; Gasse, F., Fontes, J.Ch., Van Campo, E. and Wei, K., 1996. Holocene environmental changes in Bangong Co basin (Western Tibet). Part 4: discussion and conclusions. *Palaeogeography, Palaeoclimatology, Palaeoecology*, **120**, 79–92; Lehmkuhl, F., 1997. Late Pleistocene, late-glacial and Holocene glacier advances on the Tibetan plateau. *Quaternary International*, **38–39**, 77–83; and others. The generalised trend of the eustatic sea-level rise is after Fairbanks, R.G., 1989. A 17,000-year glacio-eustatic sea level record: influence of glacial melting rates on the Younger Dryas events and deep-ocean circulation. *Nature*, **342**, 637–642. The beginning and end of periods approximate

precipitation conditions and document floods, droughts, storms and lake levels (Jingtai *et al.*, 1991; Zhang, 1991; Fang, 1993). Fang (1993) has identified three major periods of lake expansion – BC 500 to 0 AD, AD 650 to 950, and AD 1250 to 1650 in eastern China. The flood record on Huanghe extends back to the second century AD, and major floods on the Changjiang occurred in 1153, 1227, 1520, 1560, 1788, 1796, 1860 and 1870 AD, the last being the largest in the last millennium (cf. Baker *et al.*, 1987).

Apart from minor climatic and sea-level fluctuations, anthropogenic alterations in the drainage basins have been the dominant characteristics of the late Holocene in monsoon Asia (Singh, 1971; Kealhofer and Penny, 1998; Jiang and Piperno, 1999; Hori *et al.*, 2001). Consequently, the palaeoclimatic and palaeohydrological signals are hard to decipher because the available signals are interwoven with human impacts.

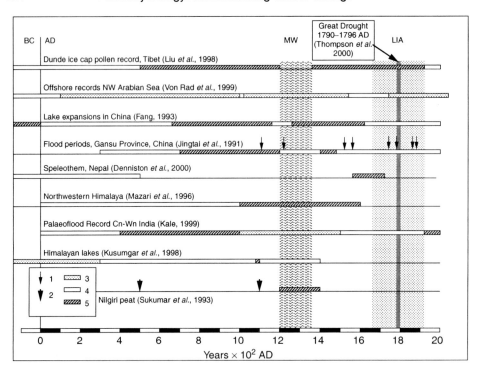

Figure 13.3 Palaeohydrology of the past 2,000 years. Key: 1 – Major floods on the Changjiang (cf. Baker *et al.*, 1987); 2 – shift towards wetter climate; 3 – more or less same as present; 4 – drier than present; 5 – wetter than present. MW – Medieval Warm period and LIA – Little Ice Age (after Liu, K.B., Yao, Z. and Thompson, L.G., 1998. A pollen record of Holocene climatic changes from Dunde ice cap, Qinghai-Tibetan plateau. *Geology*, **26**, 135–138.) Thin line represents no information. The beginning and end of periods approximate

5 DISCUSSION AND CONCLUSIONS

The Asian monsoon system is possibly the major component of the global atmospheric circulation. It is clear from multi-proxy palaeoclimatic records that the strength of the monsoon is linked to glacial–interglacial cycles, the orbital forcing and the El Niño Southern Oscillation (ENSO) events (Duplessy, 1982; Thompson *et al.*, 1989; Whetton *et al.*, 1990; Sirocko *et al.*, 1993; Schultz *et al.*, 1998; Yang *et al.*, 2000a; Thompson *et al.*, 2000). It has also been postulated that the summer monsoon initiates, amplifies and terminates climatic cycles in the northern hemisphere, by influencing the conditions that produce the greenhouse effect by injecting a large amount of water vapour into the atmosphere and by affecting snow accumulation rates (Kudrass *et al.*, 2001). Understanding the linkage between monsoons and global climates is, therefore, an important aspect in reconstructing a framework for Quaternary climatic and hydrologic changes on the global scale.

The physical, hydrological and biological environment of monsoon Asia is closely linked to the rhythm of the two monsoons. Changes in the intensity of monsoons connected with global cooling and warming during the Quaternary have strikingly affected the hydrological characteristics of the rivers of monsoon Asia. Although the poor resolution of terrestrial records precludes any firm derivation of hydrological parameters, virtually all the palaeoclimatic proxies suggest changes in temperature, precipitation, wind pattern, vegetation and sea level, which in turn caused large amplitude changes in the runoff, discharge, and sediment load across monsoon Asia. The LGM (21.5 cal kyr BP) was characterised by drier conditions with erratic and reduced fluvial activity. The summer monsoon intensified during the early Holocene (ca 10–5 ^{14}C kyr BP or 11.5–5.0 cal kyr BP), when maximum monsoon precipitation and an associated increase in fluvial activity appear to have occurred. The glacial–interglacial transition was abrupt and this had a catastrophic impact on the Asian rivers. About 5 to 3 ^{14}C kyr BP, a reduction in monsoon precipitation to a minimum occurred. Except in coastal areas and lower reaches of the rivers where sea-level changes intervened, cool, dry phases were associated with aggradation and loess deposition, and the warm, wet phases with large-scale erosion and enormous amounts of sediment transport.

The response of fluvial systems to late-Quaternary climatic changes is broadly known but the reasons for lags and leads in the fluvial response, lake levels, sedimentation and soil formation are less well understood. The beginning and the end of increased or decreased monsoon episodes, recognized on the basis of ice- and deep-sea cores, do not exactly match those derived from evidence of increased or decreased precipitation on land (Denniston *et al.*, 2000; Sarkar *et al.*, 2000). It is intriguing to note that while the 3.5 ^{14}C kyr BP event is recognizable from both terrestrial and marine records, the YD event is less well represented in the land records. Further, although the monsoon intensity exerts a first-order control on erosional rates in the Himalaya and other highlands, there is little doubt that neotectonic activity has also played an equally important role throughout the Quaternary. However, there seems to be little information that specifically relates to the contribution of neotectonic activity. One of the important themes in future studies should be separating changes caused by tectonic activity from changes that would have occurred primarily because of climatic variations.

Therefore, developing chronologies with improved time-resolution is critical to achieve a better understanding of the spatio-temporal differences in the character of the monsoons, palaeohydrological changes and the role of neotectonic activity in one of the most densely populated regions of the world.

ACKNOWLEDGEMENT

This work was carried out in connection with research projects supported by the Indian Department of Science and Technology, New Delhi. This paper was prepared while Vishwas S. Kale was visiting the Physical Research Laboratory, Ahmedabad.

REFERENCES

Ahmed, S.M., Patil, D.J., Rao, P.S., Nath, B.N., Rao, B.R. and Rajagopalan, G., 2000. Glacial-interglacial changes in the surface water characteristics of the Andaman sea. Evidence from stable isotopic ratios of planktonic foraminifera. *Proceedings of the Indian Academy of Science (Earth and Planetary Science)*, **109**, 153–156.

An, Z., Kutzbach, J.E., Prell, W.L. and Porter, S.C., 2001. Evolution of Asian monsoons and phased uplift of the Himalaya-Tibetan plateau since late Miocene times. *Nature*, **411**, 62–66.

An, Z., Kukla, G.J., Porter, S.C. and Jule, X., 1991. Magnetic susceptibility evidence of monsoon variation on the loess plateau of central China during the last 130000 years. *Quaternary Research*, **36**, 29–36.

An, Z., Porter, S.C., Kutzbach, J.E., Wu, X., Wang, S., Liu, X., LI, X. and Zhou, W., 2000. Asynchronous Holocene optimum of the East Asian monsoon. *Quaternary Science Reviews*, **19**, 743–762.

An, Z., Porter, S.C., Zhou, W., Lu, Y., Donahue, D.J., Head, M.J., Wu, X., Ren, J. and Zheng, H., 1993. Episode of strengthened summer monsoon climate of younger Dryas age on the loess plateau of Central China. *Quaternary Research*, **39**, 45–54.

Baker, V.R., Ely, L.L., O'Connor, J.E. and Partridge, J.B., 1987. Paleoflood hydrology and design applications. In V.P. Singh (ed.), *Regional Flood Frequency Analysis*. D. Reidel, Boston, 339–353.

Banerjee, P.K., 2000. Holocene and late Pleistocene relative sea level fluctuations along the east coast of India. *Marine Geology*, **167**, 243–260.

Bhattacharyya, A., 1989. Vegetation and climate during the last 30,000 years in Ladakh. *Palaeogeography, Palaeoclimatology, Palaeoecology*, **73**, 25–38.

Bishop, P. and Godley, D., 1994. Holocene palaeochannels at Sisatchanalai North-Central Thailand: ages, significance and palaeoenviornmental indications. *The Holocene*, **4**, 32–41.

Bryson, R.A. and Swain, A.M., 1981. Holocene variations of monsoon rainfall in Rajasthan. *Quaternary Research*, **16**, 135–145.

Burbank, D.W., Leland, J., Fielding, E., Anderson, R.S., Brozoric, N., Reid, M.R. and Duncan, C., 1996. Bedrock incision, rock uplift and threshold hillslopes in Northwest Himalayas. *Nature*, **379**, 505–510.

Caratini, C., Bentaleb, I., Fontugne, M., Marzadec-Kerfourn, M.T., Pascal, J.P. and Tissot, C., 1994. A less humid climate since ca 3500 yrs BP from marine cores off Karwar, Western India. *Palaeogeography, Palaeoclimatology, Palaeoecology*, **109**, 371–384.

Cayre, O. and Bard, E., 1999. Planktonic foraminiferal and alkenone records of the last deglaciation from the Eastern Arabian Sea. *Quaternary Research*, **52**, 337–342.

Chen, Z., 1999. Geomorphology and coastline change of the Lower Yangtze Delta Plain, China. In A.J. Miller and A. Gupta (eds), *Varieties of Fluvial Form*. Wiley, Chichester, 427–443.

Chen, F.H., Bloemendal, J., Wang, J.M., Li, J.J. and Oldfield, F., 1997. High resolution multi-proxy records from Chinese loess: evidence for rapid climatic changes over the last 75 kyrs. *Palaeogeography, Palaeoclimatology, Palaeoecology*, **130**, 323–335.

Colin, C., Kissel, C., Blamart, D. and Turpin, L., 1998. Magnetic properties of sediments in the Bay of Bengal and Andaman Sea: impact of rapid North Atlantic Ocean climatic events on the strength of the Indian monsoon. *Earth and Planetary Science Letters*, **160**, 623–635.

Colin, C., Turpin, L., Bertaux, J., Desprairies, A. and Kissel, C., 1999. Erosional history of the Himalayan and Burman ranges during the last two glacial interglacial cycles. *Earth and Planetary Science Letters*, **171**, 647–660.

Cornwell, K., 1998. Quaternary break-out flood sediments in Peshwar basin of Northern Pakistan. *Geomorphology*, **25**, 225–248.

Coxon, P., Owen, L.A. and Mitchele, W.A., 1996. A late Quaternary catastrophic flood in the Lahul Himalaya. *Journal of Quaternary Science*, **11**, 495–510.

Cullen, J.L., 1981. Microfossil evidence for changing salinity patterns in Bay of Bengal over the last 20000 years. *Palaeogeography, Palaeoclimatology, Palaeoecology*, **35**, 305–308.

Denniston, R.F., González, L.A., Asmerom, Y., Sharma, R.H. and Reagan, M.K., 2000. Speleothem evidence for changes in Indian summer monsoon precipitation over the last ~2300 Years. *Quaternary Research*, **53**, 196–202.

Duplessy, J.C., 1982. Glacial to interglacial contrasts in the Northern Indian Ocean. *Nature*, **295**, 494–498.

Ely, L.L., Enzel, Y., Baker, V.R., Kale, V.S. and Mishra, S., 1996. Changes in the magnitude and frequency of late Holocene monsoon floods on the Narmada river, Central India. *Geological Society of America Bulletin*, **108**, 1134–1148.

Emeis, K.C., Anderson, D.M., Doose, H., Karoon, D. and Schulz-Bull, D., 1995. Sea-surface temperatures and the history of monsoon upwelling in the Northwest Arabian sea during the last 500,000 years. *Quaternary Research*, **43**, 355–361.

Emmel, F.J. and Curray, J.R., 1982. A submerged late Pleistocene delta and other features related to the sea level changes in the Malacca Strait. *Marine Geology*, **47**, 197–216.

Enzel, Y., Ely, L.L., Mishra, S., Ramesh, R., Amit, R., Lazar, B., Rajaguru, S.N., Baker, V.R. and Sandler, A., 1999. High-resolution environmental changes in the Thar Desert, North-western India. *Science*, **284**, 125–128.

Fairbanks, R.G., 1989. A 17000 year glacio-eustatic sea level record: influence of glacial melting rates on the Younger Dryas events and deep-ocean circulation. *Nature*, **342**, 637–642.

Fang, J.Q., 1991. Lake evolution during the past 30,000 years in China and its implications for environmental change. *Quaternary Research*, **36**, 37–60.

Fang, J.Q., 1993. Lake evolution during the past 3,000 years in China and its implications for environmental change. *Quaternary Research*, **39**, 175–185.

Flenley, J., 1979. *The Equatorial Rainforest: A Geological History*. Butterworth, London, 162.

Gasse, F., Arnold, M., Fontes, J.C., Fort, M., Gilbert, E., Huc, A., Li, B., Li, Y., Liu, Q., Mé-lières, F., van Compo, E., Wang, F. and Zhang, Q., 1991. A 13000-year climate record from Western Tibet. *Nature*, **353**, 742–745.

Gasse, F., Fontes, J.Ch., Van Campo, E. and Wei, K., 1996. Holocene environmental changes in Bangong Co basin (Western Tibet). Part 4: discussion and conclusions. *Palaeogeography, Palaeoclimatology, Palaeoecology*, **120**, 79–92.

Goodbred Jr., S.L. and Kuehl, S.A., 2000. Enormous Ganges-Brahmaputra sediment discharge during strengthened early Holocene monsoon. *Geology*, **28**, 1083–1086.

Grossman, M.J., 2001. Large floods and climatic change during the Holocene on the Ara river, Central Japan. *Geomorphology*, **39**, 21–37.

Gupta, A., 1995. Magnitude, frequency and special factors affecting channel form and pro-cesses in the seasonal tropics. In J.E. Costa, A.J. Miller, K.W. Pottern and P. Wilcock (eds), *Nature and Anthropogenic Influences in Fluvial Geomorphology*. American Geophysical Union Monograph 89, Washington, DC, 125–136.

Gupta, A., Kale, V.S. and Rajaguru, S.N., 1999. The Narmada river, India, through space and time. In A.J. Miller and A. Gupta (eds), *Varieties of Fluvial Form*. Wiley, Chichester, 113–143.

Gupta, A., Rahman, A., Wong, P.P. and Pitts, J., 1987. The old alluvium of Singapore and the extinct drainage system to the South China Sea. *Earth Surface Processes and Landforms*, **12**, 259–275.

Hanebuth, T., Stattegger, K. and Grootes, P.M., 2000. Rapid flooding of the Sunda Shelf: a late-glacial sea-level record. *Science*, **288**, 1033–1035.

Hewitt, K., 1998. Catastrophic landslides and their effects on the Upper Indus streams, Karako-ram Himalaya, Northern Pakistan. *Geomorphology*, **26**, 47–80.

Hori, K., Saito, Y., Zhao, Q., Cheng, X., Wang, P., Sato, Y. and Li, C., 2001. Sedimentary facies and Holocene progradation rates of the Changjiang (Yangtze) delta, China. *Geomor-phology*, **41**, 233–248.

Huang, C.Y., Liew, P.M., Zhao, M., Chang, T.C., Kuo, C.M., Chen, M.T., Wang, C.H. and Zheng, L.F., 1997. Deep sea and lake records of the southeast Asian palaeomonsoons for the last 25 thousand years. *Earth and Planetary Science Letters*, **146**, 59–72.

Hurtado Jr., J.M., Hodges, K.V. and Whipple, K.X., 2001. Neotectonics of the Thakkhola graben and implications for recent activity on the South Tibetan fault system in the Central Nepal Himalaya. *Geological Society of America Bulletin*, **113**, 222–240.

Jiang, Q. and Piperno, D.R., 1999. Environmental and archaeological implications of a late Quaternary palynological sequence, Poyang lake, Southern China. *Quaternary Research*, **52**, 250–255.

Jingtai, W., Derbyshire, E., Xingnin, M. and Jinhui, M. 1991. Natural hazards and geological processes: an introduction to the history of natural hazards in Gansu Province, China. In I. Liu (ed.), *Quaternary Geology and Environment in China*. Science Press, Beijing, China, 285–296.

Joshi, V.U. and Kale, V.S., 1997. Colluvial deposits in Northwest Deccan, India: their significance in the interpretation of late Quaternary history. *Journal of Quaternary Science*, **12**,
391–403.

Juyal, N., Raj, R., Mauray, D.M., Chamyal, L.S. and Singhvi, A.K., 2000. Chronology of late
Pleistocene environmental changes in the Lower Mahi basin, Western India. *Journal of Quaternary Science*, **15**, 501–508.

Kale, V.S., 1998. Monsoon floods in India: a hydro-geomorphic perspective. In V.S. Kale (ed.),
Flood Studies in India. Geological Society of India, Bangalore, Memoir 41, 229–256.

Kale, V.S., 1999. Late Holocene temporal patterns of palaeofloods in Central and Western India.
Man and Environment, **24**, 109–115.

Kale, V.S. and Rajaguru, S.N., 1987. Late Quaternary alluvial history of Northwest Deccan
Upland region. *Nature*, **325**, 612–614.

Kale, V.S., Singhvi, A.K., Mishra, P.K. and Banerjee, D., 2000. Sedimentary records and luminescence chronology of late Holocene palaeofloods in the Luni river, Thar Desert, Northwest
India. *Catena*, **40**, 337–358.

Kar, A., 1990. A stream trap hypothesis for the evolution of some saline lakes in the Indian
desert. *Zeitschrift für Geomorphologie N.F.*, **34**, 37–47.

Kar, A., Singhvi, A.K., Rajaguru, S.N., Juyal, N., Thomas, J.V., Banerjee, B. and Dhir, R.P.,
2001. Reconstruction of the late Quaternary environment of the lower Luni Plains, Thar
Desert, India. *Journal of Quaternary Science*, **16**, 61–68.

Kealhofer, L. and Penny, D., 1998. A combined pollen and phytolith record for fourteen thousand years of vegetation change in Northeastern Thailand. *Review Of Palaeobotany and
Palynology*, **103**, 83–93.

Kudrass, H.R., Hofmann, A., Doose, H., Emeis, K. and Erlenkeuser, H., 2001. Modulation and
amplification of climatic changes in the northern hemisphere by the Indian summer monsoon
during the past 80 ky. *Geology*, **29**, 63–66.

Kukla, G. and An, Z., 1989. Loess stratigraphy in Central China. *Palaeogeography, Palaeoclimatology, Palaeoecology*, **72**, 203–225.

Kumar, S., Prakash, B., Manchanda, M.L., Singhvi, A.K. and Srivastava, P., 1996. Holocene
landform and soil evolution of the western Gangetic Plains: implications of neotectonics and
climate. *Zeitschrift für Geomorphologie*, **SB 103**, 282–312.

Kusumgar, S., Agrawal, D.P., Deshpande, R.D., Ramesh, R., Sharma, C. and Yadava, M.G.,
1995. A comparative study of monsoonal and non-monsoonal Himalayan lakes, India. *Radiocarbon*, **37**, 191–196.

Kutzbach, J., Gallimore, R., Harrison, S., Behling, P., Selin, R. and Laarif, F., 1998. Climate
and biome simulations for the past 21,000 years. *Quaternary Science Reviews*, **17**, 473–506.

Lehmkuhl, F., 1997. Late Pleistocene, late-glacial and Holocene glacier advances on the Tibetan
plateau. *Quaternary International*, **38–39**, 77–83.

Lehmkuhl, F., Bohner, J. and Haselein, F., 1999. Late Quaternary environmental changes and
human occupation on the Tibetan plateau. *Man and Environment*, **24**, 137–148.

Lister, G.S., Ketts, K., Zao, C.K., Yu, J.Q. and Niessen, F., 1991. Lake Qinghai, China: closed
basin lake levels and the oxygen isotope records for ostracoda since late Pleistocene. *Palaeogeography, Palaeoclimatology, Palaeoecology*, **84**, 141–162.

Liu, J., Wu, X., Li, S. and Zang, M., 1997. The last glacial stratigraphic sequence, depositional
environment and climatic fluctuations from the aeolian sand dune in Hongguang, Pengze,
Jiangxi, China. *Quaternary Science Reviews*, **16**, 535–546.

Liu, K.B., Sun, S. and Jiang, X., 1992. Environmental change in the Yangtze river delta since
12000 years BP. *Quaternary Research*, **38**, 32–45.

Liu, K.B., Yao, Z. and Thompson, L.G., 1998. A pollen record of Holocene climatic changes
from Dunde ice cap, Qinghai-Tibetan plateau. *Geology*, **26**, 135–138.

Liu, T., Ding, Z. and Rutter, N., 1999. Comparison of Milankovitch periods between continental
loess and deep sea records over the last 2.5 Ma. *Quaternary Science Reviews*, **18**, 1205–1212.

Loeffler, E., Thompson, W.P. and Liengsakul, M., 1984. Quaternary geomorphological development of the lower Mun River basin, North-East Thailand. *Catena*, **11**, 321–330.

Lu, H., Huissteden, K.V., An, Z., Nugteren, Z. and Vandenberghe, J., 1999. East Asia winter monsoon variations on a millennial time-scale before the last glacial-interglacial cycle. *Journal of Quaternary Science*, **14**, 101–110.

Maher, B.A., Thompson, R. and Zhou, L.P., 1994. Spatial and temporal reconstructions of changes in the Asian palaeomonsoon: a new mineral magnetic approach. *Earth and Planetary Science Letters*, **125**, 416–471.

Maloney, B.K., 1995. Evidence for the Younger Dryas climatic event in Southeast Asia. *Quaternary Science Reviews*, **14**, 948–958.

Marcontonio, F., Anderson, R.F., Higgins, S., Fleisher, M.Q., Stute, M. and Schlesser, P., 2001. Abrupt intensification of the SW Indian Ocean monsoon during the last deglaciation: constraints from Th, Pa, and He isotopes. *Earth and Planetary Science Letters*, **184**, 505–514.

Mazari, R.K., Bagati, T.N., Chauhan, M.S. and Rajagopalan, G., 1996. Palaeoclimatic and environmental variability in Austral-Asian Transact during the past 2000 years. *Proceedings IGBP-PAGES-PEPII Symposium*, Nagoya, Japan, 262–269.

Monecke, K., Winsemann, J. and Hanisch, J., 2001. Climatic response of Quaternary alluvial deposits in the Upper Kali Gandaki valley West Nepal. *Global and Planetary Change*, **28**, 293–302.

Naidu, P.D., 1996. Onset of an arid climate at 3.5 ka in the tropics: evidence from monsoon upwelling record. *Current Science*, **71**, 715–718.

Nigam, R., 1993. Foraminifera and changing pattern of monsoon rainfall. *Current Science*, **64**, 935–937.

Nigam, R. and Khare, N., 1992. Oceanographic evidences of the great floods in 2000 and 1500 BC documented in archaeological records. In B.U. Nayak and N.C. Ghosh (eds), *New Trends in Indian Art and Archeology*. Vol. 2, Aditya Prakashan, New Delhi, 517–522.

Nutalaya, P., Sophonsakulrat, W., Sonsuk, M. and Wattanachai, N., 1989a. Catastrophic flooding – an agent for landform development of the Khorat plateau: a working hypothesis. In N. Thiramongkol (ed.), *Proceedings of the Workshop on Correlation of Quaternary Successions in South, East and Southeast Asia*. Department of Geology, Chulalongkorn University, Bangkok, 117–134.

Nutalaya, P., Wanchai, S., Manit, S. and Nawarat, W., 1989b. Catastrophic flooding – an agent for landform development of the Khorat Plateau: a working hypothesis. In T. Narong (ed.), *Proceeding of the Workshop on Correlation of Quaternary Successions in South, East and Southeast Asia*. Chulalongkorn University, Bangkok, 95–115.

Overpeck, J., Anderson, D., Trumbore, S. and Prell, W., 1996. The Southwest Indian monsoon over the last 18000 years. *Climate Dynamics*, **12**, 213–225.

Owen, L.A., Bailey, R.M., Rhodes, E.J., Mitchell, W.A. and Coxon, P., 1997. Style and timing of glaciation in the Lahul Himalaya, Northern India: a framework for reconstructing late Quaternary palaeoclimatic change in the Western Himalayas. *Journal of Quaternary Science*, **12**, 83–109.

Pawar, N.J., Kale, V.S., Atkinson, T.C. and Rowe, P.J., 1988. Early Holocene waterfall tufa from semi-arid Maharashtra Plateau, India. *Journal Geological Society of India*, **32**, 513–515.

Pelejero, C., Kienast, M., Wang, L. and Grimalt, J.O., 1999. The flooding of Sundaland during the last deglaciation: imprints in hemipelagic sediments from the Southern South China Sea. *Earth and Planetary Science Letters*, **171**, 661–671.

Phadtare, N.R., 2000. Sharp decrease in summer monsoon strength 4000-3500 cal yr BP in the Central Higher Himalaya of India based on pollen evidence from alpine peat. *Quaternary Research*, **53**, 122–129.

Phillips, W.M., Sloan, V.F., Shroder Jr., J.F., Sharma, P., Clarke, M.L. and Rendel, H.M., 2000. Asynchronous glaciation at Nanga Parbat, Northwestern Himalaya Mountains, Pakistan. *Geology*, **28**, 431–434.

Porter, S.C. and An, Z., 1995. Correlation between climate events in the North Atlantic and China during the last glaciation. *Nature*, **375**, 305–308.

Porter, S.C., An, Z. and Zheng, H., 1992. Cyclic Quaternary alluviation and terracing in a nonglaciated drainage basin on the north flank of the Qinling Shan, Central China. *Quaternary Research*, **38**, 157–169.

Prasad, S., Kusumgar, S. and Gupta, S.K., 1997. A mid to late Holocene record of palaeoclimatic changes from Nal Sarovar: a palaeodesert margin lake in Western India. *Journal of Quaternary Science*, **12**, 153–159.

Prins, M.A. and Postma, G., 2000. Effects of climate, sea level, and tectonics unravelled for last deglaciation turbidity records of the Arabian Sea. *Geology*, **28**, 375–378.

Qiang, X.K., Li, Z.X., Powell, C.M. and Zheng, H.B., 2001. Magnetostratigraphic record of the late Miocene onset of the East Asia monsoon, and Pliocene uplift of Northern Tibet. *Earth and Planetary Science Letters*, **187**, 83–93.

Ramage, C.S., 1971. *Monsoon Meteorology*. Academic Press, New York.

Rangarajan, G. and Sant, D.A., 2000. Paleoclimatic data from 74KL and Guliya cores: new insights. *Geophysical Research Letters*, **27**, 787–790.

Rao, S.R., Lal, B.B., Nath, B., Ghosh, S.S. and Lal, K., 1963. Excavation at Rangpur and other explorations in Gujarat. *Ancient India: Bulletin of Archaeology Survey of India*. Number 18 and 19, 5–207.

Ruddiman, W.F., Co-ordinator and Complier, CLIMAP Project Members, 1984. The last interglacial ocean. *Quaternary Research*, **21**, 123–224.

Saito, Y., Yang, Z. and Hori, K. 2001. The Huanghe (Yellow River) and Changjiang (Yangtze River) deltas: a review on their characteristics, evolution and sediment discharge during the Holocene. *Geomorphology*, **41**, 219–231.

Sangode, S.J., Suresh, N. and Bagati, T.N., 2001. Godavari source in the Bengal fan sediments: results from magnetic susceptibility dispersal pattern. *Current Science*, **80**, 660–664.

Sarkar, A., Ramesh, R., Somayajulu, B.L.K., Agnihortri, R., Jull, A.J.T. and Burr, G.S., 2000. High resolution Holocene monsoon record from the Eastern Arabian Sea. *Earth and Planetary Science Letters*, **177**, 209–281.

Schulz, H., von Rad, U. and Erlankeuser, H., 1998. Correlation between Arabian Sea and Greenland climate oscillation of the past 11000 years. *Science*, **393**, 54–57.

Seltzer, G.O., 2001. Late Quaternary glaciation in the tropics: future research directions. *Quaternary Science Reviews*, **20**, 1063–1066.

Shackleton, N.J., Backman, J., Zimmerman, H., Kent, D.V., Hall, M.A., Roberts, D.G., Schneitker, D., Baldauf, J.G., Desprairies, A., Homrighausen, R., Huddlestun, P., Keene, J.B., Kaltenback, A.J., Krumsiek, K.A.O., Marton, A.C., Murry, J.W. and Westberg-Smith, J., 1984. Oxygen isotope calibration of the onset of ice-rafting and history of glaciation in the North Atlantic region. *Nature*, **307**, 620–623.

Shi, Y., Kong, Z., Wang, S., Tang, L., Wang, F., Yao, T., Zhao, X., Zhang, P. and Shi, S., 1993. Mid-Holocene climates and environments in China. *Global and Planetary Change*, **7**, 219–233.

Shroder, J.F., 1998. Slope failure and denudation in the Western Himalaya. *Geomorphology*, **26**, 81–105.

Singh, G., 1971. The Indus valley culture seen in the context of postglacial climatic and ecological studies in Northwest India. *Archaeology and Physical Anthropology in Oceania*, **6**, 177–189.

Singh, G., Wasson, R.J. and Agrawal, D.P., 1990. Vegetation and seasonal climatic changes since the last full glacial in the Thar Desert, Northwest India. *Review of Palaeobotany and Palynology*, **64**, 351–358.

Singh, I.B., 1996. Geological evolution of Ganga Plain – an overview. *Journal of the Palaeontological Society of India*, **41**, 99–137.

Sirocko, F., Garbe-Schönberg, D., McIntyre, A. and Molfino, B., 1996. Teleconnections between the subtropical monsoons and high-latitude climates during the last deglaciation. *Science*, **272**, 526–529.

Sirocko, F., Sarnthein, M., Erienkeuser, H., Lange, H., Arnold, M. and Duplessy, J.C., 1993. Century-scale events in monsoonal climate over the past 24,000 years. *Nature*, **364**, 322–324.

Srivastava, P., Juyal, N., Singhvi, A.K., Wasson, R.J. and Bateman, M.D., 2001. Luminescence chronology of river adjustment and incision of Quaternary sediments in the alluvial plain of the Sabarmati river, North Gujarat, India. *Geomorphology*, **36**, 217–229.

Steig, E.J., 1999. Mid-Holocene climate change. *Science*, **286**, 1485–1487.

Steinke, S., Kienast, M., Pflaumann, U., Weinelt, M. and Stattegger, K., 2001. A high-resolution sea-surface temperature record from the Tropical South China Sea 16,500–3000 yr BP *Quaternary Research*, **55**, 352–362.

Sugai, T., 1993. River terrace development by concurrent fluvial processes and climatic changes. *Geomorphology*, **6**, 243–252.

Sukhija, B.S., Reddy, D.V. and Nagabhushnam, P., 1998. Isotopic fingerprints of palaeoclimates during the last 30,000 years in deep confined groundwaters of Southern India. *Quaternary Research*, **50**, 252–260.

Sukumar, R., Ramesh, R., Pant, R.K. and Rajagopalan, G., 1993. A δ ^{13}C record of late Quaternary climate change from tropical peat in Southern India. *Nature*, **364**, 703–706.

Tandon, S.K., Sareen, B.K., Someshwar Rao, M. and Singhvi, A.K., 1997. Aggradation history and luminescence chronology of late Quaternary semi-arid sequences of the Sabarmati basin, Gujarat, Western India. *Paleogeography, Palaeoclimatology, Palaeoecology*, **128**, 339–357.

Thamban, M., Rao, V.P., Schneider, R.R. and Grootes, P.M., 2001. Glacial to Holocene fluctuations in hydrography and productivity along the southwestern continental margin of India. *Palaeogeography, Palaeoclimatology, Palaeoecology*, **165**, 113–127.

Thompson, L.G., 2000. Ice core evidence for climate change in the tropics: implications for our future. *Quaternary Science Reviews*, **19**, 19–35.

Thompson, L.G., Mosley-Thompson, E., Davis, M.E., Bolzan, J.F., Dai, J., Yao, T., Gundestrup, N., Wu, X., Klein, L. and Xie, Z., 1989. Holocene-late Pleistocene climate ice core records from Qinghai-Tibetan plateau. *Science*, **246**, 474–477.

Thompson, L.G., Yao, T., Davis, M.E., Henderson, K.A., Thompson, E., Lin, P.N., Beer, J., Synal, H.A., Cole-Dai, J. and Bolzan, J.F., 1997. Tropical climate instability: the last glacial cycle from a Qinghai-Tibetan ice core. *Science*, **276**, 1821–1825.

Thompson, G.G., Yao, T., Mosley-Thompson, E., Davis, M.E., Henderson, K.A. and Lin, P.N., 2000. A high-resolution millennial record of the south Asian monsoon from Himalayan ice core. *Science*, **289**, 1916–1919.

Van Compo, E., Duplessy, J.C. and Rossignol-Strick, M., 1982. Climatic conditions deduced from 150 kyr oxygen isotope-pollen record from the Arabian Sea. *Nature*, **296**, 56–59.

Verstappen, H.T., 1980. Quaternary climatic changes and natural environments in Southeast Asia. *GeoJournal*, **4**, 45–54.

Verstappen, H.T., 1997. The effect of climatic change on Southeast Asian geomorphology. *Journal of Quaternary Science*, **12**, 413–418.

von Rad, U. and Tahir, M., 1997. Late Quaternary sedimentation on the outer Indus shelf and slope (Pakistan): evidence from high-resolution seismic data and coring. *Marine Geology*, **138**, 193–236.

von Rad, U., Schaaf, M., Michels, K.H., Schulz, H., Berger, W.H. and Sirocko, F., 1999. A 5000-yr record of climate change in varved sediments from the oxygen minimum zone off Pakistan, Northeastern Arabian Sea. *Quaternary Research*, **51**, 39–53.

Wang, L., Sarnthein, M., Erlenkeuser, H., Grimalt, J., Grootes, P., Heilig, S., Ivanova, E., Kienast, M., Pelejero, C. and Pflaumann, U., 1999. East Asian monsoon climate during the late Pleistocene: high-resolution sediment records from the South China Sea. *Marine Geology*, **156**, 245–284.

Weber, M.E., Widicke, M.H., Kudrass, H.R., Hubscher, C. and Erlenkeuser, H., 1997. Active growth of the Bengal fan during sea-level rise and highstand. *Geology*, **25**, 315–318.

Wei, K.Y., Lee, M.Y., Duan, W., Chen, C. and Wang, C.H., 1998. Palaeoceanographic change in the northeastern South China Sea during the last 15000 years. *Journal of Quaternary Science*, **13**, 55–64.

Whetton, P., Adamson, D. and Williams, M., 1990. Rainfall and river flow variability in Africa, Australia and East Asia linked to El-Niño-Southern-Oscillation events. *Proceedings, Geological Society of Australia Symposium*, **1**, 71–82.

Williams, M.A.J. and Clarke, M.F., 1984. Late Quaternary environments in North-Central India. *Nature*, **398**, 633–635.

Wohl, E.E. and Cenderelli, D., 1998. Flooding in the Himalaya mountains. In V.S. Kale (ed.), *Flood Studies in India*. Memoir 41, Geological Society of India, Bangalore, 77–99.

Yadava, M.G. and Ramesh, R., 1999. Palaeomonsoon record of the last 3400 years from speleothems of tropical India. *Gondwana Geological Magazine*, **4**, 141–156.

Yang, D., 1991a. Influence of climatic and sea level changes on alluvial process of the mid-lower reaches of the Yangtze river in the last 10000 years. In T. Liu (ed.), *Quaternary Geology and Environment in China*. Science Press, Beijing, China, 337–342.

Yang, H., 1991b. Palaeomonsoon and the mid-Holocene climate and sea-level fluctuations in China. In T. Liu (ed.), *Quaternary Geology and Environment in China*. Science Press, Beijing, China, 326–336.

Yang, M., Yao, T., He, Y. and Thompson, L.G., 2000a. ENSO events recorded in the Guliya ice core. *Climatic Change*, **47**, 401–409.

Yang, D., Yu, G., Xie, Y., Zhan, D. and Li, Z., 2000b. Sedimentary records of large Holocene floods from the middle reaches of the Yellow river, China. *Geomorphology*, **33**, 73–88.

Zhang, D., 1991. Climate changes in recent 1000 years in China. In T. Liu (ed.), *Quaternary Geology and Environment in China*. Science Press, Beijing, China, 208–213.

Zheng, Z. and Lei, Z.Q., 1999. A 400,000 year record of vegetational and climatic changes from a volcanic basin, Leizhou peninsula, Southern China. *Palaeogeography, Palaeoclimatology, Palaeoecology*, **145**, 339–362.

Zheng, Z. and Li, Q., 2000. Vegetation, climate, and sea level in the past 55,000 years, Hanjiang delta, Southeastern China. *Quaternary Research*, **53**, 330–340.

Zhou, S.Z., Chen, F.H., Pan, B.T., Cao, J.X., Li, J.J. and Derbyshire, E., 1991. Environmental changes during the Holocene in Western China on a millennial timescale. *The Holocene*, **1**, 151–156.

Zhou, W., Donahue, D.J., Porter, S.C., Jull, T.A. and Li, X., 1996. Variability of monsoon climate in East Asia at the end of the last glaciation. *Quaternary Research*, **46**, 219–229.

14 Alluvial Evidence of Major Late-Quaternary Climate and Flow-regime Changes on the Coastal Rivers of New South Wales, Australia

G.C. NANSON, T.J. COHEN, C.J. DOYLE AND D.M. PRICE

University of Wollongong, New South Wales, Australia

1 INTRODUCTION

Thermoluminescence (TL) and radiocarbon evidence of climate and flow-regime changes have been interpreted from extensive exposures in Quaternary-age deposits in the Cranebrook Terrace on the Nepean River immediately west of Sydney, New South Wales (NSW), Australia (Nanson and Young, 1987; 1988) (Figure 14.1). However, more recent luminescence dates from this location requires a significant re-evaluation of the chronology these studies were based on. This chapter presents a completely revised late-Quaternary chronology for the polycyclic Cranebrook Terrace as well as new stratigraphic and chronological evidence for patterns of late-Quaternary flow-regime change in the Nambucca and Bellinger catchments on the mid-north coast of NSW (Figure 14.1). It also compares these findings with those from other alluvial chronologies developed in southeastern Australia. The total period of investigation is the duration of the last glacial cycle [the middle of marine Oxygen Isotope Stages (OIS) 5 to late OIS1], although the record is more detailed for the latter part of this period, particularly after the Last Glacial Maximum (LGM). The result is a convincing picture of widespread and connected coastal and inland climate and flow-regime changes in southeastern Australia during the past ~100 ka.

To the west of the Great Dividing Range in NSW, a comprehensive story of climate and associated flow-regime change has been developed for the pre-Holocene part of the late Quaternary (Bowler, 1978; Bowler *et al.*, 1978; Page *et al.*, 1991; Prosser *et al.*, 1994; Page *et al.*, 1996; Page and Nanson, 1996; Bowler and Price, 1998). However, because of errors in the original TL ages obtained by Nanson and Young (1988) from the Cranebrook Terrace, it has not been possible to see if the pattern from westward and eastward flowing rivers in southeastern Australia has been similar. Furthermore, recent research on the coastal rivers, both on the south coast of NSW and Victoria (Fryirs and Brierley, 1998; Brooks and Brierley, 2002; Brooks *et al.*, 2003; Nott *et al.*, 2002), the central coast of NSW (Erskine and Peacock, 2002; Webb *et al.*, 2002) and the mid-north coast of NSW (Nanson and Doyle, 1999; Cohen, in prep; Doyle, in prep) has refined the younger part of the late-Quaternary record. Data now exists with

Palaeohydrology: Understanding Global Change.   Edited by K.J. Gregory and G. Benito
© 2003 John Wiley & Sons, Ltd   ISBN: 0-470-84739-5

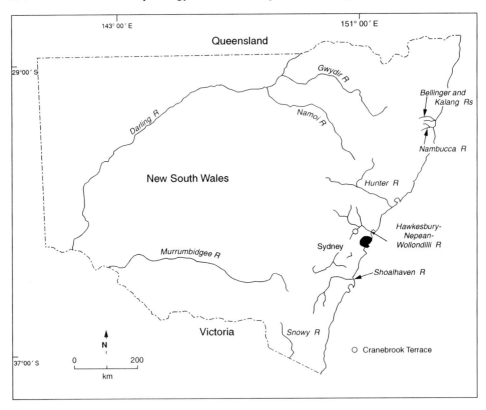

Figure 14.1   The rivers under study, coastal New South Wales

which to interpret a broad picture of climate and flow-regime change for southeastern Australia for the past 100 ka.

The first late-Quaternary alluvial chronologies for NSW coastal rivers were determined from radiocarbon dating and therefore extend back to only about 30 ka. They resulted from investigations by Walker and Hawkins (1957) and Stockton and Holland (1974) for the Nepean catchment, Walker (1962) for the lower Shoalhaven, Walker (1970) for the Macleay, Hickin and Page (1971) for the Colo, and Warner (1972) for the Bellinger River (Warner, 1970; 1972). Walker and Coventry (1976) attempted a regional correlation; however, the radiocarbon evidence available at the time was assessed by Young *et al.* (1986) as revealing no consistent pattern. Recent luminescence dating has greatly expanded the available chronological data, albeit some with bleaching problems (Wray *et al.*, 2001). The luminescence dates cited here are from sediment samples, which we are confident, from their temperature plateaux, provide sound evidence of having been well bleached at their time of burial. Throughout this paper, radiocarbon ages are presented as conventional radiocarbon years and identified as "kyr BP" or "yr BP" and luminescence ages in sidereal time as "ka".

In establishing an alluvial chronology from river terraces, a problem arises as to interpreting just what geomorphic process or event is being dated. Bull (1991) describes river terraces as being of three genetic types: tectonic, climatic and complex response. However, a fourth type that is base-level determined must commonly occur in proximity to fluctuating sea levels. River valleys along the east coast of NSW exhibit abundant evidence of Quaternary-age terrace development, but only in their lowermost reaches have they been influenced by Quaternary eustatic changes. There is evidence from the upper Shoalhaven River that clearly recognisable river terraces, well removed from eustatic effects, date back nearly 0.5 Ma (Nott *et al.*, 2002). In this chapter, only Quaternary terraces upstream of significant sea-level influence are considered. This recognises that for much of the late Quaternary, sea levels have been much lower than at present and therefore of little relevance to the position of terraces above present estuaries and deltas. Nevertheless, the question remains as to just *how* changes in climate impact to form Bull's climatic terraces. In this study, terraces standing above the contemporary floodplain are presumed to have been formed by larger and therefore more powerful flows. The thalweg is assumed in most cases to have been on or near the bedrock valley floor in the upper reaches and abrading older alluvium in the lower reaches, and the terrace is, therefore, indicative of a higher bankfull stage and thicker floodplain unit. With flow reduction, an inset floodplain forms as the smaller lower-energy river only partially reworks its previous floodplain. Clearly, such a model is strongly biased towards retaining a record of mostly successively declining flow regimes. A return to larger and more powerful flows would rework or bury earlier floodplains, and such evidence would be less obvious and therefore more difficult to find. Over relatively long periods of time, bedrock incision would have to be taken into account, but this would not be a significant factor in the face of the very low erosion rates in eastern Australia (Young, 1983). Alternative models have been proposed for terraces, including destabilised hillslopes and enhanced valley fills forming higher terraces (Walker, 1962). However, such models present complex water/sediment proportional relationships and require a degree of hillslope denudation and mobilisation that is difficult to envisage as a widespread phenomenon across the non-glaciated and non-arid coastal catchments of NSW. With the exception of swampy meadows in some headwater locations (e.g. Prosser *et al.*, 1994; Fryirs and Brierley, 1998), extensive alluvial fan, footslope and elevated valley-floor accumulations, commonly associated with such disequilibrium transport relationships, are only rarely present.

Holocene changes appear to have been less intense and therefore to have left a less-extensive record than the major OIS 5 to OIS 2 changes that have affected so substantially the alluvial surfaces of the Riverine Plain (Page *et al.*, 1996) and the Nepean and Shoalhaven rivers (data here, and Nott *et al.*, 2002). For the purpose of Quaternary study, the coastal river catchments of NSW are divisible into two broad groups. First, the relatively big catchments such as the Wollondilly–Nepean–Hawkesbury River and the Shoalhaven River have large sediment accommodation volumes and provide a relatively long-term picture of the major flow-regime change during the Quaternary. Second, the confined and steeper coastal valleys have much more limited sediment accommodation and relatively large unit stream powers. As such they are sensitive to minor changes and therefore have the potential to preserve a relatively detailed but more recent record of flow-regime change. These smaller valleys were flushed clear of most of their Pleistocene deposits and subsequent deposition has provided a resolution suitable for interpreting the less-intense and shorter-lived post-LGM and -Holocene events.

While it is beyond the scope of this chapter to cover the entire late-Quaternary literature pertaining to southeastern Australia, it is acknowledged that considerable detail has been provided by earlier work on both the Late Pleistocene and the Holocene terraces (Walker, 1962; Hickin 1970; Page, 1972; Warner, 1970; 1972). Such work has focussed on the broad chronostratigraphy of Holocene floodplains (Hickin and Page, 1971; Young, 1976), with more recent research focussing on floodplain form-process associations (Warner, 1992; Ferguson and Brierley, 1999). A number of studies have highlighted the nature and timing of sediment accumulation in upland valley fill settings and smaller basins <50 km² (Young, 1986; Melville and Erskine, 1986; Prosser *et al.*, 1994; Fryirs and Brierley, 1998; Smith, 2000).

While the various geomorphic sites within southeastern Australia contain some variability with regard to the exact timing of Holocene fluvial activity, the broad record is characterised by lateral reworking and terrace formation in the early Holocene (Page, 1972; Warner, 1970; 1972), followed by the onset of relatively stable sediment accumulation from the mid-Holocene onwards (Walker, 1962; Page, 1972; Young, 1976; Melville and Erskine, 1986; Prosser *et al.*, 1994; Nanson *et al.*, 1996; Fryirs and Brierley, 1998). The Holocene record is characterised by data from systems said to be susceptible to intrinsic thresholds, such as valley fills (Young, 1986; Prosser *et al.*, 1994; Wray *et al.*, 2001), and from larger basins where the climate signature is more apparent.

The purpose of this chapter is to combine the Quaternary stories provided by the large and small drainage basins on the seaboard of NSW, to compare this record with that obtained from the westward flowing rivers, and to construct a broad picture of climate and associated flow-regime change in southeastern Australia over the past ~100 ka. This time period is longer than that taken for most of the chapters in this book because without significant glaciation, Australia offers the opportunity to look at flow-regime changes throughout the last glacial cycle. The chapter does not attempt to review flow-regime changes for the continent as a whole, for while data exists, much of it has not yet been published. Important but less ambitious summaries of flow-regime changes on coastal NSW rivers have been undertaken by Young *et al.* (1994) and Wray *et al.* (2001).

2   THE CRANEBROOK TERRACE

The Cranebrook Terrace at latitude 33° 45′ lies immediately west of Sydney on the eastern side of the Great Dividing Range is known there as the Blue Mountains (Figure 14.1). The Terrace is a major alluvial unit and as such provides evidence of late-Quaternary changes in flow regime for the eastward flowing rivers on the central NSW coast. Furthermore, the Nepean–Wollondilly basin lies roughly east of the Murray and Murrumbidgee Rivers on the Riverine Plain (Figure 14.1), the palaeochannels of which have been described and dated in detail by Bowler (1978) and Page *et al.* (1991; 1996), and, therefore, the Cranebrook Terrace allows a direct comparison of the late-Quaternary flow regimes of eastern and westward flowing rivers of southeastern Australia.

The basin drains ~11,000 km² of rugged ranges and plateaus that reach a maximum elevation of about 1,260 m, so while not currently snow influenced it would have been at times during the Quaternary. Importantly, the basin was not glaciated in the Late Pleistocene (Barrows *et al.*, 2001). The river gradient adjacent to the 12- to 18-m-high terrace is 0.00034 and the channel carries a mean annual flood of about 2,000 m³s⁻¹. The current climate is warm and humid (Koppen, Cfb) with a uniform annual rainfall over most of the catchment of 800 to 1,000 mm.

2.1   Previous Research on the Cranebrook Terrace

The Cranebrook Terrace represents a major change in the flow regime of the Nepean River, for it stands up to 18 m in thickness from bedrock contact to its surface (and is still occasionally flooded). Termed the *Cranebrook Formation* by Jones and Clark (1991) and Smith and Clark (1991), the lower 5 to 7 m are gravels up to 300 mm in *b*-axis, much larger than anything the present river can transport, and the remainder is silty-sandy *overburden* (this term is used because while probably commonly overbank in origin, these upper fines may also have been deposited with the channel in some cases). Nanson and Young (1987; 1988) made the first detailed interpretations of late-Quaternary flow-regime changes on the Nepean River based on stratigraphic and chronological evidence available from many kilometres of exposure in a number of large gravel and sand quarries in the Cranebrook Terrace near Penrith (Figure 14.2a). Radiocarbon and TL ages suggested that the basal gravels were deposited at about 45 to 40 ka as a braid plain immediately downstream of confining sandstone gorges by a river dominated by an abundant gravel load (Figure 14.3a). A thick overburden of silty sands was believed to have been deposited initially penecontemporaneously with the uppermost gravels but continuing to become younger upwards. An eastern mud-filled palaeochannel (the Clay Band, Figure 14.2c), possibly abandoned at about 37 to 34 ka was interpreted as representing the end of the period of major terrace formation. Some stripping and replacement of the fine overburden appeared to have occurred over the western part of the terrace at ~14 ka with the modern terrace existing as a composite feature (Figure 14.3b).

In a major revision of this story, recent luminescence determinations presented here put the age of gravel deposition in the eastern part of the terrace as being significantly older than 45 ka. Furthermore, these revisions have shown the terrace to be essentially a polycyclic feature consisting of an older eastern portion and a younger two-tiered western portion. While the timing of events stands corrected, importantly, the stratigraphic and associated palaeoflow-regime interpretations from Nanson and Young's earlier work remain essentially unchanged.

In addition to interpretations of flow-regime change, Young *et al.* (1987; 1994) examined the nature of regolith weathering on the Cranebrook Terrace for use as a relative guide of depositional chronology, and Nanson *et al.* (1987), on the basis of artefacts found in the gravels (Stockton and Holland, 1974), suggested the potential archaeological significance of the terrace. The revised chronology presented below has relevance to both these studies.

2.2   Revised Chronology of the Cranebrook Terrace

The chronology compiled by Nanson and Young (1987; 1988) consisted of a series of radiocarbon dates undertaken mostly on very large but degraded logs buried near the base of the terrace sequences. A series of TL samples were analysed by the Alpha Analytic luminescence laboratory in Florida, USA. Across the terrace as a whole, five carefully pretreated wood samples analysed by the ANU lab gave ages from logs in the basal gravel of ~42 to ~30 kyr BP, not dissimilar to the Alpha Analytic TL results of ~47 to ~40 ka for the gravel (see Tables 1 and 2 in Nanson and Young, 1987; for details including the uncertainties associated with each date). The fine overburden, although averaging about 6 to 9 m in thickness and extending for kilometres, contained no samples suitable for radiocarbon dating below a depth of possible surface disturbance (about 1.5 m). However, the middle and upper overburden at Site 1 gave TL ages of ~42 and ~41 ka, respectively (site numbers in Figure 14.2a are as used by Nanson and Young, 1987; 1988).

Following this initial TL and ^{14}C dating exercise, detailed inspection of the unpublished TL laboratory data provided by the Alpha Analytic showed serious inconsistencies. The TL growth curves exhibited a lot of scatter and the temperature plateaus

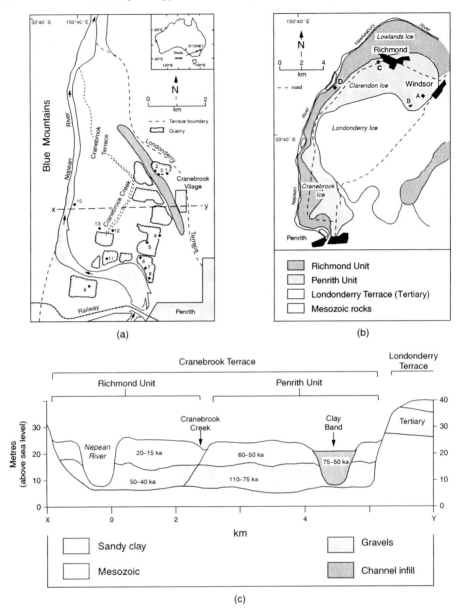

Figure 14.2   (a) The Cranebrook Terrace and sampling sites showing quarry locations at the time of sampling. X–Y marks the schematic section shown in Figure 14.2c. (b) The Cranebrook Terrace near Penrith and the Clarendon and Lowlands Terraces near Richmond and Windsor. A, B and C mark Gardiner's (1988) sampling sites. Reproduced by permission of B.H. Gardiner. (c) A schematic stratigraphic section of the Cranebrook Terrace

were unconvincing. Jerry Stipp from Alpha Analytic later agreed with this assessment (pers. comm., ca 1991). Additional sediment samples were collected in 1988 and 1989 and analysed in the University of Wollongong TL laboratory. These samples produced much more convincing laboratory data that resulted in a significantly different interpretation of the depositional chronology of the terrace (Table 14.1). At the time of second sampling, the quarry where the original samples were taken at Site 1 remained

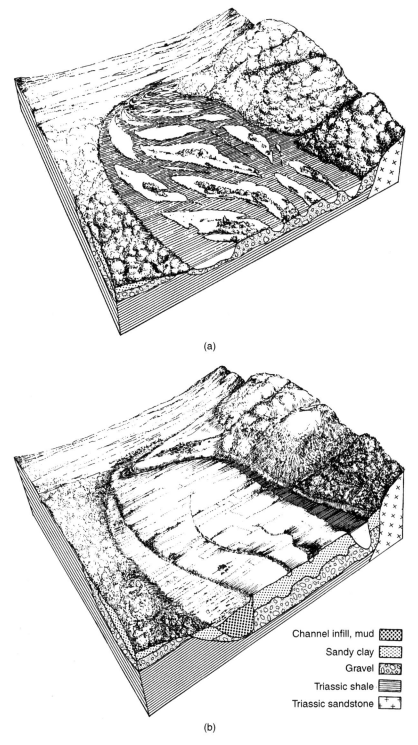

(a)

(b)

Channel infill, mud

Sandy clay

Gravel

Triassic shale

Triassic sandstone

Figure 14.3 A schematic representation of the Nepean River downstream of Penrith. (a) Formation of the Cranebrook terrace by braided river activity during Marine Oxygen Isotope Stage 5. (b) The present terrace. Flow is northwards towards the bottom right

Table 14.1 Analytical details of Thermoluminescence (TL) ages obtained from the Nepean River terraces between Penrith and Windsor

Sample number	Reference	Temp plateau region (°C)	Analysis temp. (°C)	Palaeodose (Grays)	K content (%)	Moisture content (%)	U + Th specific activity (Bq kg⁻¹)	Annual radiation (uGy yr⁻¹)	TL age (ka)
W715	Site 1 o/b 1.0 m	300–500	375	151 ± 12	1.32 ± 0.61	12 ± 5[a]	68.8 ± 4	2,627 ± 68	57.6 ± 4.7
W716	Site 1 o/b 1.8 m	300–500	375	156 ± 16	1.12 ± 0.61	12 ± 5[a]	61.6 ± 4	2,307 ± 68	67.6 ± 7.2
W717	Site 1 lower o/b	300–500	375	139 ± 18	0.88 ± 0.01	12 ± 5[a]	30.3 ± 4	1,515 ± 68	91.4 ± 12.4
W718	Site 2 u/g	300–500	375	140 ± 7	0.98 ± 0.01	15 ± 5[a]	50.8 ± 4	1,920 ± 66	73.0 ± 4.6
W719	Site 2 u/g	300–500	375	143 ± 13	0.78 ± 0.01	15 ± 5[a]	27.6 ± 4	1,331 ± 66	107 ± 11
W723	Site A o/b 3.0 m	300–450	375	182 ± 25	1.40 ± 0.01	14.2 ± 3	66.1 ± 4	2,600 ± 66	69.9 ± 9.8
W724	Site B o/b 2.0 m	275–500	375	169 ± 53	1.25 ± 0.01	1.6 ± 3	55.0 ± 4	2,586 ± 76	65.2 ± 20.6
W725	Site 3 o/b 3.5 m	275–450	375	293 ± 56	1.70 ± 0.01	7.2 ± 3	56.4 ± 4	2,911 ± 71	101 ± 19
W729	Site D o/b 5.0 m	300–450	375	81.8 ± 9.0	1.08 ± 0.01	10.3 ± 3	47.3 ± 4	2,048 ± 69	39.9 ± 4.6
W926	Site 12 o/b 1.5 m	300–375	350	129 ± 33	1.195 ± 0.005	12 ± 5[a]	69.7 ± 4	2,535 ± 68	51.0 ± 13.1
W927	Site 12 o/b 4.3 m	275–500	375	141 ± 15	1.205 ± 0.005	12 ± 5[a]	66.8 ± 4	2,489 ± 67	56.7 ± 6.0
W928	Site 12 l/g	300–425	375	108 ± 14	0.810 ± 0.005	22 ± 5[a]	27.2 ± 4	1,270 ± 61	84.7 ± 11.9
W929	Site 12 u/g	275–450	375	112 ± 18	1.050 ± 0.005	22 ± 5[a]	30.2 ± 4	1,525 ± 61	73.2 ± 12.3
W930	Site 13 C/b Ck	275–500	375	86.8 ± 11.1	1.080 ± 0.005	22 ± 5[a]	62.1 ± 4	2,072 ± 61	41.9 ± 5.5
W931	Site 12 u/g	275–500	375	105 ± 13	0.795 ± 0.005	22 ± 5[a]	38.4 ± 4	1,441 ± 61	72.9 ± 9.8
W932	Site 10 o/b 2.0 m	300–500	375	56.5 ± 5.6	1.395 ± 0.005	12 ± 5[a]	79.2 ± 4	2,891 ± 67	19.5 ± 2.0
W933	Site 10 o/b 7.3 m	275–425	375	44.9 ± 4.4	1.170 ± 0.005	12 ± 5[a]	63.7 ± 4	2,400 ± 67	18.7 ± 1.9

		Poct							
W934	Site 13 u/g	275–450	400	640 ± 9.0	0.960 ± 0.005	15 ± 5[a]	26.7 ± 4	1,490 ± 66	43.0 ± 6.3
W935	Site 13 l/g	275–500	375	55.5 ± 9.3	0.735 ± 0.005	22 ± 5[a]	20.7 ± 4	110 ± 61	50.4 ± 8.9
W936	Site 13 o/b 3.0 m	275–500	375	35.6 ± 5.2	1.205 ± 0.005	12 ± 5[a]	68.2 ± 4	2,513 ± 67	14.2 ± 2.1
W937	Site 13 u/g	275–500	375	82.9 ± 15.9	0.980 ± 0.005	15 ± 5[a]	45.7 ± 4	1,839 ± 65	45.1 ± 8.8
W991	Fine grain of W715	300–500	350–400	295 ± 24	1.32 ± 0.01	12.2 ± 3	68.8 ± 4	4,658 ± 121	63.3 ± 8.1
W1016	Fine grain of W716	300–475	350–400	239 ± 19	1.12 ± 0.01	12.1 ± 5[a]	61.6 ± 4	3,698 ± 109	64.6 ± 5.9
W856	Snowy River terrace at Orbost	300–500	350	237 ± 49	1.10 ± 0.01	0.5 ± 3	59.5 ± 4	2,555 ± 77	92.6 ± 19.5

Notes:

1. The specific activity of these specimens was measured by means of calibrated thick source alpha counting over a 42-mm scintillation screen. The values shown assume secular equilibrium for both the U and Th decay chains. The uncertainty levels indicated represent one standard deviation.
2. Potassium levels for samples W715–W729 and W856 determined by XRF, all others by AES.
3. Samples W991 and W1016 are repeat age determinations of W715 and W716, respectively, using the fine-grain polymineral additive technique. The ages indicated represent pooled mean values.

[a] Denotes assumed sample moisture content. Remaining samples assume the "as collected" moisture levels.

Key: o/b = overburden, u/g = upper gravels, l/g = lower gravels, C/b ck = Cranebrook Creek.

open but abandoned. New samples were obtained in a vertical profile from the upper gravels to near the surface of the fine overburden (standing water prevented sampling of the basal gravels). These yielded a stratigraphically consistent set of ages from 57.6 ± 4.7 ka (W715; Wollongong laboratory number) in the upper overburden 1.5 m below the surface, to 67.6 ± 7.2 ka (W716) at 2.8 m depth [these coarse-grain results are supported by fine-grain analyses on the same samples of 63.3 ± 8.1 ka (W991) and 64.6 ± 5.9 ka (W1016), respectively], to 91.4 ± 12.4 ka (W717) at the base of the fines just above the gravel (4.5 m depth) (Table 14.1). In a related study, Gardiner (1988) obtained the ages of 69.9 ± 9.8 ka (W723) (2.5 m depth), 65.2 ± 20.6 ka (W724) (2.0 m) and 101 ± 19 ka (W725) (3.5 m) for fine overburden at separate locations (A, B and C, respectively, in Figure 14.2b) overlying the morphologically and topographically similar Clarendon Terrace near Richmond, immediately downstream of the Cranebrook Terrace (Figure 14.2) (Table 14.1). Jones and Clark (1991) and Smith and Clark (1991) defined this terrace as the Clarendon Formation. It appears that the fine overburden of the Clarendon Formation (Terrace) is an equivalent feature to the overburden of the eastern portion of the Cranebrook Formation (Terrace). The upper gravels in this, the oldest portion of the Cranebrook Terrace is dated at 107 ± 11 ka (W719). Further west but just east of the Clay Band at Site 3, a sample from the upper gravels is dated at 73.0 ± 4.6 ka (W718). These results are significantly older than the 47 to 40 ka obtained for the gravels and overlying fines analysed by Alpha Analytic from the same portion of the terrace and they also indicate that the radiocarbon ages obtained from logs within the gravels must have been minimum values.

The remaining TL dates from this second sampling exercise on the Cranebrook Terrace came from a new large quarry immediately north of the east–west-oriented section of Castlereagh Road (locations named here as Sites 12 and 13) (Figure 14.2a). This quarry exposed an extensive face from the top to the base of the terrace and extended both west and east of the Cranebrook Creek several hundreds of metres. The TL results were revealing for they showed that Cranebrook Creek drains the surface of a polycyclic terrace along the boundary between the older eastern portion and the younger western portion. The gravels in the eastern portion (Site 12) dated at 84.7 ± 11.9 ka (W928) near the base and 73.2 ± 12.3 ka (W929) and 72.9 ± 12.3 (W931) near their upper limit, and consequently correspond in age to those immediately to the east of the Clay Band. The overlying fines dated at 56.7 ± 6.0 ka (W927) near their base and 51.0 ± 13.1 ka (W926) near the terrace surface, and are, therefore, of a similar age to the upper fines east of the Clay Band.

The revised ages suggest that the far eastern side of the gravels and associated lowermost fines of the Cranebrook Terrace near the Londonderry Terrace were deposited in mid-OIS 5. Overbank fines possibly continued to be deposited in OIS 3. Only slightly westward towards the Clay Band lie significantly younger gravels dating in late OIS 5 (and possibly OIS 4) with a somewhat younger overburden clearly dating in OIS 3. This difference in alluvial ages over a relatively short distance is supported from detailed multivariate soil analyses by Atkinson (1982; 1987) who recognised an older soil-geomorphic unit called the Church Lane Unit close to the Londonderry Terrace (beneath which the gravels TL dated at ∼107 ka and the fines just above the gravels at ∼91 ka) and a younger McCarthy's Lane Unit that is extensive west of, but also occurs in patches immediately east of, the Clay Band (beneath which the gravels TL dated at ∼85–73 ka) (Figure 14.2c). We have elected to name the whole of the eastern part of the Cranebrook Terrace with OIS 5 gravels at the base, the *Penrith Unit* of the Cranebrook Formation and to apply this name to the coeval Clarendon

Formation further to the north that was dated by Gardiner (1988) (Figures 14.2b and c). However, we recognise that the soil analyses by Atkinson (1982; 1987) indicate that this is polycyclic unit significantly older in the east than in the west.

To the west of Cranebrook Creek (Site 13, Figure 14.2a) the basal gravels dated at 50.4 ± 8.9 ka (W935) whereas the upper gravels gave three ages: 41.9 ± 5.5 ka (W930), 43.0 ± 6.3 ka (W934) and 45.1 ± 8.8 ka (W937). All four results are significantly younger than the gravels east of the creek. Gardiner (1988) obtained a very similar age of 39.9 ± 4.6 ka (W729) from the Lowlands Terrace at Site D (Figure 14.2b) just south of Springwood Road near Agnes Banks on the Nepean River, immediately downstream of the Cranebrook Terrace. Jones and Clark (1991) and Smith and Clark (1991) described this as the Lowlands Formation. It appears, therefore, that the Lowlands Formation (Terrace) is a comparable unit to the western part of the Cranebrook Formation (Terrace). The overlying fines immediately west of Cranebrook Creek (Site 13) gave an age of 14.2 ± 2.1 ka (W936). In the western-most extent of the Terrace at Site 10 (Figure 14.2a), along the high bank adjacent to the present Nepean River (Figure 14.2a) the fines dated at 19.5 ± 2.0 ka (W932) near the surface and 18.7 ± 1.9 ka (W933) just above the gravels. The marked change in terrace chronology west and east of Cranebrook Creek was clearly detected in Atkinson's (1982; 1987) soil study for he called the soil-geomorphic unit west of Cranebrook Creek, the Castlereagh Road Unit. We have chosen to name this western part of the Cranebrook Terrace and the coeval downstream Lowlands Terrace, the *Richmond Unit* (Figure 14.2c).

While the gravels west of Cranebrook Creek (the Richmond Unit) gave ages similar to those obtained by Alpha Analytic during the first TL dating exercise, it must be made clear that this is simply a coincidence, for none of these younger westerly gravels were exposed by quarrying and were therefore not sampled during the earlier study.

2.3   Re-interpretation of the Cranebrook Terrace

This sequence of revised TL dates for the Cranebrook Terrace fits remarkably well with the TL ages for palaeochannels on the Riverine Plain west of the Great Dividing Range (Page *et al.*, 1991; 1996; Page and Nanson, 1996). Because the eastern portion (the Penrith Unit) is the oldest part and is slightly higher than the rest, it appears that the entire terrace from the present river in the west to the base of the Tertiary age Londonderry Terrace in the east was reworked during OIS 5. Supported by independent soil studies (Atkinson, 1982; 1987), the Clay Band seems to provide an approximate boundary between the earliest OIS 5 gravels and later OIS 5 that may also include OIS 4 gravels extending westward to Cranebrook Creek.

The chronostratigraphy of the Cranebrook Terrace is summarised in an idealised section in Figure 14.2c. Most of the gravel unit east of the Cranebrook Creek (107–73 ka) probably correspond in age with the Coleambally palaeochannels (105–85 ka) on the Riverine Plain and to a previously unpublished TL age obtained from a large right bank terrace at Orbost on the Snowy River of 93 ± 20 ka (W856). Wray *et al.* (2001) obtained supporting evidence with a TL date from a small coastal valley fill on the western side of Jervis Bay of 118 ± 17 ka and a TL date of 97.0 ± 4.1 ka from Termeil Creek south of Jervis Bay. Cranebrook Terrace between the Clay Band and Cranebrook Creek, with a basal TL age of $\sim$85 ka and an age near the surface of $\sim$51 ka, appears to relate reasonably well to the Mayfield Terrace on the Shoalhaven River (Figure 14.1) dated by Nott *et al.* (2002) at $\sim$75 to 57 ka. However, there are large measurement uncertainties associated with the Mayfield ages.

Finally, Gibling (pers. comm. 1995) (cited in Young *et al.*, 1996) obtained an age from alluvium at Wogamia in the lower Shoalhaven basin of 83.7 ± 4.9 ka.

The age of the Clay Band on the Cranebrook Terrace is unknown but it was probably an active channel well prior to the two 34 to 30 kyr BP ages obtained from basal radiocarbon samples published by Nanson and Young (1987; 1988). It is situated within gravels that to the west date at ~75 ka and to the east at ~107 to 73 ka. Associated overburden deposits date near the surface at 50 ka, so unless the channel was incised into an existing terrace, it probably represents the terminal phase of deposition in the Cranebrook Formation at about 75 ka. However, it may have persisted as a channel until about 50 ka, later infilling with dark mud and organic matter.

The gravels west of Cranebrook Creek (the Richmond Unit) (50–43 ka) yield ages that correspond with the OIS 3 Kerarbury palaeochannels (55–35 ka) on the Riverine Plain. As is the Cranebrook Terrace, the Riverdale Terrace on the Shoalhaven River (Nott *et al.*, 2002) is polycyclic, and the component that dates at 50.4 ± 4.9, 45.4 ± 4.1 and 38.6 ± 11.1 ka may relate to the period of gravel deposition in the Richmond Unit and to the formation of Kerarbury palaeochannels. Significantly, Wray *et al.* (1993) obtained very similar results from three samples in a major alluvial fill at Inervary Creek in the mid-Shoalhaven basin of 56.2 ± 6.7, 52.1 ± 5.1 and 37.8 ± 4.2 ka, and Gibling (pers. comm. 1995, cited in Young *et al.*, 1996) obtained ages from weather alluvium at Wogamia in the lower Shoalhaven basin of 66.7 ± 5.7, 50.1 ± 4.3 and 38.1 ± 4.2 ka. Young *et al.* (2002) have recognised Kerarbury age large palaeochannels on the Namoi River in central western NSW dating at 56.3 ± 5.5 and 38.7 ± 2.6, well to the north of the Murrumbidgee, Shoalhaven and Nepean Rivers (Figure 14.1).

During the Yanco Phase on the Riverine Plain (20–13 ka) the fine overburden was largely stripped off the western one-third to one-half of the Cranebrook Terrace and replaced with much younger alluvium, giving ages of ~20 to ~15 ka, and leaving the basal gravels largely intact (Figure 14.2c). This period was also one of pronounced fluvial activity on the Namoi (Young *et al.*, 2002) and Gwyder Rivers (Pietsch, in prep).

In summary, there appears to have been a period of substantial construction of the Cranebrook Terrace by a braided river system in early to mid-OIS 5 (Figure 14.3a) as well as subsequent periods of reworking, each corresponding to periods of pronounced fluvial activity on the Riverine Plain during late OIS 5 and during OIS 3 and 2 (Page *et al.*, 1991; 1996). The eastern part of the Cranebrook Terrace (the Coleambally Phase Penrith Unit) is slightly higher and considerably more extensive than the Kerarbury Phase Richmond Unit (Figure 14.2c), and it is therefore assumed that the Coleambally Phase was somewhat more powerful, at least on this coastal river. There is no evidence as yet from this particular location on the Nepean River of a Gum Creek Phase (35–25 ka). There is evidence that the phases of fluvial activity on the Nepean River roughly correspond to those on the Shoalhaven River, although evidence is ambiguous due to a lack of clear differentiation on the Shoalhaven's polycyclical Riverdale Terrace that varies in age from ~50 to 17 ka (Nott *et al.*, 2002). The last period of significant alluvial reworking on the Cranebrook Terrace corresponds well with the Yanco Phase on the Riverine Plain; however, it was only capable of stripping and replacing the fine overburden on the western part of the terrace.

What is abundantly clear from the Cranebrook Terrace is that the alluvial episodes in OIS 5 and 3 were very powerful and quite unlike the relative inactivity of today's Nepean River (Figures 14.3a and 14.3b). The basal gravels throughout the terrace are very much coarser than those in contemporary bedload transport, and extensive stratigraphic exposures clearly suggest a braided system. The lack of any evidence of Holocene reworking relates mostly to the very large and compacted gravels that form the terrace boundary for the present Nepean River. However, gravel mining prior to opening up of the gravel quarries on the main terrace removed a small floodplain visible on early air photos. The present Cranebrook Terrace is a polycyclic landform containing gravels and overburden representative of several phases of climate and flow-regime change (Figures 14.2c and 14.3b).

3   NAMBUCCA AND BELLINGER CATCHMENTS

The Nambucca and Bellinger Rivers drain adjacent relatively small catchments of 1,407 km^2 and 1,100 km^2, respectively, on the mid-north coast of NSW at about 30°40′ latitude and 420 to 450 km north of Sydney and the Cranebrook Terrace (Figure 14.1). The catchments are relatively rugged and steep, reaching just over 1,500 m in elevation, with naturally forested valleys now harbouring pastured narrow terraces and floodplains. In the alluvial reaches, stream gradients vary from 0.015 to 0.001 and bedload and basal floodplain and terrace alluvium consists of cobbles in the headwaters and fine gravels in the middle and lower reaches. Climate is humid with hot summers (Koppen, Cfa). Maximum floods at Bowraville (basin area 430 km^2) are just over 2,000 m^3 s^{-1}, although zero flow characterises not infrequent droughts. Mean annual rainfall varies spatially from 1,150 to 1,600 mm, well distributed but with a summer maximum. The geology consists of Lower Permian metamorphosed sedimentary rocks of the New England Fold Belt with basalt outcrops. Compared to the Bellinger, the Nambucca catchment has a finer gravel bedload dominated by quartz in the channels and floodplains.

Because of the confined nature of most of the tributaries in the Nambucca and Bellinger catchments and the erodible nature of their sediment, it is probable that even the least erosive of the fluvial events recognised on the Riverine Plain and Nepean River was able to rework most of the older Pleistocene alluvium. However, these small steep basins have been relatively sensitive recorders of lower magnitude flow-regime changes and thereby have retained a relatively detailed late-Holocene record.

3.1   Nambucca and Bellinger Chrono-Stratigraphy

A series of terraces in the various tributaries of the Nambucca River and in the main trunk stream of the Bellinger valley have provided an important sequence of radiocarbon and TL ages (details of which are not provided here but will be contained in two Ph.D. theses soon to be completed at the University of Wollongong by Cohen and Doyle). The sampling locations are shown in Figure 14.4. These results overlap with the alluvial chronologies provided from the Cranebrook Terrace and Riverine Plain. The limited preservation has meant that just one OIS 5 TL result was obtained, an age of 78.1 ± 6.1 ka (W2732) from 6 m depth in a 12 m high (measured from the stream bed) terrace at Site 3 on Worrel Creek (Figure 14.4). The survival of this large terrace is probably due to Worrel Creek being one of the least energetic of the seven sub-catchments within the Nambucca basin and it having the most cohesive clayey sediment of them all. This late Stage 5 age agrees well with those from the Cranebrook Terrace and is not significantly different from the youngest Stage 5 ages on the Riverine Plain.

A single large terrace (10 m in height from the stream thalweg) at Site 4 on Taylors Arm in the Nambucca catchment provided three Stage 3 TL ages of 50.9 ± 11 ka (W2352), 52.9 ± 4.5 ka (W2719) and 54.6 ± 5.4 ka (W2718), a result supported by a

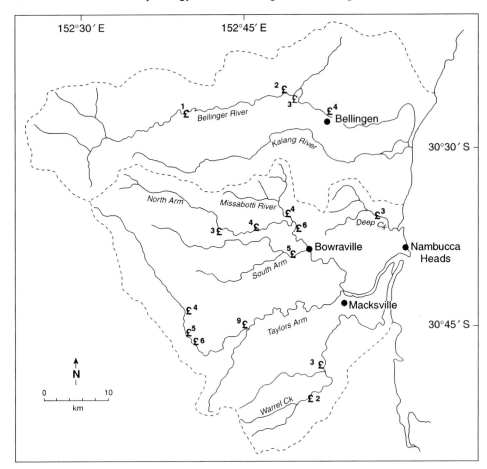

Figure 14.4   Sampling locations in the Bellinger and Nambucca catchments

basal date of 53.2 ± 6.1 ka (W2349) for the northern terrace at Site 3 on Deep Creek (Figure 14.4). This marked a very prominent period of alluviation on the Cranebrook Terrace and for Kerarbury deposits on the Riverine Plain (Page *et al.*, 1996), the Shoalhaven River (Nott *et al.*, 2002) and immediately to the west on the Namoi River (Young *et al.*, 2002). A possibly similar aged deposit on the Bellinger River was dated using AMS (accelerator mass spectrometry) radiocarbon at 48 kyr BP (OZF833) from a terrace approximately 10 m high (Site 3, Figure 14.4), and at a depth of 7 m; however, this could be a minimum age.

The upper part of the northern terrace and the base of the southern terrace at Site 3 on Deep Creek provided late Stage 3 ages of 36.1 ± 3.2 ka (W2727), 31.0 ± 3.0 ka (W2728) and 25.5 ± 2.1 ka (W2350) (Figure 14.4). These correspond well to the full age range of the Gum Creek Phase recognised on the Riverine Plain (Page *et al.*, 1996). On the Kalang River (the southern tributary in the Bellinger catchment), Warner (1970; 1972) identified an undifferentiated fluvial deposit at ~34 kyr BP that may also be from the Gum Creek Phase. Warner (1970; 1972) also radiocarbon dated a terrace on the upper Kalang River at 20 to 21 kyr BP, a result supported by an additional radiocarbon sample from a buried terrace remnant on the adjacent Bellinger River (Site 2, Figure 14.4) of ~24 kyr BP (OZF347), and preliminary OSL (optically stimulated luminescence) ages of 24 to 28 ka from the same unit. Adjustment of the radiocarbon ages using the comparisons with uranium series ages by Bard *et al.* (1990) makes them very similar to the OSL ages and puts them in OIS 3 just prior to the LGM and part of the Gum Creek Phase. They probably correspond on the Shoalhaven River to part of the Riverdale Terrace (Nott *et al.*, 2002) and to a terrace on a tributary of the Shoalhaven River, Tapitallee Creek, dated by Walker (1962). Wray *et al.* (2001) have also obtained a basal age of a polycyclic terrace on Turmeil Creek at ∼27 kyr BP.

Despite the limited preservation of older units in the confined Nambucca and Bellinger basins, there is sufficient alluvial evidence to suggest that most of the late-Quaternary fluvial episodes represented very clearly on the Riverine Plain and Nepean River are probably also represented in the narrow coastal valleys in the form of terrace fragments. However, what the Nambucca and Bellinger basins have preserved relatively well is the story of alluviation following the LGM, and particularly a radiocarbon chronology of alluviation in the mid-late Holocene.

North Arm at Site 3 in the Nambucca catchment (Figure 14.4) shows a substantial terrace unit that provided TL ages of 19.5 ± 3.6 ka (W2724) and 16.2 ± 1.2 ka (W2351) (although possibly stratigraphically reversed, their errors show them to be not significantly different, suggesting a post glacial terrace of similar age to the Yanco units on the Riverine Plain; Page *et al.*, 1996). Evidence for the survival of substantial Yanco-age deposits in the Nambucca basin is supported by alluvial deposits that yield TL ages of: 20.9 ± 4.4 ka (W2723), 16.2 ± 2.1 ka (W2351) and 10.1 ± 2.1 (W2722) at Site 6 in the North Arm; 19.3 ± 1.4 ka (W2726) at Site 5 in the South Arm; 16.6 ± 1.3 ka (W2721) and 15.4 ± 1.3 ka (W2720) at Site 9 in Taylors Arm; and 13.4 ± 1.1 ka (W2731), 12.6 ± 1.1 (W2730) and 11.6 ± 0.9 ka (W2729) at Sites 2 and 3 in Warrel Creek (Figure 14.4). In addition to the TL dates in the Nambucca basin, a radiocarbon date from a terrace 13 m above the channel in the middle to upper Bellinger valley (Site 1) yielded an age of ∼12 kyr BP (OZF836), while a terrace 11 m above the channel in the lower valley (Site 4) yielded an age of ∼17 kyr BP (OZF578). It appears that late Gum Creek and Yanco-age deposits form the major terrace units that confine the present floodplains at many locations in the Nambucca and Bellinger basins.

Large Gum Creek and Yanco-age palaeochannels, dated using TL and OSL, are present along the westward-flowing Gwyder River near Moree (Pietsch, in prep.) as are Yanco-age palaeochannels on the Namoi River (Young *et al.*, 2002). Both are rivers immediately inland of the Nambucca and Bellinger basins. Clearly, the Gum Creek and Yanco Phases of fluvial activity were characterised by much greater runoff and alluvial activity than what occurs today.

Interestingly, 13 conventional radiocarbon samples obtained from floodplains throughout the Nambucca basin failed to date any Holocene age deposits older than 3,000 yr BP. The Holocene floodplains are commonly confined between late Pleistocene terraces and appear to have been substantially reworked (flushed) during the early to middle Holocene, leaving only basal gravels from late in that period overtopped with fine overburden of even later Holocene age. We have termed this period of early to mid-Holocene fluvial activity the *Nambucca Phase*. We readily acknowledge that for earlier episodes we regard the presence of abundant alluvium as evidence of fluvial activity whereas here we are arguing that its absence is indicative of somewhat higher flows and enhanced flushing. However, while earlier episodes were clearly major events that left substantial terraces, the *Nambucca Phase* was relatively minor and was probably only able to rework floodplains between older resistant and confining terraces.

At Site 4 on Missabotti Creek in the Nambucca catchment, a gleyed organic clay within fine gravels gave an age of $2,660 \pm 50$ yr BP (Beta 101413) with the fine over-burden near the surface dating at $2,330 \pm 50$ yr BP (Beta 101412). Another example of a fine-grained floodplain is at Site 3 on Deep Creek, also in the Nambucca catchment, dated at $2,080 \pm 60$ yr BP (Beta 101421) with a clearly visible fine-grained channel fill within the floodplain dating at $1,580 \pm 70$ yr BP (Beta 101419) and $1,550 \pm 50$ yr BP (Beta 191420). In a similar situation at Site 5 on Taylors Arm, fine basal gravels dated at $1,960 \pm 100$ yr BP (Beta 101416) with sandy-silty overburden near the surface dat-ing at $1,020 \pm 60$ yr BP (Beta 101417). A well-defined palaeochannel in the floodplain at Site 6 of Taylors Arm gave a basal date of 130 ± 60 yr BP (Beta 101414) while another palaeochannel within the floodplain at Site 4 on North Arm gave a basal date of 440 ± 60 yr BP (Beta 191415). These sites show that most of the recent Holocene floodplains were formed with a fine gravel base and a vertically accreting silty over-burden from about 2,700 yr BP to the present. After about 2,500 to 2,000 yr BP, there is no sign of lateral channel migration or other forms of substantial floodplain reworking or gravel accretion that characterised the NAMof the early to middle Holocene, sug-gesting that the channels became laterally stable at this time. Relatively narrow deep palaeochannels that date from about 1,600 yr BP until the time of European settlement ($\sim$130 yr BP) indicate periodic channel abandonment by avulsion.

Additional chronostratigraphic evidence based on 30 AMS radiocarbon samples within the Bellinger catchment (Cohen, in prep) demonstrates channel stabilisation and the onset of Vertical accretion (VA) following the Nambucca Phase.This occurred at $\sim$4,000 yr BP in the upper valley and at about 3,500 to 2,500 yr BP (i.e. similar to the Nambucca catchment) in the lower valley. The chronostratigraphic data within the Bellinger valley indicates that the mid- to late Holocene has been predominantly characterised by the development of sub-horizontally laminated vertically accreted floodplains with laterally stable channels. The data, however, also demonstrate intra-reach variability. Some locations are protected by remnant terraces or bedrock spurs, whereas other sites are vulnerable to reworking through to the present due to the alignment of the main channel, localised constrictions or to the proximity to tributary junctions (Cohen, in prep.).

Once stabilised, dense rainforest would have covered the floodplains and stream banks, undoubtedly cluttering the narrow channels with large woody debris and further reducing stream power. This condition would have led to periodic avulsion (Brooks and Brierley, 2002). The result was a highly stable late-Holocene channel and floodplain system with some palaeochannel impressions, in stark contrast to the laterally more active, floodplain-reworking systems that appear to have characterised much of the late Pleistocene and early to mid-Holocene. It was these relatively recently stabilised channels, floodplains and palaeochannels, densely covered with rainforest, which the early settlers to the Nambucca and Bellinger catchments encountered when they cleared land for settlement in the mid-late nineteenth century (Nanson and Doyle, 1999).

4 INTERPRETATION OF HOLOCENE CONDITIONS ON THE NSW NORTH COAST

Our interpretation is that the Nambucca and Bellinger rivers were much more vig-orous channel systems during the Yanco Phase (20–13 ka) than they became in the late Holocene, for there is widespread evidence for thick floodplain units (substantial terraces) dating from after the LGM, followed by a reduction in the proportion of the valley floor being reworked in the Holocene. During the Yanco Phase in particular, high terraces were formed indicating that much larger flows occurred than in the late Holocene. During the very late Pleistocene and with the onset of the Holocene, it appears that the rivers reduced in discharge and load, entrenching and abandoning Yanco Phase floodplains that then acted to confine a less active river to a narrower, shallower alluvial belt (Warner, 1992). In the upper valley of the Bellinger River, this decline in activity is well preserved (Cohen, in prep.). A high terrace was formed and abandoned after about 12 kyr BP, the same happening to a lower terrace sometime after ~9 kyr BP but before ~4 kyr BP, following which the contemporary floodplain was formed. During the early to mid-Holocene (the *Nambucca Phase*), most of the flood-plains that developed were confined between Pleistocene terraces and were reworked by flows less active than those in the Yanco Phase but more active than those of today. This period of alluvial deposition has been recorded, but has limited occurrence, with radiocarbon dates by Warner (1970; 1972) and Cohen (in prep.). Essentially, quiescent conditions prevailed in most parts of the valley trough from about 3,500 to 2,500 yr BP to the present, although as early as 4,000 yr BP in the upstream reaches of the Bellinger catchment. Young *et al.* (2002) suggest a decline in the discharges on the Namoi River (immediately to the west) after about 6 ka. By the late Holocene, the rivers of the NSW north coast had become laterally stable and were supporting riparian rainforest along their banks. However, during this period the floodplains continued to build by vertical accretion. These small, laterally stable tree-lined channels with floodplains of over-bank fines are visible today in a few locations in the Nambucca and Bellinger valleys, as are isolated examples of narrow pre-European palaeochannels (Nanson and Doyle, 1999; Cohen, in prep.; Doyle, subm.).

This period of late-Holocene fluvial stability following the *Nambucca Phase* appears to have been widespread on the rivers of southeastern Australia. Illustrative of its southward extent is that Nanson *et al.* (1995) have shown a very similar sequence of Holocene lateral activity switching at 3 to 2 kyr BP to channel stability and vertical floodplain accretion on the Stanley River in western Tasmania. The data from many of the rivers of southeastern Australia suggest that the mid-late Holocene saw an increase in sediment preservation potential with the development of relatively stable, vertically accreting floodplains. The timing of this shift in channel activity appears to have been a function of local conditions as well as regional climate, and generally occurred between ~4,000 and 2,500 yr BP.

5   DISCUSSION AND CONCLUSIONS

Alluvial deposits on the Nepean, Nambucca and Bellinger rivers provide evidence of flow-regime changes in southeastern Australia from about 100 ka to the present. While the data presented in this study by no means provides a continuous or highly detailed record, they do supply evidence for understanding the nature of changing fluvial conditions in the non-glaciated coastal river basins of NSW for much of the last glacial cycle (~100 ka). As such, this evidence allows direct comparisons with late-Quaternary changes described for the westerly flowing rivers of the Riverine Plain (Page *et al.*, 1991; 1996; Page and Nanson, 1996). Figure 14.5 compares timelines of dates described here with those obtained by Page *et al.* (1991; 1996) from palaeochan-nels on the Riverine Plain and presents an inset frequency histogram of these dates combined. Using calibrations by Bard *et al.* (1990) and Stuiver *et al.* (1998), radio-carbon samples have calibrated to sidereal time for direct comparison with TL ages.

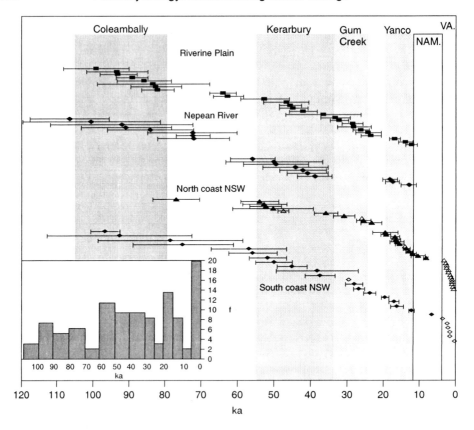

Figure 14.5 A timeline of 72 TL and 23 radiocarbon dates from the Riverine Plain, Nepean River, NSW north coast and NSW south coast. Inset: A frequency histogram of these dates plotted using a class interval of 10 ka above 30 ka and 5 ka below. Notes: (1) TL samples are presented as solid symbols and radiocarbon as open symbols. (2) Radiocarbon samples have been calibrated to sidereal time for direct comparison with TL ages, using calibrations by Bard *et al.* (1990) and Stuiver *et al.* (1998). (3) The Riverine Plain and Nepean River separate into relatively distinct fluvial episodes as only samples from channel bedload deposits or immediately overlying fines have been used. (4) Samples from the north and south coast rivers blur these groupings somewhat because these terrace exposures rarely contain much exposed bed material and therefore overburden of necessity has commonly been sampled. (5) Upland swamps, channel-less valley fills and alluvial fans have been excluded, as have terraces influenced by eustatic changes. (6) NAM is the Nambucca Phase and VA is the late-Holocene period of largely vertical accretion

The Murray and Murrumbidgee rivers on the Riverine Plain have their headwaters in the periodically glaciated Snowy mountains, with significant areas over 2,000 m in elevation (Barrows *et al.*, 2001). In contrast, the Wollondilly–Nepean system rises in country further north, unglaciated and no higher than 1,260 m, yet there is a remarkably similar pattern of fluvial activity. Coleambally Phase deposits date on the Riverine Plain between 105 and 80 ka and correspond to the Penrith Unit on the Nepean River, with basal gravels that consist of an older OIS 5 unit dating between 110 and 90 ka in the eastern portion, and a younger more extensive western unit that dates at about 85 to 73 ka (Figures 14.2b and 14.2c). Overlying soil units support our suggestion that the Stage 5 part of the Cranebrook Terrace (the Penrith Unit) is itself likely to be polycyclic. The Kerarbury Phase on the Riverine Plain dated from about 55 to

35 ka and corresponds to the Richmond Unit on the Nepean River with large and basal gravels that date from about 50 to 40 ka. However, the OIS 3 Kerarbury Phase was probably a lesser event than the Coleambally, reworking just part of the terrace and rebuilding it to a slightly lower elevation. The post LGM Yanco Phase was even less erosive, for it only reworked the overburden of the Richmond Formation in the period 20 to 15 ka (Figure 14.2c). Interestingly, the portion of the Nepean River examined in this study exhibited no recognisable unit equivalent to the Gum Creek Phase (35–25 ka) from the Riverine Plain.

The LGM appears to have been a distinctive period of very little fluvial activity. Almost no deposits of LGM age (about 24 to 20 ka) occur on the Riverine Plain (Page *et al.*, 1996; Page and Nanson, 1996), and Figure 14.5 suggests that much the same applies to the coastal rivers. Sweller and Martin (2001) recognised this as a period of devegetated periglacial activity at 1,000 m at Barrington Tops just north of the Hunter River (Figure 14.1).

Figure 14.6 has been developed from Figure 14.5 but it is also based on what is inferred from the extent and nature of the deposits. It shows in schematic form cyclical, but generally declining, episodes of fluvial activity in eastern NSW rivers during the past full glacial cycle. However, it must be appreciated that the vertical axis of this figure does not relate directly to variable stream power, but also reflects the duration and therefore, the total amount of reworking achieved by any high-energy phases.

While there were probably earlier Pleistocene terraces near Penrith during OIS 5, the Wollondilly–Nepean River was clearly a very powerful, actively migrating, probably braided river when it initially formed and then partly reworked the present Cranebrook Terrace (Figure 14.3a). It deposited an extensive unit of up to 8 m of gravels with some of the largest clasts 300 mm in "*b*" axis. On the basis of the luminescence chronology and supporting soil analyses, we contend that the Coleambally Phase (the Penrith Unit) is divided into a more powerful or longer-lived earlier period that reworked all the Cranebrook Terrace, and a lesser later phase that did not rework the far eastern portion.

The Kerarbury Phase was also either short lived or less powerful as it only reworked about one-third of the Terrace. On the basis of the extent of alluvial reworking, the Gum Creek and Yanco Phases were obviously lesser events again. The Cranebrook and Clarendon Formations on the Nepean River provide evidence that at times during OIS 5 and even OIS 3 there were periods of abundant precipitation and runoff, possibly winter-accumulated in the form of a snow-pack in the elevated Southern Tablelands and Blue Mountains. The Nepean River of today is very inactive with no prospect of currently eroding its extensive former floodplain (Figure 14.3b).

On the north coast, small catchments such as those of the Nambucca and Bellinger Rivers have experienced similar precipitation and flow-regime changes to those of the much larger rivers in eastern NSW. However, their confined nature means that they have retained only part of this record and have been particularly sensitive to Holocene changes less pronounced than those that occurred throughout much of the late Pleistocene. Small remnants of alluvium from the dominant Coleambally and Kerarbury Phases have survived, suggesting that these were significant events throughout southeastern Australia. The LGM appears to have been relatively dry and fluvially inactive. While the Yanco Phase (20–13 ka) had a relatively modest effect on the large rivers, it is widely preserved as the oldest major terrace system in the confined Bellinger and Nambucca valleys (Figure 14.7). The early to mid-Holocene (12–3 ka) (the *Nambucca*

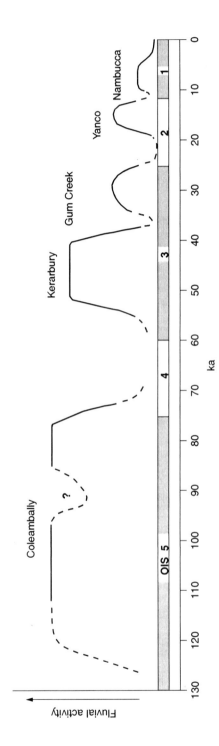

Figure 14.6 A timeline of inferred intensity of fluvial activity for the various fluvial phases (Coleambally to Nambucca) based on the extent of alluvial preservation along the NSW coastal rivers during Marine Oxygen Isotope Stages (OIS) 5 to 1. The inferred intensity of flow will result from changes in both flow magnitude and the duration of such periods of change. As a result, the vertical axis does not necessarily just represent an index of flow magnitude as reflected in stream power

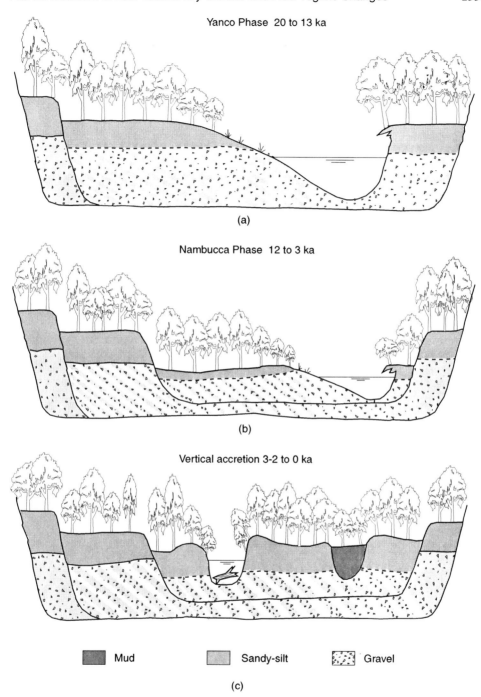

Figure 14.7 A schematic representation of floodplain and terrace formation in the Nambucca catchment. (a) Yanco phase, 20 to 13 ka. (b) Nambucca Phase, 12 to 3 ka. (c) Vertical accretion (3–2 to 0 ka) following the Nambucca Phase of lateral reworking. These sections are intended to represent the middle to lower valley, the boundaries are inferred, and the Nambucca Phase is based on very little alluvial preservation. The ages are in sidereal time for all periods

Phase) was marked by much lower flows than those of the late Pleistocene, but they were certainly more pronounced than those of today. The confining effect of the Pleistocene terraces meant that much of the early to mid-Holocene alluvium was reworked and therefore not preserved, although remnants do remain. Since about 3.5 to 2.5 ka, many of the rivers of coastal southeastern Australia have been laterally stable with floodplains vertically accreting with overbank sediment alongside well-vegetated channels (Figure 14.7). It was these low-energy, laterally stable rivers that European land clearance so dramatically destabilised within a few short years of settlement.

This broad summary of flow-regime changes on the coastal rivers of NSW in relation to changing climate is obviously open to considerable revision and refinement as more data become available, and as the significance of previous work becomes more fully appreciated. Even though some rivers have shown great stability throughout much of the period since the LGM (Brooks and Brierley, 2002), major changes such as those recorded on the Nepean River and Riverine Plain are clearly the product of climatic forcing. The more subtle changes recorded in the Holocene are in some cases difficult to interpret because of threshold and complex response effects (e.g. Nanson, 1986; Prosser *et al.*, 1994). Nevertheless, the evidence presented here shows that alluvium preserved along the rivers of coastal NSW retains a strong climatic signature (Figures 14.5 and 14.6), albeit modified by local preservation potential and intrinsic thresholds.

The findings here have important implications for understanding global climatic and hydrological changes during the Quaternary. The rivers of eastern Australia have been supplied with moisture from the adjacent Pacific Ocean. Rises and falls in sea level associated with the glacial cycles would have greatly altered the movement of ocean currents presently passing north and eastward of Australia, as well as the periodic location of an oceanic "warm pool" in the western Pacific. The palaeohydrological patterns of the coastal rivers of NSW are very likely a response to such major oceanic changes and our work continues to explore such relationships.

6 SUMMARY

Alluvial deposits on New South Wales' (NSW) coastal rivers provide evidence of flow-regime changes from about 100 ka to the present and correlate closely with fluvial phases previously described for the westerly flowing rivers of the Riverine Plain. Marine Oxygen Isotope Stage (OIS) 5 gravels date between 110 and 90 ka in the eastern portion of the Cranebrook Terrace and correspond to the Coleambally Phase deposits on the Riverine Plain that date between 105 and 80 ka. A younger and more extensive mid-western gravel unit dates at about 85 to 73 ka. Soils evidence supports the existence of possibly two periods of activity during the Coleambally Phase. The basal gravels in the most western portion of the Cranebrook Terrace were deposited at about 50 to 40 Ka and correspond to the OIS 3 Kerarbury Phase on the Riverine Plain that dates from about 55 to 35 ka. The Kerarbury Phase was probably a lesser event than the combined Coleambally, for it reworked just part of the terrace, and the Yanco Phase was even less erosive – only able to rework the upper fines on the western portion of the terrace in the period 20 to 15 ka. No recognisable equivalent to the Gum Creek Phase (35–25 ka) from the Riverine Plain has been located on the Nepean River. Clearly, the Wollondilly–Nepean River was a very powerful, actively migrating river when it formed and reworked the Cranebrook Terrace in OIS 5 and 3, yet today it is a very inactive system with no prospect of eroding its extensive former floodplain. The Cranebrook Terrace reflects a remarkable change in Quaternary flow regime for a river not affected by headwater glaciation.

On the north coast of NSW, small catchments such as those of the Nambucca and Bellinger Rivers have experienced similar precipitation and flow-regime changes to those of the much larger rivers in eastern NSW. However, their confined nature has meant that they have retained only part of their alluvial record in the form of a relatively sensitive indication of recent Holocene changes that were much less pronounced than those occurring throughout much of the late Pleistocene. A Yanco Phase flow regime (post LGM to pre-Holocene) is widely preserved as the oldest major terrace systems in the Bellinger and Nambucca valleys. The early to mid-Holocene (12–3 ka) (the *Nambucca Phase*) was marked by much lower flows than those of the late Pleistocene, but they were certainly more pronounced than those of today. The confining effect of the Pleistocene terraces meant that much of the early to mid-Holocene alluvium was reworked and not preserved, although some remnants remain. Since about 3 to 2 ka, many of the rivers of coastal southeastern Australia have been laterally stable with floodplains vertically accreting alongside well-vegetated channels. It was these low-energy, laterally stable rivers that European land clearance so dramatically destabilised.

ACKNOWLEDGMENT

We wish to thank Drs John Jansen, Ken Page and Bob (R.W.) Young for their very useful comments on an earlier draft.

REFERENCES

Atkinson, G., 1982. *Soil Survey and Erosion Control Measures for the Penrith Lakes Scheme*. Soil Conservation Service of New South Wales, Sydney.

Atkinson, G., 1987. A review of soil and geological maps of the Nepean river terraces, NSW. *Australian Geographer*, **18**, 124–136.

Bard, E., Arnold, M., Hamelin, B., Tisnerat-Laborde, N. and Cabioch, G., 1990. Radiocarbon calibration by means of mass spectrometric 230Th/234U and 14C ages of corals. An updated data base including samples from Barbados, Morurora and Tahiti. *Radiocarbon*, **40**, 1085–1092.

Barrows, T.T., Stone, J.O., Fifield, L.K. and Cresswell, R.G., 2001. Late Pleistocene glaciation of the Kosciuszko Massif, Snowy Mountains, Australia. *Quaternary Research*, **55**, 179–189.

Bowler, J.M., 1978. Quaternary climate and tectonics in the evolution of the Riverine Plain, southeastern Australia. In J.L. Davies and M.A.J. Williams (eds), *Landform Evolution in Australasia*. Australian National University Press, Canberra, 71–112.

Bowler, J.M. and Price, D.M., 1998. Luminescence dates and stratigraphic analyses at Lake Mungo: review and new perspectives. *Archaeology in Oceania*, **33**, 156–168.

Bowler, J.M., Stockton, E. and Walker, M.J., 1978. Quaternary stratigraphy of the Darling river near Tilpa, New South Wales. *Proceedings of the Royal Society of Victoria*, **90**, 78–88.

Brooks, A.P. and Brierley, G.J., 2002. Mediated equilibrium: the influence of riparian vegetation and wood on the long-term evolution and behaviour or a near-pristine river. *Earth Surface Processes and Landforms*, **27**, 343–367.

Brooks, A.P., Brierley, G.J. and Millar, R.G., 2003. The long-term control of vegetation and woody debris on channel and floodplain evolution: insights from a paired catchment study in southeastern Australia. *Geomorphology*, **51**, 7–29.

Bull, W.B., 1991. *Geomorphic Responses to Climatic Change*. Oxford University Press, New York.

Cohen, T.J., in prep. *Late Holocene Floodplain Processes and Post-European Channel Dynamics in South-Eastern Australia*, PhD thesis in preparation. School of Geosciences, University of Wollongong, Australia.

Doyle, C.J. *Fluvial Geomorphology of the Nambucca River Catchment: Late Quaternary Change, Post-Settlement Channel Degradation and Proposals for River Rehabilitation*, PhD thesis (submitted). School of Geosciences, University of Wollongong, Australia.

Erskine, W.D. and Peacock, C.T., 2002. *Late Holocene Floodplain Development Following a Cataclysmic Flood*. International Association of Hydrological Sciences, Alice Springs; in press.

Fryirs, K. and Brierley, G., 1998. The character and age structure of valley fills in Upper Wolumla Creek catchment, south coast, New South Wales, Australia. *Earth Surface Processes and Landforms*, **23**, 271–287.

Ferguson, R.J. and Brierley, G.J., 1999. Downstream changes in valley confinement as a control on floodplain morphology, Lower Tuross river, New South Wales. In A.J. Miller and A. Gupta (eds), *Varieties of Fluvial Form*. Wiley, Chichester, 377–407.

Gardiner, B.H., 1988. *The Alluvial Stratigraphy and Depositional History of the Nepean Terraces Near Penrith, New South Wales, with Specific Reference to the Clarendon Terrace*. Unpublished BSc (Hons) Thesis, University of Wollongong, Wollongong, 103.

Hickin, E.J., 1970. The terraces of the Lower Colo and Hawkesbury Drainage basins, New South Wales. *Australian Geographer*, **11**(3), 278–287.

Hickin, E.J. and Page, K.J., 1971. The age of valley-fills in the Sydney basin. *Search*, **2**(10), 383, 384.

Jones, D.C. and Clark, N.R., (eds), 1991. *Geology of the Penrith 1:100,000 Sheet 9030*. New South Wales Geological Survey, Sydney, 29–56.

Melville, M.D. and Erskine, W., 1986. *Sediment Remobilization and Storage by Discontinuous Gullying in Humid Southeastern Australia*. Publication No. 159, International Association of Hydrological Sciences, 277–286.

Nanson, G.C., 1986. Episodes of vertical accretion and catastrophic stripping: a model of disequilibrium floodplain development. *Geological Society of America Bulletin*, **97**, 1467–1475.

Nanson, G.C., Barbetti, M. and Taylor, G., 1995. River stabilisation due to changing climate and vegetation during the late Quaternary in Western Tasmania, Australia. *Geomorphology*, **13**, 145–158.

Nanson, G.C., Barbetti, M. and Taylor, G., 1996. River stabilisation due to changing climate and vegetation during the late Quaternary in western Tasmania, Australia. *Geomorphology*, **13**, 145–148.

Nanson, G.C. and Doyle, C.J., 1999. Landscape stability, Quaternary climate change and European degradation of coastal rivers in southeastern Australia. In I. Rutherfurd and R. Bartley (eds), *Proceedings of the Second Australian Stream Management Conference*, Cooperative Research Centre for Catchment Hydrology, Melbourne, Australia, 473–480.

Nanson, G.C. and Young, R.W., 1987. Comparison of thermoluminescence and radiocarbon age-determinations from late-Pleistocene alluvial deposits near Sydney, Australia. *Quaternary Research*, **27**, 263–259.

Nanson, G.C. and Young, R.W., 1988. Fluviatile evidence for a period of late Quaternary pluvial climate in coastal Southeastern Australia. *Palaeogeography, Palaeoclimatology, Palaeoecology*, **66**, 45–61.

Nanson, G.C., Young, R.W. and Stockton, E.D., 1987. Chronology and palaeoenvironment of the Cranebrook Terrace (near Sydney) containing artefacts more than 40,000 years old. *Archeology in Oceania*, **22**, 72–78.

Nott, J., Price, D. and Nanson, G., 2002. Stream response to Quaternary climate change: evidence from the Shoalhaven river catchment, southeastern highlands, temperate Australia. *Quaternary Science Reviews*, **21**, 965–974.

Page, K.J., 1972. *A Field Study of the Bankfull Discharge Concept in the Wollombi Brook Drainage Basin, New South Wales*. Unpublished M.A. (Hons) Thesis, University of Sydney, Sydney.

Page, K.J. and Nanson, G.C., 1996. Stratigraphic architecture resulting from Late Quaternary evolution of the Riverine Plain, South-Eastern Australia. *Sedimentology*, **43**, 927–945.

Page, K.J., Nanson, G.C. and Price, D.M., 1991. Thermoluminescence chronology of late Quaternary deposition on the Riverine plain of South-Eastern Australia. *Australian Geographer*, **22**, 14–23.

Page, K.J., Nanson, G.C. and Price, D.M., 1996. Chronology of Murrumbidgee river palaeochannels on the Riverine Plain, Southeastern Australia. *Journal of Quaternary Science*, **11**(4), 311–326.

Pietsch, T., in prep. *Late Quaternary Evolution of the Lower Gwydir Alluvial Fan*, PhD thesis in preparation. School of Geosciences, University of Wollongong, Australia.

Prosser, I.P., Chappell, J. and Gillespie, R., 1994. Holocene valley aggradation and gully erosion in headwater catchments, South-Eastern highlands of Australia. *Earth Surface Processes and Landforms*, **19**, 465–480.

Smith, E.A., 2000. *An Evolutionary History of Wingecarribee Swamp, NSW, and Ensuing Management Implications*. Unpublished B Envi Sci Thesis, University of Wollongong, Wollongong, 106.

Smith, V. and Clark, N.R., 1991. Cainozoic stratigraphy. In D.C Jones and N.R. Clark (eds), *Geology of the Penrith 1:100,000 Sheet 9030*. New South Wales Geological Survey, Sydney, 29–56.

Stuiver, M., Reimer, P.J., Bard, E., Beck, J.W., Burr, G.S., Hughen, K.A., Kromer, B., McCormac, G., van der Plicht, J. and Spurk, M., 1998. Intercal 98 radiocarbon age calibration, 24,000-0 cal BP. *Radiocarbon*, **40**, 1041–1083.

Stockton, E. and Holland, W., 1974. Cultural sites and their environments in the Blue Mountains. *Archaeological Phys Anthropology in Oceania*, **9**, 36–65.

Sweller, S. and Martin, H.A., 2001. A 40,000 year vegetation history and climatic interpretations of Burraga Swamp, Barrington Tops, New South Wales. *Quaternary International*, **83–85**, 233–244.

Walker, P.H., 1962. Terrace chronology and soil formation on the south coast of N.S.W. *Journal of Soil Science*, **13**(2), 178–187.

Walker, P.H., 1970. Depositional and soil history along the Lower Macleay River, New South Wales. *Journal of the Geological Society of Australia*, **16**, 683–696.

Walker, P.H. and Coventry, R.J., 1976. Soil profile development in some alluvial deposits of Eastern New South Wales. *Australian Journal of Soil Research*, **14**, 305–317.

Walker, P.H. and Hawkins, C.A., 1957. A study of river terraces and soil development on the Nepean river, N.S.W. *Proceedings of the Royal Society of N.S.W.*, **91**, 67–84.

Warner, R.F., 1970. Radio-carbon dates for some fluvial and colluvial deposits in the Bellinger Valleys, New South Wales. *Australian Journal of Science*, **32**(9), 368, 369.

Warner, R.F., 1972. River terrace types in the coastal valleys of New South Wales. *Australian Geographer*, **12**, 1–22.

Warner, R.F., 1992. Floodplain evolution in a New South Wales coastal valley, Australia: spatial process variations. *Geomorphology*, **4**, 447–58.

Webb, A.A., Erskine, W.D. and Dragovich, D., 2002. *Flood-Driven Formation and Destruction of a Forested Floodplain and In-Channel Benches on a Bedrock-Confined Stream: Wheeny Creek, Southeast Australia*. International Association of Hydrological Sciences, Alice Springs; in press.

Wray, R.A.L., Price, D.M. and Young, R.W., 2001. Thermoluminescence dating of alluvial sequences in coastal valleys of Southern New South Wales: problems and potential. *Australian Geographer*, **32**(2), 201–220.

Wray, R.A.L., Young, R.W. and Price, D.M., 1993. Cainozoic heritage in the modern landscape near Bungonia, Southern New South Wales. *Australian Geographer*, **24**, 45–61.

Young, A.R.M., 1986. Quaternary sedimentation on the Woronora Plateau and its implications for climate change. *Australian Geographer*, **17**, 1–5.

Young, R.W., 1976. Radio-carbon dates from alluvium in the Illawarra district, NSW. *Search*, **7**, 34, 35.

Young, R.W., 1983. The tempo of geomorphological change – evidence from Southeastern Australia. *Journal of Geology*, **91**, 221–230.

Young, R.W., Nanson, G.C. and Bryant, E.A., 1986. Alluvial chronology for coastal New South Wales: climatic control or random erosional events? *Search*, **17**, 270–272.

Young, R.W., Nanson, G.C. and Jones, B.G., 1987. Weathering of late Pleistocene alluvium under a humid temperate climate: Cranebrook Terrace, Southeastern Australia. *Catena*, **14**, 469–484.

Young, R.W., Short, S.A., Price, D.M., Bryant, E.A., Nanson, G.C., Gardiner, B.H. and Wray, R.A.L., 1994. Feruginous weathering under cool temperate climates during the late Pleistocene in Southeastern Australia. *Zeitshrift fur Geomorphologie*, **38**, 45–57.

Young, R.W., White, K.L. and Price, D.M., 1996. Fluvial deposition on the Shoalhaven deltaic plain, southern New South Wales. *Australian Geographer*, **27**, 215–234.

Young, R.W., Young, A.R.M., Price, D.M. and Wray, R.A.L., 2002. Geomorphology of the Namoi alluvial plain, Northwestern New South Wales. *Australian Journal of Earth Sciences*, **49**, 509–523.

PART 4   RECENT PROGRESS INTERPRETING EVIDENCE OF ENVIRONMENTAL CHANGE

15 Data Sharing in Palaeohydrology: Changing Perspectives

T. OGUCHI,[1] J. BRANSON[2] AND M.J. CLARK[2]
[1]*The University of Tokyo, Tokyo, Japan*
[2]*University of Southampton, Southampton, UK*

1 A BACKGROUND TO THE STATUS OF PALAEOHYDROLOGICAL DATA

Data are the fuel of palaeoenvironmental research – a resource that should ideally be managed in a way that provides the entire global change research community with the tools to find, access and manipulate the data needed for a particular investigation (Branson et al., 1995; 1996a; 1996b). It has been acknowledged previously that fewer sets of palaeodata are widely available in easy-to-use formats than might be expected (IGBP, 1992). There are, of course, some notable exceptions to this generalisation, including the IGBP PAGES initiative (Chapter 3, PAGES, 2002) that has assembled an impressive range of substantive collections.

In seeking to understand the differentiation between scientific fields that support effective palaeodata assembly and those that do not, one dimension of significance may be the extent to which technical issues such as data format, and indeed, the technology of data handling itself, intervene to determine the success of data sharing (e.g. Loudon, 2000). Thus, it could be argued that those fields that yield data with a good degree of consistency in data model and data format/structure (for example, tree rings or pollen), may raise fewer barriers to data aggregation than a field such as palaeohydrology in which the data model ranges across surrogates as diverse as channel dimensions, slackwater deposits, sedimentological sequence or palaeoflood levels. At a simple level, this technological constraint was recognised by Branson and Clark (1995) and Branson et al. (1996b) with respect to palaeohydrological data. Clearly, while paper-based archives are certainly a feasible basis for data sharing, they generate significant hurdles to effective use (input and output) compared with their digital counterparts. In their turn, digital data have been seen to raise technical problems when dealing with large data volumes characterised by different structures and formats. However, it would be a mistake to suggest that the balance of advantage or disadvantage in participating in a data-sharing community rests entirely on technical issues.

Clark (2003) discusses the value of data integration as the basis for comparison, identification of process drivers, change detection, hypothesis testing, impact evalua-tion and development of regional and global typologies or models. In principle, the case for data sharing is strong and broad ranging to the extent that it could be naively argued that all partners in a process of data assembly and dissemination would be (and would see themselves to be) net beneficiaries. The reality is very different. A decade

Palaeohydrology: Understanding Global Change. Edited by K.J. Gregory and G. Benito
© 2003 John Wiley & Sons, Ltd ISBN: 0-470-84739-5

after its launch, the GLOCOPH database remains as much a source of potential as of realised value and it remains vital that the drive to strengthen this position should continue. However, if this is to be successful, then the initiative will have to evolve in nature and in its promotion, since the balance of constraints and opportunities has changed substantially over the past 10 years; it has proved difficult to obtain sufficient information consistently from research investigations in order to complete the database to the originally intended extent.

The many applications of shared/integrated data proposed by Clark (2003) offer substantial potential value to the palaeohydrologist. These are, perhaps, most apparent within the research and academic communities, but it is emphasised that palaeohydrological data have substantive application outside the sphere of academic research (see Chapters 18, 21, 22). Notwithstanding this slowly developing acceptance of the value of assembling datasets within a given research or application area of interest, it is important to return to a consideration of the extent to which technological developments over the last decade have altered the balance and nature of the advantages and disadvantages of this type of exercise.

Of particular potential significance is the way in which the technical barriers to data distribution have been substantially lowered by developments in Web (Internet/Intranet) delivery such that it is now significantly more realistic to contemplate open access to data of varied types (text, numeric, image, spatial) and formats. In part, this facility has emerged as a result of software developments, including the increasing availability of open-source Web and Intranet Geographical Information Systems (GIS). However, as such technical barriers are lowered, the focus of attention shifts to the intellectual challenges of providing and using open access to data (e.g. Green and Bossomaier, 2001). In the extreme scenario, scientists who have previously been able to hide their resistance to data sharing behind a screen of technical difficulties are now exposed to the reality that the main barrier is often attitudinal. Thus, a decade after the initiation of the GLOCOPH-database initiative it is still possible to argue that the opportunities generated by data sharing remain constrained by significant problems.

We are, therefore, now faced with an opportunity to move forward into effective palaeohydrological data aggregation, freed from many of the previous technological barriers and at the same time empowered and encouraged by the expanding opportunities of Web delivery. In order to take this step, however, we need to review the scope of data sharing, both in its broader context and in the case of GLOCOPH. This will permit a focus on the issues raised by Web empowerment, and thus lead to a re-evaluation of the balance of advantage and disadvantage, as a future policy and strategy emerges for the GLOCOPH database and its spin-off activities.

1.1 Issues in Palaeoenvironmental Data Sharing

A number of international projects have begun to share datasets. These initiatives have resulted either from joint research programmes, or from studies by individual research programmes or datasets that have been "rescued" from archives. The work of programmes such as PAGES (a core programme of the International Geosphere Biosphere Project) is particularly significant (Chapter 3). This initiative has made a large number of datasets available through the World Data Center A for Palaeoclimatology, covering topics such as pollen, ice core, marine sediment, coral and plant macro fossils. The PAGES system receives over 1,000 data requests a month and it is possible to add datasets, metadata and bibliographic data to the archive from the Web site (PAGES, 2002).

Within the sphere of palaeohydrological research there are many benefits that could arise from sharing data. For example, by bringing together data from a number of researchers it would be possible to extend a single researcher's survey centred on one site to create a regional picture of hydrological change. Similarly, transects across an area could be developed, or studies in different parts of a catchment could be combined to extend the hydrological history of a river basin. The creation of datasets with a larger spatial or temporal extent than those achieved by the individual could also make the data more useful for applied purposes such as flood-impact assessment.

2 THE GLOCOPH DATABASE

The point that has been made above is that collation of the full range of research records would be highly beneficial for palaeohydrologists, but in practice compilation of the requisite data is complicated. Data would need to be acquired from all continents for a range of appropriate environmental parameters in a standard format, and this in turn requires data acquisition across the earth and environmental science disciplines. The GLOCOPH database was established in 1992 to address this challenge, with the aim of providing a data and information resource for the researchers and managers studying the palaeohydrology of the last 20,000 years. The original service was developed using the ORACLE-relational database management system. However, because of software licensing issues, the database could not be made available interactively over the Internet, so the bibliography was made available as a static Web page and information about datasets was sent to individuals on request. More recent advances in database, Internet and GIS technology have made it possible to redesign and develop the database such that it can be accessed and queried directly on the Web. The majority of the information held within the database has a spatial reference (being related to a country, river or region), and the Web site therefore offers the potential for spatial searches and queries. Related functionality is provided by the PHEIMS database (PalaeoHydrology/Environment Internet Map Server), constructed at the Center for Spatial Information Science, the University of Tokyo, Japan, which is discussed in more detail below.

The original objective of the GLOCOPH database was to provide a central data and bibliographic repository that would ensure data security and accessibility together with long-term data integrity, which, at that stage suggested a need for central data management. The priorities have since evolved such that the GLOCOPH system provision now aims to be not only a portal to a distributed palaeohydrological network but also a portal to interact with a wider palaeoenvironmental network and to be a resource for ongoing research programmes. Despite the technical limitations facing a project with limited resources, this is the development priority for the immediate future needed in order to deliver significant benefits to the palaeohydrology community. A dedicated bibliographic database was established at the beginning of the GLOCOPH research programme, which has subsequently been elaborated and continues to be extended. It includes several thousand references to palaeohydrological research investigations. The motivation for data sharing in this context focuses on familiar arguments concerning the value of undertaking comparisons between regions, of building models that extend beyond a single site or country, and of extending the time range that can be understood through the study of a single site or region. The published references represent a significant source of palaeohydrological data, and the system itself provides access to detailed results from type references that incorporate major conclusions.

The technical and functional development of the GLOCOPH Web site is discussed briefly below.

The GLOCOPH database thus potentially integrates primary and published data on environmental parameters indicative of past fluvial and hydrological processes and their drivers (such as the value of palaeoflood discharge and the techniques applied to palaeohydrological reconstruction), with the aim of

- making possible the integration of data from different sites;
- developing new indicators or information types that draw on multiple datasets (and possibly multiple disciplines);
- starting the process of standardising datasets so that they are independent of the initial analytical structures or models of their authors;
- helping to identify research gaps and opportunities, and thus fostering joint (and possibly international) research efforts.

Although the GLOCOPH project was, from the outset, established within the INQUA (International Union for Quaternary Research) professional infrastructure, it is recognised that, in the medium term, the data should be transferred to one of the major professional archiving networks. This will be a necessary evolution from the current development phase to a long-term operational phase of the database, thus maximising the maintenance of the data and the ease of access. This is one reason for the current state of development of the database, which has evolved simultaneously with the production of this volume.

Although GLOCOPH is based on published data, the task of data assembly remains complex and time consuming if a standard data model is to be achieved within a standard data structure. In the interim, therefore, it has been recognised that there is a role for another type of database with a larger number of more general data. For instance, researchers of palaeohydrology may like to access a list of existing publications that examine the Quaternary palaeoenvironment in a specific region of their interest. The database for such purposes should contain the basic information from a large number of publications, although detailed reference data for each site may be unnecessary. To meet this need, the INQUA/GLOCOPH group developed a separate Internet-accessible database called PHEIMS at the Center for Spatial Information Science, University of Tokyo, Japan. In addition, specific databases have been constructed for particular research areas of GLOCOPH, including the palaeoflood database compiled by Ely and Hirschboeck (2002) and for specific regions such as the regional database established by Diez-Herrero et al. (1998). Such thematic and regional databases have been associated with subgroups of GLOCOPH and need to be interrelated prior to integration into global databases.

3   THE PHEIMS (PALAEOHYDROLOGY/ENVIRONMENT INTERNET MAP SERVER) SYSTEM

(http://www.csis.u-tokyo.ac.jp/~oguchi/pheims.html)

The recent rapid development of Internet-related technology and GIS has facilitated the creation of Web-based user-friendly databases. In particular, the IMS technology has provided a new way of presenting geographical data (e.g. Plewe, 1997; Harder, 1998), enabling the concurrent presentation of text information and map images via the Internet. Many of the existing IMS systems have commercial purposes, but IMS is also useful for academic purposes especially in the fields of geography and geoscience (e.g. Oguchi *et al.*, 2000; 2001; 2002; Hare and Tanaka, 2001). The PHEIMS System, developed at the University of Tokyo, uses IMS technology of ArcView GIS (ESRI, USA) to facilitate on-line browsing and map visualisation of information stored in the database of palaeohydrological and palaeoenvironmental literature (Figure 15.1).

PHEIMS comprises information about publications that discuss some 6,000 sites and areas around the world (as of October 2002), and further data will be added on an ongoing basis. PHEIMS shows the location of the study sites on a digital map, so theoretical, conceptual and/or technical papers without a specific research site or area are not included. The location of an individual study is displayed either as a point or a polygon depending on the geographical extent of the research. A number of map layers are provided to give context to the study-site locations, including the country borders, major drainage networks, major lakes and the ETOPO5 Digital Elevation Model (grid interval of five arc minutes) provided by the US National Geophysical Data Center.

The main topic of each publication is classified into 18 themes (*Theme*) (Table 15.1) that are grouped into six types (*ThemeCode*). Although papers are often related to more than one of the 18 themes, PHEIMS stores just the primary theme for each paper. *Keywords* include those specified by the authors of papers, except words showing the area and the period examined by the study, which are included in *Site* and *Period*. If papers compile or reanalyse data published elsewhere, the data sources are indicated under note. The constructed data list (Table 15.2) was converted into a set of GIS point data with an attribute table. The points on the map are dense in North America and Europe, reflecting the fact that most data uploaded so far were taken from major English-language journals.

A suite of tools is available to help users access the information by changing the area of the map, the map components and the contents of the data table. Users interested in

Table 15.1 Classification of main topics of publications within the PHEIMS system

Theme group	Code	Theme	Example
Palaeohydrology	1	Flood	Palaeoflood reconstruction
		Fluvial	River-terrace development
		Glacier	Past extent of alpine glaciers
		Hydrology	Lake-level changes
Earth surface processes and materials	2	Aeolian	Reconstruction of past dune activity
		Peat land	Accumulation rate of peat
		Periglacial	Structure of ice-wedge casts
		Sedimentology	Sedimentological analysis of cave deposits
		Slope processes	Landslides induced by climatic change
		Soil	Change in humus contents in soil profiles
Palaeooceanology	3	Oceanology	Reconstruction of past ocean current
		Sea-level	Reconstruction of sea-level change
Palaeobiology	4	Archaeology	Human activities in relation to palaeoclimate
		Vegetation	Past vegetation change
		Zoology	Fossil of animals and insects
Palaeoclimatology	5	Atmosphere	Past methane content in atmosphere
		Climate	Temperature reconstruction using a diary
Chronology	6	Chronology	Application of an isotope-dating method

Table 15.2   Data list within the PHEIMS information on publications

Abbreviation	Content	Data type
ID	Identification code of the data	Text and number
Title	Title of the paper	Text
Authors	Author(s) of the paper	Text
Reference	Name of the journal or the book, volume number, year of publication	Text and number
Pages	Page number	Number
Site	Name of the site	Text
Coverage	Type of the extent of the study area	Text (*point* or *area*)
X	Longitude of the data point or the centre of the study area	Number (in decimal degrees)
Y	Latitude of the data point or the centre of the study area	Number (in decimal degrees)
Xmin	Longitude of the lower left corner of the study area	Number (in decimal degrees); not applicable to the pint coverage type
Ymin	Latitude of the lower left corner of the study area	Number (in decimal degrees); not applicable to the pint coverage type
Xmax	Longitude of the upper right corner of the study area	Number (in decimal degrees); not applicable to the pint coverage type
Ymax	Latitude of the upper right corner of the study area	Number (in decimal degrees); not applicable to the pint coverage type
Theme	Main theme of the paper[a]	Text
ThemeCode	Classification code of the theme[a]	Number
Keywords	Keywords	Text
Period	Geological/absolute age of the topic	Text and number
Note	Information about data sources, etc.	Text and number

[a] see Table 15.1.

a specific region can obtain a larger-scale map using the zoom-in tool (Figure 15.2). When a user clicks the show-attribute tool and subsequently clicks on the location of a data point on the map, an attribute table for the site or area is displayed, and details of the literature can be printed. A search tool is available, which enables users to find sites using a word such as the name of the region (Figure 15.3).

4   GLOCOPH AND THE EMERGING ROLE OF THE WEB

(http://www.geodata.soton.ac.uk/glocoph)

The GLOCOPH database itself, on the other hand, aims to facilitate analysis of environmental change and also aims to enable collaboration between researchers and with other commissions of INQUA and other organisations through direct data presentation. It is, therefore, in the course of further development in relation to an ICSU-supported project on *Past Hydrological Events related to understanding Global Change*. In addition to analysis based upon collating results from several areas, there should also be the potential to analyse results within a single area, using the GIS to

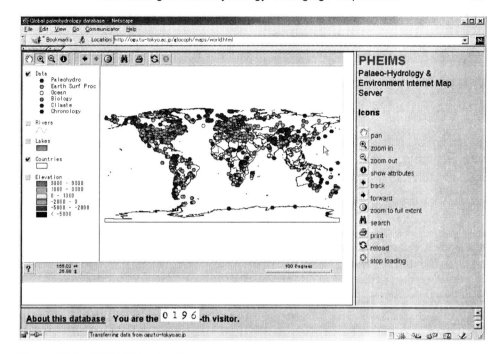

Figure 15.1   PHEIMS tool options

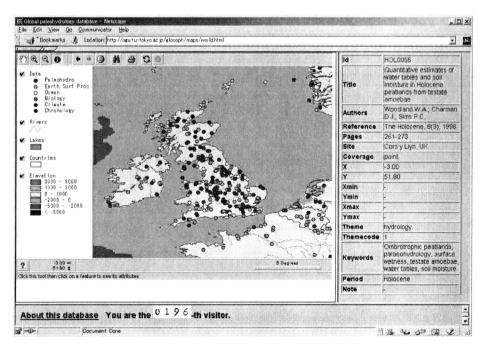

Figure 15.2   Larger-scale map with all the components, showing the area in and around the United Kingdom and the attribute table for a publication concerning palaeohydrology at Cors y Llyn, mid-Wales

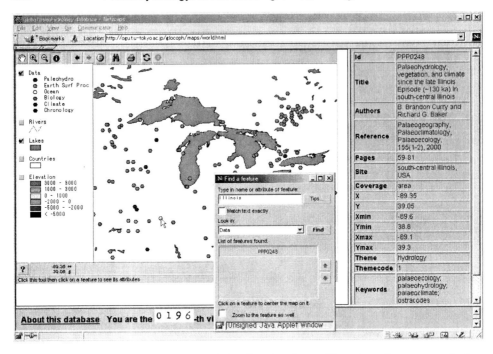

Figure 15.3 Map of the area around the Great Lakes with reduced components (no information about rivers, country borders, and elevation) and the attribute table for a site in south-central Illinois, found by a search query

obtain palaeohydrological information from selected areas and modelling hydrological responses to vegetation change within the area.

In order to facilitate information exchange on global change, a Web site has been established that includes an updated list of current researchers and projects in palaeo-hydological research, together with the literature database that can be accessed on the basis of author, area or type of approach. In making the database accessible and as stimulating as possible, a component has been designed experimentally to trial an adjunct to the database in order to obtain replies to a questionnaire from Web site users. At a time of growing widespread interest in global change, this will provide ongoing information about knowledge and understanding of the role of palaeohydrology in relation to global change.

The database continues to be developed and the Web site is being further amplified during the course of production of this book. Therefore, the continued developments extending this chapter are included on the Web site and are being further updated with particular reference to past hydrological events related to understanding of global change supported by ICSU. The ICSU research grant (for 2003) was awarded to a consortium led by GLOCOPH and composed of five other international organisa-tions: the Water Sustainability Study Commission of IGU (Dr J.A.A. Jones); Land Use and Climate Impacts on Fluvial Systems Group (LUCIFS), in PAGES and IGBP (Professor D.E. Walling); Geomorphic Challenges for the 21st Century Commission of IGU (Professor A.J. Parsons); International Commission on Continental Erosion (ICCE) of International Association of Hydrological Sciences (IAHS) (Professor Woj-ciech Froehlich); Fluvial Archives Group (FLAG) associated with INQUA and IGU (Dr Darrel Maddy). In researching "Past hydrological events related to understanding of Global Change" the group intends that the database necessary for this collaborative research activity can be developed from, and integrated with, the GLOCOPH database.

The move towards Web delivery of database and spatial services, together with the interaction that becomes possible with users, represent a major advance in the development of a true data community for palaeohydrology. While a degree of central management remains, in the form of data and metadata standards, there is an emerging sense that the information system belongs to the users rather than to the project. It is to be hoped that this will reduce the residual barriers to data sharing and will encourage individuals to deposit datasets with GLOCOPH, so that the system can become a repository rather than a point of access for published information.

5 REMAINING AND EMERGENT PROBLEMS

GLOCOPH has made great progress over the last 10 years, yet there remains much potential to be exploited. The challenge lies partly in the reserve still shown by some scientists towards the concept of sharing research data on open access. In addition, technical problems remain a significant hurdle, mainly because of the unusually diverse nature of the indicators and surrogates employed by palaeohydrologists to represent the hydrology of the past. Where the same attributes are monitored against standard scales, but in different locations (as is done in the CALM network for frozen ground phenomena, for example, Brown *et al.*, 2002), it is relatively straightforward to provide standard metadata templates for receiving data and it is easy to agglomerate datasets for distribution. In a subject such as palaeohydrology, however, where individual scientists are undertaking research on different parameters for different purposes, standardisation remains elusive, and perhaps unachievable in principle as well as in practice. In this sense, perhaps the greatest encouragement comes from the extent to which information science now recognises data diversity and uncertainty as inherent rather than aberrant properties. The continued convergence of science and information technology offers major prospects for an enterprise such as GLOCOPH.

6 THE FUTURE FOR THE GLOCOPH DATABASE SERVICE

The value of an information system lies jointly in its content and its application. As the content of GLOCOPH continues to grow, the focus and the opportunity move progressively towards the identification of application priorities. In particular, it would seem important to return now to the incentive that lay behind the initiation of GLOCOPH a decade ago, namely, the potential to undertake integrative analyses based on the assembled data. The spatial framework of such analyses could be the regional structure adopted in this volume, or it could be on the basis of spatial boundaries specified for an individual research project.

The new technologies are already releasing new potential, as is shown by recent development of both GLOCOPH and PHEIMS, but there are many opportunities yet to be fully exploited. XML (Extensible Markup Language) and open-source GIS offer further scope to broaden access to, and effective sharing of, information by removing previous technical and financial barriers. XML provides a standard and simple way of representing data (including text and images) that can be shared across a variety of languages, hardware, operating systems and software applications (Houlding, 2001). This immediately increases the scope and accessibility offered by conventional relational databases, and accords well with the "open" philosophy of a data-sharing community such as that of palaeohydrologists. Open-source programs (e.g. Neteler and Mitasova, 2002) move further along this path by providing open and free access to the source code so that functions or complete applications can be customised to provide services within initiatives such as GLOCOPH and PHEIMS. Since the open-source license allows other code writers free access allows them to build upon the code, the resulting application is freely usable by a complete community internationally. Such modification and redistribution of the source code for a piece of software (whether GIS or database) not only eases the financial burden of data access for many users but is also believed to improve the software and increase its pace of development since the re-coders and users, in effect, become the software developers. In essence, technical devices such as these contribute to the more general technical quest for data fusion and, as described by NOAA (2002), "the seamless integration of data from disparate sources. The data have been integrated across data collection 'platforms' and geographic boundaries, and blended thematically, so that the differences in resolution and coverage, treatment of a theme, character and artifacts of data collection methods are eliminated". This is a core target for information scientists as well as for palaeohydrologists. Again, the philosophy of sharing and of community exploitation of value, which is represented by open-source and data-fusion initiatives, mirrors a similar ethos of data sharing and joint information analysis and exploitation within the palaeohydrological community. This is where the technological future of GLOCOPH lies.

At the end of a decade of GLOCOPH, substantial advances in information dissemination have already provided a real service to the palaeohydrological community. Awareness of data availability has increased immeasurably, and palaeohydrologists now have a firm basis for contextualising their own work and for identifying priorities for future work. But above all, the last 10 years have seen a very significant lowering of the technological barriers to data sharing and research cooperation. Through this progress, the spotlight has moved back to individual scientists, who can no longer argue that failure to participate in data sharing reflects technological difficulties. The way is now open to create genuinely open (and in a large measure, free) access to scientific data, and this promises to empower palaeohydrologists with the ability to focus on their science rather than on the frustrations previously inherent in any attempt to integrate data and generate information.

ACKNOWLEDGEMENT

The construction of PHEIMS is supported by the Grant-In-Aid for Scientific Research, Promotion of Publication of Scientific Research Results (Database), from the Japan Society for the Promotion of Science. Special thanks are due to Miss Minako Nishikata and Mr. Yuichi Hayakawa, the chiefs of the PHEIMS construction staff at the University of Tokyo. We also acknowledge Dr. Mathias Deutsch, Universitat Erfurt, and Dr. Edgardo Latrubesse, IESA/UFG, for donating publications for PHEIMS. Development of the GLOCOPH database is supported by an INQUA grant that the authors gratefully acknowledge.

REFERENCES

Branson, J., Brown, A.G. and Gregory, K.J. (eds), 1996a. *Global Continental Changes: The Context of Palaeohydrology*. Geological Society Special Publication No. 115, The Geological Society, London.

Branson, J. and Clark, M.J., 1995. Global geocryological database: suggestions for data structure. In *International Permafrost Association Workshop on Permafrost Data Rescue and Analysis*. World Data Center A for Snow and Ice, Oslo, Norway, 1994, 21–26.

Branson, J., Clark, M.J. and Gregory, K.J., 1995. A database for global continental palaeohydrology: technology or scientific creativity? In V.R. Baker (ed.), *Global Continental Palaeohydrology*. John Wiley & Sons, Chichester, 303–321.

Branson, J., Gregory, K.J. and Clark, M.J., 1996b. Issues in scientific co-operation on information sharing: the case of palaeohydrology. In J. Branson, A.G. Brown and K.J. Gregory (eds), *Global Continental Changes: The Context of Palaeohydrology*. Geological Society Special Publication No. 115, The Geological Society, London, 235–249.

Brown, J., Hinkel, K.M., Nelson, F. and Maximov, I., 2002. *Circumpolar Active Layer Monitoring (CALM)*. Web site: http://www.geography.uc.edu/~kenhinke/CALM/.

Clark, M.J., 2003, International cooperation for international data acquisition and use. *Encyclopaedia of Life Support Systems*. UNESCO/EOLSS Publishers Company Limited; in press.

Diez-Herrero, A., Benito, G. and Lain-Huerta, L., 1998. Regional palaeoflood databases applied to flood hazards and palaeoclimate analysis. In G. Benito, V.R. Baker and K.J. Gregory (eds), *Palaeohydrology and Environmental Change*. Wiley, Chichester, 335–347.

Ely, L. and Hirschboeck, K., 2002. *Palaeoflood Database*. Compiled at the Laboratory of Tree Ring Research, University of Arizona, Tucson, AZ.

Green, D.G. and Bossomaier, T.R.J., 2001. *Online GIS and Spatial Metadata*. Taylor & Francis, London.

Harder, C., 1998. *Serving Maps on the Internet: Geographical Information on the World Wide Web*. ESRI Press, Redlands.

Hare, T.M. and Tanaka, K.L., 2001. Planetary interactive GIS-on-the-web analyzable database. Proceedings of the 20th International Cartographic Congress, Beijing, China (CD-ROM).

Houlding, S.W., 2001. XML – an opportunity for <meaningful> data standards in the geosciences. *Computers & Geosciences*, **27**, 839–849.

IGBP, 1992. Global Change Report No. 29, 71.

Loudon, T.V., 2000. Geoscience after IT Part D. Familiarization with IT applications to support the workgroup. *Computers & Geosciences*, **26**, A21–A29.

Neteler, M. and Mitasova, H., 2002. *Open Source GIS: A Grass GIS Approach*. Kluwer, Boston/Dordrecht.

NOAA, 2002. Web document: www.ngdc.noaa.gov/seg/tools/gis/fusion.shtml.

Oguchi, T., Hara, M., Saito, K., Grossman, M. and Yamamoto, S., 2002. An online database of Polish towns and historical landscapes using an Internet map server. *Geographia Polonica*, **75**, 111–117.

Oguchi, T., Saito, K., Hara, M., Kadomura, H. and Lin, Z., 2000. Alluvial fan database: presenting geographic information using an Internet map server. *Journal of Geography (Tokyo)*, **109**, 120–125 (in Japanese with English abstract).

Oguchi, T., Saito, K., Kadomura, H. and Aoki, H., 2001. Presenting geomorphological data for Japan using an Internet map server. Abstracts, Fifth International Geomorphology Conference, Tokyo, Japan, C-173.

PAGES, 2002. Web document: www.pages.unibe.ch.

Plewe, B., 1997. *GIS Online: Information, Retrieval, Mapping, and the Internet*. Onward Press, Santa Fe.

16 Fluvial Morphology and Sediments: Archives of Past Fluvial System Response to Global Change

D. MADDY,[1] D.G. PASSMORE[1] AND S. LEWIS[2]
[1]*University of Newcastle, Newcastle, UK*
[2]*University of London, London, UK*

1 INTRODUCTION

It is well established that rivers respond to a number of stimuli or forcing functions, which act over a variety of temporal and spatial scales (e.g. Chapters 5–14, Schumm and Lichty, 1965). In the long term (i.e. 10^3–10^6 years), geology and climate may play the dominant role in fluvial system behaviour, generating basin-wide sediment-landform assemblages. Landforms produced at these timescales such as river terrace flights in uplifting areas are often large and thus have good long-term preservation potential and have provided the raw data for much geological interpretation of river development. Changes in geology and in particular, climate, exert control over sediment and water supply and thus often promote system adjustment. Runoff and sediment supply, however, also vary over the short term (i.e. 10^1–10^2 years) as a result of sediment exhaustion and starvation, human activities and changes in flood frequency and magnitude. The relatively recent scientific investigation of modern river processes has underlined the immediate importance of these short-term processes but has often failed to appreciate that modern processes are frequently conditioned by long-term behaviour, that is, the system has a "memory"; such studies should not ignore the geological data that record the system history. This shortcoming is potentially significant, especially as we try to "predict" future system behaviour.

This chapter is concerned only with long-term (i.e. 10^3–10^6 years) river behaviour manifest in the sedimentary and landform records. The major controls on fluvial system behaviour over this timescale often reflect major regional shifts in climate and thus have a global trigger. Over the past two million years, rivers have been constantly forced to adjust to changing climatic conditions, many river basins have been repeatedly glaciated or seen dramatic changes in their sediment budgets resulting from slope destabilisation/stabilisation as vegetation cover contracts and expands (Vandenberghe, 1995). Superimposed upon these changes are tectonic factors that promote sediment budget changes such as uplift and earthquake triggered landslides. The picture is complicated further in some cases by factors inherited from earlier in geologic time in the way that some rivers may still be adjusting to Tertiary uplift.

Palaeohydrology: Understanding Global Change. Edited by K.J. Gregory and G. Benito
© 2003 John Wiley & Sons, Ltd ISBN: 0-470-84739-5

This chapter will concentrate on the most widespread of fluvial sediment-landform assemblages, the river terrace, in particular depositional terraces that archive substantial former floodplain fragments and extend over large proportions of the river basin.

2   TERRACE DEVELOPMENT

The formation of large-scale river terraces require interpretation not only of the aggradation phase and the sedimentary stack but also as to why the floodplain was abandoned as a terrace. At the landscape scale, we can divide the river terrace formation process into aggradational and incision phases.

2.1   Aggradation Phase(s)

Floodplains comprise horizontally and vertically stacked fluvial (channel and extra-channel), colluvial (valley slope derived) and often aeolian sediments. Each of these comprises "architectural" elements of the floodplain (cf. Miall, 1996). Field description of these elements requires extensive exposure for accurate determination; alternatively, they may be defined from borehole reconstruction of gross facies changes, or from geophysical investigation such as ground penetrating radar. Detailed descriptions of extensive exposures allow consideration of processes of interest to both the geologist and process geomorphologist.

Fluvial sediments are often subdivided into channel and extra-channel types. Within-channel sediments, transported via traction currents, often manifest as characteristic bedform assemblages. Bedforms vary depending on flow conditions and sediment type (Allen, 1982; Ashley, 1990; Collinson, 1986), thus providing some means of retrodiction of palaeoflows. Similarly, extra-channel sediments and overbank deposits often display recognisable sedimentary signatures. Combinations of these structures, together with their bounding surfaces, help define the familiar river planform types (Bridge, 1985) of straight, meandering (e.g. Bluck, 1971), braided (Best and Bristow, 1993) and anastomosing (Smith, 1983), each of which may be associated with characteristic environmental conditions.

During the construction of a floodplain, relatively short-term changes in sediment budget, resulting from either intrinsic or extrinsic triggers, can induce both planform change and spatially variable aggradation/degradation of the river bed. These changes often lead to a complex cut-and-fill stratigraphy. However, prediction of the nature of response is problematic. Fluvial systems need only respond if thresholds for change are exceeded (Schumm, 1979); however, thresholds may often represent the combined effects of changes in state of any number of variables so that different permutations may or may not result in thresholds being passed because thresholds themselves may be dynamic.

Extrinsic change may, for example, be triggered by regional climate change. High-frequency (sub-Milankovitch) climate changes have been recorded in the Greenland ice cores (Dansgaard *et al.*, 1993) and North Atlantic Ocean records (Bond *et al.*, 1993; 1997) but the precise nature of the fluvial system response in NW Europe (Chapter 8) to these climatic changes is variable with system response being widespread, for example, in the Younger Dryas, but with little obvious response to similar magnitude changes at earlier times. Intrinsic fluvial system changes such as short-term sediment starvation can result in bed degradation, or sudden large-scale debris flows onto the valley floor can promote bed aggradation. Such localised adjustments are difficult to resolve in the stratigraphic record but the identification of simultaneous system response across large areas of a basin may point to a more regional trigger.

2.2 Incision Phase(s)

Basin-wide fluvial incision events can generally be considered to reflect response to changes in external variables including climate, tectonics or base level, although significant reach-scale incision, in response to localised catchment changes, may have been superimposed upon basin-wide events. Significant localised incision leading to localised terrace formation can, for example, be caused by the shortening of rivers through diversionary events such as river capture and glacial diversion (Bishop, 1995). Additional localised incision can result from fault movement (Krzyszkowski *et al.*, 2000; Krzyszkowski and Stachura, 1998) and glacio-isostatic effects (Maddy and Bridgland, 2001).

Climate control on Quaternary fluvial incision is exerted through its influence on the sediment/discharge ratio. Incision is promoted when low sediment availability is concurrent with high discharge. Such conditions occurred for example, at the beginning of warm events when slope stabilisation by vegetation reduced sediment supply (Maddy *et al.*, 2001). Merritts *et al.* (1994) suggest that the underlying upstream controls on the sediment-discharge ratio are most likely to be climate-influenced. Incision produced upstream may migrate downstream through the process of discharge-controlled or "kinetic incision" (Leeder and Stewart, 1996).

Basin-wide incision can also result from crustal uplift (e.g. Bull and Knuepfer, 1987; Bull, 1991). Crustal uplift can result from a number of mechanisms including direct plate-tectonic stress at, or near plate margins; intraplate stress (e.g. Cloetingh *et al.*, 1990); or glacio-isostatic adjustment (e.g. Lambeck, 1993; 1995). Indeed, in NW Europe, Neogene regional uplift has been demonstrated as an important factor in valley development illustrated by the Maas (Van den Berg, 1996), the Thames (Maddy, 1997) and the rivers draining the Sudetic Foreland (Krzyszkowski *et al.*, 2000).

Finally, base-level lowering will lead to incision in the lower reaches of valleys where the lowering is not balanced by lengthening of the rivers' course onto the continental shelf. Such incision could, given enough time and a readily erodible substrate, lead to basin-wide terrace formation as the downstream incision is transmitted upstream through knickpoint recession (Begin *et al.*, 1981). Knickpoint recession may lead to variability in the absolute magnitude of incision as the knickpoint attenuates upstream, or the process may be halted before complete recession has occurred leading to terrace formation only downstream of the maximum point of recession. Although, for example, glacial low sea-level stands have been demonstrated to result in downstream incision of the Colorado River, Texas (Blum, 1993), the Mississippi (Tornqvist and Blum, 1998) and the Rhine/Maas (Tornqvist, 1998), both Schumm (1993) and Leopold and Bull (1979) have suggested that the effects of these base-level changes will tend not to be propagated far upstream, particularly in large rivers that accommodate adjustment via other channel variables.

The terrace archive, therefore, has a large number of attributes that potentially can yield important information concerning past fluvial history, providing some insight into fluvial system response to environmental change. Below, we consider first the utilisation of the gross sediment body geometry in order to reconstruct changing patterns of sediment storage over the past 14 ka in the glaciated South Tyne valley in northern England. Second, we attempt to look deeper into causes of within-terrace sedimentary architecture changes within the Northmoor Member (which spans the last glacial–interglacial cycle) of the Upper Thames valley, a valley not glaciated for several hundred thousand years.

3   LATEGLACIAL AND HOLOCENE TERRACE FORMATION IN THE SOUTH TYNE VALLEY, NORTHERN ENGLAND

Terraced Lateglacial and Holocene fluvial sediment-landform assemblages are particularly well developed in upland river basins of northern and western Britain that were wholly or partly overrun by Late Devensian Dimlington Stadial ice. Holocene fluvial activity in these environments has been the focus of extensive study over the past three decades and it has been shown that valley floors have responded in a sensitive and complex manner to Holocene climate and land-use change (e.g. Tipping, 1995; Macklin et al., 1992; Passmore, 1994; Passmore and Macklin, 1997; 2000; Rumsby and Macklin, 1996; Merrett and Macklin, 1999). These studies also acknowledge the legacy of glaciation to be manifested in the highly variable morphology of valley floors, with local constrictions formed by till and moraine deposits and tributary stream alluvial fans (see for example, Howard et al., 2000), and frequently thick deposits of glaciofluvial and glaciolacustrine sediments infilling valley floors and mantling valley sides. The tendency towards Holocene valley floor incision that is evident in many documented fluvial histories has also been attributed, at least in part, to the combination of isostatic uplift and markedly reduced rates of sediment supply following regional deglaciation (Macklin, 1999).

In general, however, while all these factors are widely recognised to have established the boundary conditions for subsequent Holocene fluvial activity, the chronology and pattern of fluvial system response to the Late Devensian glacial–interglacial transition in these areas remains poorly documented. Despite the increasing interest in Holocene valley floor development and environmental change in these environments (see Macklin, 1999), there is little information on the long-term ($10^2 – 10^4$ years) residence time of stored sediment and the rates and patterns of valley floor incision and sediment transfer (Merrett and Macklin, 1999) in deglaciated valleys.

Progress in this respect continues to be hindered by the fragmentary survival (and limited exposure) of valley fills, and the difficulties in establishing robust dating control for periods of terrace formation, although the potential of luminescence techniques has yet to be fully exploited in these contexts. However, reach-scale morphological sediment budgeting techniques have emerged as a useful means of assessing the role of fluvial sediment storage in determining the response of river systems to environmental change over millennial timescales (e.g. Huisink, 1999). Application of this approach to Lateglacial and Holocene fluvial sediment storage in a reach of the South Tyne, a gravel-bed river in northern England, has been used to quantify the volume of fluvial sediment stored and subsequently exported from the valley floor since deglaciation, and also the linkages between valley boundary conditions and long-term patterns of sediment transfer and storage (Passmore and Macklin, 2000; 2001).

This study has focussed on a 10.5-km reach of the South Tyne valley located where the northerly flowing river emerges from the North Pennine massif and turns east towards Tynedale and the North Sea (Figure 16.1). Valley morphology is characterised by alluvial basins up to 600 m wide that are separated by narrow reaches confined by steep till-mantled and bedrock bluffs (Figure 16.1). Six fluvial terraces (designated T1–6) that predate the mid-nineteenth century AD floodplain (T7) have been identified in the study reach (Figures 16.2 and 16.3). These lie between 2 to 8 m above the present

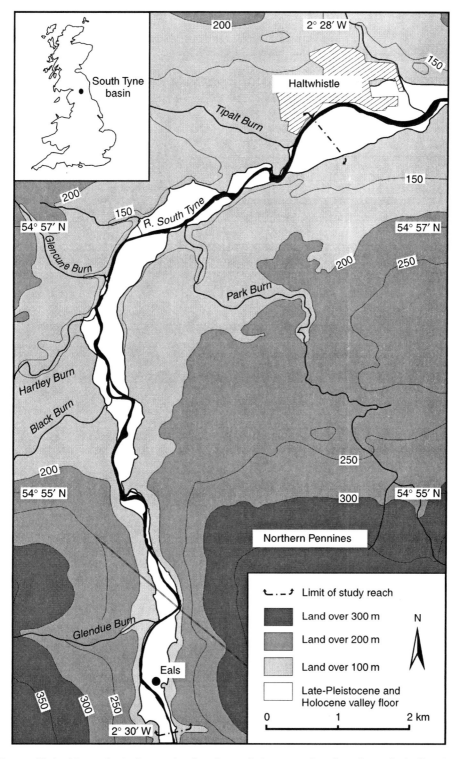

Figure 16.1 Map of study reach showing relief, area of valley floor (including late Pleistocene and Holocene fluvial terraces and alluvial fans), present river channel and location of South Tyne basin

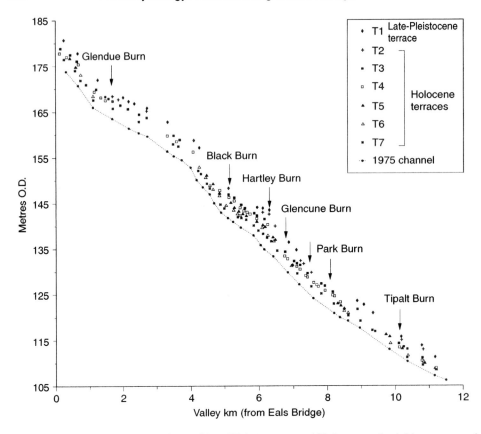

Figure 16.2 Longitudinal profiles of late Pleistocene and Holocene alluvial terraces and present channel bed in the study reach

channel bed with younger fluvial units inset below older surfaces (Figure 16.2). No direct dating controls are available for terraces T1 to T3, although T1 is believed to be equivalent to similarly high-elevation, gravel terraces recorded in river valleys elsewhere in northeast (Passmore and Macklin, 1997) and northwest England (e.g. Harvey, 1985; Tipping, 1995; Howard *et al.*, 2000) that are commonly interpreted as outwash deposits formed during deglaciation. Deposition of alluvial unit T3 is bracketed to the period 1275 BC to AD 1160 while subsequent cycles of floodplain construction and valley incision have been dated to ca AD 1160 to 1380 (T4), ca AD 1380 to 1660 (T5), ca AD 1660 to 1850 (T6) and from the mid-nineteenth century to the present (T7). The architecture of late Holocene alluvial fills in the study reach indicates that floodplain construction for the past 3,000 years has been dominated by coarse-grained migratory bed and bar assemblages, associated with episodic sub-reach scale channel braiding and bank erosion, with only thin veneers of overbank sediment.

Estimates of sediment storage and export from the study reach have been derived from a digitised database that combines the results of geomorphological mapping, topographic survey and geochronological analyses of Late Devensian and Holocene fluvial terraces. The study reach was subdivided into 0.5-km sub-reaches and the database interrogated for terrace areas and the area of the active channel zone for the period ca AD 1865 to present. This information was then merged with values of alluvial unit thickness determined either directly from lithostratigraphic analyses

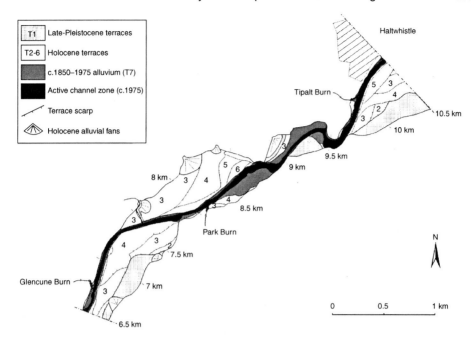

Figure 16.3 Geomorphological map of valley km 8–10.5 showing late Pleistocene and Holocene alluvial terraces, alluvial fans, the active valley floor between ca AD 1850–1975 and the active channel zone ca AD 1975

or, where unavailable, extrapolated from the nearest adjacent sub-reach. Estimates of sediment volume of each terrace and for the post-1860 active channel zone were calculated for each sub-reach.

Results show that the total present fluvial sediment storage in the study reach amounts to 7.8 million m^3 (Table 16.1; Figure 16.4) while net sediment export by fluvial processes since incision of T1 is estimated at nearly 8 million m^3 of sediment (Figure 16.5). The majority of remaining sediment is stored in relatively wide valley floors in the vicinity of valley km 0 to 2 and 5.5 to 8.5, which, respectively, accommodate ca 17% and ca 48% of the total store (Figure 16.4). Terrace T1 sediments account for 37% of total storage, and exceed 50% of storage in some sub-reaches, while only valley km 7.5 to 8.5 lack sediment of this age (Figure 16.4). Although these terraced alluvial units are at present largely decoupled from the modern active channel by inset Holocene fluvial terraces (Figures 16.3 and 16.5), it seems likely that, for much of the Holocene period, locally emplaced Late Devensian glaciofluvial deposits have contributed a significant proportion of fluvial sediment yields from the study reach. It is also likely that locally derived coarse sediment comprises a high proportion of Holocene fluvial deposits recorded in the study reach.

Valley floor morphology inherited from Pleistocene glaciation (and controlled also by bedrock geology) has also acted to condition rates and patterns of Holocene fluvial sediment transfer. Wider alluvial basins form the largest fluvial sediment stores in the study reach and have experienced the highest rates of Holocene sediment reworking and export. Conversely, relatively narrow valley reaches constricted by steep bedrock and till bluffs offer comparatively little accommodation space for fluvial sediment storage and have functioned as zones of sediment transfer over the Holocene (Figures 16.4 and 16.5; Passmore and Macklin, 2001).

Table 16.1 Volumetric estimates of fluvial sediment storage and % of total sub-reach sediment for late Pleistocene (T1), Holocene (T2–T6) and post-1850 (T7) fluvial terraces throughout the study reach

Reach (valley km)	Alluvial fans, Late Pleistocene (T1), Holocene (T2–T6) and post-1850 (T7) fluvial terraces: Sediment volume (% of total sub-reach alluvium)																	
	Alluvial fans		T1		T2		T3		T4		T5		T6		Sum (T1–T6)	Post-1850		Sum (all)	
0.5	9,300	3	78,900	26	26,100	9	124,400	41	21,800	7	9,100	3	0	0	269,600	34,600	12	304,200	4
1	36,200	7	252,900	47	0	0	106,700	20	34,900	7	42,000	8	15,600	3	488,300	45,700	9	534,000	7
1.5	1,200	0	57,000	17	0	0	189,100	55	67,100	20	400	0	0	0	314,800	28,800	8	343,600	4
2	10,100	6	29,000	18	72,200	44	26,200	16	6,300	4	0	0	0	0	143,800	20,700	13	164,500	2
2.5	30,900	8	203,800	52	17,500	4	97,200	25	5,200	1	0	0	0	0	354,600	33,900	9	388,500	5
3	15,500	9	87,500	51	34,600	20	3,400	2	0	0	0	0	0	0	141,000	31,100	20	172,100	2
3.5	0	0	58,600	51	0	0	16,100	14	4,700	4	0	0	0	0	79,400	35,300	31	114,700	1
4	0	0	46,100	25	0	0	50,400	27	12,800	7	7,300	4	0	0	116,600	71,200	38	187,800	2
4.5	56,400	13	252,300	59	0	0	0	0	6,800	2	15,800	4	9,500	2	340,800	87,200	23	428,000	5
5	7,600	2	27,500	9	0	0	57,700	18	41,200	13	67,700	21	64,600	20	266,300	51,400	17	317,700	4
5.5	21,600	4	261,700	43	0	0	151,200	25	27,300	5	13,500	2	24,000	4	499,300	106,300	18	605,600	8
6	0	0	352,800	52	10,800	2	94,100	14	100,400	15	25,600	4	32,500	5	616,200	58,100	9	674,300	9
6.5	0	0	173,800	36	39,700	8	92,400	19	77,500	16	44,100	9	14,400	3	441,900	44,400	9	486,300	6
7	21,600	4	269,900	52	21,800	4	128,900	25	24,700	5	19,500	4	0	0	486,400	31,800	6	518,200	7
7.5	10,300	2	142,500	26	84,300	15	134,100	24	151,700	28	0	0	0	0	522,900	27,600	5	550,500	7
8	36,900	8	0	0	0	0	306,100	69	51,489	11	21,322	5	0	0	415,900	28,600	7	444,500	6
8.5	69,300	17	0	0	0	0	68,200	17	201,700	50	22,100	5	7,800	2	369,100	34,700	10	403,800	5
9	18,500	10	48,800	25	0	0	11,200	6	11,500	6	2,400	1	17,400	9	109,800	83,400	48	193,200	2
9.5	20,400	7	182,300	61	0	0	25,300	8	0	0	0	0	0	0	228,000	73,100	26	301,100	4
10	0	0	96,900	54	2,400	1	23,300	13	1,700	1	0	0	6,700	4	131,000	50,200	28	181,200	2
10.5	0	0	278,500	58	27,800	6	39,800	8	51,100	11	22,400	5	14,800	3	434,400	46,700	10	481,100	6
Totals (% total alluvium)	365,800	5	2,900,800	37	337,200	4	1,745,800	22	921,300	12	291,900	4	207,300	3	6,770,100	1,024,800	13	7,794,900	100

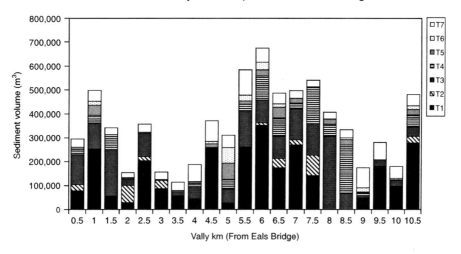

Figure 16.4 Volume (m^3) of late Pleistocene and Holocene alluvial terraces for each 0.5 km sub-reach of the South Tyne valley between Eals and Haltwhistle

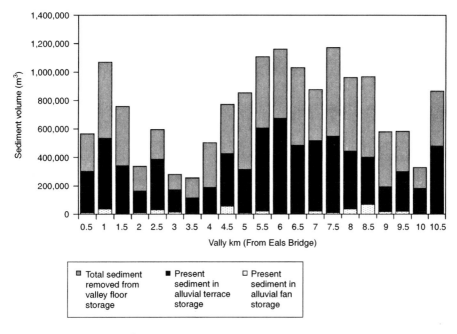

Figure 16.5 Volumes (m^3) of net Holocene fluvial sediment export and present fluvial sediment storage (including alluvial fans) for each 0.5 km sub-reach of the South Tyne valley between Eals and Haltwhistle

These findings are consistent with the widespread assumption that Pleistocene glacial and glaciofluvial drift deposits constitute the most pervasive and important source of coarse sediment to upland British rivers (e.g. Harvey, 1985; Brown and Quine, 1999; Macklin, 1999), and they provide further evidence to suggest that Holocene rivers are still in the process of adjusting to boundary conditions inherited from glaciation. Similar observations in modern and formerly glaciated river basins have been considered as evidence for a paraglacial cycle of upland geomorphic response (after Church and

Ryder, 1972) that extends well into postglacial timescales (e.g. Church and Slaymaker, 1989; Ashmore, 1993; Warburton, 1999).

4   FLUVIAL SYSTEM CHANGE IN THE UPPER THAMES VALLEY DURING THE DEVENSIAN LATEGLACIAL

In the fluvial sedimentary record, establishing the relative importance of forcing func-tions that act at different time and spatial scales can be extremely difficult, especially when the changes observed within the sedimentary record are very similar (see Lewin *et al.*, 1995). Recent work on the Upper Thames sequence has tried to establish a methodology for identifying the causal mechanisms for fluvial system change, in par-ticular, identifying criteria for linking observed sedimentological change to a known extrinsic climate change (Maddy *et al.*, 1998; Lewis *et al.*, 2001a). Lewis *et al.* (2001a) outlined five criteria that they considered important. The first two criteria concern specifically the identification of fluvial system change:

1. *Sedimentological changes*:   Changes in sediment calibre and/or facies associa-tions (groupings of genetically related lithofacies) most probably reflect changes in discharge regime in combination with changes in the nature of sediment supply. Changes in sediment calibre reflect changing energy status, but may also reflect changing sediment availability.
2. *Inferred change in fluvial style*:   Changes in planform geometry and/or tendency towards aggradation or incision may be a significant indicator of changing sedi-ment supply/discharge conditions.

   Although attributes 1 and 2 allow changes in fluvial system behaviour to be identified, such changes may result from mechanisms intrinsic to the fluvial sys-tem. A third criterion is required to establish a regional rather than localised change perhaps suggesting an extrinsic trigger is more likely.
3. *Nature of the lower bounding surface*:   The identification of regionally important erosion surfaces (bounding surfaces) may provide critical evidence of extrinsic driven change in the fluvial system. Miall (1996) described an appropriate hierar-chical classification of bounding surfaces. In his classification, fifth-order (channel changes indicative of long-term geomorphic process; $10^3 – 10^4$ years) and sixth-order (channel belt sequence change operating at the scale of Milankovitch cycles; $10^4 – 10^5$ years) surfaces represent significant changes at an appropriate spatial and temporal scale to exclude an intrinsic mechanism. Although bounding surfaces are identified at individual sites, as a network of sites becomes available over large parts of the basin it may be possible to recognise common features in successions at different sites. Aided by biostratigraphical and geochronological data (see Cri-teria 4 and 5 given below), it may be possible to correlate the more extensive bounding surface, thus indicating their regional extent.

   If significant sedimentological change is indicated by Criteria 1 to 3, it may be possible to link these changes to an extrinsic trigger. Most commonly, it is assumed that the main extrinsic trigger is climate change, but in order to establish this, two further criteria may be critical.
4. *Signal of environmental change*:   A signal of changes in the climate or other envi-ronmental parameters may be established from one or more proxy indicators such as pollen, plant macrofossils, *Coleoptera* and *Mollusca*, which can be associated with the observed sedimentological changes identified by the facies association

transitions. Given the erosive nature of fluvial deposition, it is rare that such an environmental record spans the facies association transition, but it may provide evidence relating to the period of deposition immediately prior to, or after, the sedimentological change.

5. *Chronological control*: Establishing the chronology of the sequence is essential in order to relate fluvial activity to an independent framework of environmental change and to establish the scale (localised or basin-wide) and causal mechanisms underlying the recorded response. In particular, it provides the necessary constraints to reconstruct the rates of change in fluvial activity and in some cases provides a framework for detailed assessment of lead–lag effects in landscape evolution triggered by major environmental change evidenced by associated faunal/floral changes.

Using the above criteria, evaluation of exposures at Ashton Keynes (Lewis *et al.*, 2001a) and Cassington (Maddy *et al.*, 1998) provides detailed information concerning fluvial system response to environmental changes during the last glacial–interglacial cycle. Here we consider the changes associated with the Devensian Lateglacial (Younger Dryas) climatic oscillation.

4.1 Ashton Keynes

The succession at Ashton Keynes (Grid Reference SU 057934) is divisible into four facies Associations A to D, overlain by Holocene fine-grained alluvium. A schematic representation of these associations is shown in Figure 16.6a and the characteristics of the major facies association transitions is summarised in Table 16.2 (from Maddy and Lewis *et al.*, 2001b). Here, we are concerned only with Associations B to D.

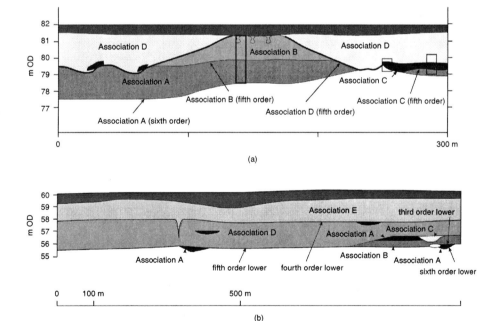

Figure 16.6 Schematic profiles showing generalised stratigraphy and disposition of facies associations at (a) Ashton Keynes (after Lewis *et al.*, 2001a) and (b) Cassington (after Maddy *et al.*, 1998)

Table 16.2   Facies associations at Ashton Keynes and Cassington. Facies coding follows Miall (1996)

	A	B	C	D	E
Ashton Keynes	Gh, Sh, Sr, Fl, Fsm	Gh, Sh	Peat, Fl	Gt, Gp	
Cassington	Gm	Sm, Sl, Sr, Fl, Fm, Fcf	St, Sm	Gm, Gp, Sp	Gm, Sm, Ss

Association B comprises laterally persistent, crudely bedded gravel sheets with minor sand-filled scours that Lewis *et al.* (2001a) consider to represent deposition in a multi-thread braided river. Cryogenic involutions within the upper sections of Association B confirm the presence of permafrost conditions prior to deposition of the overlying Association C. The transition from Association B to Association C represents a complete change in fluvial activity on this part of the valley floor. Association C comprises a peat unit, together with fine-grained organic muds intercalated with thinly bedded sands and silts, interpreted as overbank sediments. Although no fluvial channel deposits are present at the site, it is likely that the river would have adopted a single channel thread, with meandering assisted by cohesive banks formed in the peat. A pollen assemblage from the lower part of the peat unit suggests slight climatic amelioration, sufficient for scattered tree growth, mainly of juniper and birch (Lewis *et al.*, 2001a). Radiocarbon age estimates from the peat of $11,400 \pm 800$ BP (Beta-115384) and $11,690 \pm 800$ BP (Beta-115385) suggest formation within the Lateglacial Interstadial.

The B to C transition is interpreted as a fifth-order bounding surface representing a major adjustment in the fluvial regime, from a multiple channel system through to a single thread meandering channel. A reduction in sediment supply could affect such a transition in channel type. The palaeoecological evidence for climatic amelioration, supported by the age estimates, suggests this change may be climate-driven.

A further marked bounding surface separates Association C from Association D. Association D comprises lateral accretion gravels that lie within marked asymmetrical channels cut into the underlying associations that often contain collapsed and rotated blocks of Association C. These asymmetric channels are believed to result from localised scour interpreted as representing the onset of braiding as discharge and sediment supply conditions changed (Lewis *et al.*, 2001a).

A pollen assemblage from one of the rotated blocks suggests climatic deterioration at the end of the interstadial; thus, the gravels of Association D were probably deposited during the Lateglacial Stadial. Thus, a climate trigger seems appropriate, with increased sediment supply resulting in the transition to braiding.

4.2 Cassington

The succession at Cassington (SP 477108) can be divided into five facies Associations (A to E) shown schematically in Figure 16.6b, and the characteristics of the major facies association transitions is summarised in Table 16.2. Here we are concerned only with Associations D to E.

The sedimentology of Association D is typical of braided river sediments and faunal remains from Association D suggest a cold arctic climate regime. Association E is similar to that of Association D and is thus interpreted as braided river deposition. The two associations can be distinguished in the field by the presence of a marked

Table 16.3 Interpretation of facies association changes at Ashton Keynes and Cassington

		Bounding surface	Criterion					Conclusion
			1	2	3	4	5	
Ashton Keynes	B/C	5	y	y	?y	y	y	Climate-driven event indicated but not found at Cassington
	C/D	5	y	y	?y+	n	y	Time equivalent to even at Cassington but no unequivocal climate change signal
Cassington	D/E	4	y	y	?y+	n	y	Time equivalent to even at Ashton Keynes, but no unequivocal climate change signal

fourth order bounding surface, which is marked by the presence of an ice-wedge cast and the occurrence of minor scours, infilled with plant remains and molluscs. The ice-wedge cast suggests that the active channel zone must have abandoned this part of the floodplain for some time, allowing the ice wedge to form before the river returned and the gravels of Association E were deposited. Pollen and macrofossil debris from a scour fill at this boundary is dominated by birch, and the molluscan fauna is very restricted, including only two arctic/alpine species, which suggests cold climate conditions. A radiocarbon age estimate on birch twigs from one of these scours of $11,060 \pm 80$ BP (Beta-115381) places these deposits in the Devensian Lateglacial.

Although there is an indication of cold climate conditions from the flora and fauna, the magnitude or direction of any climate change cannot be inferred from the available information. However, the chronological data suggest that deposition of Association E took place after $11,060 \pm 80$ BP, most probably during the Lateglacial (Loch Lomond) Stadial (ca $11-10$ kyr BP). Thus, the Association D to E transition perhaps records a pattern of fluvial adjustment during the Devensian Lateglacial. Although there is no sedimentary record of the lower energy conditions of the interstadial at Cassington, later reworking of its flora is evident.

Table 16.3 evaluates these data against the criteria set out above, illustrating the difficulties of firmly establishing causal links. The lack of unequivocal independent evidence of climate change, together with the restricted geochronology allows only inference of a climate change trigger for these recorded changes. Unless future studies can establish more tightly constrained chronologies, then making links to triggering mechanisms will remain elusive, even though the evidence for change is contained within the archive. Preservation of independent palaeoecological data for confirmation of climate change may yet prove to be the most difficult problem to overcome.

5 DATING CHANGE IN FLUVIAL SYSTEMS

Establishing sufficiently high-resolution chronologies in order to consider causal link-ages has, until recently, proved an intractable problem. New dating techniques and the more rigorous testing of chronostratigraphies using multiple age estimation present new hope for addressing issues of cause and effect.

The most useful dating techniques within a fluvial context are those that can assess the age of the fluvial sediments themselves or the fossils contained within them. Arguably, the method with the greatest potential for constraining the ages of fluvial sequences is that of optically stimulated luminescence (OSL). The method has been particularly successful in dating fluvial sequences from the last glacial–interglacial cycle (Fuller *et al.*, 1998; Mol, 1997; Jain *et al.*, 1999) but recent studies by Rhodes (pers. comm.) suggest that successful dating of materials of much greater antiquity (up to 400 ka) may be possible in areas where radiation dose rates are low.

Despite the successes of the luminescence method, there is still a perceived wisdom that the most reliable dating method is that of radiocarbon dating and, notwithstanding the limited timescale applicability of this method, it remains the yardstick by which other techniques are measured. There are many examples of radiocarbon-dated fluvial (organic) sediments from the last glacial cycle. Often, however, these are isolated, single-age estimates of which the reproducibility, and thus reliability, is uncertain. Recent studies have attempted to offset the bias that may be introduced by the reliance on one technique by applying multiple dating methods: for example, Mol (1997) applied both luminescence and radiocarbon dating to the fluvial sedimentary record of the Neiderlausitz, Germany. Not only did this provide extended timescale coverage but also, where the two techniques overlap, some cross-checks on the validity of the age model was possible.

6 CONCLUSIONS

The approach described above, from understanding river response to a possible future-change scenario, is to look at the record of past long-term behaviour, especially in response to climate change. The only record of past response to change is archived within the fluvial sediment-landform assemblages. Their investigation requires consideration of both the morphological and sedimentological attributes of this archive. Current studies have focussed on the terrestrial record but further advances may come from investigating the vast fluvial sequences now below sea level. Large-scale morphological information can yield first approximation data for sediment storage changes associated with major climate-driven change. Furthermore, linking detailed sedimentological description with palaeoecological evidence for climate change may lead to the establishment of plausible causality. However, this methodology relies on establishing firm chronologies, a situation that has yet to be achieved in many river systems, and a point illustrated by the inadequacies evident within the studies outlined above. Notwithstanding the costs, in order to maximise the information extracted from these fluvial archives with its potential for informing rational discussion of potential system behaviour, large age-estimation programmes allied to more widespread and intensive study of Quaternary river systems are required.

ACKNOWLEDGEMENTS

This work was supported by the UK NERC (grants GR9/01656, GR3/11231 and GR3/12150). This paper is a contribution to the INQUA project "'*Northern Hemisphere fluvial response to rapid environmental change during the last two interglacial-glacial cycles (200 ka)*' and to IGCP449 '*Global correlation of Late Cenozoic fluvial deposits*'".

REFERENCES

Allen, J.R.L., 1982. *Sedimentary Structures: Their Character and Physical Basis*. Vol. 1, Elsevier, Amsterdam.

Ashley, G.M., 1990. Classification of large-scale sub-aqueous bedforms: a new look at an old problem. *Journal of Sedimentary Petrology*, **60**, 160–172.

Ashmore, P.E., 1993. Contemporary erosion of the Canadian landscape. *Progress in Physical Geography*, **17**(2), 190–204.

Begin, Z.B., Meyer, D.F. and Schumm, S.A., 1981. Development of longitudinal profile of alluvial channels in response to base-level lowering. *Earth Surface Processes and Landforms*, **6**, 49–68.

Best, J. and Bristow, C., (eds), 1993. *Braided Rivers: Form, Process and Economic Applications*. Special Publication No. 75. Geological Society of London, London.

Bishop, P., 1995. Drainage rearrangement by river capture, beheading and diversion. *Progress in Physical Geography*, **19**, 449–473.

Bluck, B.J., 1971. Sedimentation in the meandering River Endrick. *Scottish Journal of Geology*, **7**, 93–138.

Blum, M.D., 1993. Genesis and architecture of incised valley fill sequences: a late Quaternary example from the Colorado River, Gulf Coastal Plain of Texas. In P. Weimer and H.W. Posamentier (eds), *Siliciclastic Sequence Stratigraphy: Recent Developments and Applications*, American Association of Petroleum Geologists, 259–283.

Bond, G., Broeker, W., Johnsen, S., McManus, J., Labeyrie, L., Jouzel, J. and Bonani, G., 1993. Correlations between climate records from North Atlantic sediments and Greenland ice. *Nature*, **365**, 143–147.

Bond, G.C., Showers, W., Cheseby, M., Lotti, R., Almasi, P., De Menocal, P., Priore, P., Cullen, H., Hajdas, I. and Bonani, G., 1997. A pervasive millennial-scale cycle in North Atlantic Holocene and glacial climates. *Science*, **278**, 1257–1266.

Bridge, J.S., 1985. Perspectives: paleochannel patterns inferred from alluvial stratigraphy. *Journal of Sedimentary Petrology*, **55**, 579–706.

Brown, A.G. and Quine, T.A., 1999. Fluvial evidence of the medieval warm period and the late medieval climatic deterioration in Europe. In G. Benito, V.R. Baker and K.J. Gregory (eds), *Palaeohydrology and Environmental Change*. John Wiley, Chichester, 43–52.

Bull, W.B., 1991. *Geomorphic Responses to Climate Change*. Oxford University Press, Oxford.

Bull, W.L. and Knuepfer, P.L.K., 1987. Adjustments by the Charwell river, New Zealand, to uplift and climatic changes. *Geomorphology*, **1**, 15–32.

Church, M. and Ryder, J.M., 1972. Paraglacial sedimentation: a consideration of fluvial processes conditioned by glaciation. *Geological Society of America Bulletin*, **83**, 3059–3072.

Church, M. and Slaymaker, O., 1989. Disequilibrium of Holocene sediment yield in glaciated British Columbia. *Nature*, **337**, 452–454.

Cloetingh, S., Gradstein, F.M., Kooi, H., Grant, A.C. and Kaminski, M., 1990. Plate reorganisation: a cause of rapid late Neogene subsidence and sedimentation around the North Atlantic. *Journal of the Geological Society of London*, **147**, 495–506.

Collinson, J.D., 1986. Alluvial sediments. In H.G. Reading (ed.), *Sedimentary Environments and Facies*. Blackwell Scientific Publications, Oxford, 20–62.

Dansgaard, W., Johnson, S.J., Clausen, H.B., Dahl-Jensen, D., Gundestrup, N.S., Hammer, C.U., Hvidberg, C.S., Steffensen, J.P., Sveinbjörnsdottir, A.E., Jouzel, J. and Bond, G., 1993. Evidence for general instability of past climate from a 250-kyr ice-core record. *Nature*, **364**, 218–220.

Fuller, I., Macklin, M.G., Lewin, J., Passmore, D.G. and Wintle, A.G., 1998. River response to high frequency climate oscillations in Southern Europe over the last 200 k.y. *Geology*, **26**, 275–278.

Harvey, A.M., 1985. The rivers systems of North-West England. In R.H. Johnson (ed.), *The Geomorphology of North-West England*. Manchester University Press, Manchester, 122–142.

Howard, A.J., Macklin, M.G., Black, S. and Hudson-Edwards, K., 2000. Holocene river development and environmental change in Upper Wharfdale, Yorkshire Dales, England. *Journal of Quaternary Science*, **15**, 239–252.

Huisink, M., 1999. Lateglacial river sediment budgets in the Maas valley, The Netherlands. *Earth Surface Processes and Landforms*, **24**, 93–109.

Jain, M., Tandon, S.K., Bhatt, S.C., Singhvi, A.K. and Mishra, S., 1999. Alluvial and aeolian sequences along the River Luni, Barmer district: physical stratigraphy and feasibility of luminescence chronology methods. *Memoir Geological Society of India*, **42**, 273–295.

Krzyszkowski, D., Przybylski, B. and Badura, J., 2000. The role of neotectonics and glaciation on terrace formation along the Nysa Klodzka river, in the Sudeten Mountains (Southwestern Poland). *Geomorphology*, **33**, 149–166.

Krzyszkowski, D. and Stachura, R., 1998. Neotectonically controlled fluvial features, Walbrzych Upland, Middle Sudeten Mts, Southwestern Poland. *Geomorphology*, **22**, 73–91.

Lambeck, K., 1993. Glacial rebound of the British Isles I: preliminary model results. *Geophysical Journal International*, **115**, 941–959.

Lambeck, K., 1995. Late Devensian and Holocene shorelines of the British Isles and North Sea from models of glacio-hydro-isostatic rebound. *Journal of the Geological Society of London*, **152**, 437–448.

Leeder, M.R. and Stewart, M.D., 1996. Fluvial incision and sequence stratigraphy: alluvial responses to relative sea-level fall and their detection in the geological record. In: S.P. Hesselbo and D.N. Parkinson (eds), *Sequence Stratigraphy in British Geology*. Geological Society Special Publication 103, Geological Society of London, London, 25–39.

Leopold, L.B. and Bull, W.B., 1979. Base level, aggradation and grade. *Proceedings of the American Philosophical Society*, **123**, 168–202.

Lewis, S.G., Maddy, D. and Scaife, R.G., 2001a. Fluvial system response to abrupt climate change during the last cold stage: the Upper Pleistocene River Thames fluvial succession at Ashton Keynes, UK. Global and Planetary Change, **28**, 341–359.

Lewis, S.G., Maddy, D. and Scaife, R.G., 2001b. Fluvial response to rapid environmental change. A case study from Ashton Keynes, Upper Thames valley, UK. *Global and Planetary Change*, **28**, 341–359.

Macklin, M.G., 1999. Holocene river environments in prehistoric Britain: human interaction and impact. *Journal of Quaternary Science*, **14**, 521–530.

Macklin, M.G., Passmore, D.G. and Rumsby, B.T., 1992. Climatic and cultural signals in Holocene alluvial sequences: the Tyne Basin, Northern England. In S. Needham and M.G. Macklin (eds), *Alluvial Archaeology in Britain*. Oxbow Monograph 27, Oxford, 123–140.

Maddy, D., 1997. Uplift driven valley incision and river terrace formation in Southern England. *Journal of Quaternary Science*, **12**, 539–545.

Maddy, D. and Bridgland, D.R., 2001. Accelerated uplift resulting from Anglian glacioisostatic rebound in the Middle Thames valley, UK? Evidence from the river terrace record. *Quaternary Science Reviews*.

Maddy, D., Bridgland, D.R. and Westaway, R., 2001. Uplift-driven valley incision and climate-controlled river terrace development in the Thames Valley, UK. *Quaternary International*, **79**, 23–36.

Maddy, D., Lewis, S.G., Scaife, R.G., Bowen, D.Q., Coope, G.R., Green, C.P., Keen, D.H., Rees-Jones, J., Hardaker, T., Parfitt, S. and Scott, K., 1998. The Upper Pleistocene deposits at Cassington, near Oxford, UK. *Journal of Quaternary Science*, **13**, 205–231.

Merrett, S.P. and Macklin, M.G., 1999. Historic river response to extreme flooding in the Yorkshire Dales, Northern England. In A.G. Brown and T.A. Quine (eds), *Fluvial Processes and Environmental Change*. John Wiley, Chichester 345–360.

Merritts, D.J., Vincent, K.R. and Wohl, E.E., 1994. Long river profiles, tectonism, and eustasy: a guide to interpreting fluvial terraces. *Journal of Geophysical Research*, **99**, 14031–14050.

Miall, A.D., 1996. *The Geology of Fluvial Deposits*. Springer-Verlag, Berlin.

Mol, J., 1997. Fluvial response to Weichselian climate changes in the Niederlausitz (Germany). *Journal of Quaternary Science*, **12**, 43–60.

Passmore, D.G., 1994. *River Response to Holocene Environmental Change: The Tyne Basin, Northern England*. Unpublished Ph.D. Thesis, University of Newcastle upon Tyne, Tyne.

Passmore, D.G. and Macklin, M.G., 1997. Geoarchaeology of the Tyne basin: Holocene river environments and the archaeological record. In C. Tolan-Smith (ed.), *Landscape Archaeology in Tynedale*. Tyne-Solway Monograph No. 1, University of Newcastle upon Tyne, Tyne, 11–27.

Passmore, D.G. and Macklin, M.G., 2000. Late Holocene channel and floodplain development in a wandering gravel-bed river: The river South Tyne at Lambley, Northern England. *Earth Surface Processes and Landforms*, **25**, 1237–1256.

Passmore, D.G. and Macklin, M.G., 2001. Holocene sediment budgets in an upland gravel bed river: the river South Tyne, northern England. In D. Maddy, M.G. Macklin and J.C. Woodward (eds), *River Basin Sediment Systems: Fluvial Archives of Environmental Change*. Balkema, Amsterdam, 423–444.

Rumsby, B.T. and Macklin, M.G., 1996. River response to the last neoglacial (the 'Little Ice Age') in northern, western and central Europe. In J. Branson, A.G. Brown and K.J. Gregory (eds), *Global Continental Changes: The Context of Palaeohydrology*. Special Publication No. 115, Geological Society of London, London, 217–233.

Schumm, S.A., 1979. Geomorphic thresholds: the concept and its applications. *Transactions of the Institute of British Geographers, New Series*, **4**, 485–515.

Schumm, S.A., 1993. River response to baselevel change: implications for sequence stratigraphy. *Journal of Geology*, **101**, 279–294.

Schumm, S.A. and Lichty, R.W., 1965. Time, space and causality in geomorphology. *American Journal of Science*, **263**, 110–119.

Smith, D.G., 1983. Anastomosed fluvial deposits: modern examples from Western Canada. In J.D. Collinson and J. Lewin (eds.), *Modern and Ancient Fluvial Systems*. Blackwell, London, 155–168.

Tipping, R., 1995. Holocene evolution of a lowland Scottish landscape: Kirkpatrick Fleming. Part III, fluvial history. *The Holocene*, **5**, 184–195.

Tornqvist, T.E., 1998. Longitudinal profile evolution of the Rhine-Meuse system during the last deglaciation: interplay of climate change and glacio-eustasy. *Terra Nova*, **10**, 11–15.

Tornqvist, T.E. and Blum, M.D., 1998. Variability of coastal onlap as a function of relative sea-level rise, floodplain gradient, and sediment supply: examples from late Quaternary fluvial systems. In J. Canaveras, M. Angeles Garcia del Cura and J. Soria (eds), *Sedimentology at the Dawn of the Third Millennium. Proceedings of the 15th International Sedimentological Congress*, 765.

Van den Berg, M.W., 1996. *Fluvial Sequences of the Maas: A 10 Ma Record of Neotectonics and Climate Change at Various Time-Scales*. Ph.D. Thesis, University of Wageningen, Wageningen.

Vandenberghe, J., 1995. Timescales, climate and river development. *Quaternary Science Reviews*, **14**, 631–638.

Warburton, J., 1999. Environmental change and sediment yield from glacierised basins: the role of fluvial processes and sediment storage. In A.G. Brown and T.A. Quine (eds), *Fluvial Processes and Environmental Change*. John Wiley, Chichester, 385–408.

17 Palaeohydrological Modelling: from Palaeohydraulics to Palaeohydrology

J.B. THORNES

University of London, London, UK

1 INTRODUCTION

In this chapter, the author demonstrates the development and applications of selected palaeohydrological models. These range from descriptive models used to infer the responses of rivers to climate changes and their use as chronological tools, to contemporary models in which the emphasis is on the evolution of the fluvial system itself under the influence of intrinsic feedback that generate changes *within* the system and do not depend exclusively on changing external forces such as climate or human activity. The work of Schumm is frequently referred to, reflecting the enormous contribution that he has made over the last half century to fluvial geomorphology in general, and to palaeohydrology in particular, through his capacity to conceptualise processes freshly, providing new fundamental insights and to demonstrate their relevance by perceptive field and laboratory observation.

The history of palaeohydrological modelling is divided into six main phases: historic, empirical, analytical, evolutionary, global and contemporary.

1.1 Historic Palaeohydrology

Roughly up to 1950, this phase sought to account for historically documented, cartographically observed or chronologically defined river-channel changes, mainly in terms of palaeoclimates and anthropogenic changes of the landscape, in the absence of a well-developed fluvial theory and in the "geological" style of early geomorphology, with its emphasis on stratigraphy and hydraulics. It largely comprised attempts to differentiate phases of cut and fill from sections in river gravels, often dated archaeologically or by morphology. This was very much in the spirit of denudation chronology and largely dominated by inferences about the impacts of the Quaternary glacial and non-glacial sequences and stratigraphic correlations.

1.2 Empirical Palaeohydrology

This phase, roughly up to the 1980s, used empirical relationships between channel characteristics, sediment transport and discharge to infer the effects of past changes in runoff and sediment yields and the channel responses. It eventually took its source in the hydraulic geometry paradigm developed by Leopold and Maddock (1953) and

Palaeohydrology: Understanding Global Change. Edited by K.J. Gregory and G. Benito
© 2003 John Wiley & Sons, Ltd ISBN: 0-470-84739-5

extensively propagated as a result of its clear exposition in *Fluvial Processes in Geomorphology* (Leopold *et al.*, 1964). Reinforced by the quantitative wave in geomorphology, it eventually foundered in the early stages of the non-linear movement in hydrology and the shift of hydrology away from a preoccupation with discharge to a better appreciation of hillslope hydrology and channel hydraulics ushered in by Kirkby and Chorley (1967), Shen (1971) and Graf (1998).

Again the work of Schumm (1965) led to the development of a firm link between palaeohydrology and climate change, through his practical attempts to link sediment concentrations to sedimentary forms and deposits on the one hand and to mean annual temperature and effective precipitation on the other. Starting with the relationship of rainfall to runoff curves for different temperatures (Langbein and Schumm, 1958), he then adopted the curve of sediment yield and mean annual effective precipitation at a given temperature for the co-terminus United States. This "standard" curve was then translated to reflect different mean annual temperatures at 10°C intervals. It was possible then, for any given rainfall, to examine the likely effect of temperature changes on sediment yield. Although the "raw" curve itself, developed from empirical data, reflected 2,000 years of humanly changed land use, it was later demonstrated theoretically by Kirkby and Neale (1986) to be of the right basic form, with its high sediment concentration in seasonally controlled dryland environments, such as the Mediterranean. There have been no serious attempts to validate these outcomes, but the extensive compilation by Walling and Webb (1981) appears to support the general argument that seasonality is a vital component of the relationships.

1.3 Analytical Palaeohydrology

In this phase, hydrological modelling was pursued through the differential equation models of the dominant processes, such as sediment transport throughout the system either in the form of particles or as sediment waves. The forcing functions in these models have mainly been runoff and hence reflected, more adequately, the impact of changes in hillslope hydrology and allowed the links with climate to be made much more effectively through the vegetation cover. The seminal work of *Hillslope form and process* (Carson and Kirkby, 1972) is of primary importance in this phase.

1.4 Evolutionary Palaeohydrology

In this modern approach, the models attempt to account for the evolution of, and interactions between, network development and catchment morphological evolution. It was Schumm's experiments at Perth Amboy (Schumm, 1956) and in the Drainage Evolution Research Facility at Fort Collins (Schumm and Khan, 1972) that initiated evolutionary palaeohydrology through the threshold concept. Eventually, it was the development of the Geomorphic Instantaneous Unit Hydrograph (IUH) of Nash (1957) by Rodriguez-Iturbe and Valdes (1979), Rodriguez-Iturbe *et al.* (1982) and Rodriguez-Iturbe (1993) that provided a comprehensive theoretical underpinning of how the drainage network led to its own evolution, through its control on the timing and characteristics of runoff events in the evolving channel system. This is discussed at length below.

1.5 Global Palaeohydrological Models

The global climate change resulting from greenhouse gas warming assumed greater significance at the end of the twentieth century. As the Global Circulation Model

(GCM) results are treated with a higher degree of confidence in the new millennium, the potential problems in water resources in the coming century have led to a revitalisation of attempts to enable predictions of the runoff amounts and groundwater recharges to be made. It now seems much more important to develop the capacity to predict the impacts of hydrological change from climatic inputs. This has been stimulated by Eagleson's (1978a; 1978b) modelling of the components of hydrological systems using derived probability functions, driven essentially by the magnitude of rainfall, the magnitude and frequency of storm events and the optimum vegetation response to climate change. This work, first published in 1978, should carry both predictive hydrology and palaeohydrology well into the twenty-first century.

1.6 Contemporary Palaeohydrology

As computing capacity advanced through massively parallel-computing machines, the ability to model by simulation has advanced accordingly, bypassing some of the exceptionally complicated analytical approaches. So too the breakthrough in the understanding of the nature of turbulence (Sidorchuk, 1996) and its control on bedforms is overturning some of the doubtful preconceptions of both hydraulic geometry and stratigraphy. Recently, Coulthard et al. (2002) have applied this approach to Holocene changes in a small upland river basin and the corresponding alluvial fan evolution. The same technique has also been used to model, in detail, the redistribution of lead-mining pollutants of the Swaledale region of northern Yorkshire, and to the interpolation of the effects of extensive flooding on soil pollution through the redistribution of mining waste spoil by the floods. This new generation of models has immediate and obvious implication for the study of the impact of climate change at the local and regional scale. As this type of modelling advances, it is becoming possible to confirm that climate change can indeed create the changes that were attributed to it in the historical phase of palaeohydrology, or to separate climatic forces from alternative causes, such as tectonic changes (Rebeiro-Hargrave, 2000).

Because the historical and empirical modelling phases have been well reviewed in earlier volumes in this series (since Gregory, 1983), this chapter will concentrate mainly on the phases mentioned in Sections 1.4 and 1.5.

2 EMPIRICAL PALAEOHYDROLOGY

This approach relied largely on using relationships between sediment, channel characteristics and channel behaviour and various characteristics of discharge expressed as regression equations. These were then used either to infer the magnitude of the flows that formed relict geomorphological features or, even today, to infer the expected consequences of runoff changes resulting from future climate changes. The dangers of these approaches are well known (Maizels, 1983; Dury, 1981; Rotnicki, 1983; Ferguson, 1986) and can be summarised as follows.

1.   The methodology assumes that, at the time of their formation, the palaeochannels and their sediments were in fact in dynamic equilibrium with the flows forming them. However, extreme events may have been responsible for the morphological development of the channel cross section, leaving behind boundary sediments that could not be moved by successive lower flows.
2.   Unless the correlation is close to unity, the errors of prediction can be very large indeed, especially where log–log relations are involved. Sometimes the inferred flows could be between zero and impossibly large flows for a given catchment.

3. The relationships only hold over the range from which the data were collected and cannot and should not be extended to different channel morphologies or even to different environments, that is, they are not "transportable".

4. Although linear relationships between rainfall and runoff are often assumed, this is rarely substantiated. The cumulative evidence since the appreciation of hillslope hydrology in the 1960s and 1970s is that rainfall–runoff relations are commonly non-linear as a result, in part, of the importance of the vegetation cover in setting thresholds and bifurcations to runoff.

5. The importance of human activity on rainfall–runoff and sediment yield relationships has been rather ignored in inferring the potential of future climates to produce future runoffs. Even if our models can correctly derive the expected vegetation covers for a given climate or topography, they are unlikely to foresee the effects of global economic or local land management changes. Ecological models that may be used as inputs to hydrological forecasting will require more than plant production functions driven by rainfall, because competition, predation and patch dynamics are likely to be equally important in determining the plant cover component. So too is the spatial configuration in the catchment. It emerges that the vertical and horizontal arrangement of the vegetation cover will be most critical in determining future hydrologies. This was a key assumption in historical palaeohydrology that was not much explored either in the catchment, or in the channel itself, despite its recognition by Schumm (1968) as the main control of terrestrial sedimentation throughout geological time.

6. Despite these shortcomings, it would be unfair and imbalanced not to recognise the major contribution made to this approach by the authors cited above, in the shift from the qualitative historical approach to quantitative attempts to put palaeohydrology on a much firmer footing. Nevertheless, the regression approach is both unsuitable and undeliverable for palaeohydrological inferences and future projections, especially when based on simple rainfall–runoff relationships, even if they are based on more reliable scenarios of future changes. Physically based models appear to offer the main alternative, given the adequate specification of the uncertainties and methods of handling them (Bevan, 1989). These uncertainties arise spatially and temporally from the scale of resolution of the GCM output variables (see Chapter 1). Local regionalised models provide more hope in this respect than global scale models. Either downscaling of the GCM or upscaling of the physical catchment parameters has to be undertaken and, in the last few years, there has been a real effort to provide what are called "effective parameters" for large-scale models. This approach has been adopted for the System Hydrologique Europeen (SHE) model, as developed by Civil Engineering at Newcastle University (e.g. in Bathurst and O'Connell, 1992). An important parallel development is the attempt to provide a methodology for validation. This is the reverse of the same coin. Effective parametrisation seeks to optimise parameters, normally measured on a scale of centimetres for use in a model with a grid scale of the order of 1 km (Binley et al., 1989). In validation, the effort is designed to compare model outputs at coarse time and space scales with real data for natural systems at the integrated basin scale. As the deployment of large-scale models in palaeohydrology becomes increasingly common, developments in these two methodologies will become more and more critical. Distributed modelling is not the panacea for all problems of global climate change prediction. In an important critical review, Bevan (1989) concluded

"Most hydrological models are invalidated by detailed comparisons with field observations. However, there are problems that require 'physically based' predictions, in particular the problems of predicting the future hydrological effects of land use and climatic change when no data is available for model calibration or re-calibration. Thus it is necessary to improve the hydrologists available modelling tools in this area."

The modelling work of Eagleson and Rodriguez-Iturbe described below do exactly what is required.

3 GLOBAL HYDROLOGICAL MODELS

Eagleson's coherent attack on this problem (Eagleson, 1978a and b) sought to establish the understanding of climate–land surface in a form that provides insight into the physical basis for man-induced changes in both climate and water balance. He recognised that such insight can only come through the retention of physical determinism, but that uncertainty plays such a large role that his approach must (and does) deal with probability distributions. The input variables are considered stochastic and their probability distributions are transformed into the probability distributions of the output processes by using models of the deterministic physical process.

Eagleson uses a one-dimensional annual water balance based on physically realistic models of hydrological sub-processes. The model is simple enough to allow analytical derivations of the probability distributions of the critical hydrological variables from parametrised distributions of the climatic input variables.

For each component of the water balance (groundwater, runoff, evapotranspirational losses) the basic generating processes are defined. These are

1. Precipitation processes
2. PE-transpiration (Potential Evapotranspiration) processes
3. Soil moisture processes.

For example, precipitation is generated phenomenologically by Poisson arrivals of independent and identically distributed rectangular pulses of characteristic duration separated by arrival times and storm depths.

Atmospheric temperature has a small coefficient of variation and is, therefore, replaced by its mean over the rainy season. Potential evapotranspiration is also replaced by its seasonal average value and the average annual potential evapotranspiration is calculated from the climate and land surface parameters by using the Penman model (Penman, 1948). The model treats the bare and vegetated fractions of the soil as two separate components with the assumption that the vegetation cover is just enough to consume the available moisture. This "optimum" vegetation cover percent is derived for any environment on the earth and for the main soil types, so that by varying the distribution parameters of the key environmental variables, the so-defined *optimal* vegetation cover can be obtained. This, in turn, is embedded in other components of the water balance (such as the runoff/groundwater partition). The relevant soil properties are the intrinsic permeability, the soil suction, pore disconnectedness index and pore size distribution.

One product of this probability modelling (Figure 17.1) is the decomposition of the annual water balance into its main components at different places and different times, using parameters that are accessible to climate change modellers from historical records. This decomposition is expressed in terms of the annual precipitation (as the

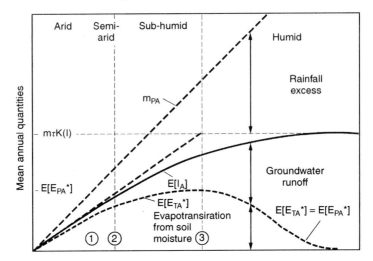

Figure 17.1 Climatic regions according to the Eagleson (1978b) model. The figure also shows the partition of the water balance according to the region. Reproduced by permission of the American Geophysical Union

sum of all the event precipitations). For palaeohydrology and its impact, the importance of this model over existing potential models rests in the following:

- The derivation of the optimal vegetation cover.
- The potential for allowing the vegetation to evolve with the climatic conditions in terms of the water-use efficiency coefficient.
- The incorporation of the intrinsic uncertainty in inputs and their realistic represen- tation in model outputs. Hydrologists have for some years stressed the need for proper identification of the errors involved in modelling (Binley and Bevan, 1991; Bevan and Binley, 1992). In these papers, Eagleson meaningfully incorporates the errors in the probabilistic specification of the models.
- The modelling suite clearly has very great potential for dealing with erosion and, ultimately, with sediment yields and concentrations, if one can accept that these are mainly controlled by climate, soil and vegetation cover (see Kosmas *et al.*, 2000).

4 EVOLUTIONARY PALAEOHYDROLOGY

Until the 1980s, global change palaeohydrology tended to overlook the problem that the basin itself was evolving. This is largely because it was assumed that evolution of the network itself was so slow as to be relatively unimportant in determining the runoff, despite the early pioneering work of Leopold and Miller (1956) that showed that the main source of sediments in a semi-arid gully system could be accounted for by channel changes. If cut and fill dominates sediment yields, then in the long term (of the order of 10^3 years), the downstream flow and sediment characteristics must be affected by the evolving network growth and network characteristics.

A concerted effort has been made by Rodriguez-Iturbe and his co-workers (e.g. Rodriguez-Iturbe and Rinaldo, 1998) to address three questions:

1. How does the channel network affect the runoff pattern?
2. Can the runoff pattern be predicted (or postdicted) from the network characteristics?

3. How is this set of relationships affected by the climate that prevails in the runoff
    generating processes?

Although the results are scattered across some 15 papers published in Water Resources
Research, an accessible and readable synthesis is available in the paper by Rodriguez-
Iturbe (1993), which the following text largely follows:

Starting with the general proposition that the climate is both the cause and effect
of network development, Rodriguez-Iturbe used Nash's (1957) concept of the IUH
to understand the effect of the network structure. The IUH is the equivalent of the
unit impulse response function–the algorithm that transfers rainfall into runoff. The
emphasis is on the volume of discharge over time at the basin outlet. This is derived
from a unit amount of rainfall over the catchment that is routed across the hillslopes
and through the channels. In a network, the rain falling on the streams further from
the outlet will take longer to reach the outlet and the response function [now called
the Geomorphic Unit Hydrograph (GUH)] reflects the travel times through the system
and hence the network configuration. This is analogous to people arriving at a main
line station from different origins, if they all leave home at the same time. The number
of people arriving at any time reflects their aggregate travel times.

The arrivals of water are conceptualised as a Markov process as Thornes and Morley
(1970) modelled vehicular movements in the road network of Dartmoor National
Park, and as envisaged (Thornes, 2002) for the movement of sediment transport when
designing the optimum location of check dams in a drainage network. In the Rodriguez-
Iturbe approach, the links of the networks are of different orders and the water moves
to a link of the next highest order with a given probability. Each order is a state of
the system. By pre-multiplying the matrix of probability transitions between different
orders by the existing volumes in each order, the state of the system at the next time
step can be predicted. A partial description of the Markovian methodology is found
in Thornes and Brunsden (1977). To avoid a step-like transition between orders, the
transition takes place at the end of the time spent in reaches of a given order, which is
averaged over all stream links of the same order. Rodriguez-Iturbe and Valdes (1979)
found that this mean waiting time can be described by an exponential distribution
and used the reciprocal of the mean waiting time to obtain the probability matrix. By
powering up the transition matrix and by pre-multiplying by the previous time vector,
the distribution of the water in the system at any discrete time is obtained (Figure 17.2).
Shannon et al. (2002) explain this idea for a very simple network. If this is plotted
for the highest order through time, it furnishes the Instantaneous Geomorphic Unit
Sedigraph (IGUS) at the mouth of the basin. The sediment yield through time is given
by the successive occupancies of the last link.

The main task is to estimate the probability matrix. If it is (reasonably) assumed
that the velocity of the stream is approximately the same throughout the system, then
the elements of the transition matrix are derived from the lengths of the links of
different orders. Proceeding analytically with the Markov process, Rodriguez-Iturbe
and Valdes (1979) were able to express the peak and time to peak in terms of the
network characteristics as

$$q_p = 0.36 R_L^{0.43} v L_\Omega^{-1} \tag{1a}$$

$$t_p = 1.58 [R_B/R_A]^{0.55} R_L^{-0.38} L_\Omega v^{-1} \tag{1b}$$

where q_p is the peak discharge of the GUH, t_p is the time to peak of the GUH, v is
the average stream velocity in the catchment, R_L is the Horton length ratio, R_B the

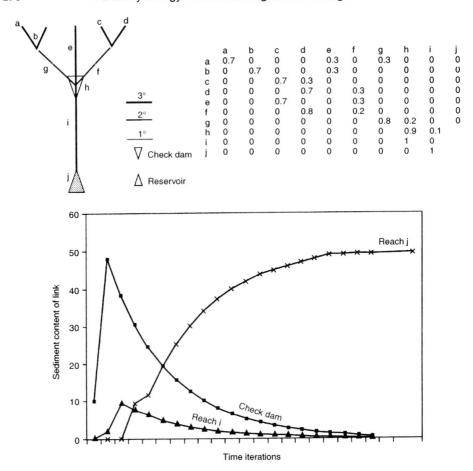

Figure 17.2 Results from Thornes' Markov model of sediment transport. The channel network is shown top left, and the state-transition matrix top right. The lower diagram shows the sediment content of particular links against time

Horton bifurcation ratio and L_Ω the length of stream of order Ω. R_A is the Horton area bifurcation ratio.

Eventually, they were able to use any waiting time distributions and also to dispense with the Markovian hypothesis.

The methodology uses lengths to define the state-transition probabilities and expresses length as a function of the network geometry.

In a further step, Rodriguez-Iturbe, Gonzalez-Sanabria and Bras (1982) used kinematic wave theory to relate mean travel times to the kinematic wave parameters and thence to the rainfall intensity and duration characteristics, thus providing a climatic control on the instantaneous geomorphic unit hydrograph (IGUH) that Rodriguez-Iturbe (1993) then termed the *geomorphoclimatic unit hydrograph*. Rainfall intensity and duration lead to the probability density functions for q_p (peak discharges) and t_p (time to peaks) and the authors applied it to four different climates and two different networks. Later, the theory was tested in the Manon River Basin in Venezuela and was found to reproduce the probability distributions and the unit hydrographs very satisfactorily.

   The short treatment given here cannot do justice to the important development of Eagleson and Rodriguez-Iturbe in putting the prediction of the impacts of climate change and network change on a firmer theoretical footing, whilst retaining the uncertainty constraints at the centre of their probabilistic formulations. They open wide avenues for future developments and provide implicit warnings against the simple regression approaches to the same problem. As and when we have better knowledge of palaeohydrological networks and climates, they could provide the tools for a better understanding of their hydrological and sedimentological responses.

   Rather than be too carried away, we have to recognise two important constraints. The first is the risk of *increasing* the non-linearity of the systems in the mathematical and digital formulations. The second is to recognise that we are dealing with topography, not simple networks and, as Rodriguez-Iturbe (1993) clearly identified, it is a three-dimensional regularity that needs to be found. Kirkby (1993) has recently made real progress in this direction by showing how contours and flow lines in the 3-D network of gully systems evolve in response to changes in storm characteristics.

   There is a third element that has to be developed in the modelling approach to palaeohydrology that emerges from Eagleson's recognition of the importance of the plant coefficient k that controls water-use efficiency. As k changes by plant evolution, the hydrological response of a catchment will also change. This comes back to the underlying issues of the intervening effects of vegetation on hydrological and sedimentological response. It is no longer possible to assume nearly instantaneous and extensive change of vegetation cover in response to climate change. Some modelling strategies simply assume that different plant functional types occupy catchments during or soon after climate change, or that the effects can be incorporated by the adoption of different densities for the different functional types (e.g. grasses or bushes). Rather, vegetation changes themselves may be slow and complex and subject to strong non-linearities. This point is developed in the following section.

5 VEGETATION STABILITY

Recent work on the stability of plant communities (Thornes, 2003) has shown that, after disturbance by climatic or anthropogenic causes, recovery is generally of a logistic nature (see Figure 17.3). This is characterised by slow growth at first, then a steeply rising growth until it is checked by limited resources (such as water stress) so that the rate of increase of biomass levels off. In systems of this type, the capacity of the vegetation to "track" the environmental change is mainly determined by the k coefficient of the logistic curve. The Mediterranean *matorral* community seems able to resist collapse despite the very large inter-annual variations of precipitation. It has a recovery time to capacity biomass and cover of the order of 13 to 15 years (Obando, 1997). The states of interest are now capacity biomass (the upper bound) and complete destruction of the vegetation (the lower bound). With fluctuations in rainfall, the vegetation exhibits a random walk between these two attractors and the interest is in finding the first passage time to either equilibrium or bust, following a climate change. This is the core of the desertification debate (Thornes and Brandt, 1993).

   Work in the eastern Cape of South Africa also shows that there are different erosional and hydrological responses to grass-dominated and bush-dominated communities and archaeological evidence shows that this boundary has been shifting over history and pre-history in response to climate and human-induced changes (Scott *et al.*, 1997). These have probably changed land surface regimes and are certainly a potential issue

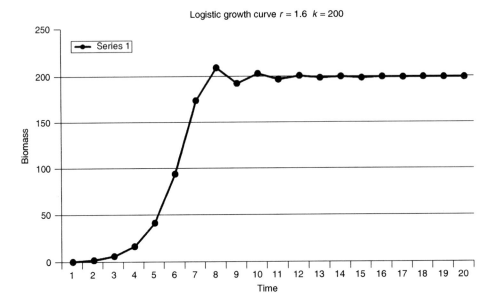

Figure 17.3 The logistic growth curve for biomass against time, when $r = 1.6$ and $k = 200$. The recovery curve stabilises after 10–12 years. Note the overshoot in Period 8

in the context of future global climate changes. The two new problems for modelling future vegetation and palaeohydrological responses are as follows. First, there is a third variable involved in the debate, that is, grazing pressure (Bosch, 1989; Thornes, 2003), and second, the shifts may be catastrophic rather than linear. The existence of two stable states with periodically oscillating noise is very similar to the problem of the stability of the states of the North Atlantic as determined by the Dansergard–Oesschger events (Ganopolski and Rahmstorf, 2001).

Because the bush–grass boundary is generally viewed as a contrast in the availability of soil moisture, the grass, which taps shallower sources, reduces the pool of deeper soil water resources that are more readily available to bushes (Walker and Noy-Meir, 1982). Thus, grazing, by removing the transpiration demand for grasses, favours bush encroachment in the finely balanced water regime of dry areas. In reality, in both space and time, there appears to be a sharp transition from bush to grass. This has variously been attributed to fire, topographic control and management treatments and as physically analogous to the grass–wood boundary of the wider savanna boundary problem (Huntley and Walker, 1982).

Along the lines indicated by Noy-Meir (1982), our field evidence points to a catastrophe theory interpretation as indicated in Figure 17.4. In this hypothesis, the three axes represent the biomass (vertical or z axis), soil moisture content (x axis) and grazing intensity (y axis). At low grazing intensities (y–z plane), there is a steady increase in biomass with available soil moisture. The actual boundary is fixed by a critical soil moisture along the continuum. The critical soil moisture itself is determined at the regional scale by the combination of precipitation and evaporation/evapotranspiration and locally, by the hillslope hydrology. As the grazing increases, the impact of soil moisture becomes more finely tuned; soil moisture is the "splitting factor" and there is a sudden shift to bush (Point A in Figure 17.4) with rising soil moisture or to grass

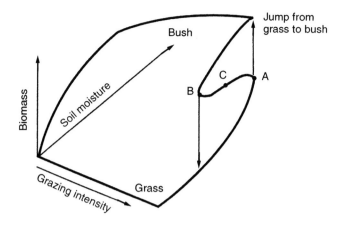

Figure 17.4 A catastrophe theory interpretation of bush–grass transition

with falling soil moisture (Point B in Figure 17.4). "Mixed" communities (Point C) are, in principle, unstable and therefore occupy less of the spatial domain than the dominant types.

This example illustrates that the shift in vegetation types, prompted by climate change, is (1) non-linear, (2) more complicated than formerly thought, (3) inexplicable without recourse to the human factors (grazing) involved, and (4) not accountable in terms of regression models of biomass productivity by rainfall. It reinforces the earlier belief (Thornes, 1988; Thornes, 1990; Thornes and Brandt, 1993) that understanding the impact of climate change on vegetation, past, present and future, involves rather complicated modelling. This is especially true when, as in the case of the Eastern Cape, the climate is not just a series of widely varying inter-annual rainfall totals but rather a strongly oscillating series reflecting periodic fluctuations in the global circulation. Figure 17.5 shows the 130-year rainfall series at Fort Beaufort, Eastern Cape, in which there are strong runs of dry and wet years that appear strongly to reflect the El Niño and La Niña oscillations (Trzaska, 2001).

6 CONCLUSIONS

Caution must be exercised in attempts to model global change impacts on hydrological systems based on regression procedures because

- they embed thresholds and other non-linearities;
- the systems are usually non-stationary, their parameters varying through time;
- statistical empirical models often incorporate high levels of uncertainty especially when applied in domains for which they are neither calibrated nor validated; in particular, the use of hydraulic geometry relationships is likely to prove unreliable as a tool for forecasting the impacts of future climate change regimes.

The prospects for distributed hydrological models, especially those that intrinsically incorporate known physical and statistical uncertainties, have improved substantially since 1980 as sounder theoretical platforms for water balance and channel network evolution processes have been developed. The acceptance of non-linear behaviour in

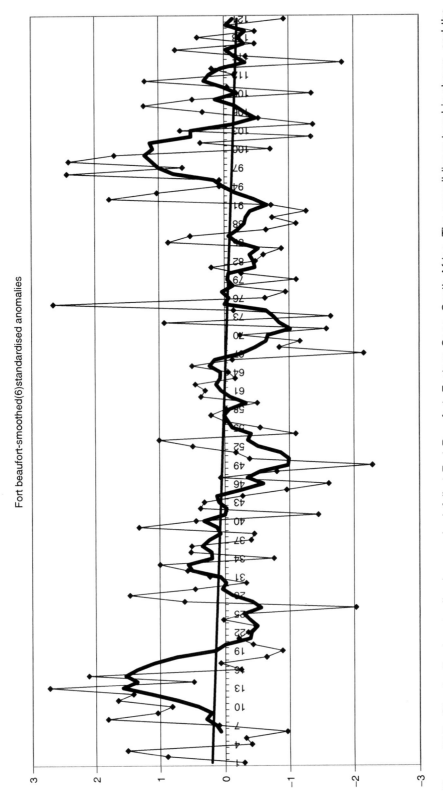

Figure 17.5 Standardised anomalies in annual rainfall at Fort Beaufort, Eastern Cape, South Africa. The overall linear trend is shown, and the annual data have been smoothed by a 6 period running mean (heavy line)

hydrological systems seriously undermines many of the conventional assumptions of both empirical and physical distributed analytical models and all modelling activities in future will have to be aware of them. The complexity of vegetation response calls for an increased effort in this area from hydrologists and palaeohydrologists to better understand the multiple stable outcomes of global climate change.

ACKNOWLEDGEMENT

This chapter was completed in 2001 while the author held the Hugh Kelly Research Fellowship at Rhodes University, Grahamstown, South Africa. The author wishes to acknowledge the support of Rhodes University and the Geography Department during his tenure of the Fellowship. Precipitation data were supplied by Mr Monde Dumas.

REFERENCES

Bathurst, J.C. and O'Connell, P.E., 1992. Future of distributed modelling: the Systeme Hydrologique Europeen. *Hydrological Processes*, **6**, 265–278.

Bevan, K.J., 1989. Changing ideas in hydrology: the case of physically-based models. *Journal of Hydrology*, **105**, 152–172.

Bevan, K.J. and Binley, A.M., 1992. The future of distributed models; calibration and predictive uncertainty. *Hydrological Processes*, **6**, 279–298.

Binley, A.M. and Bevan, K.J., 1991. Physical modelling of catchment hydrology: a likelihood approach to reducing predictive uncertainty. In D.G. Farmer and M.J. Rycroft (eds), *Computer Modelling in the Environmental Sciences*. Clarendon, Oxford, 77–88.

Binley, A.M., Bevan, K.J. and Elgy, J., 1989. A physically-based model of heterogenous hillslopes 2. Effective hydraulic conductivities. *Water Resources Research*, **25**, 1227–1233.

Bosch, O.J.H., 1989. Degradation of the semi-arid grasslands of Southern Africa. *Journal of Arid Environments*, **16**, 165–175.

Carson, M.A. and Kirkby, M.J., 1972. *Hillslope Form and Process*. Cambridge University Press, Cambridge.

Coulthard, T.J., Macklin, M.G. and Kirkby, M.J., 2002. A cellular model of Holocene upland river basin and alluvial fan evolution. *Earth Surface Process and Landforms*, **27**, 269–289.

Dury, G.H., 1981. *Environmental Systems*. Heinemann, London.

Eagleson, P., 1978a. Climate, soil and vegetation. 1. Introduction to water balance dynamics. *Water Resources Research*, **14**, 765–776.

Eagleson, P., 1978b. Climate, soil and vegetation. 2. The distribution of annual precipitation derived from observed sequences. *Water Resources Research*, **14**, 713–721.

Ferguson, R.I., 1986. Hydraulics and hydraulic geometry. *Progress in Physical Geography*, **10**, 1–31.

Ganopolski, A. and Rahmstorf, S., 2001. Rapid changes of glacial climate simulated in a coupled climate model. *Nature*, **409**, 153–158.

Graf, W.H., 1998. *Fluvial Hydraulics*. Wiley, Chichester.

Gregory, K.J., (ed.), 1983. *Background to Palaeohydrology: A Perspective*. Wiley, Chichester, 101–130.

Huntley, B.H. and Walker, B.J., 1982. *Ecology of Tropical Savannas*. Springer-Verlag, New York.

Kirkby, M.J., 1993. Long term interactions between networks and hillslopes. In K. Bevan and M.J. Kirkby (eds), *Channel Network Hydrology*. Wiley, Chichester, 255–293.

Kirkby, M.J. and Chorley, R.J., 1967. Throughflow, overland flow and erosion. *Bulletin of the International Association of Scientific Hydrology*, **12**, 5–21.

Kirkby, M.J and Neale, R.H., 1986. A soil erosion model incorporating seasonal factors. In V. Gardiner (ed.), *International Geomorphology*. Vol. II, Wiley, Chichester, 189–210.

Kosmas, C., Kirkby, M.J., and Geeson, N., (eds) 2000. *Key Indicators of Desertification and Mapping Environmentally Sensitive Areas to Desertification*. MEDALUS, Luxembourg. Office for Official Publications of European Communities, EUR18882, p. 90.

Langbein, W.B. and Schumm, S.A., 1958. Yield of sediment in relation to mean annual precipitation. *Transactions of the American Geophysical Union*, **39**, 1076–1084.

Leopold, L.B. and Maddock, T., 1953. *The Hydraulic Geometry of Stream Channels and Some Physiographic Implications*. USGS Professional Paper 252.

Leopold, L.B. and Miller, J.P., 1956. *Ephemeral Streams: Hydraulic Factors and their Relation to the Drainage Net*. USGS Professional Paper 282A, 36.

Leopold, L.B., Wolman, G. and Miller, J.P., 1964. *Fluvial Processes in Geomorphology*. Freeman, San Francisco.

Maizels, J.K., 1983. Palaeovelocity and palaeodischarge determination for coarse gravel deposits. In K.J. Gregory (ed.), *Background to Palaeohydrology: A Perspective*. Wiley, Chichester, 101–130.

Nash, J.E., 1957. *The Form of the Instantaneous Unit Hydrograph*. International Association of Scientific Hydrology, Wallingford, Publication No. 42, 114–118.

Noy-Meir, I., 1982. Stability of plant-herbivore models and possible applications to savanna. In B.H. Huntley and B.J. Walker (eds), *Ecology of Tropical Savannas*. Springer-Verlag, New York, 591–609.

Obando, J.A., 1997. *Modelling the Impact of Land Abandonment on Runoff and Soil Erosion in a Semi-Arid Catchment*. Unpublished Ph.D. Thesis, King's College, London.

Penman, H.L., 1948. Natural evaporation from open water, bare soil and grass. *Proceedings of the Royal Society of London, Series A*, **193**, 120–145.

Rebeiro-Hargrave, A., 2000. Large scale modelling of drainage evolution in tectonically active asymmetric intermontane basins, using cellular automata. *Zeitschrift fur Geomorphologie*, Supplementband **118**, 121–134.

Rodriguez-Iturbe, I., 1993. The geomorphological unit hydrograph. In K. Bevan and M.J. Kirkby (eds), *Channel Network Hydrology*. Wiley, Chichester, 43–69.

Rodriguez-Iturbe, I., Gonzalez-Sanabria, M. and Bras, R.L., 1982. A geomorphoclimatic theory of the instantaneous unit hydrograph. *Water Resources Research*, **18**, 887–903.

Rodriguez-Iturbe, I. and Rinaldo, A., 1998. *Fractal River Basins*. Cambridge University Press, Cambridge.

Rodriguez-Iturbe, I. and Valdes, J., 1979. The geomorphologic structure of hydrological response. *Water Resources Research*, **15**, 1409–1420.

Rotnicki, K., 1983. Modelling past discharges of meandering rivers. In K.J. Gregory (ed.), *Background to Palaeohydrology: A Perspective*. Wiley, Chichester, 321–354.

Schumm, S.A., 1956. Evolution of drainage systems and slopes in badlands at Perth Amboy, New Jersey. *Geological Society of America Bulletin*, **67**, 597–646.

Schumm, S.A., 1965. Quaternary paleohydrology. In H.E. Wright and D.G. Frey (eds), *Quaternary of the United States*. Princeton University Press, Princeton, 755–764.

Schumm, S.A., 1968. Speculation concerning palaeohydrologic controls of terrestrial sedimentation. *Geological Society of America Bulletin*, **79**, 1573–1588.

Schumm, S.A. and Khan, H.R., 1972. Experimental study of channel pattern. *Geological Society of America Bulletin*, **83**, 1755–1770.

Scott, L., Anderson, H.M. and Anderson, J.M. 1997. Vegetation history. In R.M. Cowling, D.M. Richardson and S.M. Pierce (eds), *Vegetation of Southern Africa*. Cambridge University Press, Cambridge, 62–78.

Shannon, J., Richardson, R. and Thornes, J.B., 2002. Modelling event-based fluxes in ephemeral streams. In L.J. Bull and M.J. Kirkby (eds), *Dryland Rivers: Hydrology and Geomorphology of Semi-Arid Channels*. Wiley, Chichester, 129–172.

Shen, H.W., 1971. *River Mechanics*. H. W. Shen Publications, Fort Collins, CO.

Sidorchuk, A., 1996. The structure of river bed relief. In P.J. Ashworth, S.J. Bennett, J.L. Best and S.J. McLelland (eds), *Coherent Flow Structures in Open Channels*. Wiley, Chichester, 397–423.

Thornes, J.B., 1988. Erosional equilibria under grazing. In J. Bintliff, D. Davidson and E. Grant (eds), *Conceptual Issues in Environmental Archaeology*. Edinburgh University. Press, Edinburgh, 193–211.

Thornes, J.B., 1990. The interaction of erosional and vegetational dynamics in land degradation: spatial outcomes. In J.B. Thornes (ed.), *Vegetation and Erosion*. Wiley, Chichester, 41–53.

Thornes, J.B., 2002. Geomorphology in the next millennium. *Proceedings of the Annual Conference of the Spanish Geomorphological Society*, Madrid, August 2000; in press.

Thornes, J.B., 2003. Exploring the grass-bush transitions in South Africa through modelling the response of biomass to environmental change. *Geographical Journal*, **169**, Part 2; in press.

Thornes, J.B., 2002b. Instability of vegetation along climatic gradients. International Workshop on Desertification, Medenine, Tunisia, 2000; submitted.

Thornes, J.B. and Brandt, C.J., 1993. Erosion-vegetation competition in a stochastic environment undergoing climatic change stochastic rainfall variations. In A.C. Millington and K.J. Pye (eds), *Environmental Change in the Drylands: Biogeographical and Geomorphological Responses*. Wiley, Chichester, 306–320.

Thornes, J.B. and Brunsden, D., 1977. *Geomorphology and Time*. Methuen, London.

Thornes, J.B. and Morley, C.D., 1970. A Markov decision model for network flows. *Geographical Analysis*, 180–191.

Trzaska, S., 2001. Regional impacts from El Niño on Southern Africa. *Medias Newsletter*, **12**, 30–33.

Walker, B.H. and Noy-Meir, I., 1982. Aspects of the stability and resilience of savanna ecosystems. In B.H. Huntley and B.J. Walker (eds), *Ecology of Tropical Savannas*. Springer-Verlag, New York, 556–590.

Walling, D.E. and Webb, B.W., 1981. Patterns of sediment yield. In K.J. Gregory (ed.), *Background to Palaeohydrology*. Wiley, Chichester, 69–100.

18 Palaeofloods and Extended Discharge Records

V.R. BAKER

The University of Arizona, Tucson, Arizona

1 INTRODUCTION

The conventional hydrological study of floods derives from the perceived need to estimate flood risk in order to reduce flood damages. In defining risk as a consequence (related to the magnitude of the event times a probability or frequency of occurrence), floods get treated statistically as populations of a single class of phenomena. This class is represented exclusively by numerical measures of magnitude (usually peak discharges) and associated probabilities of occurrence, nonoccurrence, and/or exceedance. Though logical and self-consistent for its highly limiting assumptions about the nature of "floods", conventional flood science is subject to the well-documented criticism that it fails to be associated with any regional or global reduction of flood damages (Baker, 1994; Pielke, 1999; Baker *et al.*, 2002). Despite this anomaly, an immense technical literature and most engineering practice continue to focus on the problem of estimating probabilities of flood occurrence. In the United States, this paradigm is exemplified in the much-abused, highly idealized concept of the "hundred-year flood," which will be discussed below.

Both conventional flood hydrology and palaeoflood hydrology begin with the same information: real data that indicate the properties of real floods. However, conventional flood hydrology is limited to data from stream-gauging sites. Very few stream gauges typically have record lengths that exceed a few decades. Except in some highly developed nations, in regions of low-flood variability, notably the United Kingdom, northwestern Europe, and the northeastern United States, there are very few stream gauges relative to the many streams that experience flooding. This is especially the case in arid and tropical savanna areas of extremely high flood variability. Moreover, recording at gauges is flawed, particularly for the extremely large, rare floods that destroy recording devices or prevent their safe access during the events of interest (Baker *et al.*, 2002). These limitations on conventional flood data have generally been addressed by employing a variety of modeling procedures that extend flood data from the range of actual measurements into extrapolations that are notorious for not being subject to testing against the real world (Klemes, 1986; 1987; 1989; 2000). When the conventional science of floods gets extended into the realm of human adaptation to global environmental change, the problems just noted are magnified to an alarming level. Indeed, the whole relationship of flooding to global environmental change and its associated modeling paradigm is becoming one of the most important Earth science challenges for the new millennium.

Palaeohydrology: Understanding Global Change. Edited by K.J. Gregory and G. Benito
© 2003 John Wiley & Sons, Ltd ISBN: 0-470-84739-5

2   HISTORICAL AND PHILOSOPHICAL PERSPECTIVES

Although the term "palaeoflood hydrology" was not introduced until 1982 (Kochel and Baker, 1982), this science has its origins in a long tradition of geological studies of the erosional and depositional effects of past floods (Costa, 1987; Patton, 1987). Especially influential on the development of palaeoflood hydrology were the investigations by Bretz (1923; 1925; 1929) of the cataclysmic late-Pleistocene floods of the Channeled Scabland region in the northwestern United States. Subsequent quantitative analysis of these floods (Baker, 1973) revealed methodologies that could be applied to late-Holocene palaeoflood features in a landscape currently being shaped by rare, intense floods in central Texas (Baker, 1974; 1975). Baker (1975) describes the setting as follows:

> The narrow bedrock valleys... produce constrictions of flood flows that result in large discharges being accommodated by relatively great flow depths. High flood stages on the major rivers drown the mouths of low-order tributary valleys and result in eddies at tributary junctures. The maximum height of the suspended sediment deposited at such slackwater locations might be used as a measure of the maximum stage achieved by flooding along a particular fluvial reach.

Baker (1975) calculated discharges for the indicated palaeostages and showed that they equaled or exceeded the maximum flood of record on streams that had particularly long gauge records. Obviously the resulting data are of great interest for streams that lack such records.

Palaeofloods are past or ancient floods that occurred without being studied by either (1) direct measurement using hydrological procedures, or (2) observation and recording by human observers (in which case the floods are termed *historical floods*). Palaeofloods are known to scientists via natural, physical signs or evidence of their past occurrence (Baker, 1998a). Human observation and documentation of the actual flooding does not occur, as in the case of historical floods. Instead, it is the signs and physical evidence of the floods that get documented by human observers. The archive of palaeofloods is not artificially limited by the opportunities of humans to observe floods. Instead, there is an immense archive of natural flood evidence limited only by the will and understanding to decipher it.

Despite the clarity of the above distinctions, presented numerous times in the literature on palaeoflood hydrology, much of the conventional hydrological literature continues to portray palaeofloods as a subset of historical floods, or vice versa. The latter mistake of classification is particularly prevalent in the literature on flood-frequency analysis, which commonly lumps all past floods together into the designation "historical floods." This is only one of the many confusions in the flood-frequency literature to be elaborated upon below.

The natural recording of palaeofloods is achieved via a wide variety of indices (causal signs) that indicate past flood occurrence (Baker, 1998a; 2000). At some later time, these indices can be studied by a flood scientist, who will strive to achieve an objective conceptual reconstruction of the causative past flood processes. In contrast, historical floods usually involve observations by nonscientists who are not experienced in flood phenomena. Such observations vary considerably in accuracy, relevance, and reliability. By failing to distinguish such observations from palaeofloods, some hydrologists have misinterpreted the objective recording capabilities of natural systems, and thereby misrepresented the science of palaeoflood hydrology. The problem is compounded when ignorance of fluvial geomorphology and geology leads some hydrologists to make uninformed statements about the recording potential for the natural signs of past flood processes (e.g. Hosking and Wallis, 1986; Yevjevich and Harmancioglu, 1987).

3 TYPES OF PALAEOFLOOD INVESTIGATIONS

Many techniques are available for the inference of past flood-flow parameters using principles of geomorphology and related aspects of Quaternary stratigraphy and sedimentology (Baker, 1976; Costa, 1978). The most commonly used procedures can be divided into three categories: (1) regime-based palaeoflow estimates (RBPEs), (2) palaeocompetence studies, and (3) palaeostage estimates. RBPEs utilize empirically derived relationships to relate relatively high-probability flow events, such as the mean annual flood or bankfull discharge, to palaeochannel dimensions, sediment types, palaeochannel gradients, and other field evidence for alluvial rivers (e.g. Dury, 1976; Williams, 1984). Palaeocompetence studies, in contrast, are not restricted to alluvial rivers or to flood flows of moderate frequency. They rely upon the relationships of very large sedimentary particles to their transport, usually expressed in terms of palaeoflood bed shear stress, velocity, or stream power per unit area (SS-V-SP). Although applicable to selected fluvial environments (Baker and Ritter, 1975; Costa, 1983; Williams, 1983; O'Connor, 1993), SS-V-SP studies can be of very low accuracy when local controls on sediment transport are poorly known (Maizels, 1983).

Palaeostage estimates are best exemplified by the study of SlackWater Deposits and other PalaeoStage Indicators (SWD-PSIs) in stable-boundary fluvial reaches (Baker, 1987; 2000). The list of possible palaeostage indicators is large, and includes a variety of palaeobotanical measures (Yanosky and Jarrett, 2002), scour of regolith materials marginal to a bedrock channel (Partridge and Baker, 1987), and silt emplaced on bedrock channel walls (Webb et al., 2002). Slackwater deposits may occur in many settings, including deep-in-limestone caves adjacent to river channels (Gillieson et al., 1991; Springer and Kite, 1997; Springer, 2002). The estimation of palaeoflood discharges from palaeostages has been transformed by the availability of computer-modeling procedures. In contrast to earlier reliance upon slope-area and step-backwater modeling (O'Connor and Webb, 1988), there is now a suite of modeling tools, which are best adapted to the information available for the study site of interest (Webb and Jarrett, 2002; Denlinger et al., 2002).

Closely related to palaeostage estimation is the establishment of palaeohydrological bounds, a procedure developed by the US Bureau of Reclamation for their studies of potential risk to high-hazard dams from extreme flooding (Levish et al., 1994; Ostenaa et al., 1996; Levish, 2002). A palaeohydrological bound is defined as an interval of time during which some threshold discharge has not been exceeded (Levish, 2002). Various kinds of stable geomorphological surfaces adjacent to flood-prone streams are studied for evidence of any past modification by floods. The age of the lowest surfaces not modified by flooding are used to establish the bound. Data obtained in this manner are particularly compatible with Bayesian flood-frequency analysis (O'Connell et al., in press).

Another interesting variant on palaeoflood hydrology involves the occurrence of levee-like deposits surrounding the plunge pools of tropical waterfalls that convey flood flows with relatively high suspended sand concentrations. These plunge-pool deposits can provide relatively long records of past flood occurrence (Nott et al., 1996), though it is difficult to obtain accurate absolute quantitative measurements of past flood discharges. Variations in the relative magnitudes of the floods are more easily interpreted, and indicate rather complete records of changes in the largest events over periods as long as 30 kyr (Nott and Price, 1999).

4   APPLICATIONS OF PALAEOFLOOD HYDROLOGY

4.1   Geomorphology of Bedrock Rivers

In alluvial rivers, the channel-forming discharges are generally of relatively high frequency, of statistically consistent magnitudes (e.g. bankfull), and of relative ease for direct measurement. Bedrock rivers, in contrast, achieve their infrequent, but sometimes spectacular morphological responses from rare, high-magnitude floods that are exceptionally difficult to measure during their chance occurrences (Baker, 1977). These circumstances have led to extensive use of palaeoflood hydrology for the analysis of the fluvial processes that are most significant for understanding the geomorphology of bedrock rivers (O'Connor et al., 1986; Baker and Pickup, 1987; Wohl, 2002).

4.2   Hydrology of Desert Rivers

When palaeoflood data are combined creatively with other hydrological procedures, striking advances are possible in understanding basic hydrological processes. In a study of small mountain watersheds in southern Arizona, Martinez-Goytre et al. (1994) used a rainfall-runoff model in an inverse mode to estimate rainfall intensities associated with extreme flood peaks. In regard to the generation of flood peaks, this study also demonstrated the importance of basin orientation relative to the tracks of major storms. In another study, House and Baker (2001) found that flood-frequency information could be derived from even very small drainage basins in a small region otherwise lacking in any data concerning flood characteristics.

   The extensive palaeoflood records of the Negev Desert provide an excellent demonstration for estimation of magnitude-frequency relationships for the largest floods in a hyperarid region (Greenbaum et al., 2001). Greenbaum et al. (1998) used palaeoflood data from a hyperarid catchment in the Negev Desert to analyze a three-peak flood hydrograph associated with a complex mesoscale storm pattern. The spatial pattern of the storm's rainfall distribution was characterized by radar remote sensing. This combination of real-time remote sensing of rainfall distribution and post-event palaeoflood field study illustrates the remarkable opportunity afforded by palaeoflood hydrology to advance the understanding of basic rainfall-runoff relationships in remote regions that lack adequate networks for the measurement of rainfall and streamflow. In another application, Greenbaum et al. (2002) estimated long-term transmission losses and recharge by combining the palaeoflood data with measured flood parameters in instrumented catchments.

4.3   Regional Patterns

It has long been known that patterns of Holocene fluvial aggradation and entrenchment or incision occur in cyclic patterns, probably correlating to respective phases of higher and lower flood frequency on multi-century timescales (e.g. Baker and Penteado-Orellana, 1978; Brakenridge, 1980; Starkel, 1983; Knox, 1985). The global expansion of palaeoflood studies over the past 20 years has now revealed that the century-scale variations in flood frequency extend over the past few millennia. Many studies indicate clustering of the largest floods into discrete time intervals, with time periods containing many large floods alternating with periods of few large floods. Within a given region, there are consistent patterns of large flood incidence that correlate between study sites. However, the palaeoflood records do not necessarily correlate from one region to another. Indeed, the rather extensive palaeoflood records of the southwestern United States (Ely *et al.*, 1993) and the upper Mississippi valley (Knox, 1993; 2000) seem to be out of phase with one another. Other areas showing pronounced temporal clustering of large palaeofloods are central India (Ely *et al.*, 1996; Kale *et al.*, 1997; in press); northwestern India (Kale *et al.*, 2000); northern Australia (Wohl *et al.*, 1994a; 1994b), and the Negev desert (Greenbaum *et al.*, 2000; 2001).

5 GLOBAL CHANGE

It is widely recognized that increasing atmospheric concentrations of radiatively active gases, particularly anthropogenic carbon dioxide, will lead to major changes in the global mean climate (Houghton *et al.*, 2001). Immense public expenditure is being lavished on the improvement of climate modeling as the ideal means for anticipating the presumably detrimental consequences of the indicated future climate changes. However, the current climate models are only accurate for predicting mean parameters of climate over large regions. Average seasonal temperature is much more reliably predicted than is mean regional precipitation, but extremes of temperature and precipitation cannot be reliably predicted. It is presumed that future advances in the modeling capability will eventually achieve success for accurately predicting even the hydrological consequences of these climatic extremes, but the date for achieving this goal is not known.

5.1 Modeling Global Floods

In contrast to what can be reliably inferred from the application of climate models, the most important potential impacts of global changes on society involve extreme phenomena such as hot spells, droughts, and floods (Easterling *et al.*, 2000; McCarthy *et al.*, 2001). Moreover, there is an interesting turn of logic by which the risk of these extreme phenomena is commonly cited in justification for large financial expenditures to support the climate modeling that has proven so ineffective in their prediction. While hundreds of published climate-modeling studies elaborate upon patterns of increased average temperatures of a few degrees, only a very few studies contribute anything meaningful to understanding changes in extreme floods. Given the difficulties for defining credible modeling scenarios for predicting the changes in large rainfall events that trigger flooding, McCarthy *et al.* (2001) conclude

> Global climate models currently cannot simulate with accuracy short-duration, high intensity, localized heavy rainfall, and a change in mean monthly rainfall may not be representative of a change in short-duration rainfall.

The best that can be done is to treat multiple modeling scenarios in a statistical manner. Thus, Palmer and Raisanen (2002) link the results of 19 climate models and apply a decision-making tool in a probabilistic approach for representing to policy makers the changing frequency of extreme events. Although this approach is better than that of basing decisions on a single simulation, it is still limited by systematic errors common to all models. Moreover, only the largest river basins on the planet can be analyzed this way, since model grid sizes are scaled to about 200 km. Improvement in all these aspects of modeling is possible in theory (and with much more research funding), but it will be many years before this policy can yield results that are credible (Schnur, 2002).

Both conceptual modeling (Trenberth, 1999) and intuitions from current understanding of the climate system (Meehl *et al.*, 2000) hold that warming from increasing greenhouse gases will intensify the global hydrologic cycle. Increased radiative forcing will increase surface heating and latent heating, which, in turn, will increase air temperature and evaporation. These then increase atmospheric water vapor, leading to more intense precipitation and increased storm intensity. More moisture will go into the atmosphere in the tropics, possibly enhancing tropical flood-generating phenomena, and also accelerating freshwater transport to higher latitudes. All other things being equal for a given drainage basin, the extreme flood magnitudes should increase. The limited available conventional data on recent trends in heavy precipitation does indeed show increases for some regions (southeast Australia, Russia, northeast Brazil, eastern United States, southern Canada, Norway, southern China, South Africa, and northern Japan). However, other areas (Ethiopia, east Africa, Thailand, northeastern China, and southern Japan) show decreases in heavy precipitation (Easterling *et al.*, 2000). In a global study of floods exceeding the 100-year magnitude in 29 very large river basins (areas larger than 200,000 km^2), Milly *et al.* (2002) found that the frequency of such floods increased substantially after 1950. Comparing this result with climate-modeling results, Milly *et al.* (2002) predicted a continuing trend of increasing extreme floods.

5.2 Abrupt Climate Changes and Palaeofloods

It has recently become apparent that the immense international effort to understand global change has devoted insufficient attention (and resources) to the understanding of abrupt climate changes (National Research Council, 2002). The discovery that such abrupt changes could pose a significant problem to humanity did not arise from climate modeling. Instead, it was the study of proxy records of past environmental change that revealed extreme adjustments of climate within periods of fewer than 10 years (Broecker, 1997; Alley and Clark, 1999). These proxy studies had been originally justified as a means for testing the predictive capabilities of the climate models. However, they achieved a more important purpose by showing that the whole program of climate modeling had ignored a component of climate change that is critical to the human adaptation to that change. One is tempted to ask what might be the more important science question: to test the capabilities of theoretical models, or to make significant discoveries about the natural world?

In comparing palaeoflood records with climate trends, there is mounting evidence of a general association between times of rapid climate change and more frequent occurrences of extreme floods. This is seen in palaeoflood and other fluvial records from the southwestern United States (Ely *et al.*, 1993; Ely, 1997), the upper Mississippi valley (Knox, 2000), Spain (Benito *et al.*, 1996), and much of northern, central, and western Europe (Rumsby and Macklin, 1996). Perhaps the most interesting clustering of large palaeofloods is in the period after 1950. This concentration is seen in many areas prone to incursion by tropical storms, including central India (Ely *et al.*, 1996), central Australia (Pickup *et al.*, 1988), northern Australia (Wohl *et al.*, 1994a; 1994b), and western Texas (Patton and Dibble, 1982; Kochel *et al.*, 1982). It is reasonable to hypothesize that this change may be related to enhancement of the monsoon in India (Kale, 1999) and to an increase in the frequency of occurrence and/or a change in the characteristic timescales of the El Nino Southern Oscillation phenomenon (Urban *et al.*, 2000). The latter could be related to anthropogenic warming of the Tropical Pacific Ocean (Trenberth and Hoar, 1996). At least one climate model predicts El

Nino events to become stronger and more frequent as a result of global warming (Timmermann *et al.*, 1999).

Of course, the interpretation of palaeoflood records does not rely upon inferences of flood response in relation to climate change. Palaeoflood data directly signify the properties of real floods in the same way that a weathervane indicates properties of the wind. The relationship is that of being a direct sign of the causative flood. In technical terms of logic and semiotics, this is not the same as being a substitute or "proxy" for some different phenomenon of interest (Baker, 1995; 1998a). Because palaeoflood data are direct extensions of real floods, they will tell the analyst about the behavior of those floods (i.e. various causal connections and associations), regardless of the state of understanding concerning the causal mechanisms. In this way, the need to evaluate the global flood hazard can be addressed as rapidly as data can be amassed, regardless of the state of any predictive modeling capabilities. Of course, the strategy just outlined has not been a priority for current global change science, as evidenced by the relative emphasis placed on the collection of palaeoflood data versus the emphasis accorded to the improvement of climate models.

6 FLOOD-FREQUENCY ANALYSIS

It cannot be overemphasized, particularly to engineering hydrologists, that palaeoflood hydrology merely extends the range in time for the availability of real-world information about naturally occurring floods. Thus, it is misleading (cf. Yevjevich and Harmancioglu, 1987) to compare palaeoflood hydrology to various analytical procedures, such as frequency curves, rainfall-runoff modeling, regionalization, Bayesian analysis, and probable maximum flood estimation. All these procedures stand to benefit from the introduction of more real-world information that can extend datasets in time and space, test models, or provide confidence for engineering judgments. Palaeoflood hydrology offers the means to enhance all these methods; it is not an alternative methodological procedure to be evaluated in contrast to them. The logic of the latter exercise is equivalent to criticizing the collection of hard data as an alternative to theoretical modeling pursued in the absence of such data.

There is an unfortunately common assertion that palaeoflood hydrological data inherently embody more uncertainty than do other types of flood data. More disturbingly misleading is the claim that palaeoflood hydrologic "methods of flood estimation" are inherently more uncertain than are "other methods." Aside from the confusion of data with methods, this uncertainty has been attributed (1) to large errors presumed to characterize individual flood value estimates (Hosking and Wallis, 1986), and (2) to errors in supposed "proxy" relationships between data and floods (Yevjevich and Harmancioglu, 1985). Such statements are usually made as generalities, as though they were the common wisdom of all hydrologists. They are expressed with little or no attention to the great variety of palaeoflood data sources or to the immense and growing scientific literature on the subject. The relatively high accuracy of palaeoflood data in comparison to other data on extremely large, rare floods has been demonstrated in numerous studies (e.g. Baker, 1987; 2000; Webb *et al.*, 2002). Moreover, any uncertainty is easily quantified and incorporated with great benefit into analyses of flood frequency (Stedinger and Baker, 1987; Blainey *et al.*, 2002).

6.1 Nonstationarity

Very long flood records lead to considerations of the issue of stationarity in a flood series. This issue does not arise as a consequence of scientific concern with understanding the true nature of floods. Rather, it arises when emphasis is placed on the methodology of flood-frequency analysis, which requires the treating of floods as a generalized phenomenon (Baker, 1994). Analyzed for its logic by Klemes (1987; 1989), flood-frequency methodology requires that the phenomena being analyzed be treated as a record of random samples drawn in a consistent manner from a single population. In flood-frequency analysis, the occurrence of floods in nature is considered essentially to be equivalent to the drawing of playing cards from a deck or the spinning of a roulette wheel. In the conventional application to stream-gauge records (Hazen, 1914), the presumption of population randomness must apply uniformly to flood phenomena over both the period of gauge record and the future period over which an extrapolation of this record is to hold. This leads to the notion of "stationarity" for homogeneous flood populations. The stationarity necessary for reliable flood-frequency analysis is justified by considering how well the data shows that the flooding can be analyzed as a random phenomenon with certain statistical properties. The logic of this exercise is of critically assessing data in regard to how elegantly the analyst can make statements (model) on the generalized phenomenon of "floods." It sharply contrasts the logic of criticizing the statements (models) about "floods" in the light of whatever data one can gather from the real world of floods and their effects. Klemes (2000) provides insightful and in-depth discussion of these issues in regard to the scientific nature of hydrology.

In returning to the understanding of floods in nature, as opposed to their idealized generalization as phenomena, there are three principal ways in which real floods deviate from the idealized randomness that must be presumed for conventional flood-frequency analysis. First, for a given time period, floods do not constitute a single class of phenomena. Instead, different flood magnitude/frequency characteristics are generated by different meteorological processes (Hirschboeck, 1987a; 1988). Second, in addition to such mixed-population flood series, the storm tracks generating floods may change through time with changing climate, thereby altering the incidence of large floods in particular regions (Hirschboeck, 1987b; 1988). Third, changes in land-cover and land-use patterns may alter the flood responses of a basin through time.

Resolution of these issues involves understanding the hydrology of the particular study area. In southwestern United States, for example, extreme natural flood variability and minimal human impact on the extreme flood response of large basins both preclude any significant impact for land-use change, outside urban areas. The mixed-population issue applies to all flood data, not just to palaeoflood data. The supposed violations of flood-frequency analysis assumptions derive from the true nature of floods, not from the type of data obtained from those floods. Finally, as documented for the southwestern United States in compilations by Ely et al. (1993) and Ely (1997), extreme floods and their causative flood-generating storm conditions cluster into centennial-scale periods of enhanced flood magnitudes. The past 500 years seems to be a cluster of enhanced flood magnitudes.

There are two possible reactions to the flood nonstationarity issue described above. The first reaction is exemplified by the National Research Council (1999) study of flood-frequency analyses for the American River in California. In that study, the term "bias" is used to refer to long periods of flood record that are likely to include time periods during which flood risk is different from that during the immediate planning period. (Note that here "bias" refers to the actual nature of past flooding and not to an interpretation of that flooding.) Although extensive and accurate palaeoflood data had been developed for the American River, providing a record of 3,500 years, the National Research Council (1999) chose to ignore that data in their flood-frequency analysis because of unease about climate variability over the relatively long time period. Given inadequate understanding of how floods vary in time, flood-frequency analysis must rely heavily upon judgment. It was concluded that, unless flood magnitudes have been independently and identically distributed in time during the period represented by the palaeoflood information, they cannot be incorporated with benefit into a flood-frequency analysis. It was further noted that the only way to overcome potential "bias" in the palaeoflood dataset would be to have a correct mathematical model of the variations of floods in time that would accurately predict flood parameters in order to approximately weight data from past floods. One is left to wonder how, without data on the phenomenon of interest (large, rare floods), one could ever possibly know that one had a correct mathematical model for variations of flood parameters in time.

The second response to the nonstationarity issue involves the goal of achieving understanding of the past, in contrast to the rationalizing of various presumptions of ignorance about it. Two strategies might be envisioned for this goal. In one, the clusters of extreme flooding may be taken as periods characterized by one flood distribution that would include today's flooding and reasonable extrapolations from it. The time periods for such clusters will be indicated by the palaeoflood data themselves (e.g. Ely *et al.*, 1993; Ely, 1997). These would then be used as the time periods for which the sample of palaeoflood data and gauge data could be held to be homogeneous. Another strategy would be to treat the whole problem as one of mixed populations, in which the really large floods, which dominate in the palaeoflood record, are understood in terms of their flood-causing phenomena. This would involve attempts to understand the flood hydroclimatology (Hirschboeck, 1988), but in terms of data, rather than in terms of the "correct mathematical model" envisioned for the American River by the National Research Council (1999). Of course, such attempts at achieving understanding necessarily violate the overriding goal of the first response, which is the treating of floods as a class of phenomena confined to limited properties for ease of mathematical manipulation in a flood-frequency analysis (cf. Klemes, 1987; 1989; 2000). Thus, one sees that the two responses reflect completely different goals: a limited engineering goal for the first response, exemplified by the National Research Council (1999) approach to nonstationarity and a scientific goal for the second response.

Of course, any concern with nonstationarity "bias" in flood data must apply to all data, including gauge records. The logical direction of this argument is that, because there is uncertainty about what one can learn of the past, any projections on the future must be made in total ignorance of what has ever actually occurred. Proposals for a "correct mathematical model" that would predict extreme flood phenomena with great accuracy would seem to require actual data dealing with the phenomena to be predicted. Only long palaeoflood records have the ability to include real manifestations of the rare, improbable floods that are the realizations of the predicted extreme floods. Thus, it is most unrealistic to scrutinize the relevance of data from the real world for achieving consistent functioning of a methodology that is, after all, merely presumed to represent that world. Rather, one might more productively emphasize modifications to the methodology to better represent those realities that actually threaten humans and their constructions, realities that are signified in the palaeoflood evidence of actual occurrence.

6.2 Testing of Models

In addition to providing the data for a flood-frequency analysis, palaeoflood hydrology can provide a scientific check on the reasonableness of flood-frequency estimates achieved independently of their consideration. This use is particularly important in relation to the rainfall-runoff models that have become increasingly central to modern flood-hazard evaluation. There is considerable concern that the ability to model by computer simulation is far outstripping the ability to test or validate model predictions with real data (Klemes, 1982; 1986; Pilgrim, 1986). The problem is especially acute in the study of extreme floods because of the above noted problems with applying conventional data to that issue (Klemes, 1987). Compounding this problem is the fact that certain regions of the world, particularly arid and semiarid regions (Pilgrim et al., 1988), are extremely problematic to model. This is because of (1) high spatial and temporal variability for rainfall and runoff, (2) the diversity and complexity of runoff-producing processes, and (3) the inadequacies of the conventional database.

6.3 The 'Hundred-Year Flood'

The term "hundred-year flood" constitutes an oxymoron in that the "hundred-year" period does not indicate real time, but rather is the imaginary mathematical result of inverting an annual exceedance probability. Nor does the concept deal with any real "flood" because the indicated discharge is an extrapolation, not a measurement of physical signs associated with anything that has actually happened. The presumed benefit derived from this paradigm is that the unreal predicted magnitude and associated probability can be determined algorithmically by procedures easily replicated in any engineering consulting firm, and thereby readily defended in a court of law as "best engineering practice." The latter attribute is particularly important because the whole apparatus that supports and reinforces the paradigm is a legal regulatory framework that was intended to achieve reduced flood damages (but demonstrably fails in doing so).

6.4 The 'Probable Maximum Flood'

A major area of practical application for palaeoflood hydrology is in relation to the flood design or risk numbers applied to dams, nuclear facilities, hazardous waste sites, and other critical infrastructure. In conventional engineering practice, such structures were designed in regard to the discharge of a "Probable Maximum Flood" (PMF). In what seems to be the norm for this type of nomenclature, the PMF has nothing to do with probabilities. Rather, it is a single number (not a real flood) that is determined by applying a rainfall-runoff model to the intensities and spatial distribution of a probable maximum precipitation. The latter is determined by hydrometeorological assessment of the most extreme combination of storm characteristics that could prevail over given drainage basin. Although this number is supposed to be determined in a highly objective and conservative fashion for the design of dam spillways, it turns out to not have been done consistently for many of the thousands of existing dams and other hazardous structures. Out of concern for the safety of those structures, and in order to prioritize those most at risk, it is necessary to evaluate the magnitudes to be expected for floods with probabilities in the range of about 10^{-3} to 10^{-7} chance of occurrence each year. Of course, conventional gauge data do not have records long enough for extrapolation to such small annual exceedance probabilities. The result has been the adoption of palaeoflood hydrological procedures by the US Bureau of Reclamation (Levish, 2002) and the US Geological Survey (Jarrett and Tomlinson, 2000). In these agencies, frequency estimates from the palaeoflood records are compared with the PMF in order to see which PMF determinations result in the greatest risk at various existing dams.

7   REDUCING FLOOD DAMAGES

The above considerations bring the focus back to the issue with which this essay was introduced. Hydrological procedures for estimating flood risk were originally conceived, and continue to be supported financially, for the purpose of reducing flood damages to society. That those damages continue to increase (Pielke and Downton, 2000) is an anomaly that raises serious questions about the entire scientific infrastructure that has been established for the understanding of flooding (Baker, 1994; Baker *et al.*, 2002).

In the public debate over responses to potential hazards, the most effective stimulus to the taking of action is the actual occurrence of a disaster. This is one reason politicians rush to appear before cameras at every major flood disaster. Taking advantage of the temporary suspension of the public apathy or cynicism that usually greets them, the politician can assume the role of relief agent during a crisis. The crisis often gets portrayed (in the US) as resulting from an "act of God," thereby diverting attention from the social and political factors that contributed to it. Actions taken at such times are rarely based on serious scientific reflection. The actions are inevitably taken to continue the positive impressions upon voters, at least until the next election. Scientific terms, such as the "hundred-year flood", sometimes do get employed as a basis of authority for actions taken to achieve various political goals. The usual role for science is to contribute authority, not to enlighten with understanding.

When scientific reflection is possible, generally in the long spans of time between disasters, the complacency of those inhabiting or constructing in flood-hazard zones is not readily swayed by arcane discussions of hypothetical probability distributions. Nor is public enlightenment fostered by the nearly universal confusion associated with regulatory zoning of "hundred-year" and "five-hundred-year" floodplains. These are not likely ever to resolve the continuing paradox of ineffective action to mitigate flood losses, despite immense expenditure on the science presumed to achieve that end. A proposal for breaking this deadlock is for the science to directly influence public perception. Instead of communicating the authoritative rationale of "hundred-year-flood" zones, the flood scientist will inform both the public and the decision makers of the fully documented occurrence of ancient (but very real) cataclysmic processes in the very places where public interest is greatest. This tying of flood science to public perception (Baker, 1998b) contrasts with the usual relationship of science to public policy whereby science is supported mainly for the authority that it conveys to the actions of politicians or other decision makers.

The role envisioned for flood science is more democratic than authoritarian. Informed by the realities of the natural world, science would be focused on developing better questions for policy implementation, not on providing information for the justification of policies. Relief from flood damages would not be a political end to be justified in advance, but a public good to be discovered through experience. This radical notion will entail the continual assessment of policies already implemented, which would require a role for science rather different than its current authoritarian position. Because it would also mean a completely different basis for continuing financial support, it is unlikely to achieve wide attention from those scientists who are highly invested in the current arrangements.

8   CONCLUSIONS

Palaeofloods are known to scientists through evidence (natural signs) causally connected to the responsible past or ancient floods. Evidence of palaeoflooding is limited neither by direct human observation of the flood events nor by presumed substitution ("proxy") relationships to floods. Instead, only will and understanding are necessary to access and decipher immense archives of natural flood signs.

Palaeoflood hydrology provides the means of obtaining basic data on extremely large, rare floods that are not recorded by conventional stream gauges. Palaeoflood hydrology is not a hydrological procedure appropriately contrasted with rainfall-runoff modeling, regionalization, or other methods of flood-frequency analysis. Instead, palaeoflood data constitute (1) critical extensions to the temporal and spatial ranges of input data for such modeling, (2) test sets with which to validate model predictions, and, most importantly, (3) real-world checks on model presumptions about the nature of large, rare floods.

Palaeostage evidence, most commonly slackwater deposits, is found globally to provide the most widely applicable palaeoflood information. Among the many uses for palaeoflood data are (1) advancing geomorphological process understanding of bedrock rivers, (2) providing hydrological process information for rivers lacking conventional data (including desert and monsoonal regions), (3) documenting regional patterns of large flood incidence through time, and (4) improving flood-frequency estimates. Application (4) is complicated in some nations because of arcane procedural limitations that evolved to achieve regulatory risk analyses with limited availability of conventional flood data. These procedures now impede efforts to improve flood-hazard assessment through scientific discoveries concerning the nature of large, rare floods made possible because of palaeoflood hydrology.

Patterns of large flood incidence are of immense relevance to the human consequences of global environmental change. Nevertheless, the science to identify these patterns has not been accorded an appropriately high status in international programs that purport to advance understanding of global climate change and its consequences. The overwhelming emphasis of the latter is on the improvement of climate models that are only successful in the prediction of mean conditions, not extremes. Moreover, human vulnerability to extreme floods can only be reduced when the perceptions of those at risk and of those who set policy are directed toward productive action. This will be achieved most effectively through the broad communication of direct evidence of real floods, rather than by authoritarian pronouncements of idealized symbolic representations, such as the "hundred-year flood." Instead of relying on regulatory representations of presumed flood properties, wise policy should be stimulated by the realities of flooding, and then continually reassessed and reformulated on the basis of experience. Though marginalized in current, demonstrably failed approaches to flood-hazard reduction, palaeoflood hydrology affords great potential for reducing the adverse consequences of flooding.

ACKNOWLEDGEMENT

This essay is Contribution Number 63 of the Arizona Laboratory for Paleohydrological and Hydroclimatological Analysis (ALPHA), The University of Arizona.

REFERENCES

Alley, R.B. and Clark, P.U., 1999. The deglaciation of the northern hemisphere: a global perspective. *Annual Reviews of Earth and Planetary Sciences*, **27**, 149–182.

Baker, V.R., 1973. *Paleohydrology and Sedimentology of Lake Missoula Flooding in Eastern Washington.* Special Paper 144, Geological Society of America, Boulder, CO.

Baker, V.R., 1974. Techniques and problems of estimating Holocene flood discharges. *Abstracts of the Third Biennial Meeting,* American Quaternary Association, University of Wisconsin, Madison, WI, 63.

Baker, V.R., 1975. *Flood Hazards along the Balcones Escarpment in Central Texas: Alternative Approaches to their Recognition, Mapping, and Management.* Geological Circular 75-5, University of Texas, Bureau of Economic Geology, Austin, TX.

Baker, V.R., 1976. Hydrogeomorphic methods for the regional evaluation of flood hazards. *Environmental Geology,* **1**, 261–281.

Baker, V.R., 1977. Stream channel response to floods with examples from Central Texas. *Geological Society of America Bulletin,* **88**, 1057–1070.

Baker, V.R., 1987. Paleoflood hydrology of extraordinary flood events. *Journal of Hydrology,* **96**, 79–99.

Baker, V.R., 1994. Geomorphological understanding of floods. *Geomorphology,* **10**, 139–156.

Baker, V.R., 1995. Global palaeohydrological change. *Quaestiones Geographicae,* **10**, 139–156.

Baker, V.R., 1998a. Paleohydrology and the hydrological sciences. In G. Benito, V.R. Baker and K.J. Gregory (eds), *Palaeohydrology and Environmental Change.* Wiley, Chichester, 1–10.

Baker, V.R., 1998b. Hydrological understanding and societal action. *Journal of the American Water Resources Association,* **34**, 819–825.

Baker, V.R., 2000. Paleoflood hydrology and the estimation of extreme floods. In E.E. Wohl (ed.), *Inland Flood Hazards: Human, Riparian, and Aquatic Communities.* Cambridge University Press, Cambridge, 359–377.

Baker, V.R. and Penteado-Orellana, M.M., 1978. Fluvial sedimentation conditioned by Quaternary climatic change. *Journal of Sedimentary Petrology,* **48**, 433–451.

Baker, V.R. and Pickup, G., 1987. Flood geomorphology of the Katherine Gorge, Northern Territory, Australia. *Geological Society of America Bulletin,* **98**, 635–646.

Baker, V.R. and Ritter, D.F., 1975. Competence of rivers to transport coarse bedload material. *Geological Society of America Bulletin,* **86**, 975–978.

Baker, V.R., Webb, R.H. and House, P.K., 2002. The scientific and societal value of paleoflood hydrology. In P.K. House, R.H. Webb, V.R. Baker and D.R. Levish (eds), *Ancient Floods, Modern Hazards: Principles and Applications of Paleoflood Hydrology.* Vol. 5, American Geophysical Union Water Science and Application, Washington, DC, 1–19.

Benito, G., Machado, M.J. and Perez-Gonzalez, A., 1996. Climate change and flood sensitivity in Spain. In J. Branson, A.G. Brown and K.J. Gregory (eds), *Global Continental Changes: The Context of Palaeohydrology.* Special Publication 115 of the Geological Society, Geological Society, London, 85–98.

Blainey, J.B., Webb, R.H., Moss, M.E. and Baker, V.R., 2002. Bias and information content of paleoflood data in flood-frequency analysis. In P.K. House, R.H. Webb, V.R. Baker and D.R. Levish (eds), *Ancient Floods, Modern Hazards: Principles and Applications of Paleoflood Hydrology.* Vol. 5, American Geophysical Union Water Science and Application, Washington, DC, 161–174.

Brakenridge, G.R., 1980. Widespread episodes of stream erosion during the Holocene and their climatic cause. *Nature,* **283**, 655, 656.

Bretz, J H., 1923. The Channeled Scabland of the Columbia plateau. *Journal of Geology,* **31**, 617–619.

Bretz, J H., 1925. The Spokane flood beyond the Channeled Scabland. *Journal of Geology,* **33**, 97–115.

Bretz, J H., 1929. Valley deposits immediately east of the Channeled Scabland of Washington. *Journal of Geology,* **37**, 393–427.

Broecker, W.S., 1997. Thermohaline circulation during the last deglaciation, the Achilles heel of our climate system: Will man-made CO_2 upset the current balance? *Science,* **278**, 1582–1588.

Costa, J.E., 1978. Holocene stratigraphy in flood-frequency analysis. *Water Resources Research*, **14**, 626–632.

Costa, J.E., 1983. Paleohydraulic reconstruction of flash-flood peaks from boulder deposits in the Colorado front range. *Geological Society of America Bulletin*, **94**, 986–1004.

Costa, J.E., 1987. A history of paleoflood hydrology in the United States, 1800–1970. In E.R. Landa and S. Ince (eds), *History of Hydrology*. American Geophysical Union, Washington, DC, 49–53, History of Geophysics Number 3.

Denlinger, R.P., O'Connell, D.R.H. and House, P.K., 2002. Robust determination of stage and discharge: An example from an extreme flood on the Verde river, Arizona. In P.K. House, R.H. Webb, V.R. Baker and D.R. Levish (eds), *Ancient Floods, Modern Hazards: Principles and Applications of Paleoflood Hydrology*. Vol. 5, American Geophysical Union Water Science and Application, Washington, DC, 127–146.

Dury, G.H., 1976. Discharge prediction, present and former, from channel dimensions. *Journal of Hydrology*, **30**, 219–245.

Easterling, D.R., Evans, J.L., Groisman, P.Ya., Karl, T.R., Kunkel, K.E. and Amberje, P., 2000. Observed variability and trends in extreme climate events: a brief review. *Bulletin of the American Meteorological Society*, **81**, 417–425.

Ely, L.L., 1997. Response of extreme floods in the Southwestern United States to climatic variations in the late Holocene. *Geomorphology*, **19**, 175–201.

Ely, L., Enzel, Y., Baker, V.R. and Cayan, D.R., 1993. A 5000-year record of extreme floods and climate change in the Southwestern United States. *Science*, **262**, 410–412.

Ely, L.L., Enzel, Y., Baker, V.R., Kale, V.S. and Mishra, S., 1996. Changes in the magnitude and frequency of late Holocene monsoon floods on the Narmada river, Central India. *Geological Society of America*, **108**, 1134–1148.

Gillieson, D., Smith, D.I., Greenaway, M. and Ellaway, M., 1991. Flood history of the Limestone ranges in the Kimberly region, Western Australia. *Applied Geography*, **11**, 105–123.

Greenbaum, N., Margalit, A., Schick, A.P., Sharon, D. and Baker, V.R., 1998. A high magnitude storm and flood in a hyperarid catchment, Nahal Zin, Negev Desert, Israel. *Hydrological Processes*, **12**, 1–23.

Greenbaum, N., Schick, A.P. and Baker, V.R., 2000. The palaeoflood record of a Hyperarid catchment, Nahal Zin, Negev Desert, Israel. *Earth Surface Processes and Landforms*, **25**, 951–971.

Greenbaum, N., Enzel, Y. and Schick, A.P., 2001. Magnitude and frequency of paleofloods and historical floods in the Arava basin, Negev Desert, Israel. *Israel Journal of Earth Sciences*, **50**, 159–186.

Greenbaum, N., Schwartz, U., Schick, A.P. and Enzel, Y., 2002. Paleofloods and the estimation of long term transmission losses and recharge to the lower Nahal Zin alluvial aquifer, Negev Desert, Israel. In P.K. House, R.H. Webb, V.R. Baker and D.R. Levish (eds), *Ancient Floods, Modern Hazards: Principles and Applications of Paleoflood Hydrology*. Vol. 5, American Geophysical Union Water Science and Application, Washington, DC, 1–19.

Hazen, A., 1914. Storage to be provided in the impounding reservoirs for municipal water supply. *Transactions of the American Society of Civil Engineers*, **77**, 1547–1550.

Hirschboeck, K.K., 1987a. Hydroclimatically-defined mixed distributions in partial duration flood series. In V.P. Singh (ed.), *Hydrologic Flood Frequency Modeling*. Reidel, Boston, 192–205.

Hirschboeck, K.K., 1987b. Catastrophic flooding and atmospheric circulation patterns. In L. Mayer and D. Nash (eds), *Catastrophic Flooding*. Allen & Unwin, Boston, 23–56.

Hirschboeck, K.K., 1988. Flood hydroclimatology. In V.R. Baker, R.C. Kochel and P.C. Patton (eds), *Flood Geomorphology*. Wiley, New York, 27–49.

Hosking, J.R.M. and Wallis, J.R., 1986. Paleoflood hydrology and flood frequency analysis. *Water Resources Research*, **22**, 543–550.

Houghton, J.T., Ding, Y., Griggs, D.J., Noguer, M., van der Linden, P.J., Dai, A., Maskell, K. and Johnson, C.A. (eds), 2001. *Climate Change 2001: The Scientific Basis*. Cambridge University Press, Cambridge.

House, P.K. and Baker, V.R., 2001. Paleohydrology of flash floods in small desert watersheds in Western Arizona. *Water Resources Research*, **37**, 1825–1839.

Jarrett, R.D. and Tomlinson, E.M., 2000. Regional interdisciplinary paleoflood approach to assess extreme flood potential. *Water Resources Research*, **36**, 2957–2984.

Kale, V.S., 1999. Long-period fluctuations in monsoon floods in the Deccan Peninsula, India. *Journal of the Geological Society of India*, **53**, 5–15.

Kale, V.S., Mishra, S. and Baker, V.R., 1997. A 2000-year palaeoflood record from Sakarghat on Narmada, Central India. *Journal of the Geological Society of India*, **50**, 283–288.

Kale, V.S., Mishra, S. and Baker, V.R., Sedimentary records of palaeofloods in the bedrock gorges of the Tapi and Narmada rivers, Central India. *Current Science*; in press.

Kale, V.S., Singhvi, A.K., Mishra, P.K. and Banerjee, D., 2000. Sedimentary records and luminescence chronology of late Holocene palaeofloods in the Luni river, Thar Desert, Northwest India. *Catena*, **40**, 337–358.

Klemes, V., 1982. Empirical and causal models in hydrology. In *Scientific Basis of Water Resource Management*. National Academy Press, Washington, DC, 95–104.

Klemes, V., 1986. Dilettantism in hydrology: transition or destiny? *Water Resources Research*, **22**, 177S–188S.

Klemes, V., 1987. Hydrological and engineering relevance of flood frequency analysis. In V.P. Singh (ed.), *Hydrological Frequency Modeling*. Reidel, Boston, 1–18.

Klemes, V., 1989. The improbable probabilities of extreme floods and droughts. In O. Starosolszky and O.M. Melder (eds), *Hydrology of Disasters*. James & James, London, 43–51.

Klemes, V., 2000. *Common Sense and Other Heresies: Selected Papers on Hydrology and Water Resources Engineering*. Canadian Water Resources Association, Cambridge, Ontario.

Kochel, R.C. and Baker, V.R., 1982. Paleoflood hydrology. *Science*, **215**, 353–361.

Kochel, R.C., Patton, P.C. and Baker, V.R., 1982. Paleohydrology of Southwestern Texas. *Water Resources Research*, **18**, 1165–1183.

Knox, J.C., 1985. Responses of floods to Holocene climatic change in the Upper Mississippi Valley. *Quaternary Research*, **23**, 287–300.

Knox, J.C., 1993. Large increases in flood magnitude in response to modest changes in climate. *Nature*, **361**, 430–432.

Knox, J.C., 2000. Sensitivity of modern and Holocene floods to climate change. *Quaternary Science Reviews*, **19**, 439–457.

Levish, D.R., 2002. Paleohydrologic bounds–Non-Exceedance information for flood hazard assessment. In P.K. House, R.H. Webb, V.R. Baker and D.R. Levish (eds), *Ancient Floods, Modern Hazards: Principles and Applications of Paleoflood Hydrology*. Vol. 5, American Geophysical Union Water Science and Application, Washington, DC, 175–190.

Levish, D.R., Ostenaa, D. and O'Connell, D., 1994. A non-inundation approach to paleoflood hydrology for the event-based assessment of extreme flood hazards. *American Society of Dam Safety Officials Annual Conference Proceedings*, **11**, 69–82.

Maizels, J.K., 1983. Palaeovelocity and palaeodischarge determination for coarse gravel deposits. In K.J. Gregory (ed.), *Background to Palaeohydrology*. Wiley, Chichester, 101–139.

Martinez-Goytre, J., House, P.K. and Baker, V.R., 1994. Spatial variability of paleoflood magnitudes in small basins of the Santa Catalina Mountains, Southeastern Arizona. *Water Resources Research*, **30**, 1491–1501.

McCarthy, J.J., Canziani, O.F., Leary, N.A., Dokken, D.J. and White, K.S. (eds), 2001. *Climate Change 2001: Impacts, Adaptation and vulnerability*, Cambridge University Press, Cambridge.

Meehl, G., Zwiers, F., Evans, J., Knutson, T., Mearns, L. and Whetton, P. 2000. Trends in extreme weather and climate events: issues related to modeling extremes in projections of future climate change. *Bulletin of the American Meteorological Society*, **81**, 427–436.

Milly, P.C.D., Wetherald, R.T., Dunne, K.A. and Delworth, T.L., 2002. Increasing risk of great floods in a changing climate. *Nature*, **415**, 514–517.

National Research Council, 1999. *Improving the American River Flood Frequency Analyses*. National Academy Press, Washington, DC.

National Research Council, 2002. *Abrupt Climate Change: Inevitable Surprises*. National Academy Press, Washington, DC.

Nott, J. and Price, D., 1999. Waterfalls, floods and climate change: evidence from tropical Australia. *Earth and Planetary Science Letters*, **171**, 267–276.

Nott, J., Price, D. and Bryant, E., 1996. 30,000 year record of paleofloods in tropical Northern Australia. *Geophysical Research Letters*, **23**, 379–382.

O'Connell, D.R.H., Ostenaa, D.A., Levish, D.R. and Klinger, R.E., Bayesian Flood frequency analysis with paleohydrologic bound data, *Water Resources Research*, **38**(5), noi:10.1029/-2000WR000028.

O'Connor, J.E., 1993. *Hydrology, Hydraulics and Sediment Transport of Pleistocene Lake Bonneville Flooding along the Snake river, Idaho*. Special Paper 274, Geological Society of America, Boulder, Colorado.

O'Connor, J.E. and Webb, R.H., 1988. Hydraulic modeling for paleoflood analysis. In V.R. Baker, R.C. Kochel and P.C. Patton (eds), *Flood Geomorphology*. Wiley, New York, 383–402.

O'Connor, J.E., Webb, R.H. and Baker, V.R., 1986. Paleohydrology of pool and riffle Pattern development, Boulder Creek, Utah. *Geological Society of America Bulletin*, **97**, 410–420.

Ostenaa, D.A., Levish, D.R. and O'Connell, D.R.H., 1996. *Paleoflood Study for Bradbury Dam, Cachuma Project, California*. U.S. Bureau of Reclamation Seismotectonic Report 96-3, Denver, CO.

Palmer, T.N. and Raisanen, J., 2002. Quantifying the risk of extreme seasonal precipitation events in a changing climate. *Nature*, **415**, 512–514.

Partridge, J.B. and Baker, V.R., 1987. Paleoflood hydrology of the Salt river, Arizona. *Earth Surface Processes and Landforms*, **12**, 109–125.

Patton, P.C., 1987. Measuring the rivers of the past: a history of fluvial paleohydrology. In E.R. Landa and S. Ince (eds), *History of Hydrology*, American Geophysical Union, History of Geophysics Number 3, Washington, DC, 55–67.

Patton, P.C. and Dibble, D.S., 1982. Archaeologic and geomorphic evidence for the paleohydrologic record of the Pecos River in West Texas. *American Journal of Science*, **282**, 97–121.

Pickup, G., Allan, G. and Baker, V., 1988. History, palaeochannels and palaeofloods of the Finke river, Central Australia. In R. Warner (ed.), *Fluvial Geomorphology in Australia*. Academic Press, Sydney, 177–200.

Pielke Jr., R.A., 1999. Nine fallacies of floods. *Climatic Change*, **42**, 413–438.

Pielke Jr., R.A. and Downton, M.W., 2000. Precipitation and damaging floods: Trends in the United States, 1932-97. *Journal of Climate*, **13**, 3625–3637.

Pilgrim, D.H., 1986. Bridging the gap between research and design practice. *Water Resources Research*, **22**, 165S–176S.

Pilgrim, D.H., Chapman, T.G. and Doran, D.G., 1988. Problems of rainfall-runoff modelling in arid and semiarid regions. *Hydrological Sciences Journal*, **33**, 379–400.

Rumsby, B.T. and Macklin, M.G., 1996. River response to the last neoglacials (the 'Little Ice Age') in Northern, Western and Central Europe. In J. Branson, A.G. Brown and K.J. Gregory (eds), *Global Continental Changes: The Context of Palaeohydrology*. Special Publication 115 of the Geological Society, Geological Society, London, 85–98.

Schnur, R., 2002. The investment forecast. *Nature*, **415**, 483,484.

Springer, G.S., 2002. Caves and their potential use in paleoflood studies. In P.K. House, R.H. Webb, V.R. Baker and D.R. Levish (eds), *Ancient Floods, Modern Hazards: Principles and Applications of Paleoflood Hydrology*. Vol. 5, American Geophysical Union Water Science and Application, Washington, DC, 329–343.

Springer, G.S. and Kite, J.S., 1997. River-derived slackwater sediments in caves along Cheat river, West Virginia. *Geomorphology*, **18**, 91–100.

Starkel, L., 1983. The reflection of hydrologic changes in the fluvial environment of the temperate zone during the last 15,000 years. In K.J. Gregory (ed.), *Background to Palaeohydrology*. Wiley, Chichester, 213–235.

Stedinger, J.R. and Baker, V.R., 1987. Surface water hydrology: historical and paleoflood information. *Reviews of Geophysics*, **25**, 119–124.

Timmermann, A., Oberhuber, J., Bacher, A., Esch, M., Latif, M. and Roeckner, E., 1999. Increased El Nino frequency in a climate model forced by future greenhouse warming. *Nature*, **398**, 694–697.

Trenberth, K.E., 1999. Conceptual framework for changes of extremes of the hydrological cycle with climate change. *Climate Change*, **42**, 327–339.

Trenberth, K.E. and Hoar, T.J., 1996. The 1990–1995 El Nino-Southern oscillation event: the longest on record. *Geophysical Research Letters*, **23**, 57–60.

Urban, F.E., Cole, J.E. and Overpeck, J.T., 2000. Influence of mean climate change on climate variability from a 155-year tropical Pacific coral record. *Nature*, **407**, 989–993.

Webb, R.H. and Jarrett, R.D., 2002. One-dimensional estimation techniques for discharges of paleofloods and historical floods. In P.K. House, R.H. Webb, V.R. Baker and D.R. Levish (eds), *Ancient Floods, Modern Hazards: Principles and Applications of Paleoflood Hydrology*. Vol. 5, American Geophysical Union Water Science and Application, Washington, DC, 111–125.

Webb, R.H., Blainey, J.B. and Hyndman, D.W., 2002. Paleoflood hydrology of the Paria River, Southern Utah and Northern Arizona, USA. In P.K. House, R.H. Webb, V.R. Baker and D.R. Levish (eds), *Ancient Floods, Modern Hazards: Principles and Applications of Paleoflood Hydrology*. Vol. 5, American Geophysical Union Water Science and Application, Washington, DC, 295–310.

Williams, G.P., 1983. Paleohydrological methods and some Swedish fluvial environments, I-cobble and boulder deposits. *Geografiska Annaler*, **65A**, 227–243.

Williams, G.P., 1984. Paleohydrologic equations for rivers. In J.E. Costa and P.J. Fleisher (eds), *Developments and Applications of Geomorphology*. Springer-Verlag, Berlin, 353–367.

Wohl, E.E., Fuertsch, S.J. and Baker, V.R., 1994a. Sedimentary records of late Holocene floods along the Fitzroy and Margaret rivers, Western Australia. *Australian Journal of Earth Sciences*, **41**, 273–280.

Wohl, E.E., Webb, R.H., Baker, V.R. and Pickup, G., 1994b. *Sedimentary Flood Records in the Bedrock Canyons of Rivers in the Monsoonal Region of Australia*. Engineering Research Center Water Resource Paper 107, Colorado State University, Fort Collins, CO.

Wohl, E.E., 2002. Modeled paleoflood hydraulics as a tool for interpreting bedrock channel morphology. In P.K. House, R.H. Webb, V.R. Baker and D.R. Levish (eds), *Ancient Floods, Modern Hazards: Principles and Applications of Paleoflood Hydrology*. Vol. 5, American Geophysical Union Water Science and Application, Washington, DC, 345–358.

Yanosky, T.M. and Jarrett, R.D., 2002. Dendrochronologic evidence for the frequency and magnitude of paleofloods. In P.K. House, R.H. Webb, V.R. Baker and D.R. Levish (eds), *Ancient Floods, Modern Hazards: Principles and Applications of Paleoflood Hydrology*. Vol. 5, American Geophysical Union Water Science and Application, Washington, DC, 1–19.

Yevjevich, V. and Harmancioglu, N.B., 1987. Research needs on flood characteristics. In V.P. Singh (ed.), *Applications of Frequency and Risk in Water Resources*. Reidel, Boston, 1–21.

19 Palaeohydraulics of Extreme Flood Events: Reality and Myth

P. CARLING,[1] R. KIDSON,[2] Z. CAO[1] AND J. HERGET[3]
[1]*University of Southampton, Southampton, UK*
[2]*University of Cambridge, Cambridge, UK*
[3]*Ruhr University, Bochum, Germany*

1 INTRODUCTION

Palaeohydraulics is usually defined as the science that studies and seeks to reconstruct the prehistoric, or "prior to instrumented records" water flow processes occurring in river channels, usually for exceptional unobserved floods. Often the primary aim is to delimit the discharge and mean velocity at a particular channel cross-section, based on the elevation (or palaeostage) of sediment deposits or other relict markers from one or more past extreme flood events (Jarrett and England, 2002). This objective is most commonly achieved by applying a mathematical river model and utilising palaeostage indicators as model input data to calibrate the model. Palaeohydraulics is distinct from palaeohydrology, which is concerned with the hydrology of the past including reconstruction of rainfall-runoff relationships in catchments.

However, the study of palaeohydraulics also has potential to offer insights into landscape processes and flood-hazard assessment and to make a significant contribution in palaeohydrology (Schumm, 1977; Patton *et al.*, 1979; Baker *et al.*, 1983; Baker, 1987; Jarrett, 1991; Baker *et al.*, 1993; Ely *et al.*, 1993; Baker, 1998; Enzel *et al.*, 1999). Much palaeohydraulics literature first appeared in the 1980s, when palaeoflood deposits were first utilised for reconstructing extreme flood events. In particular, river Slack Water Deposits (SWDs – Baker, 1988) were identified as "chroniclers of their own cataclysms" (Baker and Pickup, 1987). Whilst the traditional interest has been largely geomorphological, with a focus upon prehistoric and historical flood-frequency analysis (Webb *et al.*, 1988), more recently other applications have become evident. Given increasing concerns over anthropogenically induced climate change, aquatic ecological health and long-term water supply, palaeoflood studies can be applied to studies of climate change, archaeological and anthropological issues, as well as modern land use and environmental flow management.

Natural fluvial flows are hydrodynamically complicated and any simple calculations on a section, involving arithmetic formulations only for idealised steady and uniform flows (e.g. the slope-area method; Dalrymple and Benson, 1967) for modelling discharge from the palaeostage (Patton, 1977; Kochel, 1980; Baker *et al.*, 1983) can only give an estimate of flood magnitude when applied to large ungauged floods (Webb and Jarrett, 2002). Hence, reach-scale mathematical river models are needed, involving the numerical solution of a set of differential equations, given an appropriate specified

Palaeohydrology: Understanding Global Change.   Edited by K.J. Gregory and G. Benito
© 2003 John Wiley & Sons, Ltd   ISBN: 0-470-84739-5

boundary for unsteady flows and initial prescribed conditions. As the values pertaining to a variety of model parameters are not well established in basic fluid dynamics (e.g. the roughness parameter etc.), they have to be calibrated using observed data prior to the model being applied to practical situations. Even for contemporary river modelling, applications continue to be problematic (Cao and Carling, 2002a; 2002b). For palaeohydraulic processes, the uncertainty of the palaeostage data, which is essential for model calibration, makes modelling even less reliable. In brief, the attractions offered by mathematical modelling to fluvial hydraulics are seductive as rather impressive graphics can be produced easily. However, the quality and the reliability of the modelling output remain unclear. It is necessary to identify, delimit and assess the various uncertainties of models.

This chapter aims to provide a brief overview of mathematical modelling of palaeo-extreme flood events. The major issues addressed are organised according to their role in achieving the complete modelling output, that is, assumptions in model formulation, numerical scheme, model parameters and their calibration and input data quality. As there have been many reviews in the literature on contemporary hydraulic modelling (Cao and Carling, 2002a; b), particular attention is paid to those specific to palaeohydraulic modelling, that is, the palaeostage as model input for model calibration.

2 MODEL FORMULATION

Mathematical river models are built upon the mass and momentum conservation equations of the water flow, for example, the St-Venant equations (e.g. HEC-RAS, Brunner, 2001). Usually, the sole parameter to be calibrated is the roughness parameter that represents the resistance due to friction with the channel boundary.

To our knowledge, the presence and influence of a substantial sediment load on flow hydraulics and consequent hydraulic calculations, has not been implemented in existing palaeohydraulic modelling studies. However, a "moveable boundary sediment transport" module is under development for the next version of HEC-RAS. Such initiatives are timely as it must not be forgotten that during extreme floods, flows are highly competent to carry a large amount of sediment. Bedrock channels are usually considered to be supply-limited but significant quantities of sediment can originate either from the upstream sources (i.e. a sediment-laden flow entering the modelled reach) or from local scouring of the channel bed and the channel banks (Young et al., 2001). Hence, this issue can be significant even for bedrock channels, wherein there may be significant local sediment storage within the bedrock cross-section, as well as sources from bedrock erosion. The physically strongly coupled system of water flow, sediment transport and channel morphological evolution is much more complicated than the clear water flow over a non-erodible (fixed) bed. Existing models for sediment transport (mostly decoupled from the detailed hydraulic solutions) are based on a number of simplifications and appear to be invalid for extreme flood events (Cao et al., 2002). The need for further studies along this line of enquiry is evident, not least because of the implications for bed roughness. Possibly of greater importance is the fact that even if the total load is small it may be associated with significant degradation or aggradation of the bed level of any sediment filament within the bedrock-confined channel. During the passage of an unsteady flood wave, this will result in the development of SWDs at various levels along the bedrock margins leaving flood marks that do not reflect the water depth relative to a fixed bed level (e.g. Kirby, 1987). For example, Webb and Jarrett (2002) report alluvial bed level changes, in bedrock-confined channels, of between 10 m and 30 m during single flood events.

The governing equations for the fluvial water flow (e.g. the traditional St-Venant equations) and the flow-sediment-morphology system in alluvial rivers are generally hyperbolic. Perhaps in most cases, the river flows are subcritical and therefore the traditional numerical schemes are effective and efficient. However, there do exist some natural flow situations for which specially developed numerical schemes must be devised to cope with the strong hyperbolic nature, for example, mixed sub- and super-critical flow (e.g. Wohl, 2002), extreme flood-induced degradation and dam-break flow over an erodible bed (e.g. Carling and Glaister, 1987). There is scope for basic studies of advanced numerical schemes for the strong hyperbolic equations, which can be used for the mixed sub- and super-critical river flows, with or without intense sediment transport and rapid morphological evolution.

3 MODEL PARAMETERS AND CALIBRATION

The pivotal parameter in channel flood modelling is the roughness parameter (e.g. the Manning coefficient) that must be incorporated to close the resistance term in the momentum equation. Determining appropriate values is an important topic in modern fluvial hydraulics, and is recognised as an issue in recent palaeohydraulic studies (Aronica *et al.*, 1998; Wohl, 1998). The complexity of fluvial riverbed resistance stems from not only the irregular boundary but also from sediment carried by the flow. Over the last half century, there has been much controversy concerning the effect of sediment on flow resistance (increase or decrease) and its representation (Cao and Carling, 2002b). It continues to be difficult to pinpoint the friction factor of natural rivers. Often, the roughness parameter has to be tuned to reconcile the mathematical modelling outputs with measurements. Closely related to this practice is the calibration strategy to be utilised. The most popular method is to sequentially tune the roughness using experience gained from modern rivers, which may be time consuming and is intuitively unsatisfactory. More dramatically, this procedure can be subjective when distributed roughness has to be used to acquire satisfactory agreement between modelling and observation, because differing combinations of distributed roughness may lead to similar results. More discussions of this aspect can be found in Cao and Carling (2002b).

Despite the issues noted above, consideration of palaeoflows as sediment-free has proven extremely popular in the palaeohydraulic literature (e.g. Kochel and Baker, 1982; Ely and Baker, 1985; O'Connor and Webb, 1988; Partridge and Baker, 1987; Webb *et al.*, 1988; O'Connor *et al.*, 1994). The majority of studies have utilised the same method (i.e. the 1D step-backwater method, Chow, 1959; Richards, 1979; 1982; Chanson, 1999), and indeed the same modelling package (HEC-2 and its successor HEC-RAS (Brunner, 2001). A recent study by Yang *et al.* (2000) is one exception to this trend. For reasons of brevity, the issue of the appropriate level of complexity (e.g. 1D, 2D etc.) required for modelling palaeofloods cannot be addressed here. Most often the 1-D approach can be justified in as much as it is imperative to model only at the level of complexity for which palaeodata is available. However, systematic treatment of error in the two steps of (1) estimating palaeostage and (2) hydraulic modelling has been lacking in most of these studies. Error in palaeostage estimation is reduced when multiple SWDs along a reach are used. The 1D step-backwater modelling method is hence ideal for both minimising this error and placing a "probability envelope" around the estimated stage. The model procedure identifies the "best fit" discharge for a series of palaeoflow markers, as well as for identifying the range of possible discharges for the series of SWDs throughout the reach.

3.1 Input Data Quality

The issue of error in the palaeohydraulic modelling process has, with few exceptions (Freeman *et al.*, 1996; Wohl, 1998), not been addressed. This observation is in marked contrast to the hydrological literature, which has quite comprehensively addressed the issue of model error and uncertainty (e.g. Bergstrom, 1991; Beven, 2000a; Binley *et al.*, 1991; Lamb *et al.*, 1998).

Slack Water Deposits (SWDs) are a most common source of palaeostage indicators. SWDs are sediment remnants from extreme flood events. If charcoal or other organic debris is incorporated into SWDs, it is possible to date the deposits using techniques such as ^{14}C (Baker *et al.*, 1985). The ideal SWD offers two key pieces of information regarding palaeofloods: first, the stage of individual palaeofloods and second, the chronology of palaeofloods. The stage data is pivotal for estimating discharge using hydraulic models (O'Connor and Webb, 1988; Brunner, 2001). Together with a chronology, discharge estimates can then be used in flood-frequency analysis, to extend the record of extreme events for a river, beyond the gauged and historical record (Kochel and Baker, 1982; 1988). However, systematic treatment of error in estimating palaeostage has been lacking in most published studies.

Freeman *et al.* (1996) proposed three levels of error in the modelling process: natural, measurement and modelling. Relevant to the palaeostage are the natural and measurement errors. The natural error in palaeostage is the difference between the palaeostage recorded by SWDs and the actual stage of the real palaeoflood. From the General Theory of Errors (e.g. Topping, 1956), the error sources are divided into random and systematic types. Random errors are sourced from random fluctuations in conditions outside the control of the observer, whereas systematic errors are those that tend to bias results consistently in one direction, as a result, for instance, of a poorly calibrated or limited-resolution measuring instrument. Error theory provides guidance as to how these two types of error can be minimised. Random error can be reduced by replication. Systematic error can be quantified and reduced through trialing different methods of measuring or calculating a result. It is necessary to delimit how the palaeoflood discharge is sensitive to the error in the palaeostage using a mathematical river model.

The application of SWDs to estimate palaeostage was largely developed by Kochel (1980), Kochel *et al.* (1982) and Baker (1988). For the field estimation of palaeostage, the following assumptions are usually made:

1. The flood channel is bedrock confined
2. Regional aggradation and degradation are negligible
3. The heights of SWDs approximate the water surface of the flood
4. Scour and fill during floods are minimal
5. Palaeofloods occurred in the modern hydrological regime.

Each of these assumptions, if untested, represents a potential source of error in palaeostage estimation. Satisfying Assumptions 1 and 2 is notionally easy. The rate of bedrock incision and/or tectonic activity is unlikely to be of the same order of magnitude as the rate of SWD accumulation. A survey of field-based palaeohydraulic studies shows that most SWDs have been used to reconstruct flow over timescales around 1,000 years. At these timescales, this assumption is unlikely to be a significant source of error (Ely and Baker, 1985). Of course, this principle should be confirmed with independent geological evidence if available.

Assumption 3 encapsulates two processes: first, the fact that sediment is deposited at the base of a water column, and hence a palaeostage derived from a SWD is a conservative estimate of the actual stage of the palaeoflow, and second, that SWDs are records of *ineffective flow* so that they record flow conditions at the margins of channels and in backwaters that do not necessarily reflect those of the main channel (Webb and Jarrett, 2002). A key consequence of these facts is that the localised hydraulic effects (O'Connor *et al.*, 1994) can unduly influence the SWD palaeostage and not reflect those typical of flow along the thalweg. Hence, a SWD palaeostage may not be representative of the generalised flood stage that generated it.

Assumption 3 is a significant source of both random and systematic error that can be better quantified and constrained through three means:

Detailed flume experiments: It is surprising considering the number of palaeohydraulic modelling studies that there are relatively few physical flume studies to test the assumptions of SWD theory. The study by Kochel and Ritter (1987) is a rare exception. Better quantification of both process (1) and (2) would reduce this error component.

Estimation of palaeostage using several different indicators: Examples for this estimation are residual woody debris, impact scars, scour marks and silt lines, in addition to SWDs. This can aid in isolating the SWD-dependent error.

Replication of SWD palaeostage estimation along a reach: As error theory indicates, the best means of combating random error (e.g. local hydraulic effects) is to replicate the sampling. In practice, this means taking as many palaeostage measurements from different SWDs in a reach as possible. No amount of resolution within a single deposit (e.g. Jones *et al.*, 2001) can reduce error in the palaeostage. The importance of replication within a reach has been recognised by previous workers (e.g. Greenbaum *et al.*, 2000) and is obviously constrained by the SWDs available at a particular locality.

Assumption 4 is significant. Miller and Cluer (1998), for example, have highlighted the fact that debris accumulation can be substantial in bedrock-confined channels. The issue could be addressed with flume studies or through long-term monitoring of alluvial bed levels in bedrock-confined channels.

Assumption 5 is particularly problematic, as mixed populations of floods generated by different means may occur over time preventing meaningful flood-frequency analysis (Redmond *et al.*, 2002; Blainey *et al.*, 2002). One approach to manage this source of error is to incorporate independent sources of evidence, such as climatic proxies.

3.2 Model Complexity in the Palaeohydraulic Context

In model structure, error relates to inadequate characterisation of the channel geometry and roughness, issues that plague hydraulic literature (e.g. Aronica *et al.*, 1998; Wohl, 1998), and inadequate characterisation of the flow hydraulics. Traditional means of addressing these error sources in modelling is to increase the resolution of both (Beven, 2000a and b; Jakeman and Hornberger, 1993; Lane, 1998). Applied to hydraulics, the line of argument is that as river flow itself is determined by processes in several dimensions, models need also to incorporate increased complexity in order to better characterise flow (Lane and Richards, 2001). However, increasing model complexity may not be appropriate in the palaeohydraulic context.

A recent review of hydraulic literature by Miller and Cluer (1998) compared the predictive power of traditional 1-D modelling with the more sophisticated 2-D and 3-D alternatives. They concluded that 2-D modelling did produce a "slightly better fit" to a sequence of palaeostage indicators along a study reach, but that "the main features of the longitudinal profile were predicted by both models" (1D and 2D) and the "improvement may not justify the additional effort involved". They recommended choosing the more complex model (in this case the 2-D model) only where a specific question existed that could be better addressed by the more complex model (e.g. Crowder and Diplas, 2000).

There is a large literature dealing with model complexity (e.g. Elert et al., 1999; Karl et al., 2000; Stillman et al., 2000; Croke and Jakeman, 2001; de Wit and Pebesma, 2001; Wagener et al., 2001). The general consensus is that the optimal level of model complexity is a "moderate" one (Snowling and Kramer, 2001) where the variables accounting for the majority of system behaviour can be characterised, and most importantly, calibrated. This point is illustrated with the most complex models of all, General Circulation Models (GCMs), which are limited by data availability for calibration (Schnur and Lettenmaier, 1998; Xu, 1999).

Ultimately, palaeohydraulic modelling is limited by the availability of data for calibration. Whilst advances in fluid mechanics and experimental flume studies can provide insight, the absence of real data from extreme events poses a somewhat intractable problem for palaeohydraulics. Whilst opportunities to directly gauge extreme flood events are rare, they are not non-existent. Greenbaum et al. (1998) is one such exception. More consideration needs to be given to the use of remote sensing, tracer and other field studies of real extreme flood events (e.g. Adler et al., 1999). The latter requirement will entail developments in robust instrumentation.

For immediate purposes, however, it is deduced that increasing model complexity where there is both a lack of palaeo proxy data (e.g. palaeoflow indicators in several dimensions) and of contemporary real data with which to calibrate both the model and the proxy, is unlikely to increase model predictive power. It is, therefore, imperative to model at the level of complexity for which data is available. Where palaeostage is the only source of palaeoflow indicator, a simple hydraulic model is most appropriate and the limited confidence in the modelling results must be explicitly appreciated. Alternatively, given the large potential error associated with palaeostage, it may in some instances be wiser not to undertake hydraulic modelling (e.g. Jones et al., 2001). Where there is palaeoflow evidence of multi-dimensional flow characteristics (e.g. channel-scale potholes and step-pool sequences in bedrock channels), more complex modelling may be relevant. Again, the requirement for thorough and adequate field evidence is imperative. Both the palaeo (Baker, 2000) and hydrological (Beven, 2000b; 2000c) modelling literature converge at this point and it is particularly important for palaeo work in which field evidence tends to be limited.

3.3 Cumulative Error

Given the errors associated both with palaeostage estimation and the hydraulic modelling process, it is important to state the impact of this upon the final outcome, which is often a figure for downstream discharge.

The Theory of Errors provides guidance on the cumulative nature of error in the *Law of Superposition of Errors* (Topping, 1956). Where a mathematical function (i.e. model) is composed of individual variables (i.e. data sources or processes), the total error for the model is an additive function of the differentiated variables multiplied by their individual error components. Thus, although in some cases random errors due to different sources may cancel out each other, systematic errors are cumulative. Thus, it is critical to identify the error source and also to identify whether the error is systematic or random.

Applied to palaeohydraulics, this means that error from all sources (palaeostage and modelling) is cascaded through to the end result, which is the modelled discharge. If this discharge figure is then inserted into a flood-frequency analysis and combined with a chronology from ^{14}C dating (with its own set of errors to contribute), and combined overall, with a lack of calibration data, it can be seen that the potential error associated with palaeohydraulics is substantially more than that associated with contemporary hydraulics. However, systematic study of error accumulation will allow quantification of uncertainty in palaeohydraulic estimation.

In a case study on the Cache la Poudre River, Colorado, USA reported in Miller and Cluer (1998), the error margins associated with the combined palaeostage-to-discharge hydraulic modelling process were of the order of 100%. Miller and Cluer concluded that the particular site in question was not conducive to palaeohydraulic analysis. It is counterintuitive to suggest that palaeo-evidence can provide no greater insight to palaeodischarge than if the evidence did not exist. The Cache la Poudre example is an extreme one, but serves to demonstrate the importance of efforts to quantify error and reduce it where possible (e.g. choose a different site), in order for palaeohydraulics to yield meaningful results.

4 CONCLUSION

Mathematical modelling of the palaeohydraulics of many extreme flood events continues to be imprecise and possibly inaccurate owing to a lack of rigour. This occurs firstly, because of the assumptions in model formulation. Existing palaeohydraulic studies of extreme flood events have exclusively used models for clear water flows over fixed (non-erodible) bed (e.g. HEC-RAS), ignoring the fact that extreme floods are competent to carry a significant amount of sediment. This is a major shortcoming of current palaeohydraulic models, and merits detailed consideration that is technically feasible (Cao *et al.*, 2002). Secondly, the roughness parameter is an elusive quantity. Current calibration methodology is largely subjective, especially for channels with complicated geometries, where an understanding of the distributed roughness would be essential for precise modelling. Nevertheless, recent literature is exploring the concept of equi-finality in relation to the roughness parameter, wherein it is realised that the roughness value to some extent is arbitrary, depending on the scale of survey and the flow data available for a specific modelling exercise. Thirdly, specifically pertaining to palaeohydraulic modelling is the error in the input data to calibrate the model: palaeostage. The most likely source of error is the disparity between the SWD palaeostage and that of the flood that caused it. This error can be reduced, but not eliminated, through a combination of experimental flume studies, using independent palaeostage evidence and replication of SWD palaeostage estimation. A critical deduction, even with this approach, is the highly site-specific nature of palaeostage evidence, and conversely, the fact that error will increase the further one deviates from the site in question (Veldkamp and van Dyijke, 2000). In line with this observation, it is sensible to model at the level of complexity for which there is complementary data available. Bringing these two conclusions together, it is apparent that the optimal approach for palaeohydraulic modeling efforts lies at the scale of the *individual reach*. This approach has been utilised by several authors (Strupczewski *et al.*, 2001; Veldkamp and van Dyijke, 2000), and has also been applied to the management of environmental flows (Thoms and Sheldon, 2000a and b).

The future of palaeohydraulic modeling lies in adopting a multi-disciplinary approach. New contributions from fluid mechanics and flume experiments are essential to minimise palaeostage estimation error. Linking of SWD palaeostage evidence with independent forms of palaeoclimatic proxy data is essential to reconstruct the relationship between palaeoclimate and palaeodischarge. Better integration of palaeohydraulic models with traditional hydrological (e.g. Anselmo *et al.*, 1996; Rico *et al.*, 2001) and climatic models (e.g. Thomas, 2000) will help palaeohydraulic modeling realise its *raison d'être*: to understand extreme flow events of the past in order to cope better with future scenarios from a scientific, management and social perspective.

REFERENCES

Adler, M.J., Stancalie, G. and Raducu, C., 1999. Integrating tracer with remote sensing techniques for determining dispersion coefficients of the Dambovita river, Romania. *IAHS-AISH Publication*, **258**, 75–81.

Anselmo, V., Galeati, G., Palmieri, S., Rossi, U. and Todini, E., 1996. Flood risk assessment using an integrated hydrological and hydraulic modelling approach: a case study. *Journal of Hydrology*, **175**, 533–554.

Aronica, G., Hankin, B. and Beven, K.J., 1998. Uncertainty and equifinality in calibrating distributed roughness coefficients in a flood propagation model with limited data. *Advances in Water Resources*, **22**, 349–365.

Baker, V.R., 1987. Paleoflood hydrology and extraordinary flood events. *Journal of Hydrology*, **86**, 79–99.

Baker, V.R., 1988. Magnitude & frequency of palaeofloods. In K.J. Beven and P. Carling (eds), *Floods: Hydrological, Sedimentological & Geomorphic Implications*. Wiley, Chichester, 171–183.

Baker, V.R., 1998. Paleohydrology and the hydrological sciences. In G. Benito, V.R. Baker and K.J. Gregory (eds), *Palaeohydrology and Environmental Change*. Wiley, Chichester, 1–10.

Baker, V.R., 2000. South American paleohydrology: future prospects and global perspective. *Quaternary International*, **72**, 3–5.

Baker, V.R., Benito, G. and Rudoy, A., 1993. Paleohydrology of late-Pleistocene superflooding, Altay Mountains, Siberia. *Science*, **259**, 348–350.

Baker, V.R. and Pickup, G., 1987. Flood geomorphology of the Katherine Gorge, Northern Territory, Australia. *Geological Society of America Bulletin*, **98**, 635–646.

Baker, V.R., Pickup, G. and Polach, H., 1983. Desert paleofloods in central Australia. *Nature*, **301**, 502–504.

Baker, V.R., Pickup, G. and Polach, H., 1985. Radiocarbon dating of flood events, Katherine Gorge, Northern Territory, Australia. *Geology*, **13**, 344–347.

Bergstrom, S., 1991. Principles and confidence in hydrological modelling. *Nordic Hydrology*, **22**, 123–136.

Beven, K.J., 2000a. *Rainfall-Runoff Modelling: The Primer*. Wiley, Chichester.

Beven, K.J., 2000b. Uniqueness of place and non-uniqueness of models in assessing predictive uncertainty. In L.R. Bentley, J.F. Sykes, C.A. Brebbia, W.G. Gray and G.F. Pinder (eds),

Computational Methods in Water Resources. Vol. 2 Computational-Methods in Surface-Water Systems and Hydrology. Foundation for Research and Technology, Institute of Chemical Engineering and High Temperature Chemical Processes, Hellas, 1085–1091 *12th International Conference on Computational Methods in Water Resources*. 1998, Crete, Greece.

Beven, K.J., 2000c. Uniqueness of place and process representations in hydrological modeling. *Hydrology and Earth System Sciences*, **4**(2), 203–213.

Binley, A.M., Beven, K.J., Calver, A. and Watts, L.G., 1991. Changing responses in hydrology: assessing the uncertainty in physically based model predictions. *Water Resources Research*, **27**, 1253–1261.

Blainey, J.B., Webb, R.H., Moss, M.E. and Baker, V.R., 2002. Bias and information content of paleoflood data in flood-frequency analysis. In P.K. House, R.H. Webb, V.R. Baker and D.R. Levish (eds), *Ancient Floods, Modern Hazards: Principles and Applications of Paleoflood Hydrology*. American Geophysical Union, Washington, DC, 161–174.

Brunner, G.W., 2001. *Users Manual for HEC-RAS River Analysis System*. Vol. 3, US Army Corps Of Engineers, Hydrologic Engineering Centre, CA.

Cao, Z. and Carling, P.A., 2002a. Mathematical modelling of alluvial rivers: reality and myth. Part I: general overview. *Water and Maritime Engineering*, ICE, **154**, 207–219.

Cao, Z. and Carling, P.A., 2002b. Mathematical modelling of alluvial rivers: reality and myth. Part II: special issues. *Water and Maritime Engineering*, ICE, **154**, 297–307.

Cao, Z., Day, R. and Egashira, S., 2002. Coupled and decoupled numerical modelling of flow and morphological evolution in alluvial rivers. *Journal of Hydraulic Engineering, ASCE*, **128**(3), 306–321.

Carling, P.A. and Glaister, M.S., 1987. Reconstruction of a flood resulting from a moraine-dam failure using geomorphological evidence and dam-break modeling. In L. Mayer and D. Nash (eds), *Catastrophic Flooding*. The Binghamton Symposia in Geomorphology International Series, Vol. 18, Allen and Unwin, Boston, 181–200.

Chanson, H., 1999. *The Hydraulics of Open Channel Flow: An Introduction*. Arnold, London.

Chow, V.T., 1959. *Open Channel Hydraulics*. McGraw-Hill, New York.

Croke, B.F.W. and Jakeman, A.J., 2001. Predictions in catchment hydrology: An Australian perspective. *Marine and Freshwater Research*, **52**, 65–79.

Crowder, D.W. and Diplas, P., 2000. Using two-dimensional hydrodynamic models at scales of ecological importance. *Journal of Hydrology*, **230**, 172–191.

Dalrymple, T. and Benson, M.A., 1967. Measurement of peak discharge by the slope-area method. *U.S. Geological Survey Techniques Water Resources*, book 3, Chap. A2, USGS, Reston, VA, USA, 1–12.

Elert, M., Butler, A., Chen, J., Dovlete, C., Konoplev, A., Golubenkov, A., Sheppard, M., Togawa, O. and Zeevaert, T., 1999. Effects of model complexity on uncertainty estimates. *Journal of Environmental Radioactivity*, **42**, 255–270.

Ely, L.L. and Baker, V.R., 1985. Reconstructing paleoflood hydrology with slackwater deposits – Verde River, Arizona. *Physical Geography*, **6**, 103–126.

Ely, L.L., Enzel, Y., Baker, V.R. and Cayan, D., 1993. A 5000-year record of extreme floods and climate-change in the Southwestern United States. *Science*, **262**, 410–412.

Enzel, Y., Ely, L.L., Mishra, S., Ramesh, R., Amit, R., Lazar, B., Rajaguru, S.N., Baker, V.R. and Sandler, A., 1999. High-resolution Holocene environmental changes in the Thar Desert, Northwestern India. *Science*, **284**, 125–128.

Freeman, G.E., Copeland, R.R. and Cowan, M.A., 1996. Uncertainty in stage-discharge relationships. In K.S. Tickle, I.C. Goulter, X. Chengchao, S.A. Wasimi and F. Bouchart (eds), *Stochastic Hydraulics*. Balkema, Rotterdam, 601–608, *Proceedings Symposium*, Mackay, Australia.

Greenbaum, N., Margalit, A., Schick, A.P., Sharon, D. and Baker, V.R., 1998. A high magnitude storm and flood in a hyperarid catchment, Nahal Zin, Negev Desert, Israel. *Hydrological Processes*, **12**, 1–24.

Greenbaum, N., Schick, A.P. and Baker, V.R., 2000. The palaeoflood record of a hyper-arid catchment, Nahal Zin, Negev Desert, Israel. *Earth Surface Processes and Landforms*, **25**, 951–971.

Jakeman, A.J. and Hornberger, G.M., 1993. How much complexity is warranted in a rainfall-runoff model? *Water Resources Research*, **29**, 2637–2649.

Jarrett, R.D., 1991. Paleohydrology and its value in estimating floods and droughts. In Paulson, R.W., Chase, E.B., Roberts, R.S. and Moody, D.W. (Compilers), *National Water Summary 1988–1989 Hydrologic Events & Floods & Droughts*. US Geological Survey Water Supply Paper 2375, 105–116.

Jarrett, R.D. and England, J.F., 2002. Reliability of paleostage indicators for paleoflood studies. In P.K. House, R.H. Webb, V.R. Baker and D.R. Levish (eds), *Ancient Floods, Modern Hazards: Principles and Applications of Paleoflood Hydrology*. American Geophysical Union, Washington, DC, 91–109.

Jones, A.P., Shimazu, H., Oguchi, T., Okuno, M. and Tokutake, M., 2001. Late Holocene slackwater deposits on the Nakagawa river, Tochigi prefecture, Japan. *Geomorphology*, **39**, 39–51.

Karl, J.W., Heglund, P.J., Garton, E.O., Scott, J.M., Wright, N.M. and Hutto, R.L., 2000. Sensitivity of species habitat-relationship model performance to factors of scale. *Ecological Applications*, **10**, 1690–1705.

Kirby, W.H., 1987. Linear error analysis of slope-area discharge determinations. *Journal of Hydrology*, **96**, 125–138.

Kochel, R.C., 1980. *Interpretation of Flood Palaeohydrology Using Slackwater Deposits, Lower Pecos and Devils Rivers, Southwestern Texas*. Unpublished Ph.D. Thesis, University of Texas, Austin, Texas.

Kochel, R.C. and Baker, V.R., 1982. Palaeoflood hydrology. *Science*, **215**, 353–361.

Kochel, R.C. and Baker, V.R., 1988. Palaeoflood analysis using slackwater deposits. In V.R. Baker, R.C. Kochel and P.C. Patton (eds), *Flood Geomorphology*. Wiley, New York, 357–376.

Kochel, R.C., Baker, V.R. and Patton, P.C., 1982. Paleohydrology of Southwestern Texas. *Water Resources Research*, **18**, 1165–1183.

Kochel, R.C. and Ritter, D.F., 1987. Implications of flume experiments on the interpretation of slackwater palaeoflood sediments. In V.P. Singh (ed.), *Regional Flood Frequency Analysis*. Reidel, Boston, 371–390.

Lamb, R., Beven, K.J. and Myrabo, S., 1998. Use of spatially distributed water table observations to constrain uncertainty in a rainfall-runoff model. *Advances in Water Resources*, **22**, 305–317.

Lane, S.N., 1998. Hydraulic modelling in hydrology and geomorphology: a review of high resolution approaches. *Hydrological Processes*, **12**, 1131–1150.

Lane, S.N. and Richards, K.S. 2001. The 'Validation' of hydrodynamic models: some critical perspectives. In M.G. Anderson and P.D. Bates (eds), *Model Validation: Perspectives in Hydrological Science*. Wiley, Chichester, 413–438.

Miller, A.J. and Cluer, B.L., 1998. Modeling considerations for simulation of flows in bedrock channels. In K. Tinkler and E. Wohl (eds), *Rivers Over Rock: Fluvial Processes in Bedrock Channels*. Geophysical Monograph 107, American Geophysical Union, Washington, DC, 61–104.

O'Connor, J., Ely, L.L., Wohl, E.E., Stevens, L.E., Melis, T.S., Kale, V.S. and Baker, V.R., 1994. A 4500-year record of large floods on the Colorado river in the Grand Canyon, Arizona. *Journal of Geology*, **102**, 1–9.

O'Connor, J. and Webb, R.H., 1988. Hydraulic modelling for palaeoflood analysis. In V.R. Baker, C. Kochel and P.C. Patton (eds), *Flood Geomorphology*. Wiley, New York, 393–402.

Partridge, J. and Baker, V.R., 1987. Paleoflood Hydrology of the Salt River, Arizona. *Earth Surface Processes and Landforms*, **12**, 109–125.

Patton, P.C., 1977. *Geomorphic Criteria for Estimating the Magnitude and Frequency of Flooding in Central Texas*. Unpublished PhD thesis, University of Texas, Austin, Texas.

Patton, P.C., Baker, V.R. and Kochel, R.C., 1979. Slackwater deposits: a geomorphic technique for the interpretation of fluvial paleohydrology. In D.D. Rhodes and G.P. Williams (eds), *Adjustments Of The Fluvial System*. Kendall/Hunt, Dubuque, Iowa, 225–253.

Redmond, K.T., Enzel, Y., House, P.K. and Biondi, F., 2002. Climate variability and flood frequency at the decadal to millennial time scales. In P.K. House, R.H. Webb, V.R. Baker and D.R. Levish (eds), *Ancient Floods, Modern Hazards: Principles and Applications of Paleoflood Hydrology*. American Geophysical Union, Washington, DC, 21–45.

Richards, K.S., 1979. Simulation of flow geometry in a riffle-pool stream. *Earth Surface Processes*, **3**, 345–354.

Richards, K.S., 1982. *Rivers: Form & Process in Alluvial Channels*. Methuen, London, 1–74.

Rico, M., Benito, G. and Barnolas, A., 2001. Combined palaeoflood and rainfall runoff assessment of mountain floods (Spanish Pyrenees). *Journal of Hydrology*, **245**, 59–72.

Schnur, R. and Lettenmaier, D.P., 1998. A case study of statistical downscaling in Australia using weather classification by recursive partitioning. *Journal of Hydrology*, **212-213**(1-4), 362–379.

Schumm, S.A., 1977. *The Fluvial System*. Wiley, New York.

Snowling, S.D. and Kramer, J.R., 2001. Evaluating modelling uncertainty for model selection. *Ecological Modelling*, **138**, 17–30.

Stillman, R.A., McGrorty, S., Goss-Custard, J.D. and West, A.D., 2000. Predicting mussel population density and age structure: the relationship between model complexity and predictive power. *Marine Ecology Progress Series*, **208**, 131–145.

Strupczewski, W.G., Singh, V.P. and Feluch, W., 2001. Non-stationary approach to at-site flood frequency modelling I. Maximum likelihood estimation. *Journal of Hydrology*, **248**, 123–142.

Thomas, M.F., 2000. Late Quaternary environmental changes and the alluvial record in humid tropical environments. *Quaternary International*, **72**, 23–36.

Thoms, M.C. and Sheldon, F., 2000a. In-channel processes and environmental flow requirements for the Barwon-Darling river. *Rivers for the Future*, **12**(4), 10–19.

Thoms, M.C. and Sheldon, F., 2000b. Water resource development and hydrological change in a large dryland river: the Barwon-Darling River, Australia. *Journal of Hydrology*, **228**, 10–21.

Topping, J., 1956. *Errors of Observation & their Treatment*. Institute of Physics, London.

Veldkamp, A. and van Dijke, J.J., 2000. Simulating internal and external controls on fluvial terrace stratigraphy: a qualitative comparison with the Maas record. *Geomorphology*, **33**, 225–236.

Wagener, T., Boyle, D.P., Lees, M.J., Wheater, H.S., Gupta, H.V. and Sorooshian, S., 2001. A framework for development and application of hydrological models. *Hydrology and Earth System Sciences*, **5**, 13–26.

Webb, R.H. and Jarrett, R.D., 2002. One-dimensional estimation techniques for discharge of paleofloods and historical floods. In P.K. House, R.H. Webb, V.R. Baker and D.R. Levish (eds), *Ancient Floods, Modern Hazards: Principles and Applications of Paleoflood Hydrology*. American Geophysical Union, Washington, DC, 111–125.

Webb, R.H., O'Connor, J.E. and Baker, V.R., 1988. Palaeohydrologic reconstruction of flood frequency on the Escalante River, South-Central Utah. In V.R. Baker, R.C. Kochel and P.C. Patton (eds), *Flood Geomorphology*. Wiley, New York, 403–418.

de Wit, M.J.M. and Pebesma, E.J., 2001. Nutrient fluxes at the river basin scale II: the balance between data availability and model complexity. *Hydrological Processes*, **15**, 761–775.

Wohl, E.E., 1998. Uncertainty in flood estimates associated with roughness coefficient. *Journal of Hydraulic Engineering*, **124**, 219–223.

Wohl, E.E., 2002. Modeled paleoflood hydraulics as a tool for interpreting bedrock channel morphology. In P.K. House, R.H. Webb, V.R. Baker and D.R. Levish (eds), *Ancient Floods,*

Modern Hazards: Principles and Applications of Paleoflood Hydrology. American Geophysical Union, Washington, DC, 345–358.

Xu, C.Y., 1999. Climate change and hydrologic models: a review of existing gaps and recent research developments. *Water Resources Management*, **13**, 369–382.

Yang, D., Yu, G., Xie, Y., Zhan, D. and Li, Z., 2000. Sedimentary Records of Large Holocene Floods from the Middle Reaches of the Yellow River, China. *Geomorphology*, **33**, 73–88.

Young, W.J., Olley, J.M., Prosser, I.P. and Warner, R.F., 2001. Relative changes in sediment supply and sediment transport capacity in a bedrock-controlled river. *Water Resources Research*, **37**, 3307–3320.

20 Short-term Hydrological Changes

L. STARKEL

Institute of Geography, Polish Academy of Sciences, Kraków, Poland

Earlier chapters have indicated how palaeohydrological changes can often be short term and very significant and this is explored further in terms of the character of short-term hydrological changes, their frequency and zonal differences, prior to considering causes and mechanisms that are responsible.

1 THE CHARACTER OF SHORT-TERM HYDROLOGICAL CHANGES

Short-term hydrological changes arise resulting from changes in precipitation, evaporation and storage, taking place during periods of centuries up to 1,000 to 2,000 years and can be caused by fluctuations in solar radiation (Stuiver *et al.*, 1991; Bond *et al.*, 2001; Mayewski *et al.*, 1997), changes in oceanic circulation (Broecker *et al.*, 1989; Street-Perrott and Perrott, 1990), clustering of volcanic eruptions (Bryson, 1989; Zielinsky *et al.*, 1996) and other causes. Such short-term changes are superimposed upon long-term cyclic fluctuations on the scale of 10 to 20 thousands of years, controlled by variations in orbital parameters (CLIMAP, 1976; Kutzbach, 1992).

The long-term cycle of the transition from the last cold stage to the Holocene, which was expressed in general humidification in low latitudes and aridisation in the higher ones, since the mid-Holocene has been succeeded by a reversed tendency (COHMAP, 1988; Stuiver and Braziunas, 1993). Superimposed upon this trend are shorter rhythmic fluctuations recorded in advances and retreats of montane glaciers, and lake-level changes (Grove, 1979; Nessje and Johannessen, 1992; Magny, 1993; Starkel, 1983). Some of these fluctuations were recognised as Dansgaard–Oeschger events (Dansgaard and Oeschger, 1988) or Maunder and other minima in the solar radiation (Damon *et al.*, 1998; Shindell *et al.*, 2001). The character of such short-term changes may vary; in general, during the Holocene, alternate wetter and drier phases were several centuries long. The transition from one phase to another may be gradual or abrupt, even effected in several decades, like the Younger Dryas–Pre-Boreal limit (Merkt *et al.*, 2001; Goslar *et al.*, 1995).

Characteristic features of changes have been observed in hydrological records during the last 200 years as well as reconstructed from written sources for longer timescales (Pfister, 1992; Benito *et al.*, 1996; Starkel, 1998b). On the millennial scale of the Holocene, hydrological changes can be reconstructed from sediments, morphological evidence and biotic changes (Starkel *et al.*, 1991; Berglund, 1986; Wohl and Enzel, 1995). The most precise records for this timescale are found in undisturbed sequences of annually laminated lake sediments or tree rings and in ice cores in which the variability of stable isotopes can help reconstruct the course of changes in temperature

Palaeohydrology: Understanding Global Change. Edited by K.J. Gregory and G. Benito
© 2003 John Wiley & Sons, Ltd   ISBN: 0-470-84739-5

and, with lower accuracy, in precipitation (Dansgaard and Oeschger, 1988; Ralska-Jasiewiczowa *et al.*, 1998; Schettler *et al.*, 1999). In the normal sequences of lake, fluvial and other sediments, changes are registered in the chemistry and sediment facies, and in relief forms, so that by dating such changes we can recognise the time of changes in lake-level fluctuations, in flood phases or in glacier advances.

Two features of registered changes must be considered: spatial metachroneity of change and delay in the response of the system to a climatic signal. Metachroneity at the local scale is expressed on the flat shores of lake basins. In the littoral facies, rising water level is recorded by a change from peat to gyttya and then to lacustrine chalk (Digerfeldt in: Berglund, 1986). To evaluate the true timing of the change in rainfall regime and of lake transgression, the littoral records should be correlated with analysis of diatoms and cladocera from the central part of the lake basin (cf. Berglund, 1986). Metachroneity is shown at the regional scale in Africa, expressed in shifting of the savanna–steppe, steppe–semidesert ecotone (Servant, 1983; Kadomura, 1995), where the rate of such shift is difficult to measure and at particular sites is expressed as an abrupt change in lacustrine, fluvial or aeolian deposits.

The response of various systems to abrupt change in temperature or precipitation may be delayed in time or lagged. The clearest examples are demonstrated by the decay of ice sheets (Teller, 1995) because the retreat of the ice margin occurs much later than a gradual decline of its volume. Lakes in densely forested areas respond to short-term changes in parallel with the rise in groundwater level and after meeting evapotranspiration demand by plant communities. The advances of mountain glaciers may be delayed by several centuries (Anderson *et al.*, 2001), depending on the particular size and topography of ice fields (Patzelt, 1977). Delay is also shown in the gradual eastward expansion of the westerlies in Boreal time: the distinct increase in precipitation recorded in central Europe between 8.7 and 8.5 ^{14}C kyr BP reached western Siberia after 8 kyr BP (Zubakov, 1972).

Considering the detailed patterns of lake-level fluctuations, glacial advances and retreats, or phases with high flood frequency (Patzelt, 1977; Starkel, 1983; Magny, 1993; Hormes *et al.*, 2001) leads to the question: What do we register to identify short-term changes? It could be extreme positions, recognised as wetter (cooler) and drier (warmer) phases, or it could be the alternation of phases with a tendency to a wetter or a drier climate (Figure 20.1). Both instances may relate to different time intervals, depending on a range of factors, including the source of the information, and may be affected by a delay in response. Therefore, some limnologists concentrate more on recognition of lake water level rise (Digerfeldt, 1988; Street-Perrott *et al.*, 1985) as palaeobotanists focus on the rise of the upper tree line (Burga, 1988), rather than on determination of the time of the highest (or lowest) position. The highest levels can be very approximate especially when we concentrate our reconstructions on three-year time slices (cf. COHMAP, 1988; Harrison *et al.*, 1993). Therefore, to avoid misunderstanding in correlations we should recognise an essential requirement: the necessary parallel examination of various types of records including terrestrial, limnic and glacial (Figure 20.2).

As the significance of short-term palaeohydrological changes has come to be recognised in many parts of the world (e.g. Chapters 5–14), it has been increasingly appreciated that short-term changes may or may not be synchronous from one area to another. Some events occur at different times in the Quaternary. This situation arises because there are different rates of change of environmental parameters as originally expressed in the model by Knox (1972), but there are also changes, which

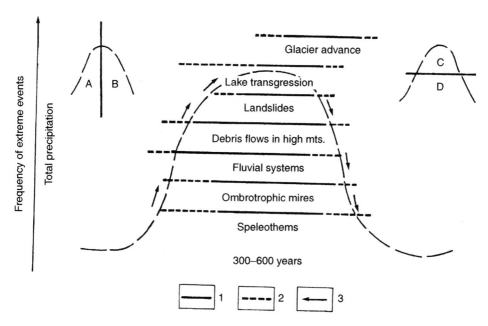

Figure 20.1 Conceptual model of reflection of phases with higher frequency of extreme events in various sediments and forms. Key: 1. main period of reflection; 2. less distinct reflection; 3. groundwater level rise and fall. In the top left – the phase of rising (A) and lowering (B) lake level; in the top right – the phase of high (C) and low (D) lake level

are synchronous at the global or hemispheric scale (Goslar *et al.*, 1995; Starkel and Basu, 2000).

1.1 Short-term Changes Realised by Event Frequency

Every change in the water budget, especially abrupt ones, usually involves a shift to a higher or lower frequency of extreme events (Starkel, 1984; 1996a; 1998a, b), most of such changes being related to disturbances in air-mass circulation including El Niño Southern Oscillation (ENSO) activity or extratropical blocking (Waylen, 1995).

Among various types of changes in event sequences these may be distinguished (cf. Starkel, 1976; 1998b):

- Heavy downpours, mostly of local extent (up to several tens of km²) with totals rarely exceeding 100 mm, but with intensity rising to several mm min⁻¹. In areas without dense vegetation cover, such rains can cause rapid overland flow, wash, gullying and, on steeper slopes, induce earth and debris flows.
- Continuous rains can affect large areas of at least several thousands of square kilometres characterised by long duration, and in spite of low intensity the precipitation totals can reach 300 to 1,000 mm and more in 2- to 5-day periods. The relevant thresholds relate to saturation of the ground and to the start of subsurface runoff causing piping, landsliding and the formation of flood waves in larger catchments, accompanied by high rates of bedload and transformation of river channels. The incidence of heavy downpours within a period of prolonged precipitation can cause a simultaneous transformation of slope and channel systems, exemplified by the October 1968 event in the Darjeeling Himalaya (Starkel, 1976).

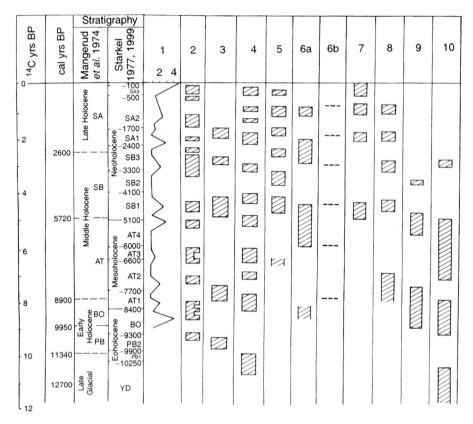

Figure 20.2 Phases of high frequency of extreme rainfalls and wetter climate. Key: 1. global acid fallout (H_2SO_4 + Ax) ($\times 10^8$ t) (Hammer *et al.*, 1980); 2. glacial advances in the Alps (Patzelt, 1977); 3. high lake water level in North Poland (Ralska-Jasiewiczowa 1989; Ralska-Jasiewiczowa *et al.*, 1998); 4. higher flood frequency in upper Vistula (Starkel *et al.*, 1996a); 5. higher activity of British rivers (Needham and Macklin, 1992); 6a. higher activity of Wisconsin rivers (Knox, 1983); 6b. major river discontinuities in the northern part of the United States (Knox, 1983); 7. frequent floods in southwestern United States (Enzel *et al.*, 1996); 8. higher level of Dead Sea (Framklin *et al.*, 1991); 9. higher lake level over Sahara (Petit-Maire and Guo, 1996); 10. higher lake levels in humid tropics of Africa (Kadomura, 1995)

- Snowy winters followed by snowmelt floods, well known from Siberia and eastern Europe, are characterised by intense slope wash over frozen soil and by snowmelt floods, which are accentuated by rainstorms and by ice jams. Such situations were frequent all over Europe during the Little Ice Age (Starkel, 2001).

- Rainy seasons (phases) take place when the rains continue for several months and the ground becomes fully saturated with water. Every occurrence of additional rain or snowmelt facilitates the crossing of the stability threshold for sliding and flooding (Gil and Starkel, 1979).

- Rainy years – several consecutive years with water surplus are reflected in lake-level rises, in faster peat growth and, in the case of repeated floods, in the stabilisation of a new phase of channel transformation.

- Droughts – in addition to the arid zones their incidence can contribute to the role of extremes. If continuing for consecutive years, droughts result in the transformation of vegetation cover, the lowering of groundwater levels and also in the desiccation of rivers and distinct reductions in lake storage.

Changes of these types of event sequences affect the water introduced to the system and so are reflected in the runoff and water storage through crossing of various thresholds (Figure 20.2). For instance, the wetter climate of the early Subboreal caused the expansion of ombrotrophic mires (Barber, 1985), flood phases are registered in the cut-fill sequences of valley floors (Starkel, 1983) and saturation of the substratum can cause the inception of landsliding (Starkel, 1997).

In the course of stabilisation of new trends in the water balance and in transformation of natural geo-ecosystems (fluvial, slope, lacustrine etc.), clustering of events can be of great importance: if several high order events follow rapidly one after another, there is insufficient time for recovery and stabilisation of the system in the intervening period. After such a clustering, the system can either return to the previous stage (equilibrium) or may shift to the new equilibrium (Starkel, 1999b). This is well exemplified by changes of channel parameters for the Gila River in Arizona (Burkham, 1972; Hooke, 1996; Figure 20.3). Such clusters of floods over several years are evident in the twentieth-century flood history (Figure 20.4) and similar examples have been identified in Europe during the decades of the Little Ice Age (Rumsby and Macklin, 1996; Brazdil et al., 1999; Starkel, 2001; Figure 20.5). Several clusterings may collectively form a phase with a higher frequency of extremes up to several hundred years in duration (Starkel, 1999b) or just accentuate a shift to new climatic conditions, such as the rapid rise in frequency of extremes about 3 ka BP that is observed in the deposits of Wisconsin rivers (Knox, 1993). A significant problem is to differentiate, using sedimentary and morphological evidence, a single extreme event from the clustering of events comprising a short-term phase, which is not of local character. Such phases can be safely recognised at the regional scale, by detailed examination

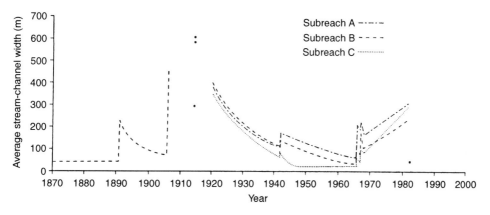

Figure 20.3   Changes of channel within three reaches of the Gila River indicating the role of single floods and their clustering in 1905–1920 and in the 1970s (after Burkham, D.E., 1972. Channel Changes of the Gila River in Safford Valley, Arizona 1846–1970. Geol. Survey Water – Supply Papers (U.S.) 655-G, 1–24; and Hooke, J.M., 1996. River responses to decadal-scale changes in discharge regime: the Gila River, SE Arizona. In J. Branson, A.G. Brown and K.J. Gregory (eds), *Global Continental Changes: The Context of Palaeohydrology*. Geological Society of Special Publication No. 115, Geological Society, London, 191–204)

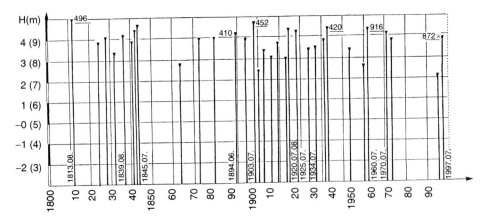

Figure 20.4 High flood levels on the Vistula River in Cracow during the last 200 years (based on Bielański, 1984). There are distinct clusterings and about 20-year-long breaks without floods. The years and months with the most extremes events are indicated. On the left, the elevation after change of water level "O" in 1941 is shown in brackets

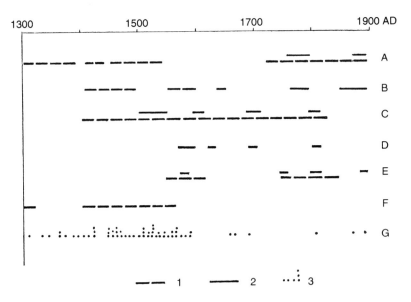

Figure 20.5 Flood records in several European countries mainly since the fifteenth or sixteenth centuries, based on chronicles, sediments and other data. Key: 1. phases of floods; 2. clustering of floods or heavy rains; 3. years of extreme rains and floods (in five-year intervals). A. English rivers (Rumsby and Macklin, 1996); B. Spanish rivers (Benito *et al.*, 1996); C. Tiber River (Camuffo and Enzi, 1995); D. Switzerland – extreme rains (Pfister, 1992); E. Czech republik – rainy years (Brazil *et al.*, 1999); F. Vistula valley (after dendrochronologically dated subfossil oaks), (Krąpiec, 1992; Starkel *et al.*, 1996a); G. Vistula valley (floods recorded in chronicles)

of various records presented by lake, bog, fluvial, slope and other systems (Gregory *et al.*, 1995; Starkel, 1996b).

In areas of water balance surplus, some features, such as the flooding of bogs and over bank, may suggest an increase in frequency of extreme events whereas the evidence for a single event is weak. In contrast, in areas of water deficit with sparse vegetation cover, a single large event (or clustering of events) may be responsible for an abrupt change in the channel shape as shown by many rivers in foothill areas of the southwestern part of the United States (Baker *et al.*, 1988). Analysis of slackwater deposits in rocky canyons provides the most exact information about consecutive events (Baker, 1987; Enzel *et al.*, 1996) although in a single profile it is only subsequent higher flood levels that are recorded. During wet phases, with a series of flood events, lake levels may rise several tens of metres in one century as in the case of the Dead Sea (Frumkin *et al.*, 1991; Netser, 1998). Lateglacial cataclysmic floods, caused by the draining of large ice-dammed lakes, being restricted in space and time, are rather different in character (Baker, 1987; Baker *et al.*, 1993).

In order to distinguish events and their clusters from longer phases, a detailed examination of continuous undisturbed profiles is necessary, concentrating on records of extreme events. Lakes and bogs, especially those with annually laminated sediments, reflect long-term variations in their granulometry (with external input components), and salinity, for example, so that only very rare extremes may be registered in such sequences (Dearing and Zolitschka, 1999). In contrast, fluvial sequences can evidence layers representing separate events so that counting them per century is possible (Enzel *et al.*, 1996; Starkel *et al.*, 1996b). In the Nahal Zin valley of the Dead Sea catchment, it was possible to detect a rise from 0.5 to 2.5 events per century in one millennium (Frumkin *et al.*, 1998). Similar clusterings of buried subfossil oak trunks from the sixth or the eleventh century at particular localities in the Vistula valley (Kalicki and Krąpiec, 1992; Krąpiec in: Starkel *et al.*, 1996b) record high flood frequency during single decades (Figure 20.6).

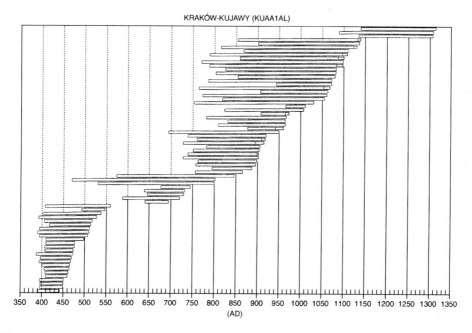

KRAKÓW-KUJAWY (KUAA1AL)

350 400 450 500 550 600 650 700 750 800 850 900 950 1000 1050 1100 1150 1200 1250 1300 1350
(AD)

Figure 20.6 Subfossil oaks dated by dendrochronological methods from the Kraków–Kujawy site in the Vistula valley (after Krąpiec, M., 1992. Skale dendrochronologiczne późnego holocenu poludniowej i centralnej Polski. *Geologia*, **18**(3), 37–119, and in print). Distinct clusterings occur ca 450 to 550 AD, 900 to 920 AD, 950 to 970 AD and 1050 to 1120 AD. Reproduced by permission of Komitet Badan Czwartorzedu

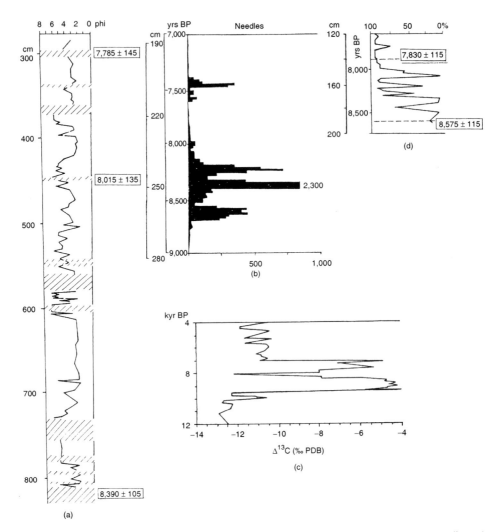

Figure 20.7 Extreme events and their clustering reflected in various sediments from the wetter phase between 8,600 and 7,800 ^{14}C yr BP (after Starkel, L., 1999a. 8,500 – 8,000 yrs BP humid phase-global or regional? *Science Reports of Tohoku University, 7th Series, Geography.* **49**, 2, 105–133); (a) Change of m/z in the alluvial fan in Podgrodzie, south Poland (after Czyżowska, E., 1997. Zapis wydarzeń powodziowych na pograniczu borealu i atlantyku w osadach stożka naplywowego w Podgrodziu. *Dokumentacja Geograficzna*, IGiPZ PAN 5, 74); (b) Macrofossils in Lago Basso, Alps (after Wick L., and Tinner W., 1997. Vegetation changes and timberline fluctuations in the Central Alps as indicators of Holocene climatic oscillations. *Arctic and Alpine Research*, 29, 445–458); (c) Fluctuations of ^{13}C in cave sediments, Soreg cave, Israel (Bar-Matthews *et al.*, 1997); (d) Mineral horizons in the peat sequence recorded in the curve of loss of ignition at Rambjorgeboten, south Norway (Torske, 1996)

Knowledge of present-day conditions helps the recognition of several Holocene phases with higher frequencies of extreme events at the regional and even at the global scale (see Chapter 7). Especially widespread and well documented are Phase 8.5 to 8.0 ^{14}C kyr BP (Starkel, 1999a) and the Atlantic–Subboreal transition (Starkel, 1995). These show a simultaneous high frequency of heavy downpours, reflected in debris flows and alluvial fans, with continuous rain recorded by new alluvial fills and avulsions of channels and the presence of rainy years recorded by lake-level rises, advances of glaciers and deep landslides (Figures 20.2 and 20.7). This pattern, during several Holocene phases, indicates great climatic instability, frequent shifts of air masses and, in the temperate zone of prevalent westerlies, a change to N–S longitudinal circulation. In lower latitudes, during such wetter phases, including that from 8.5 to 8.0 yr BP (Adamson et al., 1980), shorter drier periods are recorded, with the lowering of Nile discharges (Paulissen and Vermeersch, 1989) also indicating great instability.

Independent of these longer phases, several abrupt climatic changes have been recognised manifested by rapid alterations in frequency usually accompanied by change to warming (Younger Dryas–Pre-Boreal transition; Goslar et al., 1995; Figure 20.8) or to a cooler and wetter climate (ca 2,750 ^{14}C yr BP – van Geel et al., 1996). The number of examples of abrupt changes identified from past records is increasing (3,900–3,500 ^{14}C yr BP – Anderson et al., 1998; Mediaeval period – Stine, 1998) but these are probably of regional extent. In most cases, the beginning of every wetter and cooler phase

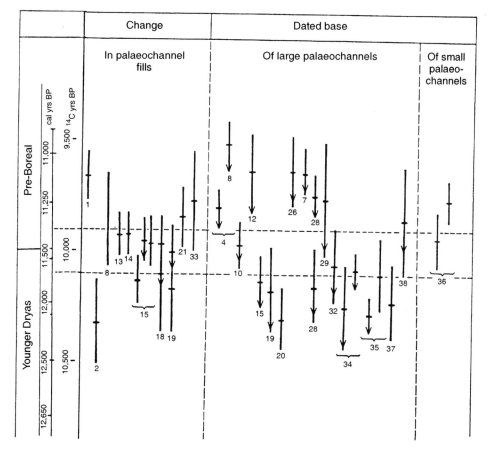

Figure 20.8 Radiocarbon-dated changes of palaeochannels at the Younger Dryas–Pre-Boreal transition in the Vistula and Warta River catchments (based on Starkel et al., 1996b and other studies). Arrows indicate that the time of abandonment is probably older than the data. Numbers refer to various localities

with a higher frequency of extremes is manifested by an abrupt change due to passing of the threshold of dominating geo- or ecosystems. It is recorded especially in the ecotonal zones and in the fluvial environment, which are not yet adapted to other frequencies (Knox, 1995; Starkel *et al.*, 1996b).

1.2 Zonal Differences in Hydrologic Regime

Considering fluctuations in the insolation regime and the position of cells steering air-mass circulation, it is possible to distinguish four latitudinal zones (a–d below) alternately having surplus and deficit in the water balance (Starkel, 1989; Waylen, 1995). In these belts, different extreme events play a leading role in short-term hydrological changes, and such changes are also affected by the shifting of ecotones. Because of the various responses to climatic change, the flood phases (wetter) are not synchronous in all climatic zones (Barnett *et al.*, 1988; Kutzbach, 1992; Baker *et al.*, 1995; Flohn, 1993).

(a) The Polar and subpolar zone (See Chapters 5 and 6) is characterised by long-term water storage in ice sheets and permafrost, seasonal storage in snow cover and seasonal runoff with snowmelt and ice-jam floods (Woo, 1986; Frenzel *et al.*, 1992). Short-term hydrological changes are more evident in the second half of the Holocene, when there was a tendency to cooling and a distinct advance of permafrost (Baulin *et al.*, 1984), with fluctuations of water storage in ice and snow as well as in flood frequency (Starkel, 1989; Maizels, 1995). In the rate of transformation of the hydrological regime, an important role was played by a feedback mechanism connected with the change of albedo during extension or reduction of the area with snow cover (Kukla, 1969).

(b) The Temperate zone (see Chapters 7 and 8) offers a great variety of extremes of changing frequency. During wetter phases, with distinct surplus of water, there are frequent extreme rainfalls of various character including heavy downpours, continuous rainstorms and rainy seasons (Starkel, 1996a). The last two are especially common at the western fringe of Europe (British Isles) as recorded in the frequency of winter floods (Rumsby and Macklin, 1997) and the extension of blanket bogs (Barber, 1985). Seasonality is also well expressed in the winter floods of the Mediterranean region (Benito *et al.*, 1996; Camuffo and Enzi, 1995). In the case of the upper Vistula basin, the channel forming discharges have a recurrence interval of 10 to 20 years, but further east, instead of floods caused by rainstorms, snowmelt floods (Starkel, 1998b) become more frequent and are the norm in eastern Europe (Lvovich, 1979). Flood phases are well expressed in alluvial fills containing clusterings of subfossil oak trunks (Becker, 1982; Krąpiec, 1992) and by frequent channel avulsions (Starkel *et al.*, 1996b).

(c) Arid zones (see Chapters 9, 14) with a water deficit occur in the latitudinal belt 15 to 30° N, extending to 50° N. Their characteristic features are long-lasting droughts disrupted by rare heavy downpours and very rare continuous rains (Waylen, 1995). Short wetter phases, connected with expansion of the monsoonal circulation or westerlies, are manifested in the increasing frequency of extreme floods, which, in conditions of sparse vegetation and thin soil cover, cause rapid runoff and storage in lakes formed in closed basins (Servant, 1983; Petit-Maire, 1986; Baker *et al.*, 1995). After one heavy rainstorm in Rajasthan, a rise in the level of the Sambhar Lake by 15 m (Starkel, 1972) was observed. Wetter phases are manifested in the clustering of heavy rainfalls that are recorded in the Dead

Sea basin (Frumkin *et al.*, 1998) and at Lake Abhe (Fontes *et al.*, 1985). Such pluvial phases caused simultaneous shift of vegetation belts over Africa towards the north (Servant, 1983), due to the rise of mean annual precipitation, which in the case of the Chad basin exceeded 600 mm, compared with 350 mm at present (Kutzbach, 1980).

A totally inverse pattern of short-term fluctuations is observed in the SW part of the United States, where during most parts of the Holocene (before 2 yr BP) is recorded the desiccation of closed depressions and the only flood phases are registered between 5.0 and 4.4 ka BP, ca 2.2 and 1.8, 1.2 and 0.8 and after 0.5 ^{14}C kyr BP (Baker *et al.*, 1995).

(d)   The humid tropics (Chapters 11, 12, 13) with a distinct surplus of water are differentiated into two types: those areas with rainfall distributed throughout the whole year and those areas with a distinct rainy season (monsoonal). Heavy rains are common throughout this system, but extreme rainstorms are not so frequent everywhere. In the renowned Cherrapunji area, continuous heavy rains are recorded every year, when the monthly precipitation exceeds 3,000 to 5,000 mm. (Starkel *et al.*, 2002). On the contrary, in the Darjeeling Himalaya threshold, rainfall with totals between 600 and 1,500 mm, capable of transforming slopes and river channels, occur 2 to 4 times per century (Starkel, 1972; Starkel and Basu, 2000). Variations in humidity in tropical Africa have also been high as shown by the millennial fluctuations of the east African lake levels, reaching up to 100 m (Hastenrath and Kutzbach, 1983; Kadomura, 1995). During phases such as those occurring between 8.4 and 8.1 ^{14}C kyr BP, high discharges of rivers such as the White Nile (Adamson *et al.*, 1980) were recorded.

Fluctuations in the rainfall regime and its totals have been reflected in the shifting of ecotones especially well recognised at the Sahara–Sahel and savanna tropical forest. At particular sites, this transition is expressed in an abrupt change, evident in a change of sediment facies, or lake salinity or the occurrence of drainage from closed depressions and vice versa (Jäkel, 1979; Kadomura, 1995; Lezine, 1998). The eastward shift ca 6 ^{14}C kyr BP of the ecotone between forest and prairie was well recognised at the Minnesota–Wisconsin transect in pollen diagrams and speleothems (Denniston *et al.*, 1999) and was explained at an early stage by the shifting position of atmospheric frontal zones (Bryson and Wendland, 1967).

1.3   Causes and Mechanisms of Short-term Hydrologic Changes

Short-term hydrologic changes are closely connected with variations of climate and air-mass circulation, which are modified again by other factors. Changes in the orbital parameters control the main climatic variations during the last 18 kyr (CLIMAP, 1976; Kutzbach and Guetter, 1984), which includes, during the early Holocene, the warming of higher latitudes of the northern hemisphere and the synchronous activisation of monsoonal circulation in lower latitudes, and finally, since 7 to 5 kyr BP, with declining irradiation, the gradual weakening monsoonal circulation and aridisation of the subtropics combined with a trend towards wetter and cooler conditions in higher latitudes (Barnett *et al.*, 1988; Issar, 1998; Steig, 1999).

Second-order (short-term) fluctuations have been correlated with variations in solar radiation, which show distinct periodicities. The reverse picture is reflected in ^{14}C and ^{10}Be content curves, but these interrelations are not simply related to solar oscillations due to the superposition of other factors (Stuiver, 1995). Damon *et al.* (1998)

distinguished cycles 2,300 years long, whereas Bond *et al.* (1997), Stuiver *et al.* (1997) and others found cycles of 1,470 years in duration. The latter are recorded in oceanic cores as horizons containing haematite-stained grains connected with ice rafting of expanding ice sheets, called *Dansgaard–Oeschger events* (Bluemle *et al.*, 1999). Probably of similar character are fluctuations in the growth rate of ombrotrophic peat in Britain, reflecting periodicities of 1,100 and 600 years (Hughes *et al.*, 2000).

During the last millennium, several minor ^{14}C culminations and minima in the solar radiation have been recognised (cf. Chambers *et al.*, 1999). These correspond to advances and retreats of the Aletsch glacier in the Alps (Flohn, 1993). By analysing the flood frequency during the Little Ice Age in Italy, Camuffo and Enzi (1995) concluded that such direct relation to Spörer and Maunder minima do not exist and other factors may coincide with changes in the solar radiation. Similar shorter rhythmicity of 200 to 300 years duration was especially distinct during Boreal time 8.7 to 8.0 ^{14}C kyr BP (Figure 20.8), recorded in Moon Lake in Dakota (Laird *et al.*, 1998), in Lake Basso in the Alps (Wick and Tinner, 1997), and in many other places (Starkel, 1999), in fluctuations of glaciers during the Venediger advance (Patzelt, 1977), as well as during the early Subboreal in bog and lake sequences (Aaby, 1976; Gajewski, 1988; Anderson, 1998).

Fluctuations of this kind are all expressed in the air-mass circulation, through its weakening or strengthening, as reflected in simultaneous increases or decreases of precipitation in various parts of the globe. Phases with reduced solar radiation are characterised by high and frequent fluctuations of the ^{18}O curve indicating very unstable temperatures (Thompson, 1992) and high frequency of hydrological extremes, such as the phase extending from 8.5 to 8.0 ^{14}C kyr BP or the Little Ice Age (Starkel, 1999). Most catastrophic floods seem to coincide with the initiation of distinct transformations in the air-mass circulation (Hirschboeck, 1988; Knox, 1983). Such rapid change occurred about 2,650 ^{14}C yr BP, indicating the return to more moist and cooler climate conditions (van Geel and Reussen, 1998). Besides these changes in solar radiation, other more irregular fluctuations were recognised, which amplify the solar signal and cause zonal or regional differences. These include the circulation of oceanic waters, volcanic activity and others. Recognition of the importance of North Atlantic Deep Water as a regulator of global water and energy exchange has explained various deviations, including the Younger Dryas rapid cooling caused by the diversion of melt waters of the Laurentide ice sheet from the Mississippi River to the St Lawrence River (Broecker *et al.*, 1989; Teller, 1995). Probably a similar, smaller melt water pulse before 8 ka BP facilitated cooling and glacier advances (Street-Perrott and Perrott, 1990). Changes in the thermohaline Atlantic circulation seem, therefore, to express the solar signal, coinciding with ^{14}C and ^{10}Be changes and with Dansgaard–Oeschger events (Bond *et al.*, 2001). Fluctuations in ENSO are restricted not only to the equatorial zone and the southern hemisphere, where they are responsible for variations in flood frequency (Diaz and Markgraf, 1993) but are also the probable driving force in clusterings of extremes during the Little Ice Age (Quinn and Neal, 1992).

The other factor determining a high frequency of extreme events could be the clustering of volcanic eruptions (Bryson and Wendland, 1967). The curve of volcanic aerosol from Greenland ice cores (Bryson and Goodman, 1980; Hammer *et al.*, 1980) has shown a clear correspondence with cooling and with frequent extreme rainfalls. More detailed examination demonstrated that the highest Holocene peak of volcanic activity ca 8.5 to 8.3 ^{14}C kyr BP (Zieliński *et al.*, 1996; Bryson and Bryson, 1998) coincided with the clustering of various types of extreme rainfalls at the global scale

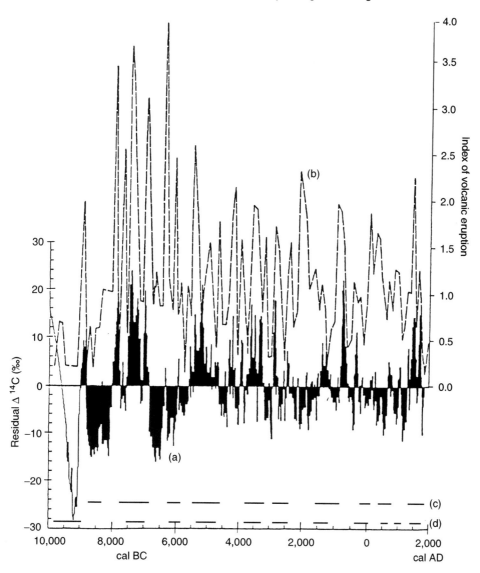

Figure 20.9 Fluctuations of residual Δ ^{14}C (Stuiver, 1995). A. indicator of volcanic erup-tions B (Bryson and Bryson, 1998) B. phases of glacier advances in the Alps (after Patzelt, G., 1977. Der zeitlicke Ablauf und das Ausmass postglazialer Klimaschwankungen in den Alpen. In B. Frenzel (ed.), *Dendrochronologie und postglaziale Klimaschwankungen in Tirol*. Veröff. des Museum Ferdinandeum, Vol. 67, Mainz, 93–123; and others) – and phases with high frequency of extreme events (after Starkel *et al.*, 1996b)

(Starkel, 1999a; Figure 20.9). Detailed examination of volcanic dust over the last mil-lennium helped recognise phases of frequent eruptions ca 1000 AD, 1055 to 1090 AD, 1285 to 1300 AD, 1340 to 1470 AD, 1570 to 1600 AD, 1630 to 1650 AD – each coin-ciding with a phase of frequent floods as recorded in sediments, clustering of subfossil oaks and in documentary sources (Bradley and Jones, 1992; Benito *et al.*, 1996; Lam-oureux *et al.*, 2001; Starkel, 2001). The direct response, by cooling, to the famous Tombola eruption in 1815 indicates a close coincidence of deviations in precipitation and temperature with volcanic activity (Flohn, 1993). The higher content of aerosols of volcanic origin and greenhouse gases superimposed on the decrease in solar radiation seems to be a cause of more distinct expression of phases with clustering of extreme events and may even have provoked the glacial advances (Starkel, 1998a; 1999).

2 CONCLUSIONS

Short-term hydrological changes are complex in character and reflect primary changes in orbital parameters and solar radiation, being transformed and adapted to the zonal and regional conditions by the air-mass circulation pattern as well as by the thermohaline circulation of oceanic water, which transmit them globally (cf. Bond *et al.*, 2001). However, such fluctuations may be amplified or reduced depending on the coincidence with volcanic activity, and on regional differences in orography and stability of ecosystems, especially in ecotonal zones. The sedimentological response to these changes depends also on the acceleration of processes caused by degradation of natural systems by man (Starkel, 1987; Gregory and Walling, 1987). Such short-term changes, increasingly evident from the research data analysed in the course of recent palaeohydrological research, exemplify major ways whereby, from reconstruction of the past, lessons can be learnt that are of value in interpreting aspects of future global change.

REFERENCES

Adamson, D.A., Gasse, F., Street, F.A. and Williams, M.A.J., 1980. Late Quaternary history of the Nile. *Nature*, **287**(5786), 50–55.

Anderson, L., Abbott, M.B. and Finney, B.P., 2001. Holocene climate inferred from oxygen isotope ratios in lake sediments, Central Brooks Range, Alaska. *Quaternary Research*, **55**, 313–321.

Anderson, D.E., Binney, H.A. and Smith, M.A., 1998. Evidence for abrupt climatic change in Northern Scotland between 3900 and 3500 calendar years BP. *The Holocene*, **8**(1), 97–103.

Baker, V.P., 1987. Palaeoflood hydrology and extraordinary flood events. *Journal of Hydrology*, **96**, 79–99.

Baker, V.R., Benito, G. and Rudoy, A., 1993. Palaeohydrology of Late Pleistocene super flooding, Altay Mountains, Siberia. *Science*, **259**, 348–350.

Baker, V.R., Bowler, J.M., Enzel, Y. and Lancaster, N., 1995. Late Quaternary paleohydrology of arid and semi-arid regions. In K.J. Gregory, L. Starkel and V.R. Baker (eds), *Global Continental Palaeohydrology*. John Wiley & Sons, Chichester, 203–231.

Baker, V.R., Kochel, R.G. and Patton, P.C., (eds), 1988. *Flood Geomorphology*. Wiley, Chichester.

Barber, K.E., 1985. Peat stratigraphy and climate change: some speculations. In M.J. Tooley (eds), *The Climate Scene*. Allen & Unwin, London, 175–185.

Bar-Matthews, M., Ayalon, A. and Kaufman, A., 1997. Late Quaternary paleoclimate in the Eastern Mediterranean region from stable isotope analysis of speleothems at Soreg Cave, Israel. *Quaternary Research*, **47**, 155–168.

Barnett, T.P., Dumenil, L., Schlese, U. and Roeckner, E., 1988. The effects of Eurasian snow on global climate. *Science*, **239**, 504–507.

Baulin, V.V., Belopukhova, Ye.B. and Danilova, N.S., 1984. Holocene permafrost in the USSR. In A.A. Velichko (eds), *Late Quaternary Environments of the Soviet Union*. University of Minnesota Press, Minneapolis, MN, 87–94.

Becker, B., 1982. Dendrochronologic und Paläoëkologie subfossiler Baumstämme aus Flussablagerungen, ein Beitrag zur nacheiszeitlichen Auenentwicklung im südlichen Mitteleuropa. *Mitteil. der Kommission für Quartärforschung*. Vol. 5, Osterreich Akad. der Wiss., Wien, 1–120.

Benito, G., Machado, M.J. and Perez-Gonzales, A., 1996. Climate change and flood sensitivity in Spain. In J. Branson, A.G. Brown and K.J. Gregory (eds), *Global Continental Changes; The Context of Palaeohydrology*. Geological Society of Special Publication No. 115, Geological Society, London, 85–98.

Berglund, B., (ed.), 1986. *Handbook of Holocene Palaeoecology and Palaeohydrology*. Wiley, Chichester.

Bielañski, A.K., 1997. *Materialy do historii powodzi w dorzeczu górnej wisly*, Politechnika Krakowska, Monografie, **217**, 1–118.

Bluemle, J.P., Sabel, J.M. and Karlen, V., 1999. Rate and magnitude of postglacial climate changes. *Environmental Geosciences*, **6**(2), 63–75.

Bond, G., Kromer, B., Beer, J., Muscheler, R., Evans, M.N., Showers, W., Hoffmann, S., Lotti-Bond, R., Hajdas, I. and Bonani, G., 2001. Persistent solar influence on North Atlantic climate during the Holocene. *Science*, **294**, 2130–2136.

Bond, G., Showers, W., Cheseby, M., Lotti, R., Almasi, P., de Menocal, P., Priore, P., Cullen, H., Hajdas, I. and Bonani, G., 1997. A pervasive millennial-scale cycle in North Atlantic Holocene and glacial climates. *Science*, **278**, 1257–1266.

Bradley, R.S. and Jones, P.D., 1992. Records of explosive volcanic eruptions over the last 500 years. In R.S. Bradley and P.D. Jones (eds), *Climate Since A.D. 1500*. Routledge, London, New York, 606–622.

Brazdil, R., Glaser, R., Pfister, Ch., Dobrovolny, P., Antoine, J.M., Barriendos, M., Camuffo, D., Deutsch, M., Enzi, S., Guidoboni, E., Kozyta, O. and Rodrigo, T.S., 1999. Flood events of selected European rivers in the sixteenth century. *Climatic Change*, **43**, 239–285.

Broecker, W.S., Kennett, J.P., Flower, B.P., Teller, J.T., Trumbore, S., Bonani, G. and Wolfli, W., 1989. Routing of meltwater from the Laurentide ice sheet during the Younger Dryas cold episode. *Nature*, **341**, 318–321.

Bryson, R.A., 1989. Late Quaternary volcanic modulation of Milankovitch climate forcing. *Theoretical and Applied Climatology*, **39**, 115–125.

Bryson, R.U. and Bryson, R.A., 1998. Application of a global volcanicity time-series on high-resolution paleoclimatic modelling of the Eastern Mediterranean. In A.S. Issar and N. Brown (eds), *Water, Environment and Society in Times of Climatic Change*. Kluwer Academic Publishers, Dordrecht, 1–19.

Bryson, R.A. and Goodman, B.M., 1980. Volcanic activity and climatic change. *Science*, **207**, 1041–1044.

Bryson, R.A. and Wendland, W.M., 1967. Tentative climatic patterns for some late-glacial and post-glacial episodes in central North America. *Life, Land and Water*. University of Manitoba Press, Winnipeg, 272–298.

Burga, C.A., 1988. Swiss vegetation history during the last 18,000 years. *New Phytologist*, **110**, 581–602.

Burkham, D.E., 1972. Channel Changes of the Gila River in Safford Valley, Arizona 1846–1970. Geol. Survey Water – Supply Papers (U.S.) 655-G, 1–24.

Camuffo, D. and Enzi, S., 1995. Climatic features during the Spörer and Maunder Minima. *Paläoklimaforschung*, **16**, 105–124.

Chambers, F.M., Ogle, M.I. and Blackford, J.J., 1999. Palaeoenvironmental evidence for solar forcing of Holocene climate: linkages to solar science. *Progress in Physical Geography* **23**(2), 181–204.

CLIMAP Project Members. 1976. The surface of the ice – age Earth. *Science*, **191**, 1131–1144.

COHMAP Members. 1988. Climatic changes of the last 18,000 years: observations and model simulations. *Science*, **241**, 1043–1052.

Czyżowska, E., 1997. Zapis wydarzeń powodziowych na pograniczu borealu i atlantyku w osadach stożka naplywowego w Podgrodziu. *Dokumentacja Geograficzna*, IGiPZ PAN 5, 74.

Damon, P.E., Eastoe, C.J., Hughes, M.K., Kailn, R.M., Long, A. and Peristykh, A.N., 1998. Secular variation of $\delta\,^{14}C$ during the medieval solar maximum, a progress report. *Radiocarbon*, **158**, 28–31.

Dansgaard, W. and Oeschger, H., 1988. Past environmental long-term records from the Arctic. In H. Oeschger and C.C. Jangway Jr. (eds), *The Environmental Record in Glaciers and Ice Sheets*. Wiley, Chichester.

Dearing, J.A. and Zolitschka, B., 1999. System dynamics and environmental change, an exploratory study of Holocene lake sediments at Holzmaar, Germany. *The Holocene*, **9**(5), 531–540.

Denniston, R.F., Gonzales, L.A., Baker, R.G., Asmerom, Y., Reagan, M.K., Edwards, R.L. and Alexander, E.C., 1999. Speleothem evidence for Holocene fluctuations of the prairie-forest ecotone, North-Central USA. *The Holocene*, **9**(6), 671–676.

Diaz, H.F. and Markgraf, V., 1993. *Palaeoclimate Aspects of El Nino – Southern Oscillation*. Cambridge University Press, Cambridge.

Digerfeldt, G., 1988. Reconstruction and regional correlation of Holocene lake-level fluctuations in Lake Bysjön, South Sweden. *Boreas*, **17**(2), 165–182.

Enzel, Y., Ely, L.L., House, P.K. and Baker, V.P., 1996. Magnitude and frequency of Holocene palaeofloods in the southwestern United States: a review and discussion of implication. In J. Branson, A.G. Brown and K.J. Gregory (eds), *Global Continental Changes: The Context of Palaeohydrology*. Geological Society of Special Publication No. 115, Geological Society, London, 121–137.

Flohn, H., 1993. Climate evolution during the last millennium: what can we learn from it? In J.A. Eddy and H. Oeschger (eds), *Global Changes in the Perspective of the Past*. John Wiley, Chichester, 295–316.

Fontes, J.Ch., Gasse, F., Camara, E., Millet, B., Saliege, J.F. and Steinberg, M., 1985. Late Holocene changes in Lake Abhe hydrology (Ethiopia – Djibouti). *Zeitschrift für Gletscherkunde und Glazialgeologic*, **21**, 89–96.

Frenzel, B., Pecsi, M. and Velichko, A.A., 1992. *Paleogeographical Atlas of the Northern Hemisphere*. Hungarian Academy of Sciences, Budapest.

Frumkin, A., Greenbaum, N. and Schick, A.P., 1998. Paleohydrology of the Northern Negev: comparative evaluation of two catchments. In A.S. Issar and N. Brown (eds), *Water, Environment and Society in Times of Climatic Change*. Kluwer Academic Publishers, Dordrecht, 97–111.

Frumkin, A., Magaritz, M., Carmi, I. and Zak, I., 1991. The Holocene climatic record of the salt caves of the Mount Sedom, Israel. *The Holocene*, **1**, 191–200.

Gil, E. and Starkel, L., 1979. Long-term extreme rainfalls and their role in the modelling of flysch slopes. *Studia Geomorphologica Carpatho – Balcanica*, **13**, 207–220.

Goslar, T., Arnold, M., Bard, E., Kuc, T., Pazdur, M.F., Ralska-Jasiewiczowa, M., Różański, K., Tisnerat, N., Walanus, A., Wicik, B. and Więckowski, K., 1995. High concentration of atmospheric ^{14}C during the Younger Dryas. *Nature*, **377**, 414–417.

Gregory, K.J. and Walling, D.E., 1987. *Human Activity and Environmental Processes*. Wiley, Chichester.

Gregory, K.J., Starkel, L. and Baker, V.R., (eds), 1995. *Global Continental Palaeohydrology*. Wiley, Chichester.

Grove, J.M., 1979. The glacial history of the Holocene. *Progress in Physical Geography*, **3**(1), 1–54.

Hammer, C.U., Clausen, H.B. and Dansgaard, W., 1980. Greenland ice sheet evidence of postglacial volcanism and its climatic impact. *Nature*, **208**, 230–255.

Hormes, A., Müller, B.U. and Schlüchter, Ch., 2001. The Alps with little ice: evidence for eight Holocene phases of reduced glacier extent in the Central Swiss Alps. *The Holocene*, **11**(3), 255–265.

Harrison, S., Prentice, J.C. and Guiot, J., 1993. Climatic controls on Holocene lake-level changes in Europe. *Climate Dynamics*, **8**, 189–200.

Hastenrath, S. and Kutzbach, J.E., 1983. Paleoclimatic estimates from water and energy budgets of East African lakes. *Quaternary Research*, **19**, 141–153.

Hirschboeck, K.K., 1988. Flood hydroclimatology. In V.R. Baker, R.C. Kochel and P.C. Patton (eds), *Flood Geomorphology*. Wiley, Chichester, 27–50.

Hooke, J.M., 1996. River responses to decadal-scale changes in discharge regime: the Gila River, SE Arizona. In J. Branson, A.G. Brown and K.J. Gregory (eds), *Global Continental Changes: The Context of Palaeohydrology*. Geological Society of Special Publication No. 115, Geological Society, London, 191–204.

Hughes, P.D.M., Manquoy, D., Barber, K.E. and Longdon, P.G., 2000. Mire-development pathways and palaeoclimate records from a full Holocene peat archive at Wolton Moss, Cumbria, England. *The Holocene*, **10**(4), 465–479.

Issar, A.S., 1998. Climate change and history during the Holocene in the Eastern Mediterranean region. In A.S. Issar and N.Brown (eds), *Water, Environment and Society in Times of Climatic Change*. Kluwer Academic Publishers, Dordrecht, 113–128.

Jäkel, D., 1979. Run-off and fluvial formation processes in the Tibesti Mountain as indicators of climatic history in the Central Sahara during the Late Pleistocene and Holocene. *Palaeoecology of Africa*. Vol. 10/11, Balkema, Rotterdam, 11–14.

Kadomura, H., 1995. Palaeoecological and palaeohydrological changes in the humid tropics during the last 20000 years, with reference to Equatorial Africa. In K.J. Gregory, L. Starkel and V.R. Baker (eds), *Global Continental Paleohydrology*. Wiley, Chichester, 177–202.

Kalicki, T. and Krąpiec, M., 1992. Kujawy Site – Subatlantic alluvia with black oaks. Excursion guide-book. *Symposium Global Continental Paleohydrology*, Kraków, Mogilany, 37–41.

Knox, J.C., 1972. Valley alluviation in southwestern Wisconsin. *Annals of the Association of American Geographers*, **62**, 401–410.

Knox, J.C., 1983. Responses of river systems to Holocene climates. In H.E. Wright Jr. (ed.), *Late-Quaternary Environment of the United States, Vol. 2, The Holocene*. University of Minnesota Press, Minneapolis, 26–41.

Knox, J.C., 1993. Large increases in flood magnitude in response to modest changes in climate. *Nature*, **361**, 430–432.

Knox, J.C., 1995. Fluvial systems since 20 000 years BP. In K.J. Gregory, L. Starkel and V.R. Baker (eds), *Global Continental Palaeohydrology*. Wiley, Chichester, 87–108.

Krąpiec, M., 1992. Skale dendrochronologiczne późnego holocenu poludniowej i centralnej Polski. *Geologia*, **18**(3), 37–119.

Kukla, J., 1969. The cause of the Holocene climate change. *Geologie en Mijnbouw*, **48**(3), 412–424.

Kutzbach, J.E., 1980. Estimates of past climate of paleolake Chad, North Africa, based on a hydrological and energy-balance model. *Quaternary Research*, **14**, 210–223.

Kutzbach, J.E., 1992. Modelling earth system changes of the past. In D. Ojima (eds), *Modelling the Earth System*. UCAR, Boulder, CO, 377–404.

Kutzbach, J.E. and Guetter, P.J., 1984. The sensitivity of monsoon climates to orbital parameter changes for 9,000 years BP, experiments with the NCAR general circulation model. In A.L. Berger *et al.* (eds), *Milankovitch and Climate*. Part 2, Reidel, 801–820.

Laird, K.R., Fritz, S.C., Cumming, B.F. and Grimm, E.C., 1998. Early-Holocene limnological and climatic variability in the Northern Great Plains. *The Holocene*, **8**(3), 275–285.

Lamoureux, S.F., England, J.H., Sharp, M.J. and Bush, A.B.G., 2001. A varve record of increased little ice age rainfall associated with volcanic activity, Arctic Archipelago, Canada. *The Holocene*, **11**(2), 243–249.

Lezine, A.M., 1998. Pollen records of past climate changes in West Africa since the Last Glacial Maximum. In A.S. Issar and N. Brown (eds), *Water, Environment and Society in Times of Climatic Change*. Kluwer Academic Publishers, Dordrecht, 295–317.

Lvovich, M.I., 1979. *World Water Resources and their Future* (English translation by R.L. Nace). American Geophysical Union, Washington.

Magny M., 1993. Holocene fluctuations of lake levels in the French Jura and sub-Alpine ranges and their implications for past general circulation pattern. *The Holocene*, **3**(4), 306–313.

Maizels, J.K., 1995. Palaeohydrology of polar and subpolar regions over the past 20,000 years. In K.J. Gregory, L. Starkel and V.R. Baker (eds), *Global Continental Palaeohydrology*. Wiley, Chichester, 259–299.

Merkt J., Müller H. and Streif H., 2001. Kurzfristige Klimaschwankungen in Quartär Klimaweissbuch, *Terra Nostra* **7**, 84–93.

Mayewski, P.A., Mecker, L.D., Twickler, M.S., Whitlow, S., Qinzhao Yang, Berry-Lyons, W. and Prentice, M., 1997. Major features and forcing of high-latitude northern hemisphere atmospheric circulation using a 110 000-year-long glaciochemical series. *Journal of Geophysical Research*, **102**, 26345–26366.

Needham, S. and Macklin, M.G., (eds), 1992. *Alluvial Archaeology in Britain*. Oxbow Books, Oxford.

Nessje, A. and Johannessen, T., 1992. What were the primary forcing mechanisms of high-frequency Holocene glacier and climatic variations. *The Holocene*, **2**(1), 70–84.

Netser, M., 1998. Population growth and decline in the northern part of Eretz-Israel during the historical period as related to climatic changes. In A.S. Issar and N. Brown (eds), *Water, Environment and Society in Times of Climatic Change*. Kluwer Academic Publishers, Dordrecht, 129–145.

Patzelt, G., 1977. Der zeitlicke Ablauf und das Ausmass postglazialer Klimaschwankungen in den Alpen. In B. Frenzel (ed.), *Dendrochronologie und postglaziale Klimaschwankungen in Tirol*. Veröff. des Museum Ferdinandeum, Vol. 67, Mainz, 93–123.

Paulissen, E. and Vermeersch, P.M., 1989. Le comportement des grand fleuves allogenes: l'example du Nil Saharien au Quaternaire superieur. *Bulletin de la Societe Geologique de France*, **5**(1), 73–83.

Petit-Maire, N., 1986. Palaeoclimates in the Sahara of Mali, a multidisciplinary study. *Episodes*, **9**, 7–16.

Petit-Maire, N. and et Guo, Z., 1996. Mise en evidence de variations climatiques holocenes rapides, en phase dans les deserts actuels de Chine et du Nord del'Afrique. *Comptes Rendus de l' Academie des Sciences Paris*, **322**(IIa), 847–851.

Pfister, C., 1992. Monthly temperature and precipitation in central Europe 1525–1979 quantifying documentary evidence on weather and its effects. In R.S. Bradley and P.D. Jones (eds), *Climate Since 1500 A.D.* Routledge, London, 118–142.

Quinn, W.H. and Neal, V.T., 1992. The historical record of El Nino events. In R.S. Bradley and P.D. Jones (eds), *Climate Since AD 1500*. Routledge, London, New York, 623–649.

Ralska-Jasiewiczowa, M., Goslar, T., Madeyska, T. and Starkel, L., (eds), 1998. *Lake Gościąż, Central Poland, A Monographic Study*. W. Szafer Institute of Botany, Polish Academy of Science, Kraków.

Rudoy, A.N. and Baker, V.R., 1993. Sedimentary effects of cataclysmic late Pleistocene glacial outburst flooding, Altay Mountains, Siberia. *Sedimentary Geology*, **85**, 53–62.

Rumsby, B.T. and Macklin, M.G., 1996. River response to the last neoglacial (The Little Ice Age) in northern, western and central Europe. In J. Branson, A.G. Brown and K.J. Gregory (eds), *Global Continental Changes: The Context of Palaeohydrology*. Geological Society of Special Publication No. 115, Geological Society, London, 217–233.

Schettler, G., Rein, B. and Negendank, J.F.W., 1999. Geochemical evidence for Holocene palaeodischarge variations in lacustrine records from the Westeifel Volcanic Field Germany: Schalkenmehrener Maar and Meerfelder Maar. *The Holocene*, **9**(4), 381–400.

Servant, M., 1983. Climatic variations in the low continental latitudes during the last 30,000 years. Abstracts of Second Nordic Symposium on Climatic Changes, Stockholm, Kobenhavn, May 16–20, 158–162.

Shindell, D.T., Schmidt, G.A., Mann, M.E., Rind, D. and Waple, A., 2001. Solar forcing of regional climate change during the Maunder Minimum. *Science*, **294**, 2149–2152.

Starkel, L., 1972. The role of catastrophic flooding in shaping of the relief of the lower Himalaya (Darjeeling Hills). *Geographical Polonica*, **21**, 103–160.

Starkel, L., 1976. The role of extreme (catastrophic) meteorological events in contemporary evolution of slopes. In E. Derbyshire (ed.), *Geomorphology and Climate*. Wiley, Chichester, 203–246.

Starkel, L., 1983. The reflection of hydrological changes in the fluvial environment of the temperate zone during the last 15,000 years. In K.J. Gregory (eds), *Background to Palaeohydrology*. John Wiley, Chichester, 213–237.

Starkel, L., 1987. Man as a cause of sedimentological change in the Holocene. *Striae*, **26**, 5–12.

Starkel, L., 1989. Global paleohydrology. *Quaternary International*, **2**, 25–33.

Starkel, L., 1995. Reconstruction of hydrological changes between 7,000 and 3,000 BP in the upper and middle Vistula river basin, Poland. *The Holocene*, **5**(1), 34–42.

Starkel, L., 1996a. Geomorphic role of extreme rainfalls in the Polish Carpathians. *Studia Geomorphologica Carpatho – Balcanica*, **30**, 21–38.

Starkel, L., 1996b. Palaeohydrological reconstruction: advantages and disadvantages. In J. Branson, A.G. Brown and K.J. Gregory (eds), *Global Continental Changes: The Context of Palaeohydrology*. Geological Society of Special Publications No. 115, Geological Society, London, 9–17.

Starkel, L., 1997. Mass movements during the Holocene: Carpathian example and the European perspective. *Paleoklimaforschung* Special Issue 19, ESF Project, European Palaeoclimate and Man 12, Vol. 19, 385–400.

Starkel, L., 1998a. Frequency of extreme hydroclimatically-induced events as a key to understanding environmental changes in the Holocene. In A.S. Issar and N. Brown (eds), *Water, Environment and Society in Times of Climatic Changes*. Kluwer Academic Publishers, Dordrecht, 273–288.

Starkel, L., 1998b. Extreme events in the last 200 years in the upper Vistula basin; their palaeohydrological implications. In G. Benito, V.R. Baker and K.J. Gregory (eds), *Palaeohydrology and Environmental Change*. Wiley, Chichester, 289–306.

Starkel, L., 1999a. 8,500–8,000 yrs BP humid phase-global or regional? *Science Reports of Tohoku University, 7th Series, Geography.* **49**(2), 105–133.

Starkel, L., 1999b. Space and time scales in geomorphology. *Zeitschrift für Geomorphologie*, Supplementband **115**, 19–33.

Starkel, L., 2001. Extreme rainfalls and river floods in Europe during the last millennium. *Geographia Polonica*, **74**(2), 69–79.

Starkel, L. and Basu, S. (eds), 2000. *Rains Landslides and Floods in the Darjeeling Himalaya*. Indian National Science Academy, New Delhi.

Starkel, L., Gregory, K.J. and Thornes, J.B. (eds), 1991. *Temperate Palaeohydrology*. Wiley, Chichester.

Starkel, L., Kalicki, T., Krąpiec, M., Soja, R., Gębica, P. and Czyżowska, E., 1996. Hydrological changes of valley floors in the upper Vistula Basin during Late Vistulian and Holocene. *Geographical Studies*, Special Issue No. 9, 7–128.

Starkel, L., Singh, S., Soja, R., Frochlich, W., Syiemlich, H. and Prokop, P., 2002. Rainfalls runoff and soil erosion in the extremely humid area around Cherrapunji, India (preliminary observations). *Geographia Polonica*, **75**, 1.

Steig, E.J., 1999. Mid-Holocene climate change. *Science*, **286**, 1485,1486.

Stine, S., 1998. Medieval climatic anomaly in the Americas. In A.S. Issar and N. Brown (eds), *Water, Environment and Change Society in Times of Climatic*. Kluwer Academic Publishers, 43–67.

Street-Perrott, F.A. and Harrison, S.P., 1985. Lake levels and climate reconstruction. In A.D. Hecht (ed.), *Paleoclimate analysis and modeling*. John Wiley, Chichester, 291–340.

Street-Perrott, F.A. and Perrott, R.A., 1990. Abrupt climate fluctuations in the Tropics: the influence of Atlantic Ocean circulation. *Nature*, **343**, 607–612.

Stuiver, M., 1995. Solar and climatic components of the atmospheric ^{14}C record. *Paläoklimaforschung*, **16**, 51–59.

Stuiver, M. and Braziunas, T.F., 1993. Sun, ocean, climate and atmospheric $^{14}CO_2$: an evaluation of causal and spectral relationships. *The Holocene*, **3**, 289–304.

Stuiver, M., Braziunas, T.F., Becker, B. and Kromer, B., 1991. Climatic, solar, oceanic and geomagnetic influences on late-glacial and Holocene atmospheric $^{14}C/^{12}C$ Change. *Quaternary Research*, **35**, 1–24.

Stuiver, M., Braziunas, T.F., Grootes, P.M. and Zielinski, G.A., 1997. Is there evidence for solar forcing of climate in the GISP 2 oxygen isotope record?. *Quaternary Research*, **48**, 259–266.

Teller, J.T., 1995. The impact of large ice sheets on continental palaeohydrology. In K.J. Gregory, L. Starkel and V.R. Baker (eds), *Global Continental Palaeohydrology*. Wiley, Chichester, 109–129.

Thompson, L.G., 1992. Ice core evidence from Peru and China. In R.S. Bradley and P.D. Jones (eds), *Climate Since AD 1500*. Routledge, London, New York, 517–548.

Torske, N., 1996. Holocene vegetation, climate and glacier histories in Jostedalsbreen region, western Norway – palaeoecological interpretations from an alpine peat deposit. In B. Frenzel (ed.), Holocene Treeline oscillations, dendrochronology and palaeoclimate, *Paläoklimaforschung*, **20**, 215–232.

van Geel, B., Baurman, J. and Waterbolk, H.T., 1996. Archaeological and palaeoecological indications of an abrupt climate change in the Netherlands and evidence for climatological teleconnections around 2650 BP. *Journal Quaternary Science*, **11**(6), 451–460.

van Geel, B. and Renssen, H., 1998. Abrupt climate change around 2650 BP in North-West Europe: evidence for climatic teleconnections and a tentative explanation. In A.S. Issar and N. Brown (eds), *Water, Environment and Society in Times of Climatic Change*. Kluwer Academic Publishers, Dordrecht, 21–41.

Waylen, P., 1995. Global hydrology in relation to palaeohydrological change. In K.J. Gregory, L. Starkel and V. Baker (eds), *Global Palaeohydrology*. John Wiley, Chichester, 61–86.

Wick, L. and Tinner, W., 1997. Vegetation changes and timberline fluctuations in the Central Alps as indicators of Holocene climatic oscillations. *Arctic and Alpine Research*, **29**, 445–458.

Wohl, E.E. and Enzel, Y., 1995. Data for Palaeohydrology. In K.J. Gregory, L. Starkel and V.R. Baker (eds), *Global Continental Palaeohydrology*. Wiley, Chichester, 23–59.

Woo, M.K., 1986. Permafrost hydrology in North America. *Atmosphere–Ocean*, **24**(3), 201–234.

Zieliński, G.A., Mayewski, P.A., Meeker, L.D., Whitlow, S. and Twickler, M.S., 1996. A 110,000 yr record of explosive volcanism from the GISP 2 (Greenland) ice core. *Quaternary Research*, **45**(2): 109–118.

Zubakov, V.A., 1972. *Palaeogeography of West-Siberian Lowland in Pleistocene and Late Pliocene (in Russian)*. Nauka, Leningrad, 1–200.

21 Palaeohydrology, Environmental Change and River-channel Management

K.J. GREGORY

University of Southampton, Southampton, UK

Earlier chapters in this volume collectively provide material that should be considered in the management of future environments: from world areas in Part 3 and on themes in Part 4 especially on palaeofloods (Chapter 18) and on short-term hydrological changes (Chapter 20). Such applications of palaeohydrology (Gregory, 1998) are particularly relevant to the management of rivers. Throughout the long experience of river management, there has been increasing emphasis upon the river *channel* because that is often the focus for river management within a basin context (Downs and Gregory, 2003). Lip service has been paid often to the importance of knowledge from, and understanding of, past fluvial systems so that such research, including palaeohydrology, should now contribute significantly to river-channel management. However, such reconciliation has seldom been treated explicitly. The statement that "The present without a past has no future" (Geoparks, 1999) cited in consideration of the progression from nature-dominated to human-dominated environmental changes (Messerli *et al.*, 2000), pertains equally well to applications of palaeohydrology.

River management is now seen in a sustainable environmental management context. A change in approach characterized as a switch from technological to ecological river management (Nienhuis *et al.*, 1998), or to sustainable management, intends that natural resources should be used without degrading their quality or reducing their quantity (Smits *et al.*, 2000). Increasingly, the impact of global change (Chapter 1) is seen as requiring sustainable adaptations to management strategies whereby in adaptive management river managers adjust their actions in response to monitoring data, which alerts them to changing environmental, economic and social conditions (see Leuven *et al.*, 2000). One effect of global change research has been the weakening of disciplinary boundaries with the growth of hybrid fields such as ecohydrology (Rodriguez-Iturbe, 2000), and greater recognition that intriguing problems are most evident at research frontiers, as exemplified by palaeohydrology. Some have taken this further by suggesting that Earth system analysis is now ushering in a second Copernican revolution (Schellnhuber, 1999) employing macroscopes to reduce rather than magnify, and to achieve "holistic" perceptions of the planetary inventory. The idea has been further extended (Lawton, 2001) by advocating a young and still emerging discipline of Earth System Science (ESS) although contributions from existing disciplines as well as from new multidisciplinary fields are still required.

Palaeohydrology: Understanding Global Change. Edited by K.J. Gregory and G. Benito
© 2003 John Wiley & Sons, Ltd ISBN: 0-470-84739-5

Paradoxically, palaeohydrology is not frequently listed amongst the disciplines contributing to global change (Gregory, 1998; Chapter 1) although the impact of global change is, and will continue to be, manifested in significant river impacts. Five potential palaeohydrology themes (Table 1.3) can relate to river-channel management in which it is necessary to know what has to be managed, how management is approached, what information is required, and how palaeohydrology can contribute. Previously, there has been some implicit and partial consideration of these aspects in disciplines, such as ecology or geomorphology, but no explicit survey of the potential of palaeohydrology has been made. No claim has also been made for a distinctive palaeohydrological perspective although there is the need to consider explicitly how the results of palaeohydrology research should inform river-channel management. A brief synopsis of management developments and requirements precedes consideration of how palaeohydrology results are utilized, and how, when collected together, they might provide additional protocols for river-channel management.

1   THE RIVER-CHANNEL MANAGEMENT CONTEXT

The history of river-channel management can be traced back to at least 4,000 years so that a context is provided by the reasons for river-channel management and the ways in which it has been approached, leading to the framework within which it is now undertaken.

Management of the river channel is necessary to contain flows that would otherwise affect adjacent floodplains and river corridors and their uses, to contain and constrain riverbank erosion, to maintain a sufficient depth of flow for navigation purposes and to support drainage schemes in valley-floor areas. Methods of river-channel management have developed sequentially in particular areas and can be summarized in terms of eight phases of river-channel management. The dates on which each of the eight phases (Table 21.1) obtained vary from one country to another, according to uses that inspired

Table 21.1   Phases of river-channel management

Chronological phase	Characteristic developments	Management methods employed
1. Control and divert river flows *Hydraulic civilizations*	Flood control Irrigation Land reclamation	Dams
2. Pre-industrial revolution	Local and small-scale river use	Drainage schemes Fish weirs Water mills Navigation Floatability
3. Industrial revolution	Industrial mills Cooling water Power generation Irrigation schemes Water supply	Dams and hydroelectric schemes
4. Twentieth-century technology	Dams and reservoirs	Large dams
5. Integrated management	Multiple river use	Basin schemes
6. Flood control	Channelization	Flood control methods
7. Alternative approaches	Restoration, renovation	Working with the river
8. Sustainability	River environment	Basin approach

river-channel management and culminating, particularly in developed areas, in major rivers that have been extensively modified, engineered and managed throughout the nineteenth and twentieth centuries.

The range of measures involved in the river-engineering or hard-engineering approach were of two major kinds: river-channel regulation (involving repeated dredging, temporary construction in the river bed, fixation of bed, elimination of obstacles from low-water channel, channel rectification and fixation, channel construction, revetments and groynes and rectification in floodplains) and discharge regulation involving dams, reservoirs and weirs (Jansen *et al.*, 1979; Petersen, 1986). This approach was successful in numerous cases by mitigating the flood or other problem addressed, and it was essential in those instances where there was no alternative to a hard-engineering solution. Four groups of factors subsequently led to questioning of the approach. Firstly, *environmental impacts* of river engineering upon the dynamic hydrosystem of channel processes, the morphology, biology including ecology, and aesthetics of river channels were progressively demonstrated by fluvial geomorphologists and ecologists (e.g. Brookes and Gregory, 1988) leading to a search for alternative approaches. Secondly, *risks* were identified where major functional problems arose from river engineering including cases of structural failure or, more commonly, where implementation of a structural-engineering solution simply moved the problem elsewhere or intensified problems in other locations. For instance, many erosion control measures tended to create greater erosion problems upstream or downstream of the "solution", and many flood control channels led to greater flood risk downstream with serious disruption to the regional water table. Thirdly, a change in economic outlook from "river engineering at all costs" to management measures that could be justified according to their *cost–benefit ratio* nullified the use of the very expensive hard-engineering techniques in all but the most extreme cases. Fourthly, and partly arising from the previous considerations, there was a change in the intellectual climate towards *sustainability*, a context in which the idea of human domination over nature had given way to a more symbiotic notion of human relations with nature.

Whereas the eight phases of river-channel management (Table 21.1) could use an approach characterized as hard or, more recently, as soft engineering, management could be either local and problem-oriented or holistic and catchment-based. The hard-engineering approach to river-channel management has been characterized as "technology will fix it" (Leopold, 1977), or as the paradigm of river engineering (Williams, 2001). The founding statement of the Institute of Civil Engineers in 1830 "to harness the great sources of power in nature for the use and convenience of man" meant that successive generations of river engineers who were trained in hydraulics and civil engineering tended to view rivers as large-scale plumbing problems (Williams, 2001). The approach worked in most cases but the four groups of factors cited above led to the quest for alternative management approaches, so that the old paradigm of river engineering was succeeded by the new paradigm of sustainable management (Williams, 2001) and in Australasia an inception phase and an engineering phase were succeeded by an environmental phase of river management (Finlayson and Brizga, 2000).

River-channel management, therefore, changed significantly towards a more sustainable approach involving softer engineering procedures wherever possible to manage river channels and embracing procedures including channel restoration. Although river-channel management has often been undertaken at the scale of the individual reach or river-channel section (Frissell and Ralph, 1998), it has increasingly been appreciated that management should be undertaken in the context of the basin as a

Table 21.2 Examples of new approaches to river management

Term	Definition	Reference
Environmental river engineering	The key requirement for sound environmental river engineering is a basic understanding of the natural processes controlling channel shape and dimensions	Hey, 1990
Restoration	Process of returning river or watershed to a condition that relaxes human constraints on the development of natural patterns of diversity. Restoration efforts constrained by lack of clear understanding of how human activities have altered processes at work within the watershed	Frissell and Ralph, 1998
Ecological river management	Recent switch from technological river management to ecological river management	Nienhuis *et al.*, 1998
Sustainable management	From an ecological point of view, sustainable development means that natural resources should be used without degrading their quality or reducing their quantity	Smits *et al.*, 2000
	Requires a carefully considered strategy for the entire catchment including (1) conservation or restoration of the natural flow regime and the hydromorphological dynamics of rivers; (2) allowing space for rivers; (3) adapting user functions to natural river dynamics	Leuven *et al.*, 2000
Adaptive management	River managers are continually adjusting their actions in response to monitoring data, which alerts them to changing environmental and economic conditions and social preferences	Leuven *et al.*, 2000
	Avoiding past mistakes on a large scale and learning from successes and failures	Frissell and Ralph, 1998

whole (Harper *et al.*, 1999), with a holistic approach completely based upon integrated basin management (IBM) (Downs *et al.*, 1991) employed such that the outcome has been a spectrum of approaches for river-channel management (Table 21.2).

2 INFORMATION REQUIREMENTS FOR RIVER-CHANNEL MANAGEMENT

All approaches to river management require particular kinds of information, often implied without being specified explicitly, and procedures, such as the 100-year flood, have been used unquestioningly (Chapter 18). River-management objectives have been broadly characterized (Knighton, 1998; Table 6.13) as hazard and resource perspectives, although the former include not only hazards (pollution, soil erosion and sediment transport) but also measures to address them (bank protection, bridge stability, flood control, floodplain zonation); and the resource aspects (e.g. navigation, water source, aesthetics) are of course susceptible to hazards. Hazards (Gares et al., 1994), potential danger or risk, relating to the river channel, as suggested for urban areas (Gregory and Chin, 2002) can provide a basis for information requirements. Twenty-eight geomorphic hazards, all associated with erosion, deposition or pattern shift, characteristic of the fluvial system were developed (Schumm, 1988; 1994) according to their association with drainage networks, slopes, channels and piedmont and coastal plains.

New approaches (Table 21.2) mean that river-channel management can now be visualized in three broad categories: engineering management or hard engineering; working with the river, usually applied to particular segments or reaches; and a sustainable approach applied holistically to the basin as a whole. Any one of these three categories may be utilized depending upon the character of the area in which management is to occur. A method of specifying river-management requirements is to identify the aspects requiring management in each category (Table 21.3, Column 2). For an engineering approach (A in Table 21.3), hazards can be itemized as discharge, sediment and solutes, debris accumulation, bank erosion and scour; whereas resources cover navigation, water supply, sediment extraction, boundary use of rivers, recreation and aesthetics. River management has traditionally embraced mitigation measures intended to minimize the impact of scour, of aggressive channel change by channelization and of flooding due to drainage schemes.

A subsequent approach of working with the river rather than against it (Table 21.2) was advocated by engineers (e.g. Winkley, 1972), often as a reaction to impacts of channelization or consequences of other management methods. Although inclined towards what we would now recognize as a "softer" approach, Winkley (1972) contended that any problem of alluvial rivers could be addressed by a blend of experience and engineering judgement, and the question arises as to whether that experience sufficiently reflected the history of the river. Adjustments consequent upon management methods include the effects of flow regulation; of dams and reservoirs extending much further downstream than the influence of below-dam scour; of urbanization similarly extending considerable distances downstream; of flow abstraction and diversion; of gravel extraction; of riparian tree removal; and of extensive catchment land-use changes and conservation by adaptive management. The group of approaches (B in Table 21.3) came to include the idea of "design with nature" which in Europe can be traced back to the mid-nineteenth century (Petts et al., 2000) although it was the subject of only sporadic action until after the mid-twentieth century. The need for an alternative to hard engineering produced this softer-management approach, including restoring the river landscape to a more "natural" character, initially involving the disciplines of landscape architecture and ecology. A book, Design with Nature (McHarg, 1969), which proposed ideas first applied to the city, suggested that environmental sciences, especially ecology, could inform the planning process as required in the 1969 National Environmental Policy Act (McHarg and Steiner, 1998). The culmination of the softer approach to management resulted in restoration of sections of river channel, spawning a variety of terms (Sear 1996; Gregory, 2000), which can be resolved into three main categories (Gregory 2002; Gregory and Chin, 2002): first, a general one, often called restoration; a second, referring to attempts to make the channel more natural,

Table 21.3   River-management requirements

Type of approach	Aspects requiring management	Information needs	Palaeo input
A. *Engineering:*			
Hazards	Discharge, flood, low flow	Long-term hydrological records	*
	Sediment, accretion, pollution	Modelling of sediment transport	(*)
	Debris accumulation	Snag build-up and frequency	(*)
	Bank erosion	Erosion history	*
	Scour	Appreciation of cut and fill	*
Resources	Navigation	Sediment delivery and movement	(*)
	Water supply	Long-term flow records	(*)
	Sediment extraction	Valley-floor sediment sequence	*
	Boundary	Channel stability and shifting	*
	Recreation	Channel stability, pollution	(*)
	Aesthetics	Channel stability, sediment	(*)
Mitigation measures	Scour below dams, structures	Channel and bed stability	*
	Channelization	Flow history, sediment delivery	*
	Drainage schemes: urban and land	Flood incidence	*
		Flood incidence and distribution	*
B. *Working with the river*			
Adjustments consequent upon management methods	Dams: channel changes Urbanization: channel changes	Location, amount and timing of change	*
Restoration: What is natural?	General restoration	Identify stable conditions	(*)
	More natural condition	Suggest appropriate natural state	*
	Restoration to prior condition	Deduce natural condition for location	*
Adaptive management	Sequential change	Likely sequence of changes	*
C. *Sustainable approach*			
Basin approach	Holistic catchment perspective	Devise holistic basin strategy understanding connectivity	*
Temporal change	History and memory of basin	Plan likely development of river in temporal sequence	*
Instability		Identify future unstable reaches	*
Sensitive reaches	Location for time(LCT)	Temporal sequence of change	*
	Location for condition (LCS?)	Basin analysis of sensitivity	*

and including approaches such as enhancement, rehabilitation, creation, naturalization and mitigation; and a third, involving restoration to some prior condition including recovery, re-establishment, reinstatement and restoration in the strictest sense. Combining concern with adjustments and restoration prompted Keller (1976) to argue that the designer of river channels should not only encompass the fact that streams are open systems but also understand the implications of convergent–divergent flow processes, of geomorphic thresholds in stream behaviour and of the complex relations implicit in erosion, deposition and sediment concentration processes. In addition to these two categories (**B** in Table 21.3), an adaptive management approach (Table 21.2) has been added, whereby management constantly improves and adjusts, on the basis of improved understanding of how the ecosystem works and on constantly updated information on the status and response of the ecosystem, obtained by monitoring critical indicators (Petts *et al.*, 2000, 505). Even unwanted interference in river-modification schemes can be used to advantage if it is treated as ecological experimentation (Boon, 1992), whereby the results can be employed, in the case of reservoir releases and river abstractions, to change the way in which schemes are managed.

The third category of approaches (**C** in Table 21.3) embracing sustainability, includes at least three aspects. Unlike restoration and working with the river approaches, which tend to concentrate on small spatial scales or on specific reaches or particular problems and often fail to take into account upstream/downstream linkages and impacts within the geomorphic system (Fryirs and Brierley, 2000), the sustainable approach adopts a broader spatial and temporal context. A catchment-based management approach (Harper *et al.*, 1999) can be employed in a sustainable manner (Sear, 1996) appropriate to the particular characteristics of the local environment. Its implementation requires use of the hierarchy of river systems (Frissell *et al.*, 1986) and their dimensions; relationships between discharge and channel dimensions; the importance of the physical and biological continua of natural rivers (Vannote *et al.*, 1980) and the role of heterogeneity in maintaining biodiversity. The entire basin as a framework for management requires that any upstream or downstream effects are examined, and that the connectivity and characteristics of the complete channel network (Gregory, 2002) are considered. Catchment-based approaches for assessment of river reaches have recently been developed to include the geomorphic approach of river styles (e.g. Brierley and Fryirs, 2000) and definitions used by managing authorities now include total catchment management (e.g. Joliffe and Ball, 1999).

The three broad approaches to river-channel management (Table 21.3) are each associated with particular aspects requiring management so that it is possible to suggest how particular information needs relate to each of these aspects.

3   CONTRIBUTIONS FROM PALAEOHYDROLOGY

Contributions to global change can be visualized in terms of data derivation, mechanics of change, spatial contrasts, coupling with Global Circulation Models (GCM) and construction of new models (Chapter 1, Table 1.3) so that against the context of types of river management (Table 21.3), it is possible to deduce specific contributions that might be made from palaeohydrology (final column of Table 21.3 indicated by ∗) with less comprehensive inputs indicated by (∗).

The context is provided for these contributions from the knowledge of temporal changes often advocated as a necessary context for river-channel management, although usually in general rather than in specific terms. Observations since 1984

(Table 21.4) stressing the geomorphological importance of the temporal view are not exclusive to palaeohydrology because a long-term perspective is multidisciplinary with three major issues evident (Table 21.4). First, affirming the need for a temporal perspective (marked **A** in column 2 of Table 21.4) because each river has a history (Leeks *et al.*, 1988), which may be reflected in what can be thought of as a landscape memory (Macklin and Lewin, 1997). Whereas engineers have been typically concerned with a particular design period, perhaps of ca 50 years, over which equilibrium is often assumed (James, 1999), varying timescales significant in shaping river history and morphology should be considered (Hey, 1990). Thus, engineering geomorphology has been interpreted as defining all strategies at the reach and watershed scale based on historical and contemporary assessment of geomorphological dynamics (Brookes, 1995). Second, reasons suggested as to why temporal change should be considered in relation to river management include (**B** in Table 21.4) the fact that variability in time as well as in space is the key to the understanding of fluvial systems (Schumm, 1991); that longer-term studies are required to explain river processes (Vandenberghe and Maddy, 2000); that the effects of climate change still affect contemporary river systems; and that Late-Quaternary environmental changes provide the "initial" conditions for present-day processes, their activity rates and the resulting morphology (Macklin and Lewin, 1997). Landscape histories and the cumulative consequences of land-use changes may assist in the prediction of geomorphological futures (e.g. Wasson, 1994); geomorphic inputs may explain the reasons for erosion and flooding problems (Poesen and Hooke, 1997); and it is long-term evolution that provides understanding of the proximity of individual reaches to threshold conditions (Gilvear, 1999). Third, there are a number of specific implications (**C** in Table 21.4) including the facts that palaeo landforms can furnish information on levels of stability (Gilvear, 1999); and that, because landforms and sediment were formed under very different conditions and have subsequently had to adjust to present conditions (Macklin and Lewin, 1997), analysis can be aided by application of multiple stable and unstable equilibria amending regime approaches, bearing in mind that instability may occur by way of some changes being cyclical, not the result of human activity (Dollar, 2000). In order to formulate sustainable solutions appreciating temporal change, it is suggested that there is a need to link change at regional, catchment, reach and hillslope scales (Brierley and Stankoviansky, 2002), to analyse locational probability of change (Graf, 2000), and to use a catchment-wide evolutionary perspective as a basis for rehabilitation (Fryirs and Brierley, 2000) because, for example, sediment exhaustion (Fryirs and Brierley, 2001) may not be understood sufficiently without a context of temporal change.

Considering these issues (Table 21.4) with the potential contributions suggested in Table 21.3 indicates four contributions of palaeohydrology associated with

- flow records, especially floods;
- the sensitivity of reaches and segments of the basin to change in a basin context;
- the temporal sequence of sediment, morphological and hydrological history;
- the question of what is natural.

Flow records and floods are perhaps the primary way in which palaeohydrology can contribute to river and river-channel management, primarily to avert the dangers of design being based upon incomplete or insufficiently long hydrological records, exemplified in New South Wales, Australia, where the underdesign of many dam spillways obtains because hydrological data was collected in below-average rainfall and flow periods between 1900 and 1945 (Riley, 1988; Tooth and Nanson, 1995).

Table 21.4 Long-term change as an input to river-channel management

Comment	Issues identified	Source
No matter of prime significance to the river engineer (and for that matter the geomorphologist) on which ignorance is so profound as that of *climate change* and how it affects river form and process	Climate change **B**	Chorley *et al.*, 1984
There can be considerable variability of fluvial system morphology and dynamics through *time*. In addition, there is great variability in *space or location* and a result of different geological, climatic and relief conditions	Variability in space and time **B**	Schumm, 1988
As interest in geomorphology shifts from the observation of equilibrium states to the recognition of multiple stable and unstable equilibria, it is clear that *regime assumptions* need setting in a more dynamic framework. To abandon regime bespeaks a replacement; dialogue between engineering and geomorphology is desirable to derive this framework	Multiple stable and unstable equilibria; more dynamic framework for regime approaches **C**	Lewin *et al.*, 1988
The concept of each river as a *system with a history* has value, (information on river-channel pattern, on sediment sources, sizes and dynamics is needed)	Each river system has a history **A**	Leeks *et al.*, 1988
The importance of Quaternary climatic history and lithology of river basins in controlling (vertically) processes that set up the relevant *thresholds; feedback* from effective floods bring about long-lasting shifts in control on morphological development	Thresholds and feedback effects **C**	Newson and Macklin, 1990
Understanding of reach and catchment-scale processes over *varying timescales* complements engineering skills and provides the physical basis for the maintenance and enhancement of the conservation value of river corridors	Varying timescales **A**	Hey, 1990
Intricacy of landscape histories and the cumulative *consequences of land-use changes*, provide a basis for realistic predictions of geomorphic futures	Cumulative consequences of land-use changes **B**	Wasson, 1994 in Brierley and Stanko-viansky, 2002

(continued overleaf)

Table 21.4 (*continued*)

Comment	Issues identified	Source
The question is not whether it is possible to re-build a self-maintaining channel reach, but rather whether we can establish a *landscape within which a river can live*	Establish a landscape within which a river can live **C**	Morris, 1995
Engineering geomorphology is defining alternative strategies at both the reach scale and the watershed scale based on both *historical and contemporary assessment of geomorphological dynamics*	Strategies based on both historical and contemporary dynamics **B**	Brookes, 1995
An understanding of the *interactions between river hydrology and geomorphology and the regeneration of floodplain vegetation* is critical to the success of any restoration initiative.	Interaction of hydrology and geomorphology **C**	Richards *et al.*, 1996
Understanding and managing the impacts of environmental change in a drainage basin... requires an appreciation of river processes that operate over longer timescales and larger spacescales than those that are generally familiar to the river engineer... the *physical landscape has a* longer *memory* than either the climate system or biosphere and it is Late-Quaternary (the last 125,000 years of Earth history) environmental changes that provide the "initial" conditions for present-day river processes, their activity rates and the resulting morphology.	Late-Quaternary environmental changes provide "initial" conditions for present processes and dynamics **B**	Macklin and Lewin, 1997, 15.
A high proportion of existing drainage-basin landforms and sediments were... formed under very different conditions to those that exist today...	Landforms and sediments formed under very different conditions and have had to adjust **B**	Macklin and Lewin, 1997,17
... the need to span a *wide range of timescales* in time and space is not a new challenge for applied geomorphologists. What perhaps is less familiar to geomorphologists is the need to assimilate the role of geomorphology fully with the needs and perspectives of project planners and team leaders, who operate outside academia and who must be convinced that spatially diffuse and long-term impacts are relevant	Range of timescales pertinent to river management **A**	Thorne *et al.*, 1997, 366

Table 21.4   *(continued)*

Comment	Issues identified	Source
... geomorphological analyses will at least help the engineer or manager to view local problems and particular parts of the fluvial system with the appropriate perspectives of time and space. These perspectives are the essential prerequisites for the formulation of *sustainable solutions*	and necessary for formulation of sustainable solutions **C**	Thorne *et al.*, 1997, 368
Four key contributions that fluvial geomorphology can make to the engineering profession with regard to river and floodplain management: (1) promote recognition of lateral, vertical and *downstream connectivity* in the fluvial system and the inter-relationships between river planform, profile and cross-section; (2) stress the importance of understanding *fluvial history and chronology* over a range of timescales, and recognizing the significance of both palaeo and active landforms and deposits as indicators of levels of landscape stability; (3) highlight the sensitivity of geomorphic systems to environmental disturbances and change, especially when close to geomorphic *thresholds*, and the dynamics of the natural systems; and (4) demonstrate the importance of landforms and processes in controlling and defining *fluvial biotopes* and to thus promote ecologically acceptable engineering.	Recognizing palaeo as well as active landforms as indicators of levels of landscape stability; **C** Proximity to thresholds; **B**	Gilvear, 1999
River managers need to understand fluvial systems as they change through time. River engineers are typically concerned with a design period of the order of 50 years; a time period over which static equilibrium and graded conditions of fluvial systems are commonly assumed. While river engineers need not be concerned with landforms evolution over cyclic time, they should recognize the value of *historical viewpoints and methods*	River engineers typically concerned with design periods of the order of 50 years but need historical view and methods **A**	James, 1999, 265, 267, 268

(continued overleaf)

Table 21.4   (*continued*)

Comment	Issues identified	Source
In addition to providing *historical background* and identifying general principles, geomorphological inputs can be focused directly into the river-management decision-making process. Many of the traditional river-management problems, such as erosion and flooding, have a geomorphological origin. Geomorphological analysis can assist in diagnosing the causes of the problems, development of appropriate management strategies that address the causes, and evaluation of the likely implications of management strategies.	Analysis of historical background can aid in diagnosing the cause of problems such as erosion and flooding **B**	Finlayson and Brizga, 2000, 9.
Fluvial geomorphology provides a *catchment-wide, evolutionary perspective* on river changes, providing an understanding of processes and direction of change to river structure that are often lacking in studies of ecological recovery. Framing river rehabilitation strategies in terms of recovery potential provides an integrative biophysical template for identifying sustainable target conditions for rehabilitation of degraded river courses.	Rehabilitation needs to be devised in terms of understanding of catchment-wide, evolutionary perspective **C**	Fryirs and Brierley, 2000
In restoration efforts, the locational probability image can be useful in identifying those places where the planting and nurturing of riparian vegetation is most likely to proceed with the least likelihood of damage from channel migration	Locational probability of change relevant to planning restoration **C**	Graf, 2000
It is clear that tectonic, climatic and environmental changes have impacted on fluvial systems throughout geological time... It is necessary that geomorphologists bring this to the attention of river managers, as it is possible to misinterpret natural *instability* in fluvial systems as being the result of human impact..., or to misdiagnose cyclical changes as channel instability... or to exaggerate human impact...	Instability may not always be the result of human impact, can be cyclical changes in instability **C**	Dollar, 2000

Table 21.4 (*continued*)

Comment	Issues identified	Source
It is important to acknowledge that channel and valley properties respond to long-term effects of climate change and tectonic activity... in the short term, *information required on* the intrinsic characteristics and evolution of rivers are to explain gradual processes	Information on evolution of rivers needed to explain river processes **B**	Vandenberghe and Maddy, 2000
Sediment exhaustion from parts of the catchment, and from river courses elsewhere, has major implications for the geomorphic recovery potential of rivers, constraining what can realistically be achieved in terms of river rehabilitation	Sediment exhaustion in context of long-term development **C**	Fryirs and Brierley, 2001
Evidence suggests that there is no singular relationship between event magnitude and landform change, and that short climatic cycles may not always produce noticeable changes in fluvial succession. The way *fluvial systems react depends on the physiographic setting of the system, which includes antecedent conditions*	Lack of singular relationships so that setting of system and former conditions need to be considered **C**	Dollar, 2002
Geomorphic insights into the nature and extent of change at regional, catchment, reach and hillslope scales have considerable implications for sustainable management of landscape. More effective *integration of these differing scales of sedimentary cascades*, moving beyond case study applications, will significantly expand the range of critical tools that geomorphologists provide for future land management planning... within catchment connectivity of biophysical processes may vary markedly from system to system.	Need to link change at regional catchment, reach and hillslope scales **C**	Brierley and Stankovian-sky, 2002

Although there are real physical limits to the size of floods (O'Connor *et al.*, 2002), a subject that palaeohydrological models could test, palaeoflood hydrology has evolved (Baker *et al.*, 2002) to overcome not only inaccuracies in estimating the ages of floods and in reconstructing flood discharges but also the lack of robust statistical methods for incorporating palaeoflood data into flood-frequency analysis and effects of climatic shifts and non-stationarity.

Convincing demonstrations by Victor Baker (Chapter 18) and other workers have shown how palaeoflood data reduce uncertainty in estimates of long return period floods, bringing very specific benefits in the design or retrofitting of dams or other floodplain structures. This is illustrated by a regional chronology of flood magnitude and frequency for small basins in western Arizona showing, from regional flood-frequency models, how large floods can be at variance with the 100-year floods so that predictive equations may not be appropriate for regulatory management and design purposes in the absence of real data on flooding (House and Baker, 2001). Palaeohydrologic bounds (Levish, 2002) are time intervals during which a particular discharge has not been exceeded, and there are limits on palaeostage provided by stable geomorphic surfaces adjacent to a stream where the age of the surface established the duration of the bound. In the past, probable maximum flood (PMF) derived from probable maximum precipitation (PMP) based on rainfall-runoff modelling may not have been the most effective approach, and consideration of the implications of palaeoflood analysis leads to the requirement for a more flexible view of palaeoflood history (Chapter 18). Other aspects of the water balance can also be illuminated by the palaeohydrological record.

Sensitivity of sections of the channel system arises because, although management has always concentrated upon sensitive problem reaches, the need for a focus in the context of the entire watershed, so linking the reach and the watershed scale (Brookes, 1995), is now appreciated. For a sustainable approach (C in Table 21.3), it is essential to be able to identify reaches that are already sensitive or may become unstable in the future. Whereas instability arising from mitigation measures and from adjustments consequent upon management methods has been recognized (B in Table 21.3), a sustainable basin approach employing knowledge of past development seeks to identify sensitive reaches throughout the basin. There are at least four possible interpretations of sensitivity (Downs and Gregory, 1995), which may be regarded as "the state of a landform or its propensity for change" (Schumm, 1988, 22). Methods are needed to identify sensitive components and segments of the fluvial system, effectively requiring prediction of geomorphic events with location, sensitivity, singularity and complexity aspects needing careful consideration (Schumm, 1991). Two methods offered to solve these four problems (Schumm, 1988) are location for time substitution (LTS) and location for condition substitution (LCS). The first, (LTS), often referred to as the ergodic method, requires arrangement of landforms in a sequence to show change through time so that it can be used to predict landform response. The second, (LCS) involves measuring the characteristics of relatively stable and unstable landforms so that a comparison allows identification of critical threshold conditions and sensitive landforms.

Identification of sensitive reaches requires isolation of thresholds (Schumm, 1979), described as conceptually elegant but methodologically difficult to locate precisely. Adoption of a basin approach allows analysis of the connectivity existing between the various components of the basin, with the connectivity between the river and the floodplain being vital and in tandem with influencing activities in the catchment (Holmes, 1998). Categorization of reaches of the fluvial system is an explicit way of implementing a basin-wide approach (Gregory and Chin, 2002) and there are a number of ways of categorizing reaches of the channel system such as the five categories suggested for the Bega catchment in New South Wales (Fryirs and Brierley, 2000) – a mixed classification of a kind that could be developed by combining threshold analysis with segment characterization in a palaeohydrological context, thus identifying sensitive reaches in a temporal as well as a spatial sense.

The controls upon specific reaches include the role of woody debris (Gregory, 2003). Along channels with riparian forest there are regular and significant inputs of Large Woody Debris (LWD) to the channels that may be chronic (frequent, but small in magnitude and occurring because of tree mortality and bank failure) and episodic (infrequent but providing large amounts of material from windthrow, ice storm, fire or flood events) (Fetherston et al., 1995). It has been suggested (Brown, 1998) that multiple channel systems characterized northwest European floodplains prior to deforestation and channelization, leading to an argument for the restoration of multiple channel systems, which involves regular flooding of parts of the floodplain, with secondary channels being allowed to exist and dead wood being left in the system even when obstructing channel flow.

The temporal sequence of sediment, morphological and hydrological history was one of the earliest focii for palaeohydrology (Leopold and Miller, 1954). Whereas river engineers are typically concerned with a design period of the order of 50 years, a time period over which static equilibrium and graded conditions are commonly assumed, there is a need to recognize the historical viewpoint (James, 1999) based upon analysis over longer timescales. Thus, Mediterranean areas require future research in order to identify dated alluvial sequences more fully and reliably, to provide more comprehensive coverage of river environments and to set future alluvial chronologies for particular sites into appropriate understanding of basin sediment dynamics, with better understanding of timing, spatial patterning, style and amounts of tectonic activity (Lewin et al., 1995), but it is clear that the impact of short-term climate variability has implications for river-basin management (Maas and Macklin, 2002). This requires the reconstruction of sediment development through cut and fill and patterns of sediment accretion and of erosion so that temporal sequences can be constructed. Four important features of recent research have included first a closer relationship between studies of basin processes and sediment accretion, admirably exemplified by the use of radionuclides to date patterns of floodplain accretion (e.g. Walling et al., 1996; Owens and Walling, 2002); second, the production of models of sediment residence in the landscape (e.g. Brown, 1987), thus documenting the extent to which, and the rates at which, sediment is transferred through the system. Consideration of the spatial basin scale cannot be isolated from temporal change because categorization of channels relates very specifically to sediment movement. Sediment exhaustion from parts of the Bega catchment, and from river courses elsewhere, has major implications for the geomorphic recovery potential of rivers, constraining what can realistically be achieved in terms of river rehabilitation (Fryirs and Brierley, 2001). Watershed scale is also related to the sedimentological characteristics of overbank alluvium (Knox and Daniels, 2002) deposited by large floods so that the spatial pattern needs to be considered with temporal variability. Such developments have progressed, thirdly, to the creation of more realistic temporal models (see Chapter 17) such as the cellular automaton approach utilized for northern England (Coulthard et al., 1999); and fourthly, to an approach specified to interpret site-specific boundary conditions linked to predictions of morphological behaviour over realistic time and spatial scales (Sear et al., 1995) allowing more complete interpretation of alluvial systems to become possible (Lewin, 2001).

What is natural is pertinent to restoration management (Tables 21.2, 21.3) particularly since the World Conservation Strategy (1980) emphasized that humanity, as part of nature, has no future unless nature and natural resources are conserved (IUCN, UNEP and WWF, 1991). Restoration is the process of returning the river or watershed to a condition that relaxes human constraints on the development of natural patterns of diversity (Frissell and Ralph, 1998); but restoration approaches have evolved very significantly (Brookes and Shields, 1996, Figure 1.2) since water quality restoration was first suggested in the mid-twentieth century. In river-channel restoration, it is necessary to consider how knowledge of environmental change illuminates what is natural; informs sustainable river-channel design with respect to extreme events, temporal change sequences, patterns of land-use change, ecological influences upon and within river channels and how interpretations are conditioned by the range of interested disciplines from engineering hydraulics to philosophy. It is now generally accepted that the water planner of the future must be able to handle many more concepts than flow equations, turbine efficiencies and flood frequencies (Jobin, 1998), because of the inclusion of sustainability in planning concepts. Some have argued that we must relinquish the notion of restoring the various components of the fluvial system to their prehistoric conditions, having instead a more modest goal, to put in place a multifunctional riparian zone that mimics natural forms to the greatest possible extent, so that the question is not really whether it is possible to re-build a self-maintaining channel reach but rather whether we can establish a landscape within which the river can live – a natural looking, multifunctional channel and floodplain that supports diverse aquatic and terrestrial plant and animal communities and requires minimal maintenance (Morris, 1995). Growing general acknowledgement of the importance of the natural environment, evident in the legislative development in almost all countries, but often requiring conservation of the flora and fauna and natural beauty without specifying what they mean, can require that the ecological basis for river management needs "wild/pristine examples", to tell us what many rivers once were (Harper *et al.*, 1995). Questions to be addressed are whether restoration is feasible, what is its precise objective and will it avoid future impacts downstream or upstream?

Such questions become increasingly significant as restoration is used more extensively and will relate to areas affected by dam decommissioning (Graf, 2001; Heinz III Center, 2002). Furthermore, because the concept of "natural" is a social construct, each community socially negotiates an appropriate mix of human and biophysical components in the local landscape (Rhoads *et al.*, 1999) so that education may be required to convey what is appropriately natural in landscapes.

4 CONCLUSION

Palaeohydrology does not readily provide analogues for the future but it can contribute background and insight into the scenarios and mechanisms that may occur. Reference to the time dimension is necessary in approaches to river-channel and drainage-basin management (Table 21.4) so that three conclusions from palaeohydrology merit explicit emphasis.

- *First*, the need, particularly in some regions, to make greater reference to palaeohydrological and environmental change research in river-channel management because the appropriate reference management timescale should be greater than the 50 years or 100 years of design flood;
- *Second*, to relate palaeohydrological information to the basin structure: because integrated basin management (IBM) is not fully integrated if it does not include sufficient geomorphological input and awareness of short-term change (Downs *et al.*, 1991);
- *Third*, as river-channel management increasingly utilizes stream restoration, more explicit attention should be given to the restoration objectives in relation to palaeohydrological history.

Recent recommendations for river-channel management have included a protocol for the restoration of regulated rivers (Stanford *et al.*, 1996), guiding principles for sustainable river restoration (Brookes and Shields, 1996), strategies for channel and basin management (Poesen and Hooke, 1997) and principles of good river management (Wharton, 2000). From these and other strategies the needs generally now agreed upon are to

- use integrated, basin-wide planning, and a holistic approach for channel and flood management; and to describe and formalize the problem of ecological restoration at the catchment scale;
- work with nature and not against it, emulating nature in river designs; restore environmental (habitat) heterogeneity but letting the river do the work;
- employ a sustainable approach;
- consider hazards created by erosion and sedimentation together with those of flood discharges, structures being designed for high sediment loads;
- adopt non-structural and do-nothing approaches wherever possible, using procedures that have least damaging effects on environment.

In undertaking such procedures, it is now generally agreed that it is necessary to

- keep areas under review by adaptive ecosystem management (Table 21.2);
- employ a detailed appraisal process; consult widely, considering all the environmental issues alongside the engineering and economic objectives;
- undertake post-project appraisal so that knowledge about impacts of river management continues to grow;
- take into account high spatial and temporal variability of floods and flood impacts, and of their feedback effects;
- stress that exact predictions are not possible by giving scientific results to decision-makers as a range of possibilities and probabilities with consequences of extremes indicated.

These points distilled from schemes previously suggested lead to a further protocol embracing palaeohydrological inputs for application to a particular area:

- obtain as much historical data as possible on floods, flood hazard and flood mitigation measures and channel behaviour for channel management;
- consider which timescale is an appropriate basis for management for channel management and seek to augment the continuous record;
- include awareness of the period of records used as the basis for previous channel-management decisions;
- adopt a basin perspective by identifying reaches and segments of channel that are unstable and sensitive, as a result of mitigation or management measures or human activity including those that may become sensitive in the future;
- set the pattern of sensitive reaches into a basin context taking account of changes in sediment history including phases of storage and exhaustion;
- consider any detectable phases in the palaeohydrology or sediment budget record to set the management period into a temporal pattern;
- when restoring channels, give careful consideration to the following:
  - Is restoration feasible for the particular channel?
  - Should restoration be to a more natural state or to some specific prior condition and, if the latter, what is the basis for the decision?

- Does the restored state present the most stable channel that will avoid impacts downstream or upstream?
- consider that "natural" in any area is a social construct that must be negotiated with the local community giving opportunity for education of that community in relation to palaeohydrology.

These additional protocol suggestions are not intended to be definitive but adapted as necessary and should be considered for application to specific areas.

ACKNOWLEDGEMENT

This chapter was developed during the tenure of a Leverhulme Emeritus Fellowship, which is gratefully acknowledged, and thanks are due to Professor D.E. Walling for his comments on an earlier draft.

REFERENCES

Baker, V.R., Webb, R.H. and Kyle House, P., 2002. The scientific and societal value of palaeoflood hydrology. In P.K. House, R.H. Webb, V.R. Baker and D.R. Levish (eds), *Ancient Floods Principles and Application of Palaeoflood Hydrology*. American Geophysical Union, Washington, DC, 1–19.

Brierley, G.J. and Fryirs, K., 2000. River styles, a geomorphic approach to catchment characterization: implications for river rehabilitation in the Bega catchment, New South Wales, Australia. *Environmental Management*, **25**, 661–679.

Brierley, G.J. and Stankoviansky, M., 2002. Geomorphic responses to land use change. Special issue of *Earth Surface Processes and Landforms*, **27**(4), 339–462.

Brookes, A., 1995. Challenges and objectives for geomorphology in UK river management. *Earth Surface Processes and Landforms*, **20**, 593–610.

Brookes, A. and Gregory, K.J., 1988. Channelization, river engineering and geomorphology. In J.M. Hooke (ed.), *Geomorphology in Environmental Planning*. Wiley, Chichester, 145–167.

Brookes, A. and Shields, D.R., (eds), 1996. *River Channel Restoration. Guiding Principles for Sustainable Projects*. Wiley, Chichester, 1996.

Brown, A.G., 1987. Long term sediment storage in the Severn and Wye catchments. In K.J. Gregory, J. Lewin and J.B. Thornes (eds), *Palaeohydrology in Practice*. Wiley, Chichester, 307–332.

Brown, A.G., 1998. The maintenance of diversity in multiple channel floodplains. In R.G. Bailey, P.V. Jose and B.R. Sherwood (eds), *United Kingdom Floodplains*. Otley, Westbury, 83–92.

Chorley, R.J., Schumm, S.A. and Sugden, D.E., 1984. *Geomorphology*. Methuen, London.

Coulthard, T.J., Kirkby, M.J. and Macklin, M.G., 1999. Modelling the impacts of Holocene environmental change on the fluvial and hillslope morphology of an upland landscape, using a cellular automaton approach. In A.G. Brown and T.M. Quine (eds), *Fluvial Processes and Environmental Change*. Wiley, Chichester, 31–47.

Dollar, E.S.J., 2000. Fluvial geomorphology. *Progress in Physical Geography*, **24**, 385–406.

Dollar, E.S.J., 2002. Fluvial geomorphology. *Progress in Physical Geography*, **26**, 123–143.

Downs, P.W. and Gregory, K.J., 1995. Approaches to river channel sensitivity. *Professional Geographer*, **47**, 168–175.

Downs, P.W. and Gregory, K.J., 2003. *River Channel Management*. Arnold, London.

Downs, P.W., Gregory, K.J. and Brookes, A., 1991. How integrated is river basin management? *Environmental Management*, **15**, 299–309.

Fetherston, K.L., Naiman, R.J. and Bilby, R.E., 1995. Large woody debris, physical process and riparian development in montane river networks of the Pacific Northwest. *Geomorphology*, **13**, 133–144.

Finlayson, B.L. and Brizga, S.O., 2000. Introduction. In S. Brizga and B. Finlayson (eds), *River Management. The Australasian Experience*. Wiley, Chichester, 1–10.

Frissell, C.A., Liss, W.J., Warren, C.E. and Hurley, M.D., 1986. A hierarchical framework for stream habitat classification: viewing streams in a watershed context. *Environmental Management*, **10**, 199–214.

Frissell, C.A. and Ralph, S.C., 1998. Stream and watershed restoration. In R.J. Naiman and R.E. Bilby (eds), *River Ecology and Management*. Springer-Verlag, New York, 599–624.

Fryirs, K. and Brierley, G., 2000. A geomorphic approach to the identification of river recovery potential. *Physical Geography*, **21**, 244–277.

Fryirs, K. and Brierley, G., 2001. Variability in sediment delivery and storage along river courses in Bega catchment, NSW Australia; implications for geomorphic river recovery. *Geomorphology*, **38**, 237–265.

Gares, P.A., Sherman, D.J. and Nordstrom, K.F., 1994. Geomorphology and natural hazards. *Geomorphology*, **10**, 1–18.

Geoparks, 1999. *UNESCO network of Geoparks (Quotation of F.Broudel, French historian 1902–1985)*. UNESCO, Division of Earth Sciences, Paris.

Gilvear, D.J., 1999. Fluvial geomorphology and river engineering: future roles utilizing a fluvial hydrosystems framework. *Geomorphology*, **31**, 229–245.

Graf, W.L., 2000. Locational probability for a dammed, urbanizing stream: Salt river, Arizona, USA. *Environmental Management*, **25**, 321–335.

Graf, W.L., 2001. Damage control: restoring the physical integrity of America's rivers. *Annals of the Association of American Geographers*, **91**, 1–27.

Gregory, K.J., 1998. Applications of palaeohydrology. In G. Benito, V.R. Baker and K.J. Gregory (eds), *Palaeohydrology and Environmental Change*. Wiley, Chichester, 13–26.

Gregory, K.J., 2000. *The Changing Nature of Physical Geography*. Arnold, London.

Gregory, K.J., 2002. Urban channel adjustments in a management context: an Australian example. *Environmental Management*, **29**, 620–633.

Gregory, K.J., 2003. The limits of wood in world rivers. In S.V. Gregory (ed.), *Wood in World Rivers*. American Fisheries Society, Bethesda, Maryland.

Gregory, K.J. and Chin, A., 2002. Urban stream channel hazards. *Area*, **34**, 312–321.

Harper, D., Smith, C., Barham, P. and Howell, R., 1995. The ecological basis for the management of the natural river environment. In D.M. Harper and A.J.D. Ferguson (eds), *The Ecological Basis for River Management*. Wiley, Chichester, 219–238.

Harper, D.M., Ebrahimnezhad, M., Taylor, E., Dickinson, S., Decamp, O., Verniers, G. and Balbi, T., 1999. A catchment-scale approach to the physical restoration of lowland UK rivers. *Aquatic Conservation: Marine and Freshwater Ecosystems*, **9**, 141–157.

Heinz III Center, 2002. *Dam Removal. Science and Decision Making*. The Heinz Center, Washington, DC.

Hey, R.D., 1990. Environmental river engineering. *Journal of the Institution of Water and Environmental Management*, **4**, 335–340.

Holmes, N., 1998. Floodplain restoration. In R.G. Bailey, P.V. Jose and B.R. Sherwood (eds), *United Kingdom Floodplains*. Otley, Westbury, 331–348.

House, P.K. and Baker, V.R., 2001. Palaeohydrology of flash floods in small desert watersheds in Western Arizona. *Water Resources Research*, **37**, 1825–1839.

IUCN, UNEP and WWF, 1991. *Caring for the Earth. A Strategy for Sustainable Living*. Earthscan, London.

James, A., 1999. Time and the persistence of alluvium: River engineering, fluvial geomorphology, and mining sediment in California. *Geomorphology*, **31**, 265–290.

Jansen, P.Ph., Van Bendegom, L., van den Berg, J., de Vries, M. and Zanen, A., 1979. *Principles of River Engineering: The Non Tidal Alluvial River*. Pitman, London.

Jobin, W., 1998. *Sustainable Development for Dams and Waters*. Lewis Publishers, Boca Raton, Boston.

Joliffe, I.B. and Ball, J.E., (eds), 1999. 8th *International Conference, Urban Storm Drainage*. Institution of Engineers, Australia.

Keller, E.A., 1976. Channelization: environmental, geomorphic and engineering aspects. In D.R. Coates (ed.), *Geomorphology and Engineering*. State University of New York, Binghamton, 115–140.

Knighton, A.D., 1998. *Fluvial Forms and Processes*. Arnold, London.

Knox, J.C. and Daniels, J.M., 2002. Watershed scale and the stratigraphic record of large floods. In P.K. House, R.H. Webb, V.R. Baker and D.R. Levish (eds), *Ancient Floods Principles and Application of Palaeoflood Hydrology*. American Geophysical Union, Washington, DC, 237–255.

Lawton, J., 2001. Earth system science. *Science*, **292**, 1965.

Leeks, G.J., Lewin, J. and Newson, M.D., 1988. Channel change, fluvial geomorphology and river engineering: The case of the Afon Trannon, Mid Wales. *Earth Surface Processes and Landforms*, **13**, 207–224.

Leopold, L.B., 1977. A reverence for rivers. *Geology*, **5**, 429, 430.

Leopold, L.B. and Miller, J.P., 1954. *Postglacial Chronology for Alluvial Valleys in Wyoming*. United States Geological Survey Water Supply Paper 1261, 61–85.

Leuven, R.S.E.W., Smits, A.J.M. and Nienhuis, P.H., 2000. From integrated approaches to sustainable river management. In A.J.M. Smits, P.H. Nienhuis, and R.S.E.W. Leuven, *New Approaches to River Management*. Backhuys, Leiden, 329–347.

Levish, D.R., 2002. Palaeohydrologic bounds: non exceedance information for flood hazard assessment. In P.K. House, R.H. Webb, V.R. Baker and D.R. Levish (eds), *Ancient Floods Principles and Application of Palaeoflood Hydrology*. American Geophysical Union, Washington, DC, 175–190.

Lewin, J., 2001. Alluvial systematics. In D. Maddy, M.G. Macklin and J.C. Woodward (eds), *River Basin Sediment Systems*. Balkema, Netherlands, 19–41.

Lewin, J., Macklin, M.G. and Newson, M.D., 1988. Regime theory and environmental change-irreconcilable concepts? In W.R. White (ed.), *International Conference on River Regime*. Wiley, Chichester, 431–445.

Lewin, J., Macklin, M.G. and Woodward, J.C., (eds), 1995. *Mediterranean Quaternary River Environments*. A.A.Balkema, Rotterdam.

Maas, G.S. and Macklin, M.G., 2002. The impact of recent climate change on flooding and sediment supply within a Mediterranean mountain catchment, Southwestern Crete, Greece. *Earth Processes and Landforms*, **27**, 1087–1105.

Macklin, M.G. and Lewin, J., 1997. Channel, floodplain and drainage basin response to environmental change. In C.R. Thorne, R.D. Hey and M.D. Newson (eds), *Applied Fluvial Geomorphology for River Engineering and Management*. Wiley, Chichester, 15–45.

McHarg, I.L., 1969. *Design with Nature*. Natural History Press, New York.

McHarg, I. and Steiner, F.R., (eds), 1998. *To Heal the Earth. Selected Writings of Ian L. McHarg*. Island Press, Washington, DC.

Messerli, B., Grosjean, M., Hofer, T., Nunez, L. and Pfister, C., 2000. From nature-dominated to human-dominated environmental changes. *Quaternary Science Reviews*, **19**, 459–479.

Morris, S.E., 1995. Geomorphic impacts of stream channel restoration. *Physical Geography*, **16**, 444–459.

Newson, M.D. and Macklin, M.G., 1990. The geomorphologically effective flood and vertical instability in river channels – a feedback mechanism in the flood series for gravel-bed rivers. In W.R. White (ed.), *International Conference on River Flood Hydraulics*. Wiley, Chichester, 123–140.

Nienhuis, P.H., Leuven, R.S.E.W. and Ragas, A.M.J., (eds), 1998. *New Concepts for Sustainable Management of River Basins*. Backhuys, Leiden.

O'Connor, J.E., Grant, G.E. and Costa, J.E., 2002. The geology and geography of floods. In P.K. House, R.H. Webb, V.R. Baker and D.R. Levish (eds), *Ancient Floods Principles and Application of Palaeoflood Hydrology*. American Geophysical Union, Washington, DC, 359–385.

Owens, P.N. and Walling, D.E., 2002. Changes in sediment sources and floodplain deposition rates in the catchment of the river Tweed, Scotland, over the last 100 years: the impact of climate and land use change. *Earth Surface Processes and Landforms*, **27**, 403–423.

Petersen, M.S., 1986. *River Engineering*. Prentice Hall, Englewood Cliffs, NJ.

Petts, G.E., Sparks, R. and Cambell, I., 2000. River restoration in developed economies. In P.J. Boon, B.R. Davies and G.E. Petts (eds), *Global Perspectives on River Conservation*. Wiley, Chichester, 493–508.

Poesen, J.W.A. and Hooke, J.M., 1997. Erosion, flooding and channel management in Mediterranean environments of southern Europe. *Progress in Physical Geography*, **21**, 157–199.

Rhoads, B.L., Urban, M., Wilson, D. and Herricks, E., 1999. Interaction between scientists and nonscientists in community-based watershed management: emergence of the concept of stream naturalization. *Environmental Management*, **24**, 331–368.

Richards, K.S., Hughes, F.M.R., El-hames, A.S., Harris, T., Pautou, G., Peiry, J.L. and Girel, J., 1996. Integrated field, laboratory and numerical investigations of hydrological influences on the establishment of riparian tree species. In M.G. Anderson, D.E. Walling and P.D. Bates (eds), *Floodplain Processes*. Wiley, Chichester, 611–636.

Riley, S.J., 1988. Secular change in the annual flows of streams in the NSW section of the Murray-Darling basin. In R.F. Warner (ed.), *Fluvial Geomorphology of Australia*. Academic Press, Sydney, 245–266.

Rodriguez-Iturbe, I., 2000. Ecohydrology: A hydrologic perspective of climate-soil-vegetation dynamics. *Water Resources Research*, **36**, 3–9.

Schellnhuber, H.J., 1999. Earth System analysis and the second Copernican revolution. *Nature*, **402**, C19–C23.

Schumm, S.A., 1979. Geomorphic thresholds. *Transactions of the Institute of British Geographers*, **4**, 485–515.

Schumm, S.A., 1988. Geomorphic hazards: problems of prediction. *Zeitschrift fur Geomorphologie*, Supplementband **67**, 17–24.

Schumm, S.A., 1991. *To Interpret the Earth: Ten Ways to be Wrong*. Cambridge University Press, Cambridge.

Schumm, S.A., 1994. Erroneous perceptions of fluvial hazards. *Geomorphology*, **10**, 129–138.

Sear, D.A., 1996. In A. Brookes and D.R. Shields (eds), *River Channel Restoration. Guiding Principles for Sustainable Projects*. Wiley, Chichester, 149–177.

Sear, D.A., Newson, M.D. and Brookes, A., 1995. Sediment-related maintenance: the role of fluvial geomorphology. *Earth Surface Processes and Landforms*, **20**, 629–647.

Smits, A.J.M., Nienhuis, P.H. and Leuven, R.S.E.W., 2000. *New Approaches to River Management*. Backhuys, Leiden.

Stanford, J.A., Ward, J.V., Frissell, W.J., Liss, C.A., Williams, R.N., Lichatowich, J.A. and Coutant, C.C., 1996. A general protocol for restoration of regulated rivers. *Regulated Rivers*, **12**, 391–413.

Thorne, C.R., Hey, R.D. and Newson, M.D., (eds), 1997. *Applied Fluvial Geomorphology for River Engineering and Management*. Wiley, Chichester.

Tooth, S. and Nanson, G.C., 1995. The geomorphology of Australia's fluvial systems: retrospect, perspect and prospect. *Progress in Physical Geography*, **19**, 35–60.

Vandenberghe, J. and Maddy, D., 2000. The significance of fluvial archives in geomorphology. *Geomorphology*, **33**, 127–130.

Vannote, R.L., Minshall, G.W., Cummins, K.W. and Sedell, J.R., 1980. The river continuum concept. *Canadian Journal of Fisheries and Aquatic Sciences*, **37**, 130–137.

Walling, D.E., He, Q. and Nicholas, A.P., 1996. Floodplains as suspended sediment sinks. In M.G. Anderson, D.E. Walling and P.D. Bates (eds), *Floodplain Processes*. Wiley, Chichester, 399–440.

Wasson, R.J., 1994. Living with the past: uses of history for understanding landscape change and degradation. *Land Degradation and Rehabilitation*, **5**, 79–87.

Wharton, G., 2000. *Managing River Environments*. Cambridge University Press, Cambridge.

Williams, P.B., 2001. Stuck between two paradigms: river management versus river engineering. Keynote address to American Society of Civil Engineers, *Wetlands Engineering & River Restoration Conference 2001*, Reno, Nevada.

Winkley, B.R., 1972. River regulation with the aid of nature. *International Commission Irrigation and Drainage, Eighth Congress*, 433–457.

PART 5 PROSPECT

22 Concluding Perspective

K.J. GREGORY[1] AND G. BENITO[2]
[1]*University of Southampton, Southampton, UK*
[2]*CSIC-Centro de Ciencias Medioambientales, Madrid, Spain*

Palaeohydrology has become an important key to assist understanding of the potential effects of global change by using analogues from the past. New multidisciplinary methods developed during the last decade assisting the study of palaeohydrological change have brought new perspectives in the detailed reconstruction of palaeoenvironments, which can now be incorporated in conceptual and numerical macroscale and mesoscale models.

At present, future predictions for world climate systems are based on highly sophisticated computer models, supported by the vast amount of available remotely sensed data that is limited by the short period of records. Once most of the scientific community is persuaded that climate is changing, new efforts should focus on how change impacts on the different components of the hydro-biosphere, such as hydrology, as well as on the adaptations to change. However, downscaling from global models is not easily achieved (Chapter 1, this volume), and regional hydrological projections from global models are still uncertain, sometimes contradictory, and are frequently speculative (e.g. flood-forecasted scenarios). Recent efforts in assessing the hydrological impacts of global change are carried out by different international programs on global change such as IGBP, International Hydrological Programme of UNESCO, WCRP (Chapter 3). Hitherto long-term palaeohydrological records have not been considered sufficiently, and only the IGBP project on Past Global Changes (PAGES) deals with palaeoenvironmental science. A human-dominated environment, with increasing vulnerability of societies and economies to extreme events and natural variability (Messerli *et al.*, 2000), demands detailed information on how climate change can impact on different regions of the World. Long-term records for diverse circumstances of climatic variability can substantially improve future predictions of hydrological extremes, including floods and droughts, and of water resources.

Advances in palaeohydrology were related in five categories to global change (Chapter 1). The first, *coupling Global Climate Change Models to hydrological models* incorporating palaeohydrological data is already occurring to some limited extent. In the last decade, there has been a real effort to include palaeohydrological results either in the downscaling of Global Circulation Models or in the upscaling of the physical catchment parameters (Chapters 1 and 17). Palaeohydrological *quantification* of extreme events (palaeofloods) in terms of magnitude and frequency can be used to assign weighted probabilities for specific hydrologic events on given climate scenarios (Chapters 10 and 18). It is clear that short- and long-term observations of climatic change are pushing hydrologists to include non-stationarity in flood-frequency

Palaeohydrology: Understanding Global Change. Edited by K.J. Gregory and G. Benito
© 2003 John Wiley & Sons, Ltd ISBN: 0-470-84739-5

analysis, including analysis performed with long-gauge records. Because palaeoflood records include manifestations of rare floods, they have the ability either to solve the problems associated with non-stationarity in flood-frequency analysis or to emphasize modifications to the flood-hazard methodology so that they can better represent those realities indicated by the palaeoflood data (Chapter 17).

Learning from past palaeohydrological conditions can provide a more accurate projection of the hydrological response to climate episodes and changes than that inferred from climatic modelling. In subarctic fluvial environments such as the Usa river, a temperature 3°C higher during the Holocene optimum resulted in a reduction of the mean annual flood by some 30%, and an equivalent reduction in specific stream power, inducing changes in the sediment grain size being transported and in river pattern from braided to meandering (Chapter 5). Knowledge of palaeohydrological conditions created during the Holocene climatic optimum may be projected to other high-latitude areas susceptible to potential future climate changes.

Conclusions from most chapters focus on the derivation of data to complement periods of continuous hydrological records relating to water balance, hydrological extremes, water quality and sediment involving historical sequences (Chapter 21). Continuous detailed palaeohydrological continental records cannot be found at any one site, although there are some semi-continuous terrestrial proxy data from mire sequences, lake sediments and loess deposits (see references in Chapters 8 and 11). Quantitative river discharges (runoff and flood peak discharges) have been reconstructed using relict fluvial landforms (palaeochannels) and through the analysis of stratigraphical sequences (floodplain sequences and slackwater flood sediments). These fluvial records are limited by stratigraphical discontinuities and by the preservation of the landform-sediment architecture (Chapter 16). Knowledge is uneven in different parts of the world. High- and low-latitude zones contain large gaps in their palaeohydrological record whereas in the mid-latitude zone (especially the temperate regions of Europe and North America) existing palaeohydrological data allow interpretation of a broad picture of climate and flow-regime change for the last 20 kyr (Chapters 7, 8 and 10).

Available data for different world zones provide palaeohydrological figures very far removed from contemporary hydrological conditions. During the Lateglacial (cold periglacial climate with extensive permafrost), the Russian Plain and West Siberia rivers developed channel widths up to 15 times the present ones with discharges 5 to 6 times greater (Chapter 6). Most of the annual flow was presumably drained during spring floods. Holocene warming conditions affected the extension of the periglacial hyperzone, producing large variations in the spatial distribution of surface runoff, with large runoff reductions in areas of permafrost degradation. East Siberia is situated at present within the permafrost zone, and there are no indicators of major changes in river morphology and runoff during the Lateglacial and the Holocene. In Central Europe, the Holocene was characterised by humid phases (usually cooler) with frequent extreme events (300–500 years long) alternating with longer phases, which were relatively dry with less frequent floods (Starkel, 1991). Humid phases at 8.5 kyr BP (Starkel, 1999), at the Atlantic Subboreal transition (Starkel, 1995), at the late Bronze (van Geel et al., 1998) and the Little Ice Ages extended throughout western Europe (Chapter 8) and even southern Europe (Chapter 9). In the Mediterranean region, the most important changes in the Holocene palaeohydrological regime are related to hydrological extremes (flood and droughts). Palaeoflood records show periods of very frequent recurrences of large floods, the most recent ones during the late Bronze Age

(700–500 BC), AD 500 to 1000 and/or AD 1600 to 1700, alternating with periods of very few or no large floods.

Many problems remain in the interpretation of tropical palaeohydrology and there are comparatively few studies of the major rivers of Africa and South America. In Africa, almost all estimates of rainfall change have been based on boundary values for reconstructed vegetation types, or on changes in palaeolake levels (Gasse, 2000). Fluvial activity was not in phase throughout the river systems in Africa (Chapter 11), with general figures indicating a decrease in fluvial activity in response to cooling of northern climates and oceans (25–15 cal kyr BP; during the Younger Dryas, 8.2–7 cal kyr BP; 4.5–2 cal kyr BP). The unique record of the Nile floods in Egypt (Nilometer) shows episodes of high floods in early 800 AD, around 1100 AD and between 1350 and 1470 AD (Hassan, 1981). Palaeohydrological reconstructions of the large South American fluvial systems remain at an early stage of knowledge. Fluvial activity decreased during glacial and Lateglacial periods (Chapter 12), whereas the Late Holocene, and particularly the last 1,000 years, seems to have been a period of rapid recuperation of fluvial systems with high rates of sedimentation.

In southern Asia, the physical, hydrological and biological environment is closely related to the rhythm of both the southwest (Indian) and the southeast (East Asian) monsoons. The Last Glacial Maximum (LGM) was characterised by drier conditions with erratic and reduced fluvial activity, followed by a summer-monsoon intensification during the early Holocene (ca 10–5 ^{14}C kyr BP or 11.5–5 cal kyr BP) associated with an increase in fluvial activity (Chapter 13). About 5 to 3 cal kyr BP, a reduction in monsoon precipitation to a minimum was recorded. In southeastern Australia, the early to mid-Holocene was marked by much lower flows than those in the Pleistocene, but they were certainly more pronounced than those of today (Chapter 14). In the last 3,000 years, rivers have been laterally stable with floodplains vertically accreting alongside well-vegetated channels.

Palaeohydrological studies have provided information on the *mechanics of temporal change*, demonstrating their complexity from the Neolithic to the twentieth century because of the influence of human activity. In this respect, anthropogenic influence and climatic factors have controlled the palaeohydrologic evolution of most parts of Europe during the Holocene. The main activities modifying regional hydrological outputs are deforestation and cultivation, which have triggered the acceleration of runoff and, at the same time, have required increased water storage by reducing evapotranspiration. Late-Quaternary climatic variability produced seasonal changes in precipitation, which are difficult to understand from present climatic analogues. In general, it is difficult to differentiate the roles played by climate and man in determining changes observed in fluvial channels. In some particular regions, other temporal changes are described so that in many Asian rivers it is difficult to separate fluvial changes caused by neotectonic activity from changes that would have occurred primarily because of climatic variations (Chapter 13).

A major concern in palaeohydrological studies is to gain some insight into the *spatial contrasts* necessary for the understanding of regional and global hydroclimatic changes. Regional correlations in different world areas (Chapter 4) are addressed throughout this volume. Event-based stratigraphy may produce accurate records for individual floods, producing a good picture of the response of extreme events to Holocene climatic variability. It is notable that clustering of floods into discrete episodes of higher frequency events is recurrent in many parts of the World, such as in North America (Chapter 10). In the southwest of North America, floods were relatively frequent about 5,800 and 4,200 years ago, and after about 2,400 years ago, except between about 800 and 600 years ago (Ely, 1997). In the upper Mississippi River, relatively large floods occurred between about 7,000 and 5,500 years ago, followed by an episode of smaller floods between about 5,500 and 3,300 years ago, then returning to generally larger floods after 3,000 years ago (Chapter 10). Two main conclusions can be extracted from these chronologies (Chapter 10): first that extreme events (floods and droughts) tend to either "turn on" or "turn off" with shifts in the frequencies of certain patterns of large-scale ocean–atmosphere circulation, and second that strong teleconnections exist between regional flood episodes in North America, and may reflect either in-phase or out-of-phase relations depending on the prevailing hemispheric-scale ocean–atmosphere circulation patterns.

Sufficient progress has now been made to be able to contemplate constructing *new models* of environmental change. Providing uncertainty is quantified, simple hydraulic models applied at the scale of the individual reach may be appropriate (Chapter 19), but mathematical palaeohydraulic modelling is still in a premature stage, so that advances in contemporary hydraulics will permit more realistic modelling of palaeofloods. Progress has been made towards the instantaneous geomorphic unit hydrograph (IGUH) (Chapter 17) and it is such developments that will be able to link palaeohydraulic models with tried hydrological and climatic models, employing a multidisciplinary approach to achieve downscaling and to create a new generation of models.

To construct and implement such models, it will be necessary to maintain and enhance a multidisciplinary approach, to continue to acquire palaeohydrological data with robust dating controls, employing new techniques to date alluvial sequences including dating other drift organics as well as times of sediment transport, especially for periods of terrace formation or occurrence of palaeohydrological events. Spatial discontinuities continue to be problematic, and the complexity of the fluvial system means that it is still difficult to understand how different components of the basin (vegetation, land use, etc.) affect hydrological responses. The differences that exist between areas in storage and the legacy of past events in the present system need to be disaggregated. Relict landforms and sediments stored along rivers introduce timescales that are of varying importance from one area to another so that, for management purposes, reference needs to be made to longer timescales. For scientific explanations, attention needs to be given, with caution, to climatic linkages (teleconnections) across Europe, for example, and particularly to the correlation of the terrestrial (fluvial) record with available continuous palaeoclimatic records such as those from ocean-floor cores.

Global change in relation to hydrology and hydrological regime is the subject of a number of current investigations (e.g. Alverson *et al.*, 2000) many of which draw attention to the coarseness of spatial resolution and to the modelling of hydrological components, which is less reliable than temperature and pressure (e.g. Jones and Woo, 2002). In establishing palaeohydrological regimes and palaeoflow changes over the last 20,000 years, the outstanding problems relate mainly to discontinuity of alluvial sequences, poor chronological constraints, and the complexity of the intrinsic parameters of the drainage basins. Some of the themes introduced here will contribute to the development of the next research frontiers in which short-term events (Chapter 20) and triggers for change are particularly important. The role of INQUA (Chapter 2) in catalysing international multidisciplinary research should continue to foster closer links with other organisations and commissions as one significant way forward (Chapter 15).

The significance of past global changes for the future has been stressed (e.g. Bradley, 2000) but, as changes that affect rivers are expressed through the drainage basin, it remains essential to ensure that the translation of climate signals through the characteristics of the catchment are fully understood – one urgent theme for the continuing research agenda in palaeohydrology.

REFERENCES

Alverson, K.D., Oldfield, F. and Bradley, R.S., (eds), 2000. Past Global Changes and their significance for the future. *Quaternary Science Reviews*, **19**, 3–7.

Bradley, R.S., 2000. Past global changes and their significance for the future. *Quaternary Science Reviews*, **19**, 391–402.

Ely, L.L., 1997. Response of extreme floods in the southwestern United States to climatic variations in the late Holocene. *Geomorphology*, **19**, 175–201.

Gasse, F., 2000. Hydrological changes in the African tropics since the last glacial maximum. *Quaternary Science Reviews*, **19**, 189–211.

Hassan, F.A., 1981. Historical Nile floods and their implications for climate change. *Science*, **212**, 1142–1145.

Jones, J.A.A. and Woo, M.K., 2002. *Modelling the impact of climate change on hydrological regimes*. Special issue of *Hydrological Processes.*, **16**(6), 1135–1352.

Messerli, B., Grossjean, M., Hofer, T., Nunez, L. and Pfister, C., 2000. From nature dominated to human-dominated environmental changes. *Quaternary Science Reviews*, **19**, 459–479.

Starkel, L., 1991. The Vistula River Valley: a case study for Central Europe. In L. Starkel, K.J. Gregory and J.B. Thornes (eds), *Temperate Palaeohydrology*. John Wiley, Chichester, 171–188.

Starkel, L., 1995. Palaeohydrology of the temperate zone. In K.J. Gregory, L. Starkel and V.R. Baker (eds), *Global Continental Palaeohydrology*. Wiley, Chichester, 233–258.

Starkel, L., 1999. 8500–8000 yrs BP humid phase-global or regional? *Science Reports of Tohoku University*, 7th Series (Geography). **49**, 2, 105–133.

van Geel, B., van der Plicht, J., Kilian, M.R., Klaver, E.R., Kouwenberg, J.H.M., Ressen, H., Reynaud-Farrera, I. and Waterbolk, H.T., 1998. The sharp rise of $\delta^{14}C$ at ca 800 cal. BC. Possible causes, related climatic teleconnections and the impact on human environments. *Radiocarbon*, **40**, 335–350.

Appendix Discussion at the Fifth International Meeting of GLOCOPH in Pune, India, December 2002

The fifth international meeting of GLOCOPH, held in India in December 2002, was organised by Professor Vishwas Kale, University of Pune, supported by many colleagues including Dr V. Joshi, S.P. College, and Dr P. Hire, HPT/RYK College. An excellent three days of paper presentations in Pune was followed by three equally impressive days of field excursions that covered the area southwards from Pune to Goa. Although there was no one session devoted exclusively to papers from this volume, a number of the lectures were related to, or built upon, the content of particular chapters. Following the paper on *Palaeohydrology, environmental change and river channel management* by K.J. Gregory, the importance of education to communicate the significance of new approaches was stressed by R. Sinha, and A.G. Brown advocated that floodplain management be associated with river channel management. V. Baker presented a stimulating paper on *Palaeofloods and Global Change*, and the ensuing discussion led to the need for the explanation of the significance of the research to nonscientists and decision makers. *PHEIMS: A web based data base for the global palaeoenvironment* by T. Oguchi with M. Nishikata and Y. Hayakawa prompted discussion about the ways in which this excellent database could be extended either by taking in other publications or by including additional material.

The 32 papers and 7 posters presented can be classified using the scheme of Table 1.1 to show that, over the 12-year period, some 260 papers have been presented at GLOCOPH international meetings as follows:

Dominant theme	Pune 2002	Total
Glacial	2	15 (5.8%)
Technique-based including palaeoecology, historical records	7	55 (21.2%)
Modelling	3	18 (6.9%)
Processes	4	37 (14.2%)
Sediment-based	16	80 (30.8%)
Basin components including palaeochannels, lakes, planform	5	41 (15.8%)
Drainage Basin	2	14 (5.4%)
		100.1

Palaeohydrology: Understanding Global Change. Edited by K.J. Gregory and G. Benito
© 2003 John Wiley & Sons, Ltd ISBN: 0-470-84739-5

Several of the Pune papers were very clearly related to chapters in this volume, including those by

- G. Benito, V.R. Thorndycraft, M. Rico, A. Sopeira, Y. Sanchez and M. Casas *Palaeoflood hydrology applications to flood risk assessment in Spanish mediterranean rivers* (Chapter 9)
- M. Thomas *Late Quaternary sedimentary fluxes from tropical watersheds* (Chapter 11)
- E. Latrubesse, J.S. Stevaux, S.A. da Silva, M. Bayer and N.J. Pawar *Geomorphology and human induced changes of the Araguaia fluvial basin* (Chapter 12)
- V. Kale, V. Joshi and P. Hire *Palaeohydrological reconstructions based on the analysis of a palaeochannel and Toba ash associated alluvial sediments in western Deccan Trap region, India* (Chapter 13).

However, two other characteristics of the papers presented at the meeting were clearly apparent. First, the stimulus provided by the papers authored by Indian scientists, often dealing with large and complex fluvial situations but also drawing attention to issues of considerable significance. Such issues arose generally in special lectures provided by

- S.K. Tandon *Fluvial response to climate change in monsoonal areas*
- I.B. Singh *Late Quaternary evolution of Ganga plain*

And also more specifically in a number of lectures including those by

- M.R. Bhutiyanim, V.S. Kale and N.J. Pawar *Global warming effects on the Himalayan Mountains – a cause for alarm?*
- V. Jain and R. Sinha *Hydrological controls on the geomorphological variability of the Ganga plains and their implications to climate change.*

In these and in other papers, the significance of palaeohydrological research for understanding the present and also for contributing to managing the future is becoming increasingly evident.

The ongoing potential of palaeohydrology was also evident during the discussion after individual papers and was also stressed during business meetings of GLOCOPH on the 2nd and 4th of December 2002. There was a clear consensus that ongoing research is required, that this should exploit the potential of the several promising leads that have emerged during GLOCOPH 1991 to 2003, and that a strong multi-disciplinary research programme is needed to achieve the necessary international collaborative research.

The success of the meeting in India is a great credit to Professor Vishwas Kale and his colleagues and to the support received from the Geological Society of India – support that will continue as the Society has agreed to publish a monograph containing papers presented at this conference. In addition, an international context is required, admirably provided by INQUA for the period 1991 to 2003. However, as INQUA reviews its structure and may terminate all of its commissions in 2003, we hope that any subsequent structure will enable multidisciplinary international palaeohydrology research to flourish. This was why the GLOCOPH response to initial proposals by INQUA stated our belief that "any new structure must be sufficiently flexible to allow these existing successful INQUA sponsored research activities to continue although we accept that all such activities should be subject to regular periodic review...".

The success shown by the publications produced and by the spirit of GLOCOPH engendered by all of the researchers who have been involved will undoubtedly continue. One way will be through the ICSU sponsored project on past hydrological events related to the understanding of global change (see. pp. xi, 266, 268). In addition, ongoing research will certainly occur, increasingly demonstrating the significance and the potential of palaeohydrology. We hope that this book has been one clear step in that direction.

Ken Gregory and Gerardo Benito
8th December 2002

Index

Note: Page numbers in *italics* refer to Figures, those in **bold** refer to Tables